COLOUR TELEVISION

TECHNOLOGY, TRANSMISSION AND RECEPTION

COLOUR TELEVISION

TECHNOLOGY, TRANSMISSION AND RECEPTION

Second Edition

R. R. GULATI

Formerly Professor
Electrical and Electronics Engineering Department
Birla Institute of Technology and Science
Pilani

NEW AGE INTERNATIONAL (P) LIMITED, PUBLISHERS

New Delhi • Bangalore • Chennai • Cochin • Guwahati • Hyderabad
• Jalandhar • Kolkata • Lucknow • Mumbai • Ranchi
Visit us at **www.newagepublishers.com**

Published by New Age International (P) Ltd., Publishers
First Edition: 1992
Second Edition:2007
Reprint: 2008

Branches :

- 36, Malikarjuna Temple Street, Opp. ICWA, Basavanagudi, **Bangalore.** ✆ (080) 26677815
- 26, Damodaran Street, T. Nagar, **Chennai.** ✆ (044) 24353401
- Hemsen Complex, Mohd. Shah Road, Paltan Bazar, Near Starline Hotel, **Guwahati.** ✆ (0361) 2543669
- No. 105, 1st Floor, Madhiray Kaveri Tower, 3-2-19, Azam Jahi Road, Nimboliadda, **Hyderabad.** ✆ (040) 24652456
- RDB Chambers (Formerly Lotus Cinema) 106A, Ist Floor, S.N. Banerjee Road, **Kolkata.** ✆ (033) 22275247
- 18, Madan Mohan Malviya Marg, **Lucknow.** ✆ (0522) 2209578
- 142C, Victor House, Ground Floor, N.M. Joshi Marg, Lower Parel, **Mumbai.** ✆ (022) 24927869
- 22, Golden House, Daryaganj, **New Delhi.** ✆ (011) 23262370, 23262368

ISBN : 81-224-2026-5

Rs. 295.00

C-08-04-2477

2 3 4 5 6 7 8

Printed in India at Ram Printograph, Delhi.
Typeset at Goswami Associates, Delhi

PUBLISHING FOR ONE WORLD
NEW AGE INTERNATIONAL (P) LIMITED, PUBLISHERS
4835/24, Ansari Road, Daryaganj, New Delhi-110002
Visit us at **www.newagepublishers.com**

Preface to the Second Edition

This book provides a comprehensive text on modern colour television with major emphasis on PAL B & G colour system adopted by India and many other countries. The treatment is essentially non-mathematical and presented in an easy-to-read format.

Twenty chapters of the book are divided into four major sections: (1) Colour signal generation and transmission; (2) Colour Television systems and standards; (3) Colour receiver circuit analysis and trouble shooting and (4) Advances in television systems, new receivers and digital technology.

Chapter 1 to 5 are devoted to colour TV optics, signal generation, modulation and transmission. Chapters 6 and 8 describe various television systems and characteristics of PAL B & G and I versions. Chapter 7 is written to introduce functioning of various sections of the PAL-D colour television receiver.

Chapter 9 to 18 give circuit details of different subsystems of modern colour receivers with a separate chapter on synthesized and microprocessor controlled channel selection and tuning. While describing various circuit blocks, the focus is on circuit functions because state-of-the-art in television receivers has advanced so much that relatively complex functions involving numerous components and circuits have been reduced to a single integrated circuit. This is not to imply that the need for circuit diagrams does no exist, rather section blocks and associated peripheral circuits of various ICs have been given and explained to emphasize special features of circuits operation and adjustments. A separate chapter explains Switched Mode Power Supply (SMPS) techniques and associated circuits. Chapter 18 ties together all the subsystems and associated circuitry to describe complete circuit of a colour receiver. Alignment, testing and servicing of a colour receiver are also explained with reference to the circuit under investigation.

In the last section, chapter 19 is on the functioning and control of modern micro-processor controlled colour receivers, PLASMA and LCD screen TV receivers, Extended Definition Television (EDTV), High Definition Television (HDTV), and elements of Digital Television.

Chapter 20 is devoted to satellite communication system, digital signal processing, data compression, data encrypting, multiplexing and transmission. Satellite acquisition for receiving down-link satellite signals, digital receiver decoder and TV signal distribution are also explained.

In addition each chapter provides a discussion on common faults and servicing of the subsystem investigated in it. At the end of each chapter there is a set of questions to test in-depth understanding of the material presented.

This volume and my first book "MONOCHROME AND COLOUR TELEVISION are complementary texts and companion volumes for a comprehensive course in Black and White and Colour television technology.

R.R. GULATI

CONTENTS

1

OPTICS OF COLOUR TELEVISION

INTRODUCTION

For fully grasping the essentials of colour television it is necessary to be fully conversant with the basics and functions of monochrome (B & W) television. Also, in order to attain a clear concept of Colour Television (CTV), certain fundamental qualities of light must be fully understood, it is also necessary to grasp the technique of mixing colours to produce different hues (colours) and the sensitivity of human eye to perceive them. The purpose of this chapter is to briefly explain such aspects so that essentials of CTV are fully assimilated.

1.1 NATURE OF VISIBLE LIGHT

When white light from the sun is examined, it is found that the radiation does not consist of a single wavelength but it comprises of a band of frequencies. In fact, white light is a very small part of the large spectrum of electromagnetic waves which extend from very low to beyond 10^{25} Hz. The visible spectrum extends over only an octave that centres around a frequency of the order of 5×10^{14} Hz. When radiation from the entire visible spectrum directly reaches the eye we see white light. If, however, part of the range is filtered out, and only the remainder of the visible spectrum reaches the eye, we see a colour. The entire electromagnetic spectrum is shown in Fig. 1.1 where the visible spectrum has been expanded and shown separately to demonstrate the range of colours it contains. Note that the various colours merge into one another with no precise boundaries.

1.2 PERCEPTION OF COLOURS

All objects that we observe are focused sharply by the lens system of the eye on its retina. The retina which is located at the back side of eye has light sensitive organs which measure visual sensations. The retina is connected with the optic nerve which conducts light stimuli as sensed by the organs to the optical centre of the brain.

According to the theory formulated by Helmholtz the light-sensitive organs are of two types—*rods* and *cones*. The rods provide brightness sensation, and thus perceive objects only in various shades of grey from black to white. The cones that are sensitive to colour are broadly in three different groups. One set of cones detects the presence of blue colour in the object focused on the retina, the second set perceives red colour and the third is sensitive to the green range. Each group of cones can thus be thought of as a set 'tuned' to only a small band of frequencies and so absorbs energy from a definite

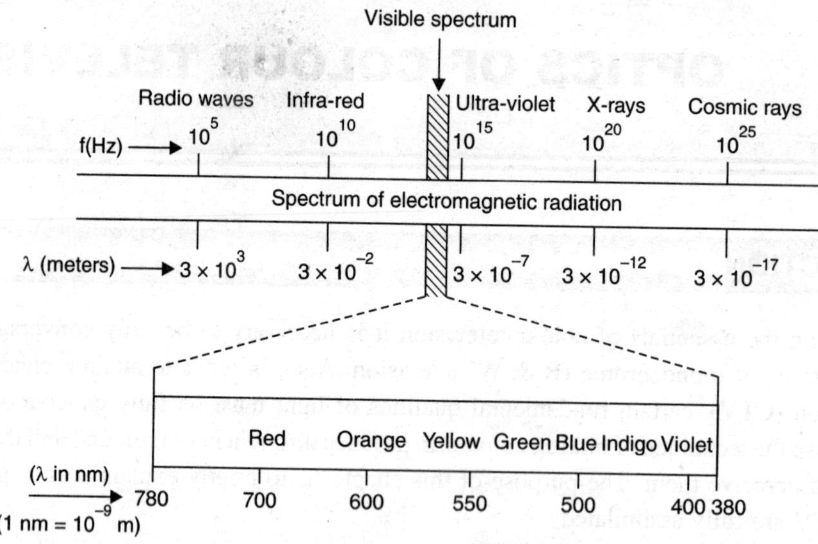

Fig. 1.1 Region of sunlight in the electromagnetic spectrum.

range of electromagnetic radiation to convey the sensation of corresponding colour or range of colours. The combined relative luminosity curve showing relative sensation of brightness produced by individual spectral colours radiated at a constant energy level is shown in Fig. 1.2. It will be seen from the plot that

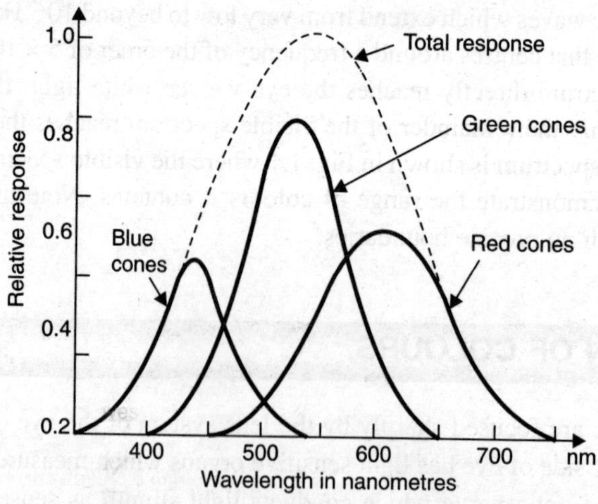

Fig. 1.2 Approximate relative response of the eye to different colours.

sensitivity of the human eye is greatest for green light, decreasing towards both the red and blue ends of the spectrum. In fact the maximum is located at about 550 nm, a yellow green, where the spectral energy maximum of sunlight is also located.

1.3 MIXING OF COLOURS

The radiation of each wavelength in the visible spectrum is perceived by the eye as a certain hue or colour in its pure and fully saturated form. These colours cannot be exactly reproduced by conventional methods like printing or painting. However, the infinite variety of colours that we see around can be produced either by subtractive mixing using various pigments or by additive mixing of light of different colours.

Subtractive mixing. A sensation of colour is produced by opaque objects or materials when white light falls on them. It is so because any surface does not reflect all the wavelengths of the incident light uniformly. In fact, only some out of these are reflected and the rest get absorbed at the surface of the object. Thus, any object is seen in a colour corresponding to the spectrum of radiated light which is not absorbed but instead gets reflected. For example, a red apple absorbs light of all colours except red, which it reflects and thus looks red. An object that appears black absorbs all the visible light. Similarly, a surface looks grey when the entire light spectrum is reflected uniformly but less, *i.e.,* a part of the incident light being absorbed by the object. Based on the above facts special chemicals called *pigments* are used for printing and painting. A yellow pigment absorbs violet, blue, green, orange and red but reflects the remaining yellow. Similarly, blue pigments reflect only blue and absorb all other colours. When pigments of two or more colours are mixed they reflect wavelengths of only those colours which are common to both and these combine to give the sensation of a new colour. However, when no wavelength of the light spectrum is common in the pigments which are mixed, the entire incident light is absorbed and the object looks black. Thus, pigments create a given colour by subtracting parts of the spectrum of incident white light and reflecting the remaining which gives the objects its characteristic colour. Making colours by mixing paint pigments is therefore described as 'Subtractive Mixing', since each added pigment subtracts more from white light and leaves less to be reflected to the eye. The formation of various colours including black by subtractive mixing is shown in Fig. 1.3 (*a*).

Additive mixing. This is altogether a different method of producing various colours. When lights of two or more different colours are mixed and projected on a white screen, the eye perceives it as a combined colour. In effect, the light reflected from the screen seems to the eye as if it is coming from a new source of light which is different from any of the sources actually projected.

As explained earlier, all light sensations to the eye are divided into three main groups. The optic nerve system then integrates different colour impressions in accordance with the curves shown in Fig. 1.2 to perceive actual colour of the object being seen. A yellow colour, for example, can be distinctly seen by the eye when red and green groups of cones are excited at the same time with corresponding intensity ratio. Similarly, any colour other than red, green and blue will excite different sets of cones to generate cumulative sensation of that colour. Thus, a white colour is perceived by additive mixing of sensations from all the three sets of cones.

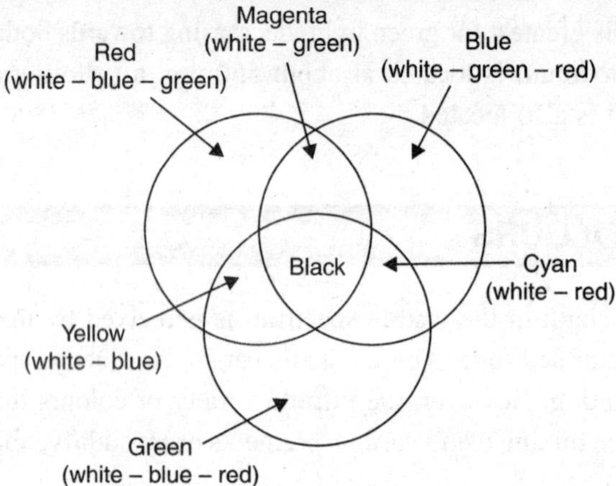

Fig. 1.3 (*a*) Subtractive colour mixing: The diagram shows the effect of mixing colour pigments under white light.

Hence, in additive mixing which forms the basis of colour television, light from two or more colours obtained either from independent sources or through filters can create a combined sensation of a different colour. Note that different colours are created by mixing pure colours and not by subtracting parts from white. The additive colours mixing is illustrated in Fig. 1.3 (*b*).

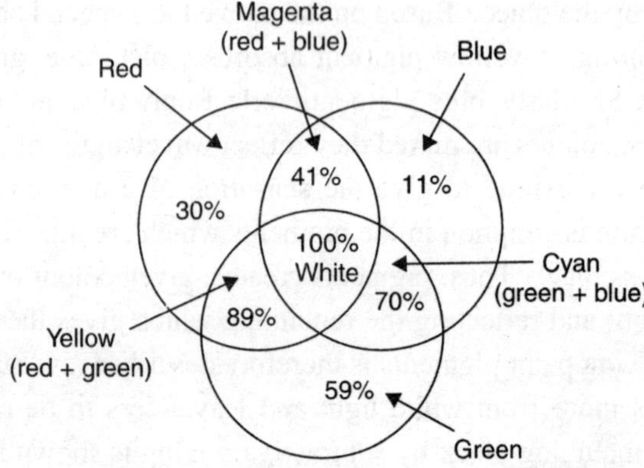

Fig. 1.3 (*b*) Additive colour mixing. The diagram shows the effect of projecting green, red and blue light beams on a white screen in such a way that they overlap.

1.4 THREE COLOUR THEORY

The number of different colours that can be formed by additive mixing depends on the number of coloured lights used for this purpose. The more the number of such lights employed in the mixing scheme, the wider can be the reproducible range of colours. However, the number of different lights has to be limited to a few to avoid system complexity and cost.

Experimental results have established that lights of red, green and blue colours when combined with each other in different proportions, produce a wide range of gamut of colours. The perception of

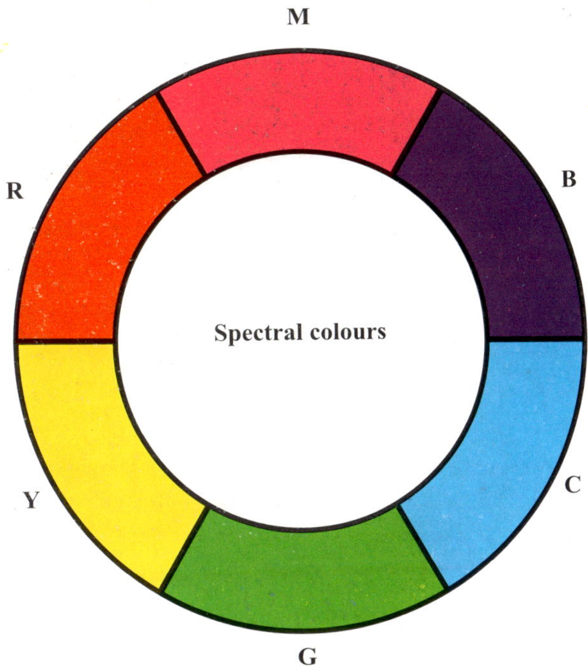

Colour Plate 1: Colour circle with primary and complementary colours

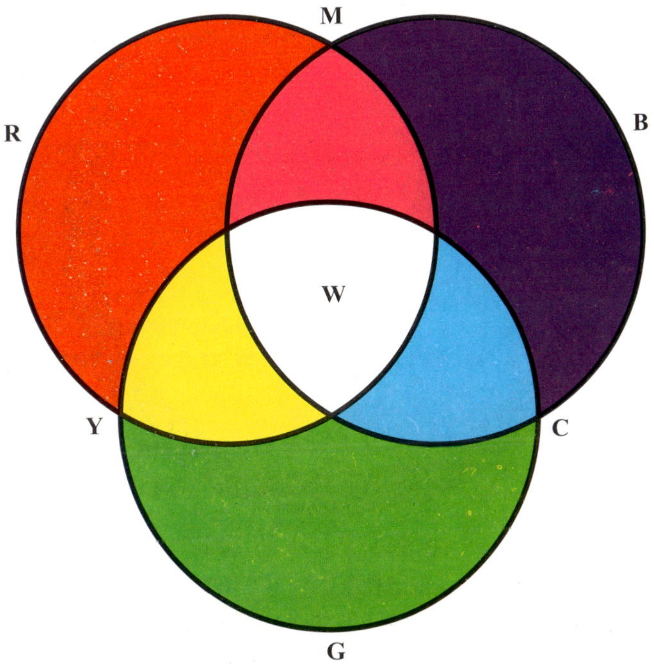

Colour Plate 2: Additive colour mixing

Colour Plate 3: Colour bar pattern with colours in order of decreasing brightness: white, yellow, cyan, green, magenta, red and blue

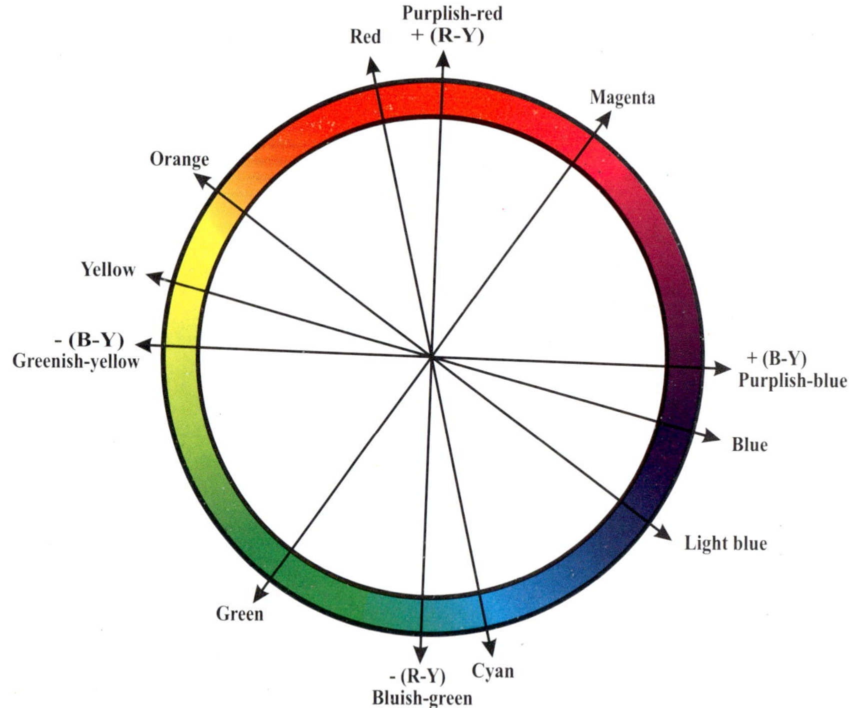

Colour Plate 4: Colour circle diagram showing different hues corresponding to phase angles of the chrominance signal

colours thus produced is enhanced by the fact that the three sets of cones that are associated with the optic nerve have greatest sensitivity to red, green and blue. Therefore, red, green and blue have been, standardized as basic colours for colour television and are known as Primary Colours.

By pair-wise additive mixing of primary colours, the following complementary colours can be produced:

Red + Green = Yellow

Red + Blue = Magenta (purplish red shade)

Blue + Green = Cyan (greenish blue shade)

Colour Plate 1 depicts the location of primary and complementary colours on the colour circle. If a complementary colour is added in an appropriate and fixed proportion to the primary which it itself does not contain, white is produced. Colour plate 2 shows formation of complementary colours by mixing of primary colours. Note that additive mixing of the three primary colours in defined proportions produces white.

Grass Man's Law

As already explained the eye is not able to distinguish each of the colours that mix to form a new colour but instead perceives only the resultant colour. Thus, the eye behaves as though the output of three types of cones are additive. The subjective impression which is gained when green, blue and red lights reach the eye simultaneously, may be matched by a single light source having the same colour. In addition to this, the brightness (luminance) impression created by the combined light source is numerically equal to the sum of brightnesses (luminances) of the three primaries that constitute the single light. This property of the eye of producing a response which depends on the algebraic sum of red, green and blue inputs is known as Grass Man's law. White has been seen to be reproduced by adding red, green and blue lights. The intensity of each colour may be varied. This enables simple rules of addition and subtraction.

1.5 TRISTIMULUS VALUES AND CHROMATICITY OF SPECTRAL COLOURS

Equal quantities of primary colours do not appear to be equally bright nor do they involve equal amounts of light energy. As an illustration, if inputs to a colour picture are adjusted to make the raster white and later primary colours are viewed separately in turn, the green phosphor on excitation by the green signal will appear to be brightest while the blue will appear to be least bright. This is expected because of the nature of response of human eye to different colours as explained earlier. Therefore, in order to obtain quantitative values of various colours which must be mixed together, all such factors have to be taken into account.

Based on these considerations and extensive tests with a large number of observers, the primary spectral colours and their intensities required to produce different colours by mixing have been standardized. The red, green and blue have been fixed at wavelengths of 700 nm, 546.1 nm and 435.8 nm, respectively. The component values (or fluxes) of the three primary colours to produce various other colours have also been standardized and are called *tristimulus values* of different spectral colours. The reference white for CTV has been chosen to be a mixture of 30% red, 59% green and 11% blue. These percentages of light fluxes are based on the sensitivity of eye to different colours.

Thus, for example, 0.3 lumen of red + 0.59 lumen of green + 0.11 lumen of blue will produce one lumen of white light. Furthermore, in accordance with the law of colour additive mixing, one lumen of white light (see Fig. 1.3. (*b*)) will also be produced by 0.89 lm of yellow + 0.11 lm of blue or 0.70 lm of cyan + 0.30 lm of red or 0.41 lm of magenta + 0.59 lm of green.

It may be noted that if the concentration of luminous flux is reduced by a common factor from all the constituent colours, the resultant colour will still be white, though its level of brightness will decrease. The brightness of different spectral colours is associated with that of white. Yellow for example, (see Fig. 1.3 (*b*)) appears 89% as bright as the reference white, reflecting the addition of 59% brightness of green and 30% brightness of red. Similarly, any other combination of primary colours will produce a different colour with a different relative brightness reference to white which has been taken as 100 percent.

Chromaticity Diagram

The factors stated above are summed up in the chromaticity diagram shown in Fig. 1.4. As seen, it is a convenient space-coordination representation of all the spectral colours and their mixtures based on the tristimulus values of the primary colours contained by them.

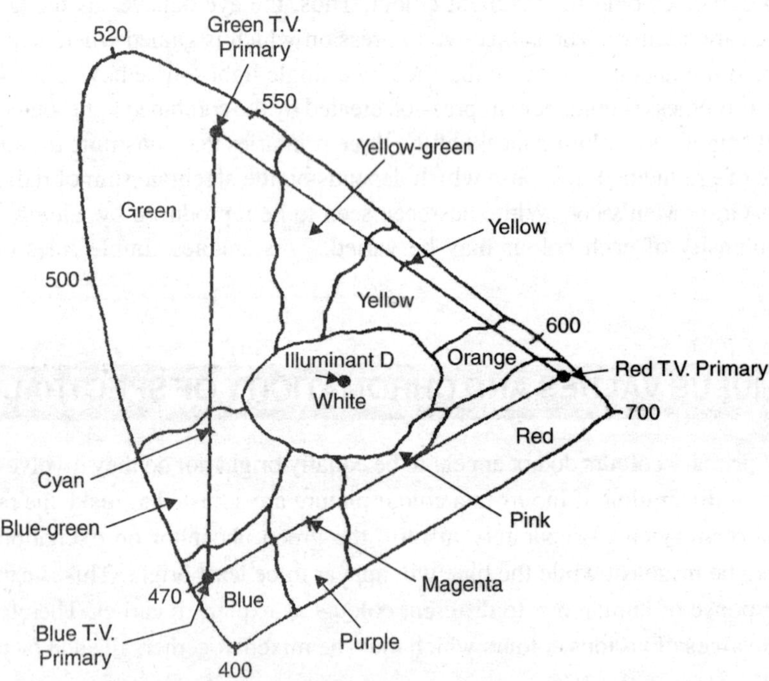

Fig. 1.4 The chromaticity diagram. The inner triangle rests on the TV primaries and the area enclosed by the triangle represents the range of colours in a TV display

1.6 HUE, SATURATION AND LUMINANCE

Any colour has three characteristics to specify its visual information. These are: (*i*) Hue or Tint, (*ii*) Saturation and (*iii*) Luminance.

(*i*) ***Hue or Tint.*** This is the predominant spectral colour of the received light. Thus, the colour of any object is distinguished by its hue or tint. The green leaves have green hue and red tomatoes have red hue. Different hues result from different wavelengths of spectral radiation and are perceived as such by the sets of cones in the retina.

(*ii*) ***Saturation.*** This is the spectral purity of the colour light. Thus, saturation may be taken as an indication of how little the colour is diluted by white. A fully saturated colour has no white. As an example, vivid green is fully saturated and when diluted by white it becomes light green. The hue and saturation of a colour put together is known as *Chrominance*. Note that it does not contain the brightness information. Chrominance is also called *Chroma*.

(*iii*) ***Luminance.*** Brightness and luminance are often taken to mean the same information but actually it is somewhat different. It is difficult to define brightness for technical purposes because it only gives an objective assessment of sensation perceived by the eye due to the effect of illumination of surroundings. Thus, to overcome such a difficulty, luminance is taken to mean the subjective visual sensation of brightness of light or colour. To be more specific, luminance is a measure of the light emitted from a surface. Light may be emitted by direct radiation from a source at a surface or it may be reflected by a surface which is bathed in incident light. As an illustration, in monochrome TV camera, the video signal developed describes the amount of light falling on its target plate. Since this in turn depends upon the amount of light being reflected from the scene, the camera output represents luminance of the scene element by element. For a given colour, luminance is the fundamental brightness of the colour impression and is the amount of light intensity as perceived by the eye regardless of the colour. In black and white pictures, better lighted parts have more luminance than the dark areas. Different colours also have shades of luminance in the sense that though equally illuminated they appear differently bright as indicated by the relative brightness response curve of Fig. 1.2. Thus, on a monochrome TV screen, dark red colour will appear as near black, yellow, close to white and light blue colour as light grey.

1.7 PERCEPTION OF COLOURS VERSUS BLACK AND WHITE

The fundamental characteristics of a CTV system are based on perception of colours by the human eye. Black and White vision is more acute than colour vision. An important aspect of this distinction is the inability of human eye to perceive very small coloured areas in a scene. Detailed studies have shown that perception of colours produced by the three primaries is limited to relatively large areas. Further, for medium size objects or areas, only two primary colours are needed. This is so, because for finer details the eye fails to distinguish purple (magenta) and green-yellow hues from greys. As the coloured area becomes very small, the red and cyan hues also become indistinguishable from greys. Thus, for very fine colour details all persons with normal vision are colour blind and see only changes in brightness even for coloured areas. Its technical importance is that while the TV system should be capable of reproducing fine details of black, white and grey, coloured areas need not be sharply defined.

The eye is also more critical of some hues than others. For example, the viewer tends to be very critical of variations in flesh tones whereas similar changes in brown and green are likely to be accepted. In other words, orange tints need to be reproduced with minimum departure from the original scene whereas green hues may vary somewhat without recognition by the viewer. Technically, this means that orange hues and flesh tones must be picked up, transmitted and reproduced with the minimum possible error. Special circuits are often included in colour receivers to achieve such a performance.

REVIEW QUESTIONS

1. What is the ratio of visible electromagnetic spectrum to the total electromagnetic spectrum?
2. Explain how the human eye perceives brightness and colour sensations. Comment on the spectral response of the human eye.
3. What do you understand by additive and subtractive mixing of colours? Why is additive mixing suitable for colour television?
4. Name the primary and complementary colours used in colour television. What is Grass-Mans' Law? Explain with an example how the primary and complementary colours can be combined to produce white?
5. What are tristimulus values of spectral colours? What is their significance in colour television?
6. Explain how hue, saturation and luminance describe fully any colour scene. What do you understand by the chrominance of a colour?
7. Explain limitations of the human eye to perception of colours when compared with black, grey and white? Is black the complement of white?

2 LUMINANCE AND COLOUR SIGNAL GENERATION

INTRODUCTION

The black and white picture is described by its brightness element by element. Its electrical representation is the luminance signal. The colour content of any scene is represented by its hue and saturation. This is contained in the chrominance signal. Thus, to ensure compatibility, the final signal from a colour camera has two parts, a luminance signal and a chrominance signal which are combined suitably to form the complete colour signal.

The ultimate need in a monochrome receiver is a voltage corresponding to the brightness of any colour scene. Similarly, in a colour receiver, the final requirement at the input of colour picture tube is three voltages proportional to red, green and blue contents of the picture element by element. However, compatibility requirements dictate that instead 'colour-difference' signals be generated and transmitted. A suitable matrix is then employed in the receiver to recover red, green and blue voltage signals. Compatibility requirements and generation of luminance and colour-difference signals are described in this chapter.

2.1 COMPATIBILITY

Compatibility implies that (*i*) the colour television signal must produce a normal black and white picture on a monochrome receiver without any modification of the receiver circuitry, and (*ii*) a colour receiver must be able to produce a black and white picture from a normal monochrome signal. This is referred to as *reverse compatibility*.

To achieve this, that is, to make the system fully compatible the composite colour signal must meet the following requirements:-

(*i*) It should occupy the same bandwidth as the corresponding monochrome signal.

(*ii*) The location and spacing of picture and sound carrier frequencies should remain the same.

(*iii*) The colour signal should have the same luminance (brightness) information as would a monochrome signal, transmitting the same scene.

(*iv*) The composite colour signal should contain additional colour information together with the ancillary signals needed to allow this to be decoded.

(*v*) The colour information should be carried in such a way that it does not affect the picture reproduced on the screen of a monochrome receiver.

(*vi*) The system must employ the same deflection frequencies and sync signals as used for monochrome transmission and reception.

In order to meet the above requirements, it becomes necessary to encode the colour information of the scene in such a way that it can be transmitted within the same channel bandwidth of 7 MHz and without disturbing the brightness signal. Similarly, at the receiving end, a decoder must be used to recover the colour information back in its original form for feeding it to the tricolour picture tube.

2.2 COLOUR TV CAMERA

Figure 2.1 shows a simple block schematic of colour TV camera. It essentially consists of three camera tubes in which each tube receives selectively filtered primary colours. Each camera tube develops a signal voltage proportional to the respective colour intensity (luminance) received by it. Light from the scene is processed by the objective lens system. The image formed by the lens is split into three images by means of glass prisms. These prisms are designed as diachroic mirrors. A diachroic mirror passes one wavelength and rejects other wavelengths (colours of light). Thus, red, green and blue colour images are formed. The rays from each of the light splitters also pass through colour filters called *trimming filters*. These filters provide highly precise primary colour images which are converted into video signals by image-orthicon or vidicon camera tubes. Thus, the three colour signals are generated. These are called Red (R), Green (G) and Blue (B) signals.

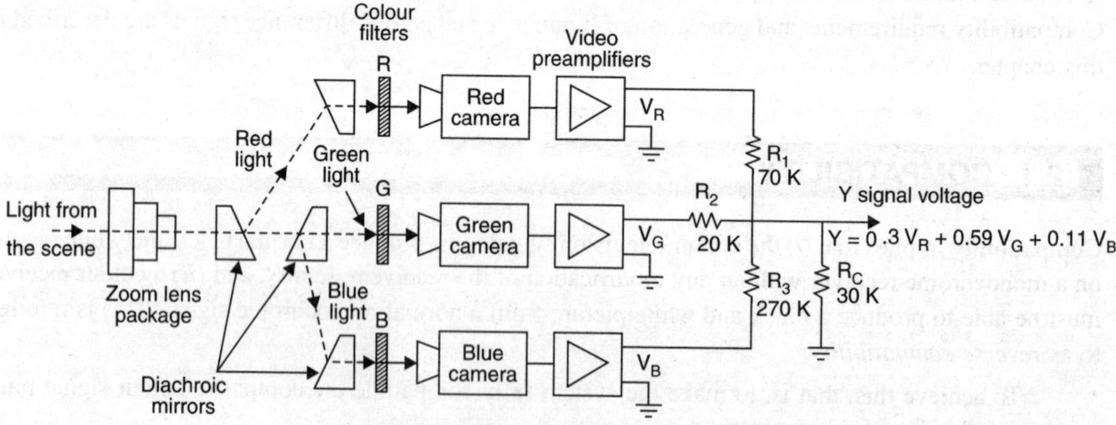

Fig. 2.1 Plan of a colour TV camera showing generation of colour voltages and Y matrix for obtaining the luminance signal.

Simultaneous scanning of the three camera tubes is accomplished by a master deflection oscillator and sync generator which drives all the three tubes. The three video signals produced by the camera represent three primaries. By selective use of these signals, all colours in the visible spectrum can be reproduced on the screen of a special (colour) picture tube. This is explained in Chapter 3.

Single Tube Camera

The method of obtaining three video outputs with a single vidicon varies from make to make. In one method, specially designed colour filters are bonded on the face plate of pick-up tube. As the beam scans it develops colour video outputs but one at a time depending on the colour of filter it encounters. A somewhat complicated system of multiplexing is then employed to collect separately each colour output voltage. In another type of camera developed by 'Hatachi' of Japan, multiplexing has been avoided by using a striped target instead of the conventional uniform target. The face of the vidicon is bonded with a filter having three stripes of red, green and blue that are separated from each other. The target of the pick-up tube also has three sections. Each section is located in such a way that it is just behind (in parallel) the corresponding colour filter section. Thus, the three targets develop red, green and blue outputs depending on the intensity of colour light falling on them.

Normalization of Colour Voltages

Since tristimulus values of the chosen red, green and blue are standardized to have the same maximum value for saturated colours, output voltages of the three camera tubes are set to have equal values with a white card placed in front of the camera. The signals reach absolute maximum values because of the uniform spectral energy distribution of white light. The normalization or equal voltage adjustment of V_R, V_G and V_B is made with maximum illumination that the system is designed to handle.

In a studio, the usual practice is to normalize the outputs with the maximum white luminance available at the camera. Normalization *i.e.,* setting of $V_R = V_G = V_B$ ensures that colour-difference signals, which are actually transmitted, disappear on whites. It is an aid to making the colour and black and white systems compatible. This aspect is fully explained in a subsequent section.

2.3 GAMMA (γ) CORRECTION

The overall television system must be linear for distortion free reproduction of pictures on the receiver screen. This implies that besides other factors, the light emitted from the picture tube should be directly proportional to the light falling on the camera target plate. However, it does not happen because the beam current of a picture tube is not linearly related to the driving voltage applied between its grid and cathode. The relationship between luminance (L) of the screen against grid-to-cathode voltage (Vg) can be expressed as $L \propto (Vg)^{\gamma}$ where (γ) varies between 2.0 and 2.6 from tube to tube.

In order to compensate for this non-linearity at the receiver end *i.e.,* picture tube, an opposite distortion (non-linearity) referred to as 'Gamma Correction' is introduced at the transmitting end. For this, the camera output voltages are passed through corrective networks having input-output characteristics just opposite to that at the receiving end. Thus, the light output becomes directly proportional to the light falling on the camera target plate.

It should be noted that gamma correction is necessary both in colour and monochrome TV systems. In some textbooks, the corrected camera outputs are written as V_R', V_g' and V_B' to indicate that these are gamma-corrected values. Each output is separately corrected before normalization and hence before forming luminance and chrominance signals.

Since gamma correction to colour voltages is applied on generation and not altered in any way during encoding and decoding of the colour signal, there is no need to use separate symbols for the uncorrected and corrected values of camera outputs. Hence, in this text we will use V_R, V_G and V_B for the gamma-corrected values unless specified otherwise.

2.4 CAMERA OUTPUTS ON DIFFERENT COLOURS

The camera outputs are normalized after gamma correction with white light incident on the lens turret. The maximum white light that the system can handle is said to cause 100% brightness. The corresponding voltages after gamma correction and normalization are taken in most texts equal to *one volt for convenience of manipulations while explaining various processes. Thus, for 100% brightness $V_R = V_G = V_B = 1V$ is used in this text. As the level of brightness decreases, the camera outputs also decrease and become zero on darkness, say when there is no illumination.

Assuming maximum brightness, Fig. 2.2 shows the nature of output voltages when a picture having red, green and blue bars is scanned. Note that at any one instant only one camera tube delivers output depending on the colour being scanned. The three voltages are equal and will decrease uniformly as brightness on the bar pattern is reduced.

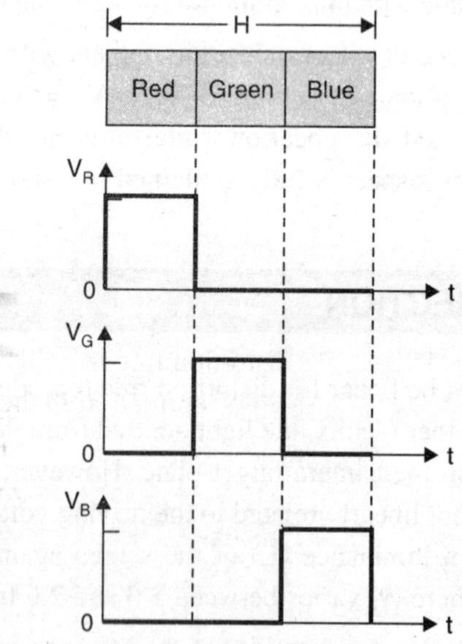

Fig. 2.2 Camera output voltage for the red, green and blue colour bars. Note that 'H' indicates one horizontal scanning line.

In order to illustrate camera outputs for a complementary colour and white, Fig. 2.3 shows output voltages obtained on scanning a strip having yellow and white bands besides the three pure colour bars. Notice the outputs for yellow. Since it includes red and green voltages, V_R and V_G are

*The actual value of any of the colour signals depends on the point in the system at which they are measured. Thus a peak luminance signal which can be one volt when it is formed at the camera may be represented by a signal less than 1 mV at the receiver antenna, by a few volts at the detector, and by hundred or so volts at the picture tube.

produced simultaneously. The magnitude of V_B is zero because there is no blue in yellow. Similarly, for the white bar, all the three colour outputs are produced when it is scanned. The magnitude of V_R, V_G and V_B in each case depends on the intensity of brightness. It will vary from 1 V (100%) to zero volt as illumination varies from 100% brightness to zero at darkness.

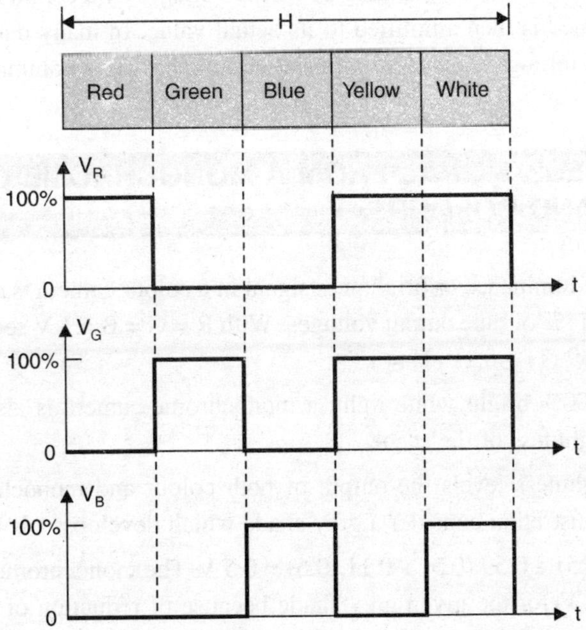

Fig. 2.3 Red, Green and Blue camera outputs for the bar pattern shown above the signal waveforms.

2.5 THE LUMINANCE SIGNAL

Equal quantities of primary colours as defined for colour television do not appear to be equally bright nor do they involve equal amounts of energy. If the colour picture tube is adjusted to display white and the three primary emissions (colours) are then viewed separately in turn, the green appears to be the brightest and thus has the highest luminance. At the other end, blue appears to be least bright and so has the least luminance. With luminance of white taken as 'one' (1) relative luminance values as perceived by the eye have been found out to be 0.299 for red, 0.587 for green and 0.114 for blue. Therefore, for constituting the luminance (Y) signal, colour voltages developed by the camera must contribute signal amplitude in these ratios. Accordingly Y signal that carries brightness information of the scene can be formed by adding fractions of the developed red, green and blue video signals in proportion to the coefficients stated above. However, for most practical purposes it is sufficiently accurate to form the luminance signal as: $V_y = 0.3\ V_R + 0.59\ V_G + 0.11\ V_B$

'Y' Signal Matrix

The reduction and subsequent addition of the three primary signals V_R, V_G and V_B can be effected by feeding them into a matrix network, the basic circuit of which is shown in Fig. 2.1 along with camera details. The values of R_1, R_2, R_3 and R_c are so chosen that V_y which develops across the summing resistance R_c is equal to $0.3\ V_R + 0.59\ V_G + 0.11\ V_B$. Generally, the prefix 'V' is omitted and only Y, R, G and B are used to represent corresponding voltages. Thus, $Y = 0.30\ R + 0.59\ G + 0.11\ B$.

As stated earlier, R, G and B are the gamma corrected and normalized values of colour camera outputs. Usually matrixing is done by scaling down the voltages by a common factor to avoid crosstalk. The Y signal thus formed is then amplified to its actual value. In many matrixing circuits, the colour voltages are combined through transistor amplifier circuits having a common load resistance.

2.6 BRIGHTNESS SIGNAL FROM A MONOCHROME CAMERA ON WHITES AND COLOURS

As explained above the luminance or brightness signal in a colour camera is obtained by adding 30% of red, 59% of green and 11% of blue output voltages. With R = G = B = 1 V set for 100% bright white we get: $Y = 0.3\ (1) + 0.59\ (1) + 0.11\ (1) = 1V$

For the same 100% bright white light, a monochrome camera is also set to develop one volt output to represent brightness of the scene.

At reduced brightness levels the output of both colour and monochrome cameras fall by the same amount. As an illustration consider a grey shade which develops: $R = G = B = 0.5\ V$

Then $Y = 0.3\ (0.5) + 0.59\ (0.5) + 0.11\ (0.5) = 0.5\ V$. The monochrome camera will also develop the same output of 0.5 V for the given grey shade because of reduction of brightness by 50 percent. Thus, the luminance signal amplitude for any shade of white from both colour and monochrome cameras remains the same.

In general, if R = G = B, which is only possible for any shade of white, the magnitude of $Y = (0.3R + 0.59G + 0.11B) = R = G = B$ volts and is also equal to the voltage developed by the black and white camera.

Next consider a colour object, say a pure red colour bar and assume that it is scanned by a colour and a monochrome camera simultaneously. As shown in Fig. 2.3, R = (IV (100%) for pure red at 100% brightness. Since G = B = 0 V for red, $Y = 0.3\ (1) + 0.59\ (0) + 0.11\ (0) = 0.3\ V$. The monochrome camera output will also be 0.3 V because red is 30% as bright when compared to white light. As the brightness level on the colour bar is reduced to become zero, R will change from 1 V to 0 V and monochrome camera output will vary from 0.3 V to 0 V. Such a behaviour is true for any colour which may be pure, complementary or otherwise desaturated.

Y Signal Amplitude on Colours

The colours in the bar pattern of Fig. 2.4 have been chosen in order of decreasing relative luminance. The magnitude of V_R, V_G, V_B and V_Y as obtained on scanning the bar from left to right with a colour camera are shown below the bar pattern. Notice that the amplitude of Y as obtained for each colour by adding set fractions of R, G and B has a staircase pattern. It is maximum for white since all the camera tubes develop voltage when it is scanned.

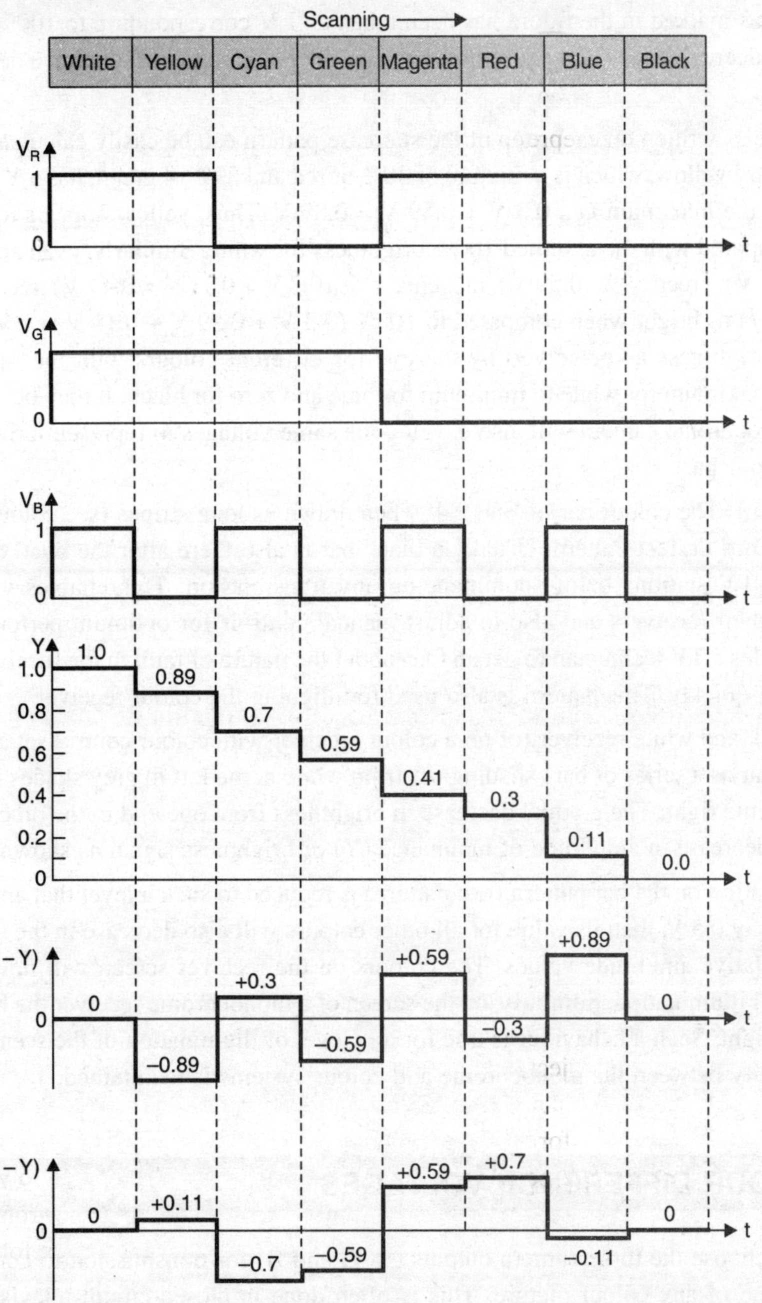

Fig. 2.4 Formation of luminance (Y) and colour-difference video signal with the resistance matrix from colour camera outputs. Note that yellow appears 89% (.3R + 59G) bright as compared to 100% for white. Similarly, cyan, green, magenta, red and blue appear 70%, 59%, 41%, 30% and 11% bright respectively as compared to assumed 100% brightness for white. Note the values marked are unweighted. (See colour plate 3)

Its value as marked in the figure has been taken as 1 V corresponding to 100% brightness. The magnitude of Y decreases for succeeding bars because the colours that follow have decreasing relative luminance.

The value as written on each step of the staircase pattern can be easily calculated. For example, the complementary yellow, which is a mixture of 30% of red and 59% of green, has a Y signal amplitude equal to 89% of the maximum *i.e.*, 0.3 V + 0.59 V = 0.89 V. Thus, yellow appears to the eye 89% as bright when compared with the assumed 100% brightness for white. Similarly, cyan appears 70% (0.59 V + 0.11 V = 0.7 V), green 59% (0.59 V), magenta 41% (0.3 V + 0.11 V = 0.41 V), red 30% (0.3 V) and blue 11% (0.11 V) as bright when compared to 100% (0.3 V + 0.59 V + 0.11 V = 1 V) for white. It is thus clear that brightness as perceived by the eye for different colours with the same illumination decreases from maximum for white to minimum for blue and zero for black. It may be emphasized once again that a monochrome camera will also develop the same voltages to represent brightness levels on scanning the colour bar.

Test pattern. The colour bar of Fig. 2.4. when drawn as long stripes (see colour plate No. 3) is called the Test Card or Test Pattern. Usually a black bar is also there after the blue. Such a pattern is telecast by all CTV stations before commencing any transmission. This enables viewers to check performance of their receivers and also to adjust manual controls for optimum performance. The test pattern also enables a TV technician to get an idea about the nature of fault in the receiver thus enabling him to localize it quickly. This pattern is also used for aligning the colour receiver.

On a black and white receiver (or on a colour receiver with colour control set at zero) the same pattern will appear as a series of bars shading-off from white at the left to grey shades in the centre and black at the extreme right. The gradual decrease in brightness from one end to the other corresponds to the progressive decrease in amplitude of luminance (Y) or brightness signal as shown in Fig. 2.4.

If illumination on the bar pattern (test pattern) is reduced to such a level that amplitude of Y for white becomes, say 0.5 V, then its value for all other colours will also decrease in the same proportion, thus retaining relative amplitude values. The colours on the receiver screen will now appear half as bright as with full illumination. Similarly, on the screen of a monochrome receiver the bar shadings will appear half as bright. Such a behaviour is true for any level of illumination of the scene being telecast. Thus, compatibility between the monochrome and colour systems is maintained.

2.7 COLOUR DIFFERENCE VOLTAGES

It is possible to choose the three camera outputs (R, G and B) for transmission to convey luminance, saturation and hue of any colour picture. This is often done in closed-circuit television where three coaxial cables are used, one for each of the three colour voltages between the transmitter and receiver. However, the transmission of R, G, and B is not suitable if the system is to be compatible. This problem is solved by choosing Y and any two of the three colour-difference signals. Therefore, it becomes necessary to convert colour output voltages to corresponding values of the colour-difference signals. This is naturally done before transmission. The three possible colour difference signals are (R – Y), (G – Y) and (B – Y). These are so called because they are obtained by subtracting the Y signal form each of the primary colour voltages. In practice, (R – Y) and (B – Y) are used. (G – Y) is not considered suitable for transmission for reasons to be explained later. The amplitudes of (B – Y) and (R – Y) for the colour bar pattern of Fig. 2.4 are shown below the R, G, B and Y amplitude values.

Colour Difference Signal Amplitudes

The colour difference signals can be represented in terms of R, G and B as under:

The red colour-difference signal

$$(R - Y) = R - 0.3R - 0.59G - 0.11B \text{ (as } Y = 0.3R + 0.59G + 0.11B)$$
$$= 0.7R - 0.59G - 0.11B$$

Similarly, the green colour-difference signal

$$(G - Y) = G - 0.3R - 0.59G - 0.11B$$
$$= - 0.3R + 0.41G - 0.11B$$

and the blue colour-different signal

$$(B - Y) = B - 0.3R - 0.59G - 0.11B$$
$$= - 0.3R - 0.59G + 0.89B$$

If R, G and B change by the same amount, which means that if brightness and saturation of the colour are changed by adding or subtracting white light, it is clear that the colour-difference signals will remain the same.

For pure colours, only one voltage *i.e.*, either R, G or B is present and the remaining two are zero. Assuming unit value (1V) for the colour voltages in turn, the magnitudes of colour-difference voltages for the three primary colours will be as tabulated below:

Colour	(R – Y)	(G – Y)	(B – Y)
Red	0.7	– 0.3	– 0.3
Green	– 0.59	0.41	– 0.59
Blue	– 0.11	– 0.11	0.89

If V is a voltage which is not less than the largest of V_R, V_G and V_B, then $(V - V_R)$, $(V - V_G)$ and $(V - V_B)$ represent complementary colours. For example, $(V - V_G)$ represents cyan which is the complementary of green. It may also be noted that colour-difference signals for complementary colours are negative of the corresponding primary colours. This can be seen from the waveforms shown in Fig. 2.4.

2.8 CHOICE OF (R – Y) AND (B – Y) FOR TRANSMISSION

The luminance signal Y = 0.3R + 0.59G + 0.11B and also = 0.3Y + 0.59 Y + 0.11Y = 1 Y. Equating the above two expressions for Y we get, 0.3 (R – Y) + 0.59 (G – Y) + 0.11 (B – Y) = 0.

This is true for any colour.

CHAPTER 2

As is obvious from the derived equation, it is only necessary to specify two of the three colour-diffference signals as the third can be derived from them. The signals (R – Y) and (B – Y) are transmitted and the missing (G – Y) is obtained in the receiver. Since 0.30 (R – Y) + 0.59 (G – Y) + 0.11(B – Y) = 0.

therefore,
$$(G-Y) = -\frac{0.30}{0.59}(R-Y) - \frac{0.11}{0.59}(B-Y)$$

i.e.,
$$(G-Y) = -0.51\,(R-Y) - 0.186\,(B-Y)$$

Thus, a suitable matrix can be set up in the receiver circuit to obtain (G – Y) from the other two colour-difference signals.

Unsuitabilitiy of (G – Y) for Transmission

As shown above, (G – Y) = – 0.51 (R – Y) – 0.186 (B – Y).

Since the required amplitudes of both (R – Y) and (B – Y) for obtaining (G – Y) are less than unity, these may be derived using simple resistor attenuators across respective signal paths. However, if (G – Y) is to be one of the two transmitted signals then (*i*) if (R – Y) is the missing signal, its matrix would have to be based on the expression:

$$(R-Y) = -\frac{0.59}{0.3}(G-Y) - \frac{0.11}{0.3}(B-Y)$$

The factor 0.59/0.3 (= 1.97) implies gain in the matrix and thus would need an extra amplifier.

(*ii*) Similarly, if (B – Y) is not transmitted, the matrix formula would be:

$$(B-Y) = -\frac{0.59}{0.11}(G-Y) - \frac{0.3}{0.11}(R-Y)$$

The factor 0.59/0.11 = 5.4 and 0.3/0.11 = 2.7, both imply gain and two extra amplifiers would be necessary in the matrices. This shows that it would be technically less convenient and uneconomical to use (G – Y) as one of the colour-difference signals for transmission.

In addition, since the proportion of G in Y is relatively large in most cases, the amplitude of (G – Y) is small. It is either the smallest of the three colour difference signals, or is atmost equal to the smaller of other two. The smaller amplitude together with the need for gain in the matrix would make S/N ratio problems more difficult then when (R – Y) and (B – Y) are chosen for transmission.

2.9 COMPATIBILITY WITH COLOUR-DIFFERENCE SIGNALS

The following points bring out the suitability of colour-difference signals as an aid to compatibility:

(*i*) For any picture element having no colour R = G = B = Y

Assuming these amplitudes equal to 1 V we get,

(R – Y) = 0V and (B – Y) = 0V

Thus, the colour-difference signals disappear on all shades of white. The only signal that gets transmitted is the Y signal. Hence, there is no interference from colour-difference signals to the reproduction of black and white parts of the picture.

(*ii*) The colour-difference signals are smaller in amplitude as compared to both colour and luminance signals. Thus, their transmission is easier and causes least interference to adjoining signals. In practice, for 90% of the time it is found that colour-difference signals have less than one-third of their maximum values.

(*iii*) The Y signal is transmitted with full frequency bandwidth of 5 MHz for maximum brightness details in monochrome. However, such a large frequency spectrum is not necessary for colour-difference video signals because the eye cannot perceive very fine details in colour. Therefore, colour-difference signals are reduced in bandwidth before final transmission. The receiver has thus to derive a low definition colouring picture to superimpose on the sharp-luminance picture. This results in simplified receiver design.

2.10 PRODUCTION OF COLOUR DIFFERENCE SIGNALS

The circuit of Fig. 2.1 is reproduced in Fig. 2.5 to explain the generation of (B – Y) and (R – Y) voltages. The voltage V_Y as obtained from the resistance matrix is low because R_c is chosen to be small to avoid crosstalk. Hence, it is amplified before it leaves the camera sub-chassis, Also, the amplified Y signal is inverted to obtain – Y as the output. This is passed on to the two-adder circuits. One adder circuit adds the red camera output to – Y to obtain (R – Y) signal. Similarly, the second adder combines blue camera output to – Y and delivers (B – Y) as its output. This is illustrated in Fig. 2.5. The difference signals thus obtained bear information both about the hue and saturation of different colours.

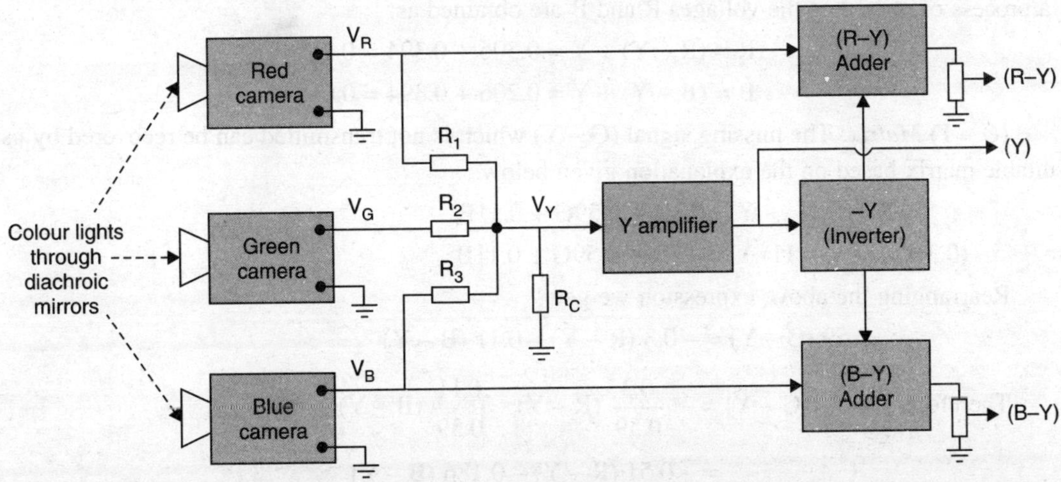

Fig. 2.5 Production of luminance and colour-difference signals.

Note that $Y = 0.3\,R + 0.59G + 0.11B$

$(R - Y) = 0.7R - 0.59G - 0.11B$

and $(B - Y) = 0.89B - 0.59G - 0.3R$

2.11 VALUES OF LUMINANCE (Y) AND COLOUR DIFFERENCE SIGNALS ON COLOURS

While televising colour scenes even when voltages R, G and B are not equal, the 'Y' signal still represents monochrome equivalent of the colour because the proportions 0.3, 0.59 and 0.11 taken of R, G and B, respectively still represent the contribution which red, green and blue lights make to the luminance. This aspect is illustrated by considering some specific colours.

(*i*) *Desaturated purple.* Consider a desaturated purple colour, which is a shade of magenta. Since the hue is magenta (purple) it implies that it is a mixture of red and blue. The word desaturated indicates that some white light is also there. The white light content will develop all the three *i.e.*, R, G and B voltages, the magnitudes of which will depend on the intensity of desaturation of the colour. Thus, R and B voltages will dominate and both must be of greater amplitude than G. As an illustration let R = 0.7, G = 0.2 and B = 0.6 volts. The white content is reperesented by equal quantities of the three primaries and the actual amount must be indicated by the smallest voltage of the three, that is, by the magnitude of G.

Thus, white is due to 0.2R, 0.2G and 0.2B. The remaining 0.5R and 0.4B together represent the magenta hue.

The luminance signal Y = 0.3R + 0.59G = 0.11B. Substituting the values or R, G and B we get Y = 0.3 (0.7) + 0.59 (0.2) + 0.11 (0.6) = 0.394 (volts).

The colour-difference signals are:

$$(R - Y) = 0.7 - 0.394 = +\ 0.306 \text{ (volts)}$$

$$(B - Y) = 0.6 - 0.394 = +\ 0.206 \text{ (volts)}$$

At the receiver after demodulation, the signals Y, (B – Y) and (R – Y), become available. Then by a process of matrixing the voltages R and B are obtained as:

$$R = (R - Y) + Y = 0.306 + 0.394 = 0.7\text{V}$$

$$B = (B - Y) + Y = 0.206 + 0.394 = 0.6\text{V}$$

(G – Y) Matrix: The missing signal (G – Y) which is not transmitted can be recovered by using a suitable matrix based on the explanation given below:

$$Y = 0.3R + 0.59G + 0.11B$$

also $$(0.3 + 0.59 + 0.11)\ Y = 0.3R + 0.59G + 0.11B$$

Rearranging the above expression we get:

$$0.59\ (G - Y) = -\ 0.3\ (R - Y) - 0.11\ (B - Y)$$

Therefore, $$(G - Y) = -\ \frac{0.3}{0.59}\ (R - Y) - \frac{0.11}{0.59}\ (B - Y)$$

$$= -\ 0.51\ (R - Y) - 0.186\ (B - Y)$$

Substituting the values of (R – Y) and (B – Y)

$$(G - Y) = -\ (.51 \times 0.306) - (0.186 \times 0.206) = -\ 0.156 - 0.038 = -\ 0.194$$

Therefore,

$$G = (G - Y) + Y = -\ 0.194 + 0.394 = 0.2$$ and this checks with the given value.

Reception on a monochrome receiver. Since the value of luminance signal Y = 0.394 V and peak white corresponds to 1 volt (100%) the magenta will show up as a fairly dull grey in a black and white picture. This is as would be expected for this colour.

(2) *Desaturated orange.* A desaturated orange having the same degree of desaturation as in the previous example is considered now. Taking R = 0.7, G = 0.6, and B = 0.2, it is obvious that output voltages due to white are R = 0.2, G = 0.2 and B = 0.2. The red and green colours which dominate and represent the actual colour content with, R = 0.5 and G = 0.4 give the orange hue. Proceeding as in the previous example we get:

Luminance signal Y = 0.3R + 0.59 G + 0.11B

Substituting the values of R, G and B we get Y = 0.586V.

Similarly, the colour-difference signal magnitudes are:

$$(R - Y) = (0.7 - 0.586) = + 0.114$$

$$(B - Y) = - (0.2 - 0.586) = - 0.386$$

and

$$(G - Y) = - 0.51 (R - Y) - 0.186 (B - Y) = 0.014$$

At the receiver by matrixing we get

$$R = (R - Y) + Y = 0.7$$

$$G = (G - Y) + Y = 0.6$$

$$B = (B - Y) + Y = 0.2$$

This checks with the voltages developed by the three camera tubes at the transmitting end.

Reception on a monochrome receiver: Only the luminance signal is received and, as expected, with Y = 0.586 the orange hue will appear as bright grey.

Polarity of the Colour Difference Signals

As has been demonstrated by the above two examples, both (R – Y) and (B – Y) can be either positive or negative depending on the hue they represent. The reason is that for any primary, its complement contains the other two primaries. Thus, a primary and its complement can be considered as opposite to each other and hence the colour-difference signal turn out to be of opposite polarities. This is illustrated by the colour phasor diagram of Fig. 2.6. Observe that a purplish-red hue is represented by + (R – Y) while its complement, a bluish-green hue corresponds to – (R – Y). Similarly, + (B – Y) and – (B – Y) represent purplish-blue and greenish-yellow hues, respectively (see colour plate 4). Note that green colour is obtained by a combination of – (R – Y) and – (B – Y) while cyan is obtained by a combination of – (R – Y) and + (B – Y) signals. Furthermore, any one of the three primaries or their complementaries can be obtained by a combination of two of the above four signals. It may also be noted that the colour-difference video signals have no brightness component and represent only different hues.

2.12 CONSTANT LUMINANCE

It is expected that at the receiving end luminance of each element of a transmitted picture remains the same irrespective of whether colour signal is received on a black and white or colour receiver. Therefore,

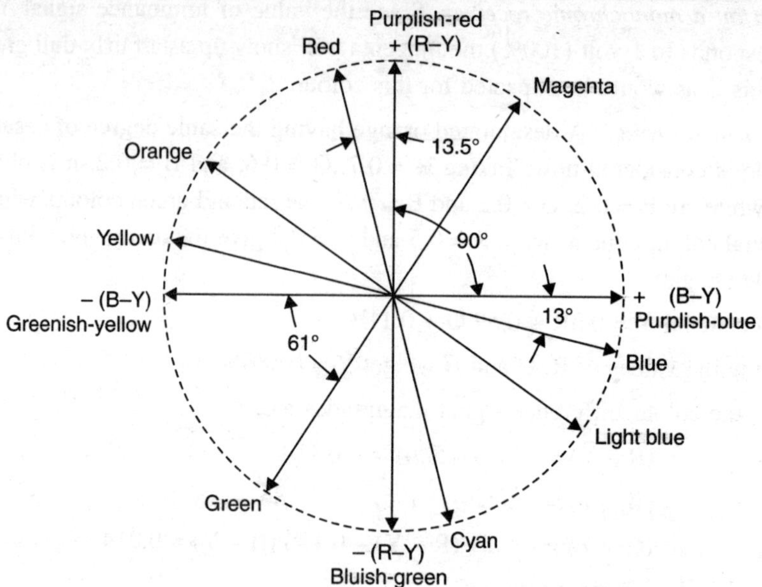

Fig. 2.6 Colour circle showing location and magnitude (100%) of primary and complementary colours (See colour plate 4).

in order to produce correct luminance the picture tube must be fed with voltages that are gamma corrected to offset non-linearity of the tube characteristics. But on colour transmission, the luminance signal is not truly gamma corrected because the camera output voltages are subjected to gamma correction before Y signal is formed. Actually the Y signal on gamma correction should become.

$$(Y)^{1/\gamma} = (0.3R + 0.59G + 0.11B)^{1/\gamma}$$

where R, G and B are the colour voltages before gamma correction. However, in practice it is:

$$Y^{1/\gamma} = 0.3R^{1/\gamma} + 0.59G^{1/\gamma} + 0.11^{1/\gamma}$$

because of prior correction. Thus, the two expressions are not same mathematically except when R = G = B *i.e.*, when grey or white elements are transmitted. For voltages on colours, when R, G and B are not equal, substitution of arbitrary values in the above expressions shows that amplitude of the transmitted luminance signal is smaller than it should be. Such an exercise for pure red colour (assuming γ = 2.0) yields the true gamma-corrected value of Y equal to

$$(0.3R)^{1/\gamma} = (0.3(1))^{1/2} = (0.3)^{1/2} \text{ (since R = 1 V).}$$

However, because of gamma correction prior to formation of Y signal, the value produced will be $0.3(R)^{1/2} = 0.3(1)^{1/2} = 0.3$.

At the receiving end where light output from the picture tube is proportional to square of input voltage; the luminance produced by the truly gamma corrected Y signal will be $[(0.3)^{1/2}]^2 = 0.3$, whereas the actual luminance produced will be $(0.3)^2 = 0.09$. The reproduced luminance is thus less than it should be.

Such an exercise for other primary colours would show that the error is greatest on blue and least on green. To sum up, colour areas of the scene will appear darker on a black and white receiver.

However, in actual practice it is not so because the discrepancy gets mostly corrected in a very interesting way. The chrominance signal which is present in the video on a colour transmission is applied to the picture tube input as a sinewave superimposed upon the luminance level. Due to non-linearity of the tube's I_b/V_g characteristics, the increase in Ib due to one-half cycle of the sinewave is more than the decrease due to the other half. The signal is thus partially rectified and causes a net increase in the beam current because of the added dc component due to rectification of the chrominance signal. The greater the level of saturation more is the dc component that gets added to the otherwise lower luminance (dc) level and thus raises the beam current to almost the same level as it should be. Thus, the brightness level of coloured areas on a black and white screen is not much affected and compatibility between the two systems is maintained.

There is no such problem with a colour receiver because luminance signal is eliminated in the final matrix. For example, the blue gun is excited by a voltage = (B – Y) + Y = B which is gamma corrected and Y does not get applied as such. Thus, brightness details of colours are not affected when colour transmission is received on a colour receiver.

The effect of any noise voltage on the chrominance signal is diminished because an increase in (R – Y) or (B – Y) or both due to noise causes a decrease in (G – Y) which is equal to [– 0.51 (R – Y) – 0.186 (B – Y)] and hence the effect is partly compensatory. The use of colour-difference signals thus helps to achieve constant luminance by making luminance more independent of arbitrary changes in the chrominance signal than it would be if all the three colour-difference signals are transmitted. This also justifies the use of colour-difference signals instead of R, G and B.

REVIEW QUESTIONS

1. What do you understand by compatibility in TV transmission? Enumerate essential requirements that must be met to make a colour system fully compatible with black and white transmission.
2. Describe with a diagram the construction of a colour TV camera and its optical system. Why are the outputs of all the three camera tubes set equal when standard white light is made incident?
3. What do you understand by gamma correction of camera output voltage? How is it connected with the beam current/drive voltage characteristics of a picture tube?
4. Draw camera output waveshapes for V_R, V_G and V_B for primary and complementary colours. What is the effect of desaturation on camera output voltages? Illustrate your answer by sketching colour voltage output for a desaturated magenta hue.
5. Explain how the 'Y' and colour difference signals are developed from camera outputs. Why is the 'Y' signal set = 0.3R + 0.59G + 0.11B?
6. What is the significance of colour-difference signals? How are they suitable as an aid to compatibility? Write expressions for (R – Y), (B – Y) and (G – Y) in terms or R, G and B.
7. Why is the (G – Y) colour-difference signal not chosen for transmission? Explain how it is obtained in the receiver for modulating corresponding beam of the picture tube.
8. Explain how colour-difference signals disappear at the output of the signal combining matrix on white and grey shades. What is the significance of 'Y' signal in colour transmission and reception?
9. While televising a static desaturated colour scene, the camera outputs were found to have the following amplitudes:

$$V_R = 0.7V, V_G = 0.6V \text{ and } V_B = 0.3V$$

What is the basic hue of the scene? Compute values of (G – Y) and Y signals and establish that true hue and brightness will be reproduced in the receiver. What shade will such a signal produce in a monochrome receiver?

3 COLOUR PICTURE TUBES AND ASSOCIATED CIRCUITS

A colour picture tube has three separate electron beams. The inside of its screen is coated with three different phosphors, one for each of the chosen red, green and blue primaries. The three phosphors are physically separate from one another and each is energized by an electron beam of intensity that is proportional to the respective colour voltage reproduced in the television receiver. The object is to produce three coincident rasters which produce red, green and blue contents of the transmitted picture. While seeing from a normal viewing distance the eye integrates the three colour information to convey sensation of the hue at each part of the picture. Based on the gun configuration and the manner in which phosphors are arranged on the screen, three different types of colour picture tubes have been developed. These are:

1. Delta-gun colour picture tube
2. Guns-in-line or Precision-in-line (P-I-L) colour picture tube
3. Single gun or Trinitron Colour Picture tube

3.1 DELTA-GUN COLOUR PICTURE TUBE

This tube was first developed by the Radio Corporation of America (R.C.A). It employs three separate guns (see Fig. 3.1(*a*)), one for each phosphor. The guns are equally spaced at 120° interval with respect to each other and tilted inwards in relation to the axis of the tube. They form an equilateral triangular configuration.

As shown in Fig. 3.1(*b*), the tube employs a screen where three colour phosphor dots are arranged in groups known as *triads*. Each phosphor dot corresponds to one of the three primary colours. The triads are repeated and depending on the size of the picture tube, approximately 1,000,000 such dots forming nearly 333,000 triads are deposited on the glass face plate. Abcut 1-cm behind the tube screen (see Figs. 3.1(*b*) and (*c*)) is located a thin perforated metal sheet known as the *shadow mask*. The mask has one hole for every phosphor dot triad on the screen. The various holes are so oriented that electrons of the three beams on passing through any one hole will hit only the corresponding colour phosphor dots on the screen. The ratio of electrons passing through the holes to those reaching the shadow mask is only about 20 percent. The remaining 80 percent of the total beam current energy is dissipated as a heat loss in the shadow mask. While the electron transparency in other types of colour picture tubes is more, still, relatively large beam currents have to be maintained in all colour tubes compared to monochrome tubes. This explains why higher anode voltages are needed in colour picture tubes than are necessary in monochrome tubes.

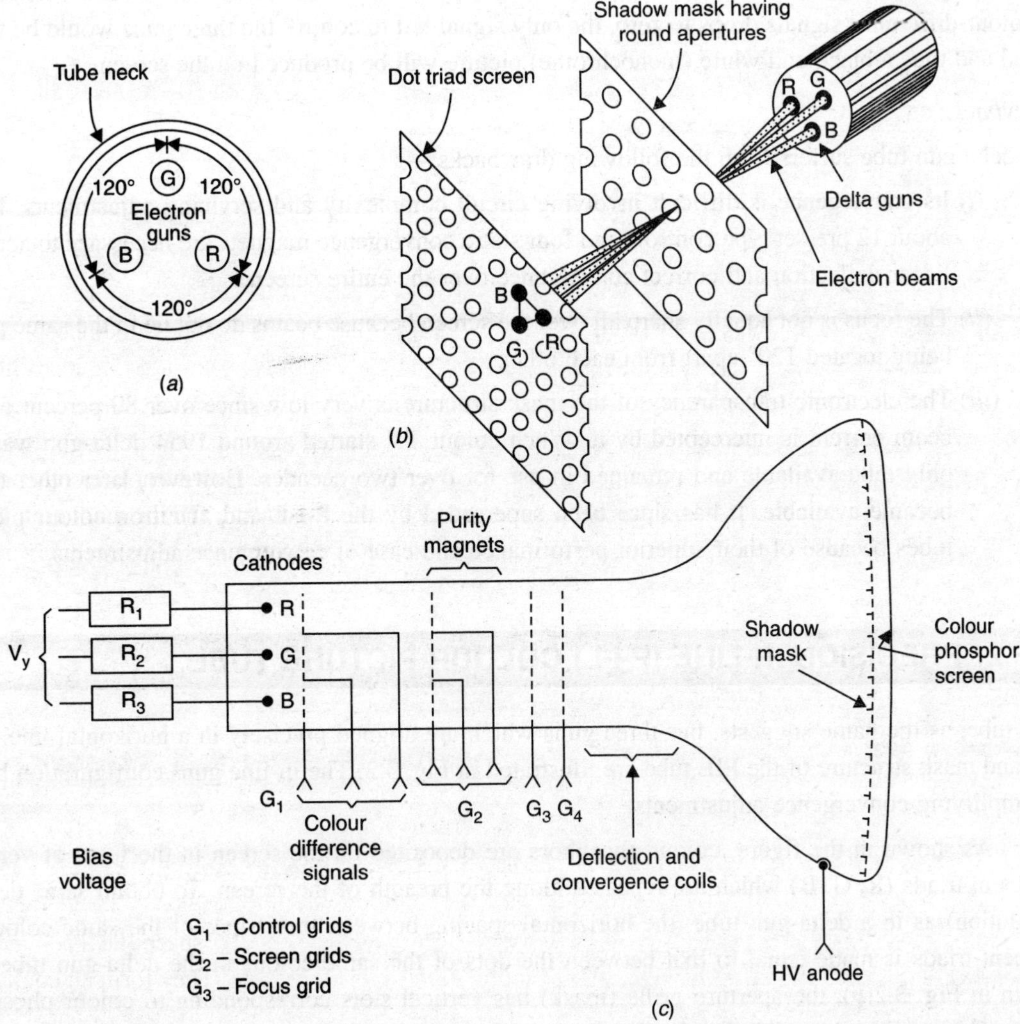

Fig. 3.1 Delta-gun colour picture tube. (*a*) guns viewed from the base, (*b*) electron beams, shadow mask and dot-triad phosphor screen, (*c*) schematic diagram showing application of 'Y' and colour-difference signals between the cathodes and control grids.

Generation of Colour Rasters

The overall colour seen is determined both by the relative intensity of each beam and the phosphors which are being bombarded. If only one beam is 'on' and the remaining two are cut-off, dots of only one colour phosphor get excited. Thus, the raster will be seen to have only one of the primary colours. Similarly, if one beam is cut-off and the remaining two are kept on, the rasters produced by excitation of the phosphors of two colours will combine to create the impression of a complementary colour. The exact hue will be determined by relative strengths of the two beams. When all the three guns are active simultaneously, lighter shades are produced on the screen. This is so because red, green and blue combine in suitable amounts to form white, and this combines with whatever colours are present to desaturate them. Naturally, intensity of the colour produced depends on the intensity of beam currents. Black in a picture is just the absence of excitation when all the three beams are cut-off. If the amplitude of colour-difference signals drops to zero, the only signal left to control the three guns would be the Y signal and thus a black and white (monochrome) picture will be produced on the screen.

Drawbacks

The delta gun tube suffers from the following drawbacks:—

(*i*) Its convergence is difficult involving circuit complexity and servicing adjustments. In all about 12 pre-set type controls and four static convergence magnets are necessary to achieve linear deflection and correct convergence over the entire screen.

(*ii*) The focus is not equally sharp all over the screen because beams do not lie in the same plane being located 120° apart from each other.

(*iii*) The electronic transparency of the mask structure is very low since over 80 percent of the beam current is intercepted by it. When colour TV started around 1954 delta-gun was the only tube available and remained in use for over two decades. However, later other tubes became available. It has since been superseded by the P-I-L and Trinitron colour picture tubes because of their superior performance and ease of convergence adjustments.

3.2 PRECISION-IN-LINE (P-I-L) COLOUR PICTURE TUBE

This tube, as the name suggests, has three guns which are aligned precisely in a horizontal line. The gun and mask structure of the PIL tube are illustrated in Fig. 3.2. The in-line guns configuration helps in simplifying convergence adjustments.

As shown in the figure, colour phosphors are deposited on the screen in the form of vertical stripes in triads (R, G, B) which are repeated along the breadth of the screen. To obtain same details (resolution) as in a delta-gun tube, the horizontal spacing between the stripes of the same colour in adjacent triads is made equal to that between the dots of the same colour in the delta-gun tube. As shown in Fig. 3.2(*b*), the aperture grille (mask) has vertical slots corresponding to colour phosphor stripes. One vertical line of slots is for one group of fine stripes of red green and blue phosphors. Since all the three electron beams are on the same plane, the beam in the centre (green) moves along the axis of the tube. However, because of inward tilt of the right and left guns, the blue and red beams travel at an angle and meet the central beam at the aperture grille mask. The slots in the mask are so designed that each beam strikes its own phosphor and is prevented from landing on other colour phosphors.

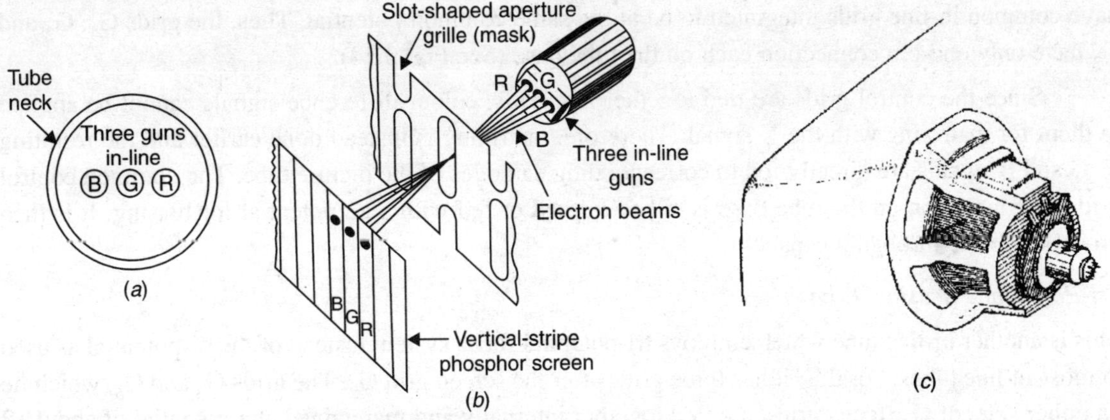

Fig. 3.2 Precision-in-line (P-I-L) or cathodes in-line colour picture tube. (*a*) in-line guns, (*b*) electron beams, aperture grille and striped three colour phosphor screen (*c*) yoke and magnetic assembly.

Beam Focusing

A bi-potential beam focusing system employing four grids is commonly used in P-I-L tubes. The grid structures are cylindrical metal tubes of various lengths which are closed at one end with an apperture disc that has a small hole in the centre. These cylinders establish necessary electrostatic field to produce an electron beam. The aperture discs are used to select electrons that can reach the phosphor screen from those moving along axis of the tube.

The grid (G_1) may have positive, negative or zero potential with the condition that net difference of potential (bias) between the cathode and control grid is such that it makes the control grid negative with respect to cathode. On application of video signal the instantaneous bias changes to control flow of electrons (beam current) moving on towards the phosphor-striped screen.

The purpose of screen grid (G_2) is to accelerate electrons towards phosphor screen and also to provide focusing action. Modern colour tubes use low-voltage focus technique with about 500V at G_2. Following the screen grid is the focus grid G_3. As the name suggests the electrostatic field due to potential on it enables focusing of the beam to a fine point on the screen. The potential of G_3 is set around 2 KV by a pre-set to obtain correct focusing action.

The accelerating anode (G_4) consists of a cylinder that is connected to the graphite (aquadag) coating on the inside of tube bell. The coating extends up to the phosphor screen but does not touch it. It is connected to the positive terminal of a high voltage supply which is typically 25 KV in colour picture tubes. This EHT (high voltage) establishes a very strong electric field along the funnel of the tube causing beam electrons to remain focused on deflection all along the screen surface. Extra high voltage is obtained from the line-output (horizontal) deflection circuit and connected to the inner conductive coating through a circular metal cap imbedded in the wall of the tube shell.

Integrated Gun Structure

Earlier in-line colour picture tubes employed three separate guns with three independent control, screen and focus grids. This meant separate pre-sets for each of these grids. However, modern P-I-L tubes have common in-line grids integrated to be at the same common potential. Thus, the grids G_1, G_2 and G_3 have only one-pin connection each on the tube base (See Fig. 3.14).

Since the control grids are tied together internally, colour-difference signals cannot be applied to them for matrixing with the Y signal. Therefore, matrixing is instead done earlier and the resulting R, G and B signals are directly fed to corresponding cathodes of the picture tube. The common control grid-pin connection on the tube base is either earthed or fed with a dc potential for biasing. It is then grounded for ac through a capacitor.

Tri-Potential Focusing Tube

This is another in-line tube which employs tri-potential focus system instead of the bi-potential as used in most-in-line tubes. For this, it has three grids after the screen grid G_2. The grids G_3 and G_5, which lie on either side of G_4 (focus grid), are tied together internally and maintained at a potential of about 12 KV. The focus grid (G_4) has nearly 7 KV on it for providing double-lens action. It is claimed that such a focus arrangement enables a very sharp beam spot on any location of the screen and is in particular useful for screen sizes larger than 51 cm.

3.3 PURITY AND CONVERGENCE

The three beams on deflection must produce pure colours all over the screen. This implies that if green and blue guns are turned off, a pure red raster should be produced. Similarly, it should be possible to obtain green and blue rasters when only corresponding guns are switched on one at a time. These rasters are then said to be of good PURITY. If colour light levels of the three rasters are properly set and all the beams switched 'on' simultaneously, a pure white raster will be perceived. The purity requirement is thus satisfied.

Loss of purity, which may be due to either wrong entry angles of the beams, or incorrect point of deflection under the yoke or strong magnetic fields, can cause discolouration. This can happen in any one or all the three colour rasters. Such a loss of purity causes patches of colour in one or more areas of an otherwise black and white picture. Poor purity is more of a problem while receiving black and white transmission than with colour reception where it is somewhat comouflaged by the colour programme material.

For obtaining colour purity each beam should land at the centre of corresponding phosphor dot irrespective of the location of beams on the raster. This needs precise alignment of the colour beams and is carried out by a circular magnet assembly known as the *purity magnet.* It is mounted externally on the neck of the tube and close to the deflection yoke. The purity magnet assembly consists of several flat washer like magnets held together by spring clamps in such a way that these can be rotated freely. The tabs on the magnets can be moved apart to reduce resultant field strength. This is illustrated for a two pole magnet in Fig. 3.3. As shown in the same figure, the tabs when moved together change the direction of magnetic field. Two, four and six pole magnet units are employed to achieve individual and collective beam deflections. Thus, to affect purity and static convergence the beams can be deflected up or down, right or left and diagonally by suitably orienting the purity magnets.

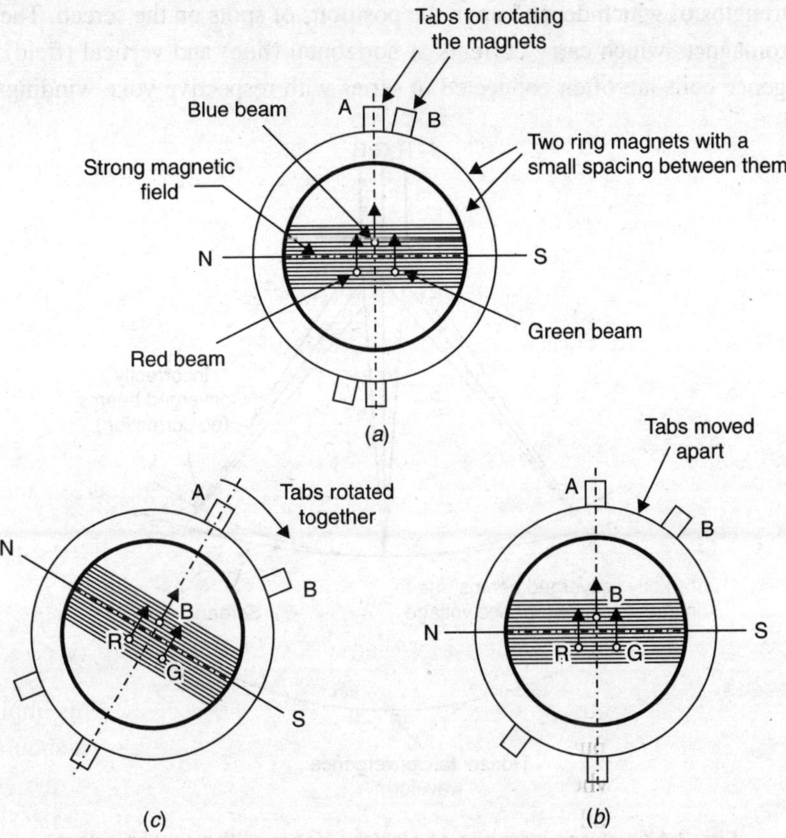

Fig. 3.3 Two pole purity magnet assembly (*a*) strong magnetic field when tabs (A and B) are nearly together, (*b*) spreading the tabs reduces magnetic field, (*c*) rotating the magnets together rotates the magnetic field to cause change in the direction of beam deflection.

Yoke Position

The position of yoke on the tube neck determines location of deflection centre of the electron beams. A wrong setting will result in poor purity due to improper entry angles of beams into the mask openings. Since deflection due to yoke fields affect landing of beams on the screen more towards edges of the tube, the yoke is moved along neck of the tube to improve purity in those regions.

Convergence

The three colour rasters should not only be pure but also coincident *i.e.,* they must fully overlap each other for correct reproduction of both colour and monochrome pictures. The technique of bringing the beams together so that they hit the same part of the screen at the same time to produce three coincident rasters is referred to as convergence. Convergence errors are caused by (*i*) non-coincident convergence planes, (*ii*) non-uniformity of the deflection field and (*iii*) flat surface of the picture tube screen. Figure 3.4 illustrates correct and incorrect convergence of beams. Proper convergence is achieved by positional adjustment of the individual beams. It falls into two parts referred to as (*i*) static and (*ii*) dynamic convergence. Static convergence involves movement of the beams by permanent magnetic fields which, once correctly set, bring the beams into convergence in the central area of the screen. Convergence over rest of the screen is achieved by continuously varying (*i.e.,* dynamic) magnetic fields, the instantaneous strengths of which depend upon the positions of spots on the screen. These fields are set up by the electromagnets which carry currents at horizontal (line) and vertical (field) frequencies. In practice convergence coils are often connected in series with respective yoke windings.

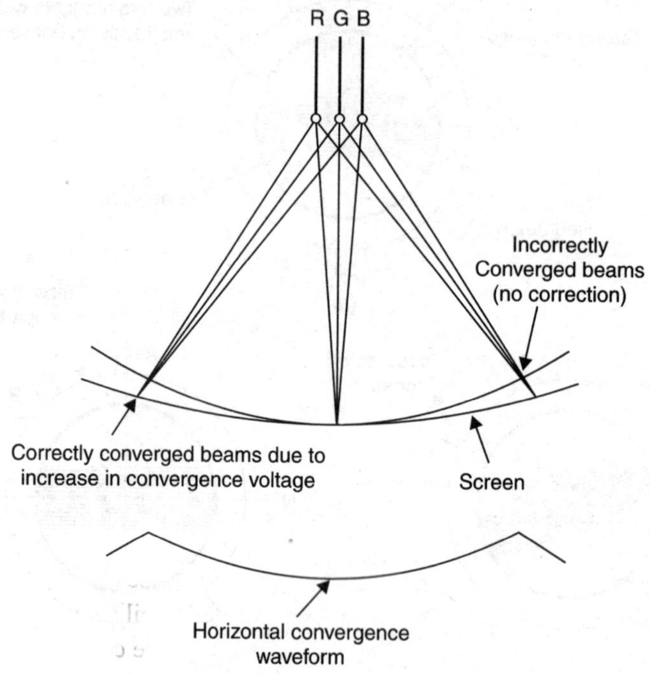

Fig. 3.4 Over-convergence of electron beam at the screen edges unless a corrective deflection field is utilized.

In an in-line picture tube, the problem of misconvergence is only in the horizontal plane. It is so because guns being in line, the distance of beams to the screen for vertical deflection remains the same for any location on the raster. Therefore, no vertical misconvergence occurs. Thus, convergence difficulties are much less in in-line tubes as compared to delta-gun where both vertical and horizontal misconvergence occurs because the three guns are not in the same plane.

The convergence problem in in-line tubes has been further simplified by perfecting yoke and tube designs. In fact, in modern in-line colour tubes practically no dynamic convergence adjustments are necessary. Any adjustments that may be necessary are made at the factory. However, for purity and static convergence adjustments, ring magnets are mounted on the neck of the tube.

3.4 THE DEFLECTION UNIT

As explained in the previous section, misconvergence of the three electron beams R, G and B is due to non-linearity of both horizontal and vertical deflecting fields. Most of the non-linearity in P-I-L picture tubes is in the horizontal plane and it is corrected by employing 2, 4 and 6-pole permanent magnets as detailed in sections 3.3 and 3.7. However, still, field astigmatic errors persist, both due to non-linearity of horizontal and vertical deflecting fields. The necessary correction is carried out by designing special shaped deflection coils, with the aim to provide deflection field distributions as shown in Fig. 3.5.

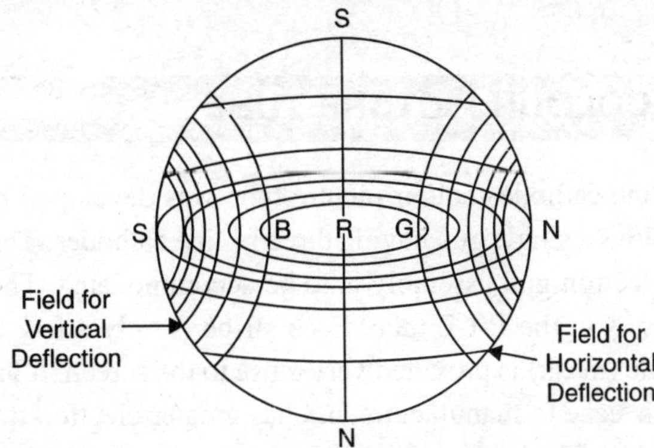

Fig. 3.5 The astigmatic deflection fields required for a self-converging in-line picture tube. The field shapes are achieved by precision construction of the deflection yoke.

Deflection Yoke

After prolonged research and development many picture tube manufactures introduced deflection units for providing automatic self-convergence. 'PHILIPS' were the first to introduce such a unit for 110° deflection and screen size up to 66 cm (≈ 26 inch). It makes use of the nature of in-line gun array in conjunction with a specially designed saddle shaped deflection coil assembly. This system, called 20 AX, needs least number of corrections for obtaining purity and true convergence. The yoke assembly is provided with a plastic ring for necessary axial alignment. Similarly, there is provision to rotate the assembly by a small angle for correct raster orientation, later with the aid of computer aided design techniques. PHILIPS introduced 30 AX yoke and deflection unit for more accurate deflection angles and large screen sizes.

In this and other similar systems marketed by other manufacturers, the three electron guns, mask and deflection yoke assembly are constructed with such precision that hardly any purity and convergence adjustments are necessary. However, small errors do occur on account of mounting tolerances, stray magnetic field effects and non-uniform field distribution. Small corrections are therefore unavoidable in such assemblies.

3.5 MERITS OF IN-LINE GUN PICTURE TUBES

Based on the above discussion merits of a in-line colour picture tube may be summarized as under:

(*i*) The main advantage is the elimination of all convergence adjustments after the tube leaves the factory. A precision wound toroid-deflection yoke is installed and cemented into place. The design and placement of this yoke is the key to the tube's "self-convergence" capability.

(*ii*) The vertical phosphor stripes are also an important feature of the preconverged tube design. This is true because with vertical phosphor stripes, there cannot be vertical mis-register of the electron beams. Also as mentioned earlier, with the three electron beams in the horizontal plane, no vertical convergence correction is needed.

(*iii*) With the introduction of integrated gun assembly, separate grid potentials are not necessary for the three guns. A common connection means same potential and hence convenient to provide and adjust electrode voltages.

3.6 TRINITRON COLOUR PICTURE TUBE

The Trinitron or three in-line cathodes colour picture tube was developed by 'SONY' Corporation of Japan around 1970. It employs a single gun having three in-line cathodes. This simplifies constructional problems since only one electron gun assembly is to be accommodated. The three phosphor triads are arranged in vertical stripes as in the P-I-L tube. Each stripe is only a few thousandth of a centimetre wide. A metal aperture grille (mask) is provided very close to the screen. It has one vertical slot for each phosphor triad. The grille is easy to manufacture and has greater electron transparency as compared to both delta-gun and P-I-L tubes. The beam and mask structure, together with constructional and focusing details of the Trinitron are shown in Fig. 3.6. The three beams are bent by an electrostatic lens assembly. Since the beams have a common focus plane a sharper image is obtained with good focus over the entire picture area. All this simplifies convergence problems and fewer adjustments are necessary.

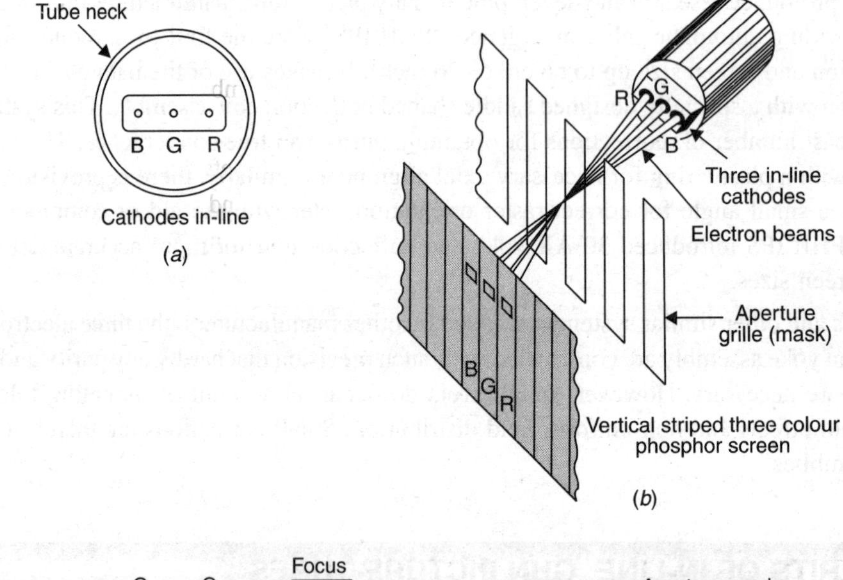

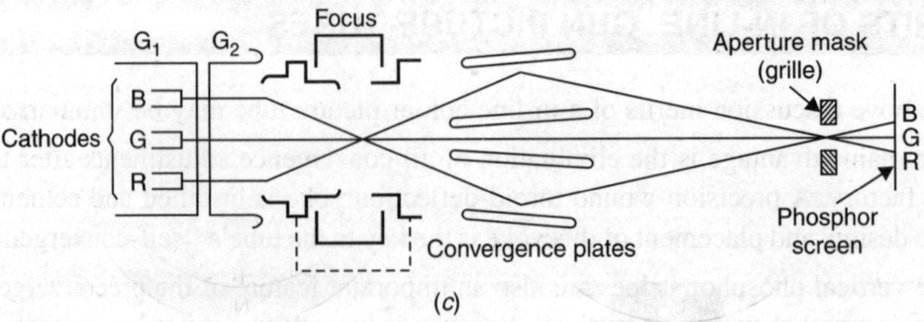

Fig. 3.6 Trinitron (cathodes in-line) colour picture tube (*a*) gun structure, (*b*) electron beams and vertical-striped three colour phosphor screen, (*c*) focus and convergence details.

Electrons emitted from the cathodes pass through small holes in control grid G_1 and are speeded up by the screen grid G_2. The electron beams pass through holes in this electrode and proceed at high speed into the focus electrode region. This is essentially an electrostatic-lens assembly unit which brings the beams to a point. Only the green electron beam proceeds in a straight line. The red and blue beams come together at a point and then diverge. The focusing action has the effect of reducing diameter of each beam. The beams are then attracted to the converging plates which operate at about 19 KV. The second electron lens arrangement brings the beams to precise focus on the aperture grille.

Field purity is obtained by proper adjustment of purity magnets and correct positioning of the deflection yoke. A green field is ordinarily used in this procedure. However, convergence at the edges of the screen requires the addition of a parabolic voltage waveform to the converging plates. The latest version of Trinitron incorporates a low magnification electron gun assembly, long focusing electrodes and a large aperture lens system. The new high precision deflection yoke with minimum convergence adjustments provides a high quality picture with very good resolution over large screen display tubes.

3.7 PURITY AND STATIC CONVERGENCE ADJUSTMENTS

Modern P-I-L tubes are manufactured with such precision that hardly any purity and static convergence adjustments are necessary. However, to correct for small errors due to mounting tolerances and stray magnetic fields, multiple permanent magnet units are provided. As shown in Fig. 3.2 (*c*), these are mounted on the neck of tube next to the deflection yoke. The assembly incorporates four ring-shaped permanent magnet units.

The magnetic rings comprise of (*i*) two pairs of 2-pole magnets, (*ii*) one pair of 4-pole magnets and (*iii*) one pair of 6-pole magnets. Each pair consists of an inner and outer ring of identical magnetic configuration. As in any purity magnet assembly, (see Fig. 3.3) rotating one of the two rings varies the resultant magnetic field strength and rotating them together varies direction of the resultant field. For mutual rotation in opposite direction, the rings are coupled by small pinion gears. Details of various adjustments for a typical multipole unit are as follows:

Horizontal Colour Purity

As illustrated in Fig. 3.7 horizontal colour purity is obtained by varying field strength of the 2-pole magnet situated between the 4-pole and 6-pole magnets.

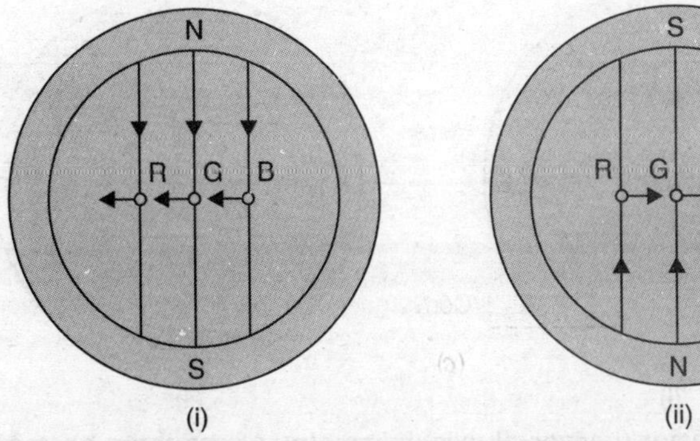

Fig. 3.7 Horizontal colour purity. All the three beams move equally in the horizontal direction.

Static Convergence

It is obtained by varying the field strength and direction of the 4-pole and 6-pole magnet pairs. The 4-pole field moves the outer electron beams (red and blue) equally in opposite directions. The 6-pole field moves the outer electron beams equally in the same direction. The central beam (green) is unaffected. Magnetic field directions and consequent deflections are illustrated in Fig. 3.8.

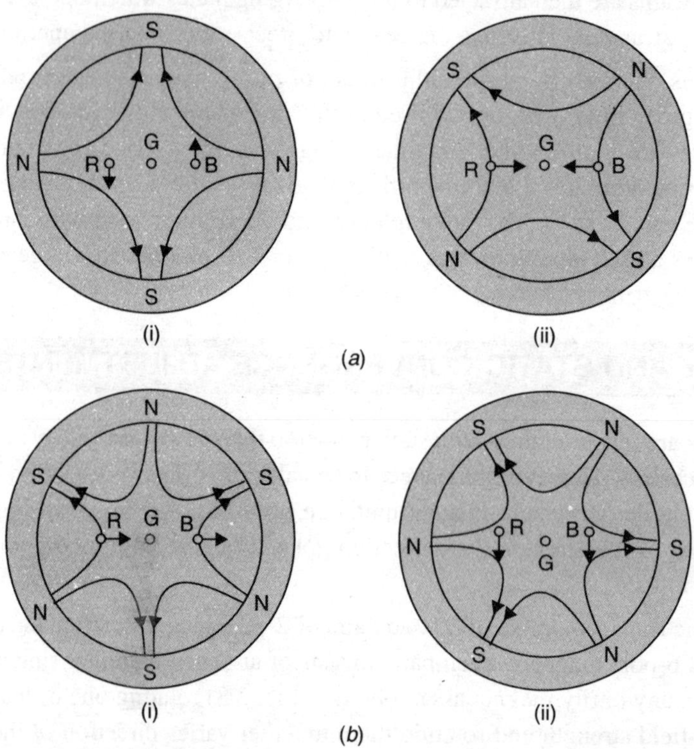

Fig.3.8 Static convergence (*a*) the outer electron beams (R&B) move equally in opposite directions, (*b*) the outer electron beams move equally in the same direction.

Raster Symmetry

Horizontal axis or raster symmetry is adjusted by varying field strength of the 2-pole magnet located at the rear of the multi-pole unit. All the three beams (Fig. 3.9) are equally moved in a vertical direction.

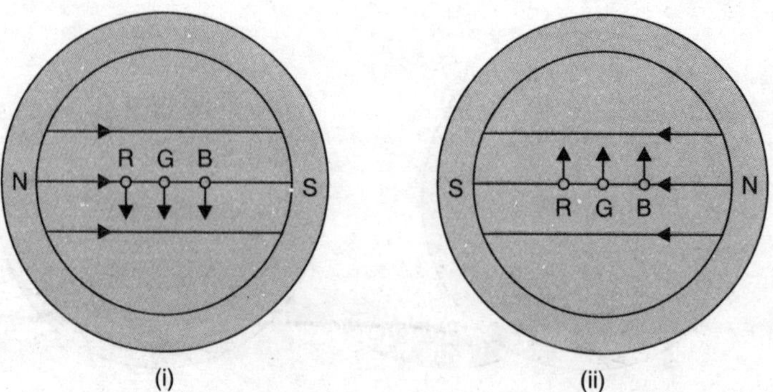

Fig. 3.9 Raster (horizontal axis) symmetry. All the three beams move equally in the vertical direction.

3.8 DYNAMIC CONVERGENCE ADJUSTMENTS

As already stated the display unit consisting of picture tube and deflection unit is inherently self-converging. No correction is necessary for deflection in the vertical direction. However, little adjustments necessary for linear horizontal deflection are provided. For this purpose two types of four pole dynamic magnetic fields are used. One is generated by additional windings on the yoke ring of the deflection unit. It is energized by adjustable sawtooth currents synchronized with scanning. The other type of dynamic field is generated by sawtooth and parabolic currents which are synchronized with scanning and flow through the deflection coils. Thus, the need that field distribution at the gun end be barrel shaped horizontally and pincushion vertically is met. Necessary adjustments are carried out at the factory and no external controls are provided.

Since the deflection yoke magnetic flux centre must align with the electron beam axis, this alignment if necessary is made by turning the yoke sideways and tilting it up and down. Wedges placed between the yoke and tube neck are provided to enable motion both in the horizontal and vertical directions. In all three wedges are used. Once overall convergence is correctly obtained, the screw of clamp holding the yoke is firmly tightened to fix it in position. The adhesive tapes are then put on the wedges to keep them in position.

3.9 PINCUSHION CORRECTION

Since magnetic deflection of the electron beams is along the path of an arc, with point of deflection as its centre, the screen should ideally have corresponding curvature for linear deflection. However, the small curvature of the tube screen does not coincide with the deflection path because distances at the four corners are greater as compared to central portion of the face plate. Therefore, electron beams travel farther at the corners causing more deflection at the edges. This results in a stretching effect where top-bottom (N-S) and right-left (E-W) edges of the raster tend to bow inwards towards centre of the screen. The result as illustrated in Fig. 3.10(*a*) is a pincushion like raster. Though the figure shows an exaggerated view of the so-called pincushion distortion, it needs correction to ensure correct registration of colours in those areas. The use of permanent magnets for the elimination of pincushion distortion is not feasible for colour receivers because the magnets would tend to introduce purity problems. Therefore, dynamic pincushion correction is used with colour picture tubes. Figs. 3.10(*b*) and (*c*) show basic circuits and associated waveforms illustrating dynamic pincushion correction techniques. Such a correction automatically increases horizontal width and vertical size in those regions of the raster that are shrunken because of pincushion distortion.

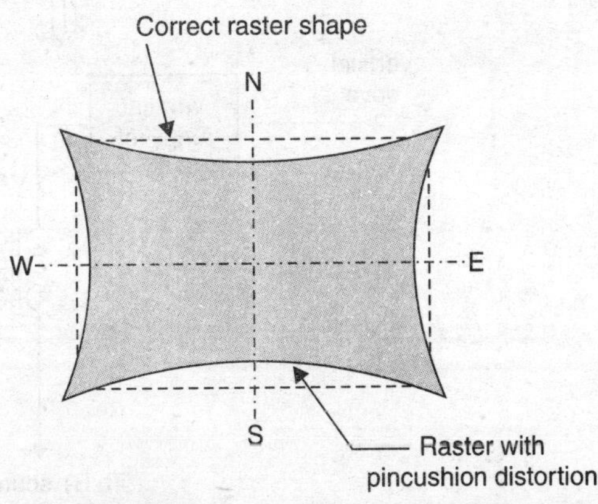

Fig. 3.10 (*a*) Pincushion distortion.

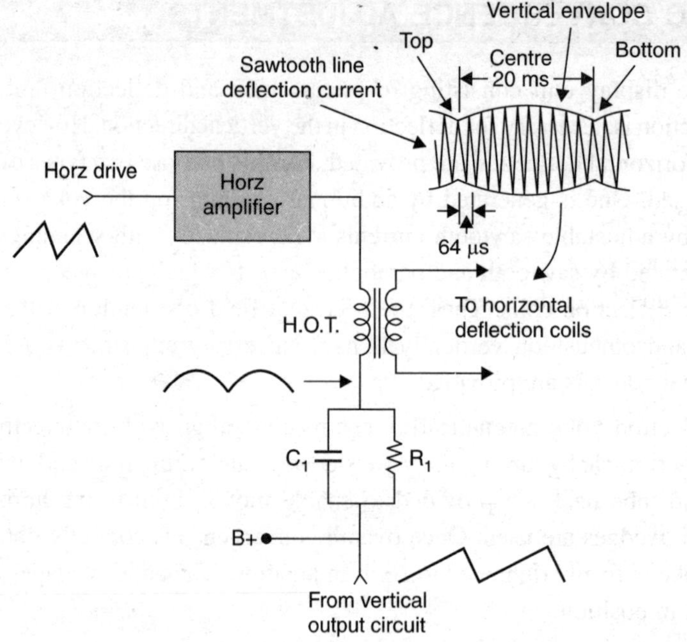

Fig. 3.10 (*b*) Horizontal (E-W) pincushion correction circuit and waveforms.

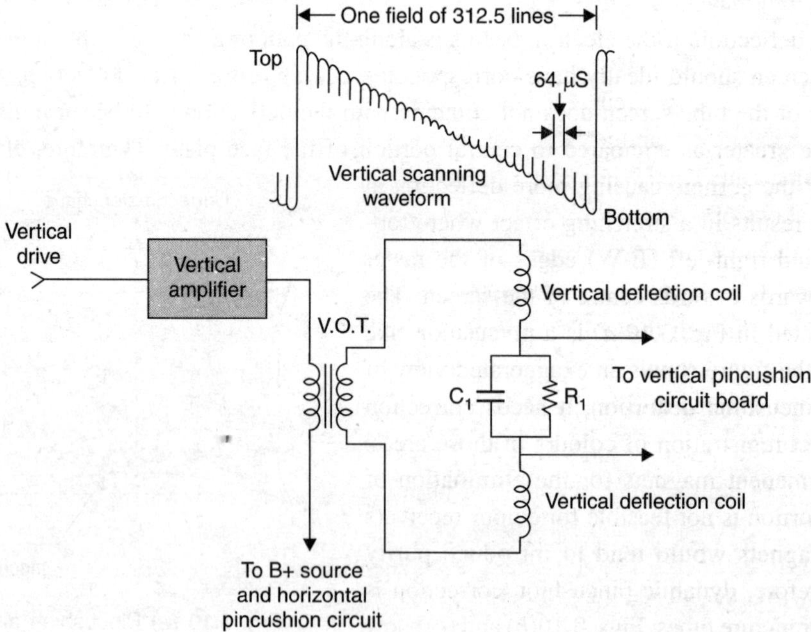

Fig. 3.10 (*c*) Vertical (N-S) pincushion correction circuit and waveforms.

3.10 DEGAUSSING AND ADG CIRCUITS

The main cause of poor purity is the susceptibility of the mask and its mounting frame of become magnetized by the earth's magnetic field and /or by any other strong magnetic fields. The effect of these localized magnetic fields is the deviation of electron beams from their normal path. Thus, the beams strikc wrong phosphors causing poor purity especially at edges of the screen. To prevent such effects the picture tubes are magnetically shielded. It is done by placing a thin silicon steel (mu-metal) housing around the bell of the tube. Since the mask structure and shield material have non-zero magnetic retentivity they get weakly magnetized by the earth's magnetic field and other extraneous fields such as from magnetic toys, domestic electrical apparatus, etc. Thus, over a time, despite initial adjustments or whenever the colour receiver is moved from one location to another, the stray field changes to effect purity. To overcome this drawback, some form of automatic degaussing is incorporated in all colour receivers. Degaussing means demagnetizing iron and steel parts of the picture tube mountings. A magnetic object can be demagnetized by placing it in an alternating magnetic field which becomes weaker over a period of time. This way the magnetized object is forced to assume the strength of the external degaussing field and becomes weaker and weaker as the degaussing field diminishes. A degaussing coil is used for this purpose. It is wrapped round the tube bowl close to the rimband of the screen. The circuit is so designed that when the receiver is first switched on, a strong mains current passes through the coil and then dies away to an insignificant level after a few moments. This way the effects of localized magnetic fields are removed each time the receiver is used.

Automatic Degaussing (ADG) Circuit

There are many degaussing circuits in use. Figure 3.11(*a*) shows details of on automatic degaussing circuit. It uses a thermistor and a varistor combination for controlling the flow of alternating current through the degaussing coil. When the receiver is turned on the ac voltage drop across the thermistor is quite high (about 60 volts) and this causes a large current to flow through the degaussing coil. Because of this heavy current, the thermistor heats up, its resistance falls and voltage drop across it decrease. As a result, voltage across the varistor decreases thereby increasing its resistance. This in turn reduces ac current through the coil to a very low value. The circuit components are so chosen that initial surge of current through the degaussing coil is close to 4 amperes and drops to about 25 mA in less than a second. Once the thermistor heats up degaussing ends and normal ac voltage is restored to the rectifier circuit.

Another ADG circuit that is commonly used is shown in Fig. 3.11(*b*). The secondary of the mains transformer is tapped at about 100V and connected to the degaussing coil through a thermistor having positive temp-coefficient characteristics. When the receiver is switched on, resistance of the thermistor is low (25 Ω) and a high ac current flows through the degaussing coil, demagnetising magnetic metal parts near the picture tube. In about a second's time the thermistor heats up and develops very high resistance (1 MΩ) and current to the degaussing coil drops to almost zero. The process repeats every time the receiver is switched on for use.

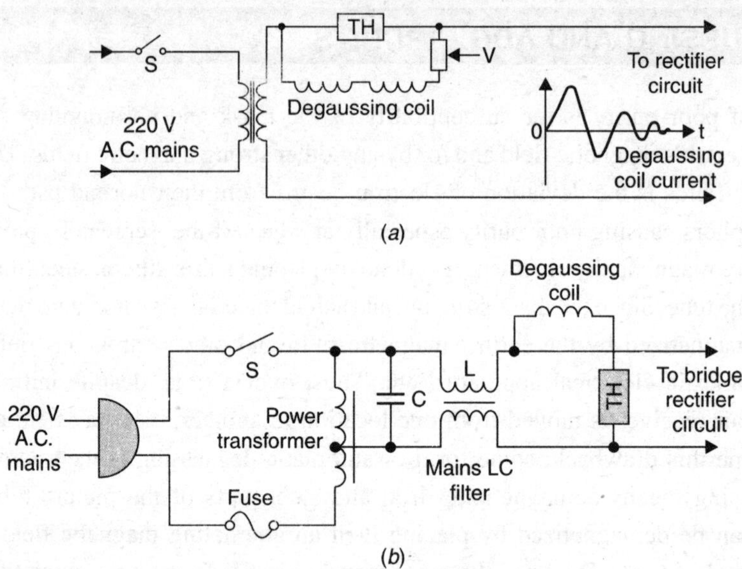

Fig. 3.11 Automatic degaussing circuits (*a*) Thermistor and varistor degaussing circuit, (*b*) Positive thermister degaussing circuit.

If the set is moved or faced in a different direction, the mains switch must be thrown off for atleast 10 minutes in order to allow the thermistor to fully cool down to enable the AGC circuit to become fully operative.

External Degaussing Coil

Portable degaussing coils are used for very bad cases of poor purity. These coils are approximately 30 cm in diameter and consist of about 1000 turns of No. 20 SWG enamelled wire forming a bundle about 1-cm thick. Degaussing is accomplished by connecting the degaussing coil to the 220V mains and then moving it over the screen face, sides, top and bottom of the tube. Care is taken not to bring the coil too close to the neck of the tube or to the speaker where permanent magnets are mounted. After about 10 seconds the coil is moved away to about 3 metres from the tube. There it is turned at right angles to the receiver before switching off the supply. If the degaussing coil is not moved away from the picture tube before turning off the mains supply to it, a strange looking colour impurity appears on the raster.

3.11 WHITE BALANCE (GREY SCALE) ADJUSTMENTS

It may be recalled that red, green and blue lights must combine in definite proportions to produce white light. The three phosphors have different efficiencies. Also the three guns may not have identical I_b/V_{gk} characteristics and cut-off points. Therefore, it becomes necessary to incorporate suitable adjustments such that monochrome information is reproduced correctly (with no colour tint) for all setting of the contrast control. In practice, this amounts to two distinct steps.

Adjustments of Low-Light Level

Non-coincident I_b/V_{gk} characteristics and consequent difference in cut-off points of the three guns results in appearance of coloured tint instead of pure dark grey shades in areas of low brightness. To correct this, it is necessary to bring the cut-off points in coincidence. This is achieved by making the screen grid (*i.e.,* 1st anode) voltages different from each other. Potentiometers are normally provided for this purpose in the dc voltage supply to the three screen grids. For colour picture tubes having a common external connection for the screen grids which in fact is more common in modern in-line tubes, black level voltages at the three cathodes are varied to make the cut-off points coincident. This is done by adjusting bias voltage at the emitters of corresponding driver amplifier transistors. For this pre-sets are provided (see Fig. 3.14) in the dc supply used for biasing the transistors. These pre-sets are called cut-off controls for the three guns of the colour picture tube. As obvious the aim of such adjustments is that all the three guns are brought to the threshold of producing a visible raster the moment a drive is applied at the cathodes.

Adjustment of High-Light Level

It is equally necessary to ensure that all other levels of white are also correctly reproduced. This amounts to compensating for the slightly different slopes and also for the different phosphor efficiencies. This is achieved by varying video (luminance) signal drives to the three guns. Since the red phosphor has the lowest efficiency, maximum red video signal is fed to the red cathode and then by the use of potentiometers video signal amplitudes to the green and blue guns are varied (See Fig. 3.14) to obtain optimum reproduction of high lights. Separate potentiometers are provided in the blue and green drive amplifier input circuits.

3.12 COLOUR PICTURE TUBE SPECIFICATIONS AND SPECIAL FEATURES

The main specifications of a colour picture tube are as under:

Picture Tube Numbering

All picture tubes are identified by a specific code scheme consisting of numbers and letters. For example a colour picture tube having the designation '510YX-B22' indicates that the diagonal measurement of the tube screen in 510 mm (51 cm) and its characteristics are in accordance with the YX code meaning of the manufacturer. The last three letters (B22) tell about the screen phosphor. Many manufacturers add their own prefix as a special identification code.

Heater Volts/MA

A colour picture tube has three heaters, one for each cathode. However, these are paralleled to one pair of plug pins in the tube base. The heater voltage is either 5 V or 6.3 V ac. Heater current drain is around 900 mA but some tube draw as much as 1800 mA. Small size tubes used in battery operated receivers have a rating of about 12.6 V/150 mA.

Size and Deflection Angle

The length of a picture tube depends on two factors—its faceplate size and deflection angle. The listed deflection angle is along diagonal of the tube and is the total angle through which beams can be deflected by the yoke magnetic field. For a given screen size the greater the deflection angle, the shorter is the overall length of the tube. Increased deflection angle requires that the yoke's magnetic field be increased. For this the tube neck diameter must be decreased to place the magnetic field close to the beams if power consumed in the yoke windings is not to be unduly increased. This can be seen from the following data:

Deflection angle	Neck diameter
70°	5.08 cm
90°	3.63 cm
100°/114°	2.2/2.90 cm

A 90° deflection angle colour picture tube has an overall length of 47 cm whereas a 110° angle tube has a length of 36 cm only.

Tube Bases

As in any vacuum tube a circular base is provided for picture tubes. A 13 or 14 pin base is commonly used for colour picture tubes. The electrode leads terminate into strong steel pins mounted on the base. The electrodes are so located that there is no possibility of any shorts or mutual interference between signals fed to them. The pins fit into the socket holes tightly. A key is provided in the base to ensure that pins and socket holes mate correctly.

Special Features

Colour picture tubes are manufactured by many companies in different countries. The performance characteristics of similar tubes are basically the same. However, all makes have certain special features which distinguish them from others. Some such features are considered.

Aluminium Coating on the Screen

All present day picture tubes have a thin aluminium coating on the phosphor screen instead of the conventional ion traps. The coating is thick enough to act as a net that prevents massive ions from reaching the phosphor screen while it is coarse enough to allow the much lighter beam electrons to pass through it. The aluminium coating being a good electrical conductor provides direct return path for the beam currents to the high voltage supply. This prevents accumulation of charge near the screen and enables higher beam currents. This in turn means more light output from the phosphors and hence a brighter picture.

In the absence of aluminium coating, 70 percent of the emitted light would travel back inside the tube thus considerably reducing brightness of the picture. The contrast is also impaired because of interference caused by light which is returned to the screen after reflections from some points on the inner walls of the tube. The use of a polished aluminium coating prevents any of the light which is generated by the phosphors from travelling back into the tube. Most of it then reaches the viewer and a very bright picture is seen.

Contrast does not improve much unless certain precautions are taken. The visible spots on the phosphor are surrounded by rings of light. These are caused by the reflected light rays which strike the fluorescent crystals and scatter. This is known as 'halation' effect. The light rings cause a hazy glow in the region surrounding the beam spot and reduce maximum possible details contrast. Infact, due to rescattering of light the areas in the picture which should be in total darkness receive some light and the result is a reduction in the contrast ratio. It is here where the phosphor crystal structure, nature of aluminum coating and overall shape of the bell and screen combine together to improve brightness and contrast of the picture. Each manufacturer uses its own know-how to produce a bright and high contrast picture tube.

Beam Focusing and Convergence

The size of the beam spot and its correct landing on the phosphor go a long way in improving overall definition (resolution) of the reproduced picture. A very sharp beam spot can be obtained by an improved design of the bi-potential lens system or by incorporating tri-potential high voltage focusing technique. Correct colour registration needs purity of the rasters and proper convergence of beams. This needs a well-designed yoke and accurately fabricated aperture grille.

Thus, the design of electrode guns, focusing system and correctness of slots in the aperture grille govern the quality of colour picture a tube can produce. Here again manufacturers differ in their approach for achieving desired results.

Black Matrix in-between Phosphor Stripes

As already explained a thin polished aluminium coating over the phosphor screen greatly improves brightness of the picture. In an attempt to further increase light output from the screen all modern in-line-colour tubes have black lines in-between the vertical phosphor stripes. These are constructed out of an opaque jet black material. The black lines tend to reduce scattering of light and thus both brightness and contrast are considerably improved. Some tube manufacturers highlight this aspect to stress improved performance of their product.

Colour Phosphors

Older colour tubes used light metals in the form of sulphide, sulphate and phosphates as phosphors for producing different colour light outputs. These were of unequal efficiency, *i.e.*, for equal amounts of electron beam current the light output of each phosphor was different. Red was the poorest in efficiency and required highest beam current. The improved phosphors used in modern colour tubes have almost the same efficiency. In general zinc sulphide is used for generating blue, zinc silicate for green and special rare-earth elements such as europium and yttrium for red. While these phosphors produce bright colours, their compositions can be varied a bit to get desired variations. Rare-earth phosphors are deposited on the screen in a dry state for optimum brightness uniformly spread over the entire screen. In addition, they reflect white light in contrast to sulphide phosphors that reflect a yellowish white light.

The combining of the red, green and blue lights in suitable proportions produces the effect of white light. Actually there is no specific white light. Sunlight, daylight, moonlight are all white lights. It is the wavelength of each colour light and its intensity which determines the nature of white light produced by their mix. Thus, almost all brands of colour picture tubes have a slightly different white and even the reproduced colours are somewhat different. The choice depends on several factors including individual liking of white and different hues. For examples, 'EURO COLOUR' means European colours and this perhaps indicates the liking of colour shades by people living in the part of that world. Similarly, HELIO-CHROM means 'sun' like colours. In fact, picture tube manufacturers incorporate some variations in colour shades and that of white in their products to meet different requirements.

Quick Start

The phosphors used in old tubes were not efficient and needed higher beam current intensity for sufficient light output. This meant larger cathode surfaces which in turn needed more time to come up to the emission temperature on switching on the receiver. In earlier receivers employing valves the heater

CHAPTER 3

chain was a series circuit across the mains supply and tube filaments took a long time to become red hot and cause emission. Even solid state receivers employing tubes for the line output and H.V. rectifier needed considerable time to produce the EHT supply. However, in modern all solid state receivers the only filament that is to be heated is that of the picture tube and is fed from an independent winding on the line output transformer. The phosphors have been considerably improved and need lesser beam currents to produce sufficient light output. Thus, the cathode structures are smaller having coatings of barium and strontium oxides which produce copious emission at a relatively low temperature. In addition, with semiconductor rectifiers high voltages for picture tube electrodes build up very fast and enable beam currents to flow the moment enough electrons get emitted at the cathodes. The raster appears within five seconds of the switching on of the receiver. This is now true for all modern picture tubes and hence, quick start is no longer a special feature of any one or a few makes of picture tubes.

3.13 PICTURE TUBE PRECAUTIONS

It is important to be conversant with the following topics for maximum reliability from TV picture tubes and safety for technicians and set owners.

Shock Hazard

The high voltage at which colour picture tubes operate is around 25 KV. The receiver must be turned off before touching the tube or the high voltage generation circuitry. The outer surface of the tube envelope also has aquadag coating. The inner and outer coatings form a capacitor with glass in-between as the dielectric. The capacitance is of the order of 2000μμf and is used to filter the EHT supply. The charge stored in this capacitor stays for many days after the receiver is switched-off because it feeds a very high impedance load (very low beam currents). Therefore, it must be discharged to chassis ground by a jumper before touching any part of the high voltage supply.

Implosion Protection

Because the picture tube has a large surface area a very strong force of air pressure exists on the glass envelope. If a crack occurs in the glass the external force due to vaccum inside the glass envelope can produce an implosion inward before the tube explodes outwards. In order to safeguard the viewer against such an implosion a flat sheet of glass possessing the same contour as the external surface of the tube screen is permanently bonded to the screen by an epoxy resin. Other methods of implosion protection include (*i*) rimband affixed around the skirt of the face plate and secured by tension bands over the rimband, (*ii*) a metal 'T' band around the skirt of the faceplate and secured by a clip and (*iii*) a resin filled steel shell affixed around the front end of the rectangular skirt of the faceplate.

X-radiation

X-rays are invisible radiation with wavelengths much shorter than visible light. Prolonged exposure to X-rays can be harmful. The X-rays are produced when a metal anode is bombarded by high velocity electrons generally with anode voltages above 16KV. Colour receivers with anode voltages of 25 KV and higher can produce soft X-rays. Television receivers are designed to limit X-radiation to a very safe level. In colour receivers, X-radiation is prevented by not allowing the high voltage to exceed recommended value. In latest receiver designs the two methods that are often used are (*i*) to cut-off horizontal scanning or (*ii*) to remove horizontal synchronization when the anode voltage becomes too high. Both methods aim at disabling the horizontal (line) oscillator circuit.

X-ray Prevention Circuit

A typical circuit that is used for disabling the horizontal oscillator when EHT exceeds a certain limit is shown in Fig. 3.12. A fraction of the flyback pulses induced during retrace intervals in the line output transformer are sampled by a small winding on it. These are rectified and filtered by diode D_1 and capacitor C_1. The dc output is divided by a resistive network consisting of R_1, R_2, R_3 and R_4. The emitter-to-base bias of the control transistor Q_1 is so set through R_5 R_6 and zener diode D_2 that it stays in cut-off during normal operation of the receiver. If excessive high voltage develops because of large horizontal drive pulses or due to any other defect in the horizontal output or EHT circuit, the corresponding increase in the pulse amplitude to diode D_1 raises the rectified dc voltage. This in turn increases the voltage that feeds the emitter of Q_1. Since voltage across D_2 does not change, the transistor is forward biased and goes into conduction. The output at the collector of Q_1 (across R_7) is smoothened by the RC filter forward by R_8 and C_2. The error voltage thus developed feeds into the horizontal oscillator circuit. As shown in this circuit, the error voltage is applied at the input of error amplifier, the output of which is used to disable the horizontal oscillator. Thus, the circuit provides an automatic control for preventing any X-radiation by limiting the high voltage to a predetermined safe value. Note that terminals 'A' and 'B' can be shorted to check that the circuit is in operation. As obvious by-passing R_2 means more current through D_1 and application of higher voltage at the emitter of Q_1 and operation of the circuit to disable the horizontal (line) oscillator. This check is usually carried out after serving the receiver and the circuit is called 'FAIL SAFE'.

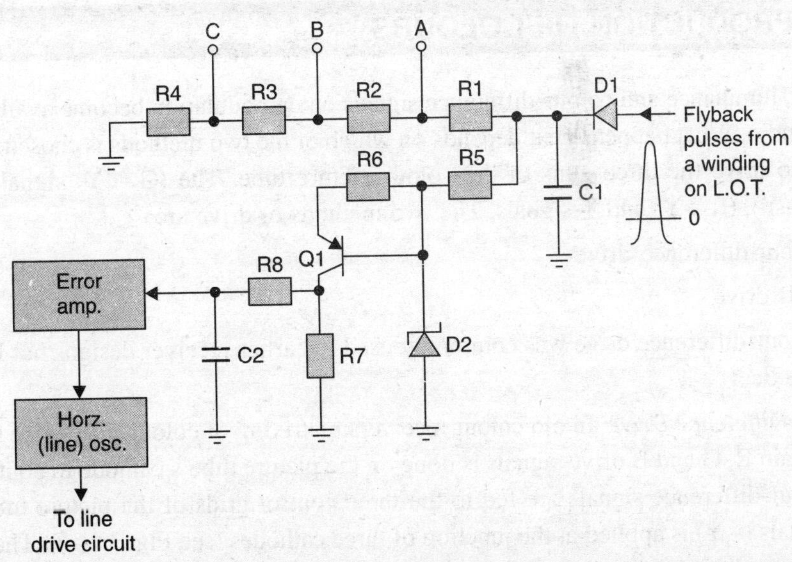

Fig. 3.12 X-radiation prevention circuit.

Excessive Beam Current Shut Down

Some receivers have excessive beam current shut-down circuits to prevent blooming of the picture, X-radiation and damage to the phosphors. In this case, it is the beam current that is sampled either at the

H.V. tripler or diode-split at the return path of beam currents. The error voltage is processed in a similar manner as explained above for X-radiation protection and finally used to shut-off the horizontal oscillator.

Spark Gaps

The picture tube electrodes operate at very high voltages and have close spacings. Any accumulation of charge can result in arcing or flashover between them and cause picture tube damage or associated transistor circuit failure. To prevent such a problem spark gaps are provided on the picture tube base socket. Spark gaps are provided for all tube elements except heaters. These are designed to have breakdown voltages slightly above the maximum expected operating potentials. Any excess charge (high voltage) is thus bypassed through the spark gap to ground. The spark gaps are moulded into the tube base socket. In an another type it is an intergral part of a disc capacitor. Yet another type consists of two electrodes mounted into a ceramic moulding. In old tubes, the spark gaps are external to the picture tube. In some tubes, neon type gas diodes are used to bypass the charge on ionization of the gas when a high voltage surge occurs.

DC Heater Biasing

In colour receivers, cathodes of the picture tube are kept at a dc potential of about 100 volts with respect to ground because the control grid is earthed and colour video signals are fed at the cathodes. In order to reduce any possibility of heater-to-cathode shorts, in some receiver designs a dc bias is placed on the heaters. The bias voltage brings the heaters close to cathode voltages, thus reducing any possibility of arcing and shorting between these elements.

3.14 REPRODUCTION OF COLOURS

In the receiver, luminance and colour-difference signals on demodulation become available in the form they left the studio. What happens next depends on which of the two methods is chosen to process them before using to drive the three guns of the colour picture tube. The (G – Y) signal is obtained by matrixing (B – Y), (R – Y) and Y signals. The two methods of drive are:

(1) Colour-difference drive

(2) RGB drive

The colour-difference drive was commonly used in earlier receiver designs but R, G, B drive is the current practice.

Colour-Difference Drive In old colour receivers matrixing of colour-difference signals with the Y signal to obtain R, G and B drive signals is done in the picture tube's cathode to control grid circuit. The three colour-difference signals are fed to the three control grids of the picture tube and inverted luminance signals (– Y) is applied at the junction of three cathodes (see Fig. 3.1(*c*)). The signal voltage subtract from each other to develop control voltages for the three guns:

$$V_{G1} - V_K = (V_R - V_Y) - (-V_Y) = V_R$$

$$V_{G2} - V_K = (V_G - V_Y) - (-V_Y) = V_G$$

and

$$V_{G3} - V_K = (V_B - V_Y) = (-V_Y) = V_B$$

Colour Reproduction

It is instructive to know how different hues are produced on the screen on application of these signals to the picture tube electrodes. The waveforms shown in Fig. 3.13(*a*) illustrate the way Y and colour-difference signals combine to produce red hue. For 100% saturated red and using unweighted values for simplicity Y = 0.3(1) + 0.59(0) + 0.11(0) = 0.3(30%), (R – Y) = + 0.7(70%), (G – Y) = – 0.3(30% and (B – Y) = – 0.3(30%).

The Y signals is inverted and applied at the cathodes by strapping them together and colour-difference signals are applied to corresponding control grids. The inverted luminance signal – Y = – 0.3 subtracts from (R – Y) = 0.7 to give a full drive equal to 1.0 [(0.7) – (– 0.30)] *i.e.*, 100% to the red gun. A red raster is thus produced. The signals (G – Y) and (B – Y) being equal to Y in magnitude subtract to cancel each other. Thus, both green and blue guns stay at cut-off and do not produce any beam currents. This is illustrated in Fig. 3.13(*a*) where as shown only red raster can be produced.

Similarly, waveshapes can be drawn for the other two primary colours to show that only green and blue colour rasters will be generated. As an illustration for the generation of a complementary colour Fig. 3.13 (*b*) shows necessary waveforms for producing magnets. For this Y = 0.3(1) + 0.59(0) + 0.11(1) = 0.41(41%) and (R – Y) = (1 – 0.41) = + 0.59(59%). Simialrly, (B – Y) = (1 – 0.41) = + 0.59 *i.e.*, 59% and (G – Y) = (0 – (0.41) = – 0.41(41%). As obvious net signal to the green gun will be zero. However, the red and blue guns will receive 100% drives to produce red and blue rasters. The eye on intergrating these rasters will perceive a magenta shade. A similar exercise will enable understanding generation of yellow and cyan. In some earlier designs matrixing is done outside the tube and colour voltages are directly applied at the corresponding control grids. The cathodes are then returned to a fixed dc voltage.

However, in modern colour picture tubes, as explained earlier, the three grids are joined together inside the tube and only one external connection is provided. It is either directly grounded or maintained at a fixed dc potential and grounded from the ac point of view. The value of dc voltage on G_1 depends on the dc potential at cathodes and tube characteristics.

Voltage amplitudes at the cathodes of colour picture tube | Waveshapes | Net drive

(*a*) 100% saturated red

(i) V_K (Red gun) 0 –(Y) 30% + – (R–Y) 70% ≡ – (R) 100% Full drive

(ii) V_K (Green gun) 0 –(Y) 30% + (G–Y) + 30% ≡ 0V No drive

(iii) V_K (Blue gun) –(Y) 30% + (B–Y) + 30% ≡ 0V No drive

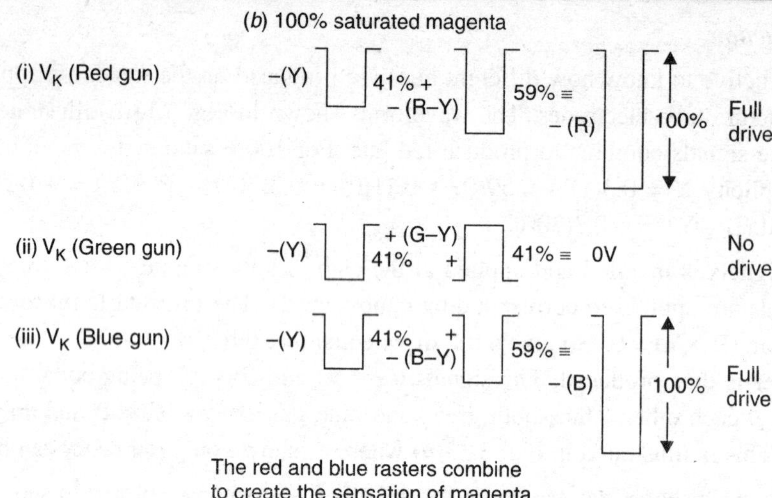

Fig. 3.13 Chroma and luminance signal waveforms corresponding to pure colour bars of hues (*a*) red, (*b*) magenta.

Matrixing in Drive Amplifiers

In not too old receivers matrixing is carried out in the colour driver amplifier circuits where colour-difference signals are fed at the base and Y signal at the emitter of amplifier transistors. As an illustration of such a matrixing arrangement video board circuits and their connections to the picture tube cathodes are shown in Fig. 3.14. The transistors Q_1, Q_2 and Q_3 are red output, green output and blue output driver amplifiers, respectively. The colour-difference signals (R – Y), (G – Y) and (B – Y) as obtained from three pins of chroma IC are fed at the base terminal of corresponding transistors. The Y signal together with vertical and horizontal blanking pulses as obtained from the last video amplifier is fed at the emitters of these transistors. Since colour-difference signals feed at the base and the Y signal at the emitter, the net signal drive to the transistor amplifier is the difference of the two. Hence, output at the collector of Q_1 is the red drive signal (R). Similarly, output signals at the collector of Q_2 and Q_3 transistor are the green and blue (G & B) drive signals, respectively.

These are dc coupled to corresponding picture tube cathodes through isolating (flashover protection) resistors R_1, R_2, and R_3. Luminance signal is directly applied at the emitter of Q_1 (red output amplifier) but through potentiometers P_1 and P_2 at the emitters of Q_2 and Q_3. Thus, while Q_1 receives full signal, the drives to Q_2 and Q_3 are adjusted with pots P_1 and P_2. These then become drive controls for obtaining optimum reproduction of high lights (pure white). The dc bias voltage to the three transistors is through potentiometers P_3, P_4 and P_5 from a 12-volt supply. These are varied to get correct cut-off for the three guns to obtain grey scale tracking at low-light levels.

RGB Drive

In this method of drive, matrixing is done within the chroma IC and R, G and B video signals become available at separate pins of the IC. These are fed to corresponding drive amplifiers to obtain –R, – G, and – B video voltages. The control grids, as explained earlier, are returned together to ground. With control grids at zero potential it is necessary to apply negative going voltages at the cathodes to drive the guns. The necessary invertion is provided by the drive amplifiers. Thus, – R, – G and – B video signals are fed at corresponding cathodes. More about RGB drive is explained in section 14.13 of the chapter that is devoted to chroma signal processing. Refer Fig. 14.21 to see how RGB drive is fed to the colour picture tube through drive amplifiers.

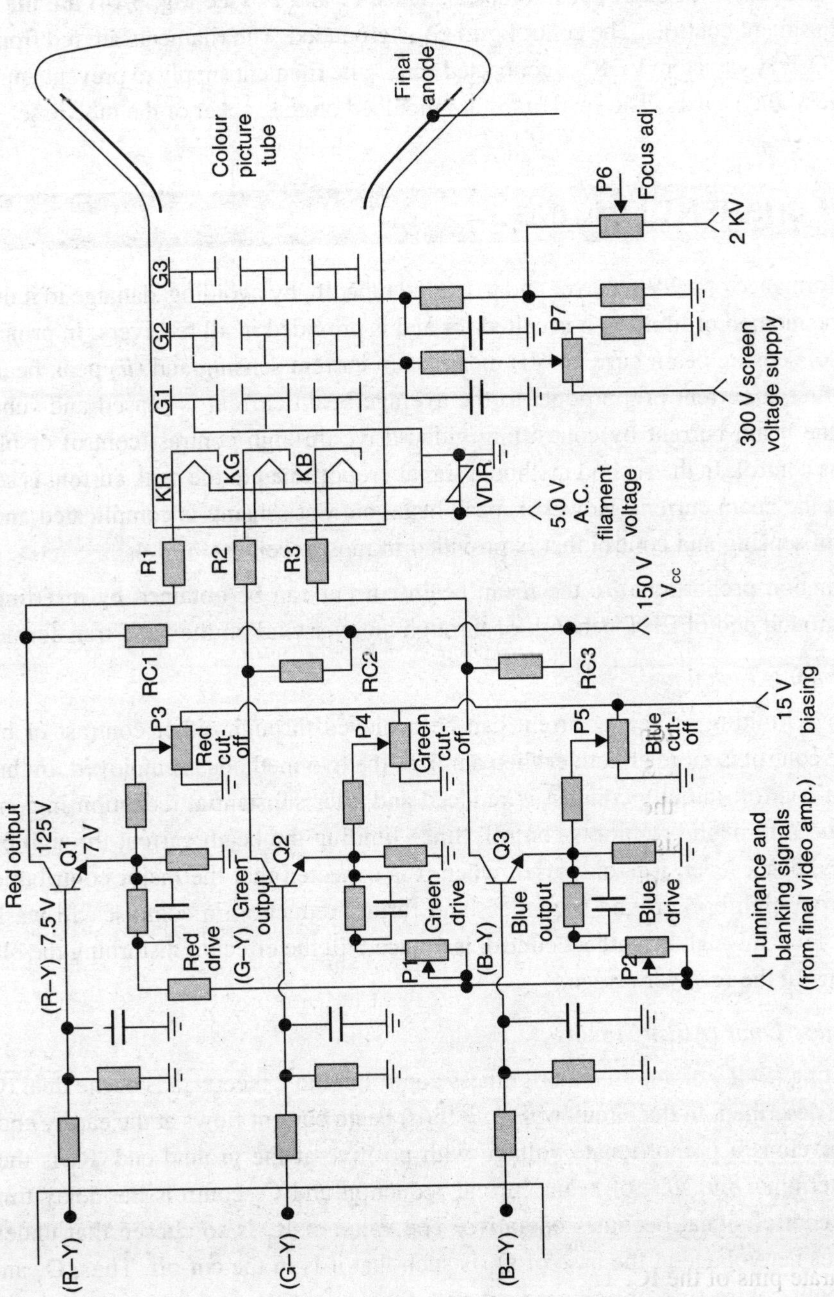

Fig. 3.14 Colour drive board and picture tube circuitry where matrixing is done in the drive amplifiers.

Picture Tube Circuit

The EHT voltage of about 25 KV developed in the line output circuit is applied through a special cable to the inner aquadog coating. The focus and screen voltages are also obtained from the same source and fed to corresponding electrodes through potentiometers. Thus, P_6 and P_7 (see Fig. 3.14) are the focus and screen voltage adjustment controls. The control grid G_1 is grounded. The filaments are fed from a 5 V ac winding on the L.O.T. A varactor (VDR) is connected across the filament supply to prevent application of any higher voltage to the heaters. The spark gaps are moulded on the socket of the tube base.

3.15 BEAM CURRENT LIMITING

For optimum performance of a TV receiver and to extend tube life by avoiding damage to it due to high beam current some method of limiting it is called for and is provided in all receivers. In principle there are two methods of sensing beam current—(*i*) mean beam current sensing and (*ii*) peak beam current sensing. For the first, a current proportional to the average beam current is sensed and subsequently utilized to limit the beam current by controlling video-drive through contrast control or black level through brightness control. In the second method a signal proportional to the peak current is sensed and later used to limit the beam current. However, peak beam current sensing is complicated and it is the mean beam current sensing and control that is provided in most receivers.

The information proportional to the mean beam current can be obtained by inserting a small resistance at the ground end of EHT network or through an overwind on the EHT transformer.

Beam Current Limiting Technique

As explained above limiting of beam current can be achieved through either contrast or brightness control. In fact the control is more effective when a mix of the two methods is employed. In this *i.e.*, the dual beam current control; initially contrast is reduced and after substantial reduction in contrast, the black level is reduced through brightness control. Since limiting the beam current through brightness control disturbs the black level, it is only used when it is expected that the major contribution to the beam current is from high background brightness and further reduction in contrast will wash-out the picture. The black level through brightness control is reduced till the effect of disturbing the black level is not too annoying on the receiver screen.

Automatic Brightness Limit (ABL) Circuits

Beam current limiting (BCL) or automatic brightness control in most receivers is of the dual type. Two typical circuits are described. In the circuit of Fig. 3.15(*a*) beam current flows at the earthy end of EHT through R_2 and develops a proportionate voltage with positive at the ground end. R_1 is the biasing resistance that determines the start of beam current reduction and C_1 controls the delay time of the circuit before the control signal becomes operative. The value of R_1 is so chosen that under normal beam current values net voltage at the base of Q_1 is such that it is in the cut-off. Thus, D_1 and D_2 are back biased and both contrast and brightness controls function in the normal manner. As the beam current exceeds, the permitted average value, voltage drop across R_2 increases making potential of point 'B' less positive. The emitter-base junction of Q_1 then becomes forward biased and it conducts. The flow of emitter current reduces potential at point 'A' permitting diodes D_1 and D_2 to become forward

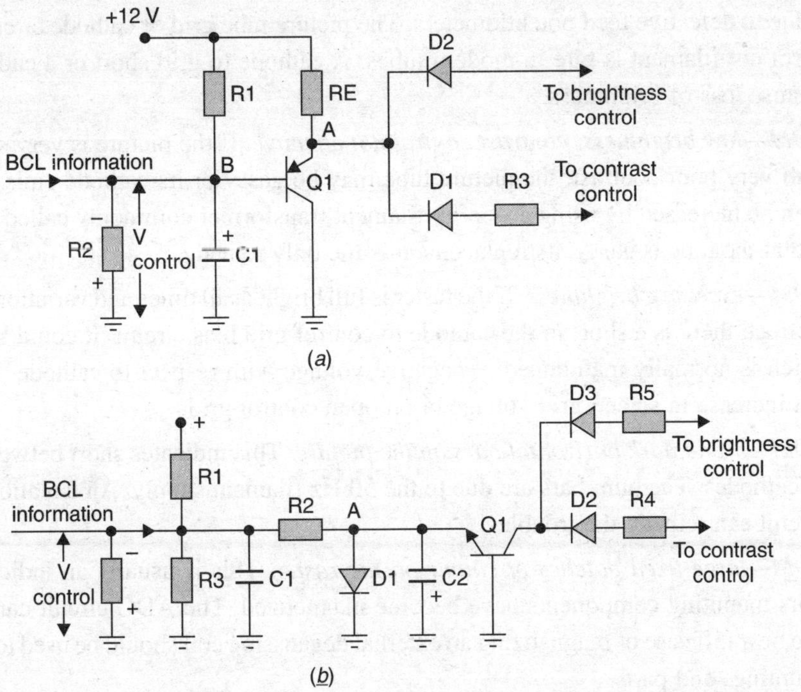

Fig. 3.15 Typical beam current limiting circuits.

biased. This affects the potentials of both contrast and brightness controls because of the now available shunt path through emitter to collector of Q_1. The resistance R_3 determines reduction in drive through contrast control before the start of reduction through brightness control. Thus, reduction in contrast and brightness control voltages reduces beam current and the circuit provides an automatic control over it.

In the circuit of Fig. 3.15(*b*) the control voltage is obtained in the same way as in the previous circuit *i.e.*, at the earthing end of EHT network. Resistance R_1 decides the threshold of beam current. Resistance R_2 controls the range of beam current or control voltage with beam current limiter in action. In typical designs it is 800 μA to 1000μA for beam current and –3.9V to – 4.9V for corresponding control voltage. The diode D_1 limits the voltage at point 'A' to + 0.7V under accidental serious overload thereby protecting associated circuits. The circuit operates in a similar way to that described with Fig. 3.15(*a*). Diode D_2 conducts first and R_4 decides the reduction in beam current before diode D_3 conducts at a control voltage close to – 4.35V. The resistor R_5 determines reduction in black level after contrast is reduced to a predetermined value.

3.16 COMMON FAULTS IN COLOUR PICTURE TUBES AND ASSOCIATED CIRCUITS

In this section, symptoms of some common faults in the picture tube and associated circuitry and their remedies are briefly described.

SYMPTOM—no brightness. There are a number of things in a television receiver that can cause complete loss of brightness. For example, the EHT section can be defective or focus and screen voltages may be absent due to defective feed potentiometers. The picture tube grid or cathode circuit can also be defective. A burnt out filament is rare in modern tubes. A cathode to grid short or a cathode to heater short can also cause loss of brightness.

SYMPTOM—low brightness, contrast control not effective. If the picture is very slow to appear and is weak with very poor contrast, the picture tube may be gassy or its cathode emission very low. The emission can be increased by using a step-up filament transformer commonly called a 'booster.' If it is confirmed that the tube is gassy, its replacement is the only remedy.

SYMPTOM—excessive brightness. If the raster is full bright at all times and variation of brightness control has no effect, there is a short in the cathode to control grid bias circuit. It could be a shortened control grid which is normally maintained at negative voltage with respect to cathode. It can also be due to too much increase in screen grid voltage or an open control grid.

SYMPTOM—broad dark horizontal bars on the picture. This indicates short between the picture tube heater and cathodes. The hum bars are due to the 50 Hz filament supply. An isolation transformer in the heater circuit can remedy the trouble.

SYMPTOM—large fixed patches of colour on the raster. This is usually an indication that the picture tube or its mounting components have become magnetized. The ADG circuit can be defective or if it happens to be a stiff case of magnetizing an external degaussing coil should be used to demagnetize the affected mountings and parts.

SYMPTOM—only two colours are available. This can be due to third gun being defective or failure of corresponding colour drive amplifier on the colour video drive board.

SYMPTOM—one colour is weak. This could be due to poor emission from corresponding cathode or defect in the associated gun. This fault is rare, and weak colour could be due to malfunctioning of the colour drive amplifier.

SYMPTOM—colour tint in the picture as brightness control is varied. The reason could be poor grey scale tracking. Readjusting colour cut-off controls and drive controls associated with the colour driver amplifiers will remedy this fault.

REVIEW QUESTIONS

1. Describe with suitable diagrams the gun arrangement and constructional details of a delta-gun colour picture tube. Why is it necessary to connect a very high voltage at the final anode of a colour picture tube?
2. What are the drawbacks of a delta-gun tube? Describe constructional details of a PIL tube and explain how it is different from a delta-gun colour tube. What are its distinguishing features?
3. Define purity and convergence and explain why elaborate static and dynamic correction become necessary to obtain colour purity and coincident rasters.
4. Describe essential features of a Trinitron colour picture tube. Explain why is it considered superior to both the delta-gun and PIL picture tubes.
5. Describe how a multipole-ring magnets assembly can be used to deflect the electron beam in different directions. Describe purity and static convergence adjustments which are normally provided with a PIL tube deflection assembly.

6. What do you understand by grey scale tracking? Explain how it is carried out both for low and high light levels of the video signal.
7. What is pincushion correction and explain how both E-W and N-S pincushion corrections are carried out?
8. What do you understand by degaussing and why is it necessary? Describe with a typical circuit how degaussing is affected each time the receiver is switched on.
9. Describe main specifications of colour picture tubes. Discuss some of the special features that distinguish different makes of P-I-L tubes from each other.
10. What are the common precautions that are taken in picture tubes for the safety of technicians and set owners?
11. Why is it necessary to prevent X-radiation from the HV circuit of a colour receiver? Describe a method with a circuit diagram that is commonly used to shut down the EHT circuit when high voltage exceeds a certain limit.
12. Describe how pure colour rasters and other colours are produced on application of appropriate colour-difference and luminance signals to the drive circuitry of the colour picture tube.
13. Why is it necesssary to limit the beam current from exceeding a certain maximum value? How does it affect the picture quality? Describe how beam current can be limited by controlling either contrast or brightness control.
14. Why is the dual control of beam current preferred? Draw such an automatic brightness limiting circuit and explain how it operates to limit the beam current.
15. Describe briefly symptoms and remedies of some of the common troubles in the picture tube and its associated circuitry.

4

CHROMINANCE SIGNAL–GENERATION AND PROCESSING

INTRODUCTION

The colour signal must be accommodated in the standard TV channel to meet compatibility requirements. For this, colour information should be within the video-passband of 5 MHz. It is desirable to fix the colour signal near the higher video frequency end because there is very little 'Y' signal energy beyond about 4 MHz. This avoids any strong beat interference between the two signals.

The bandwidth of colour-difference signals is kept within 1.5 MHz because higher colour video frequencies are not necessary due to limitation of the human eye to perceive very fine details in colour. Therefore, it becomes necessary to frequency translate the colour signals to a higher video frequency range. The carrier frequency chosen for this purpose is around 4.43 MHz and is called the colour subcarrier.

It is a difficult matter to modulate both (B – Y) and (R – Y) signals on one and the same carrier. However, this problem is solved by shifting the subcarrier phase by 90° to create two carrier frequencies —the numerical value remaining the same. This is called *quadrature modulation*. The two output signals (phasors) obtained on quadrature modulation are combined to form the chrominance or chroma signal.

The chrominance signal thus formed is combined with the Y-video signal. This then becomes the complete colour signal where information about hue and saturation is contained in the chrominance signal and brightness in the luminance (Y) signal. The combined signal together with syn pulses is modulated on the channel picture carrier frequency for transmission.

4.1 AMPLITUDE MODULATION

In general, amplitude modulation means multiplication or acting upon a carrier, say $A_c \cos \omega_c t$ ($\omega_c = 2\pi f_c$) by an information signal $A_m \cos \omega_m t$ ($\omega_m = 2\pi f_m$). The modulator output can be written as :-

$A = A_c (1 + m \cos \omega_m t) \cos \omega_c t$. Assuming 100% modulation *i.e.*, $m = \frac{A_m}{A_c} = 1$ we get $A = A_c \cos \omega_c t + A_c \cos \omega_m t \cos \omega_c t$. The second term can be expanded to get:

$$A = A_c \cos\omega_c t + \frac{A_c}{2} \cos (\omega_c + \omega_m) t + \frac{A_c}{2} \cos (\omega_c - \omega_m) t.$$

Thus, the modulated output consists of the original carrier f_c, and two side-band frequencies ($f_c + f_m$) and ($f_c - f_m$). It may be noted that if the modulating signal is complex, as is often the case, there will be a series of side-band frequencies depending on the nature of modulating signal. These groups of frequencies that lie on either side of the carrier are called *upper* and *lower sidebands*.

As an illustration of amplitude modulation Fig. 4.1 shows a sinusoidal modulating signal (*a*) and its amplitude modulated waveshape (*b*) with a suitable carrier. The waveshape (*b*) with identical upper and lower envelopes is the summation of carrier and upper and lower sideband frequencies as given by the expression for 'A' the modulator output.

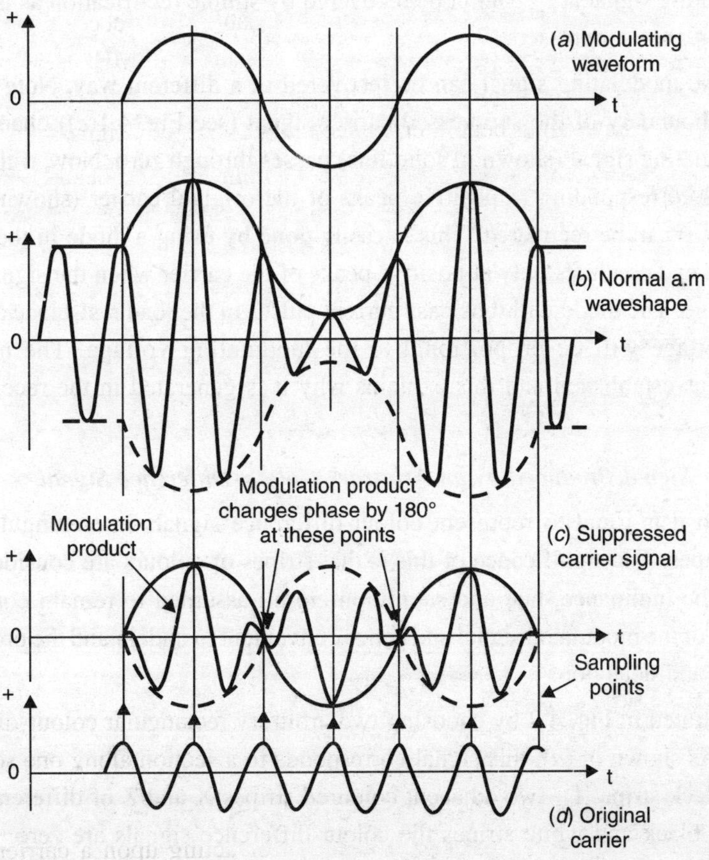

Fig. 4.1 Amplitude modulation (*a*) modulating signal (*b*) modulated carrier wave (*c*) modulated output when carrier is suppressed (*d*) carrier wave.

CHAPTER 4

Suppressed Carrier Modulation

As is obvious from the expression for 'A' both sidebands contain full information about the modulating signal. The carrier is transmitted because it is needed for easy demodulation at the receiving end. It can also be generated in the receiver and thus need not be transmitted. When carrier is suppressed during the process of amplitude modulation, it is called suppressed carrier modulation.

In colour TV, suppressed carrier modulation is used while translating colour-difference signals to the higher video frequency range. The reason for doing so is the large amplitude of the carrier as compared to sideband frequency components. If allowed to remain as part of the modulated wave, it can cause serious interference and dot patterning since the colour signal is interleaved in the luminance (Y) signal. Therefore, as a remedial measure the subcarrier is suppressed during modulation and thus not transmitted. It is again generated at the receiver for demodulating the chrominance signal.

When the carrier component is suppressed, the modulated signal looks quite different as shown in Fig. 4.1 (*c*). It represents the addition of two sinewaves of frequencies ($f_c + f_m$) and ($f_c - f_m$). When compared with the waveshape (*b*), its upper and lower envelopes are no longer sinusoidal. As a result the original modulating signal at f_m cannot be recovered by simple rectification as is possible when the carrier is present *i.e.*, not suppressed.

However, the modulating signal can be recovered in a different way. Note that the waveform within the overall boundary of the suppressed carrier signal (see Fig. 4.1(*c*)) changes phase by 180° each time the modulating signal (shown in solid line) passes through zero. Now, if the waveshape (*c*) is sampled at instants corresponding to positive peaks of the original carrier (shown at *d*) the shape of modulating signal (*a*) can be recovered. This is easily done by using a diode in the demodulator. It is kept reverse-biased and conducts only at positive peaks of the carrier when the signal at (*d*) gets added to the biasing voltage. The diode will thus pass current pulses in the load resistance at the rate of carrier and the output voltage will be proportional to the modulating voltage. The need of carrier for demodulation is thus established and this explains why it is generated in the receiver and fed to the demodulator.

Suppressed Carrier Signal Produced by an Arbitrary Colour-Difference Signal

In colour television it is usual to represent colour-difference signals in rectangular form instead of sinusoidal waveshapes. The significance of this is that stripes of colours are considered. For the width of a given stripe, the luminance, hue and saturation are all assumed to remain constant. This means simple waveshapes of the modulated signal and hence convenient to understand the process of suppressed carrier modulation and detection.

This is illustrated in Fig. 4.2 by choosing two arbitrary rectangular colour-difference signals of opposite polarity. As shown in (*a*), this signal corresponds to a section along one scanning line which passes through a black stripe 'D' two adjacent coloured stripes X and Z of different hues and a white stripe 'W'. On the black and white stripes the colour-difference signals are zero. At 'X' the chosen colour is such that the colour-difference signal is positive and of constant value. At 'Z' the colour is such that it gives rise to a negative colour-difference signal of amplitude larger than that for the coloured stripe 'X'.

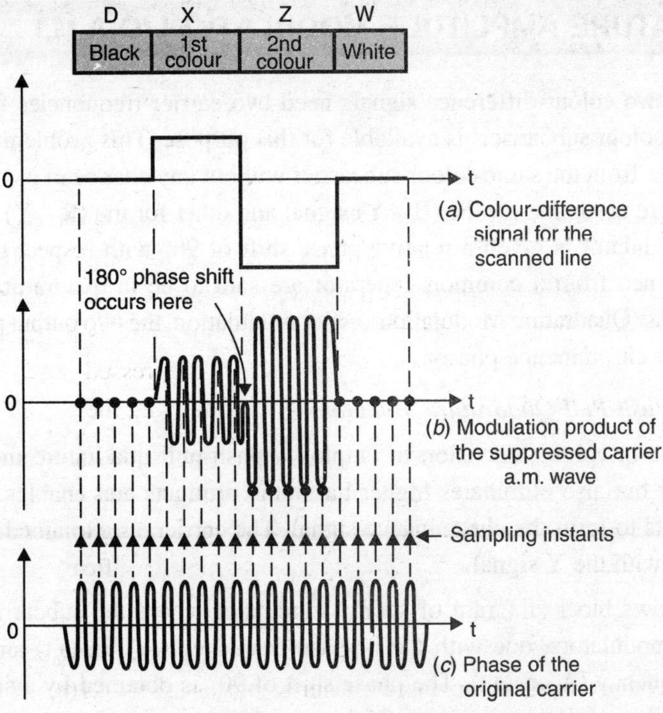

Fig. 4.2 Suppressed carrier signal produced by an arbitrary colour-difference signal.

The waveshape (*b*) shows corresponding suppressed carrier double sideband amplitude modulated signal. This is referred to as the modulation product. It is an RF signal which is processed and transmitted to convey the modulating intelligence. At 'D' and 'W' where colour-difference signal is zero, the modulation product is also zero because carrier is suppressed.

The modulation products for X and Z are of constant amplitude but directly proportional to the amplitudes of corresponding modulating signals. Note their opposite polarity and the phase shift of 180° at the point where the colour-difference signal changes from positive to become negative. Also notice carefully by comparison with the carrier wave shown at (C) that the modulation product during the stripe 'X' (where the colour-difference signal is positive) is in phase with the original (but now suppressed) subcarrier. Conversely at 'Z' where the colour-difference signal is negative, the modulation product is in antiphase with the original (now suppressed) subcarrier.

For demodulation, the process is the same as explained with reference to Fig. 4.1. The modulation product is sampled during positive peaks of the subcarrier as shown by arrows in Fig. 4.2. This will result in small conduction currents at the instants marked by black dots on the modulation product waveshape. The output voltage will thus have the waveshape of the modulating colour-difference signals. It is nearly continuous because sampling occurs at the high rate of carrier frequency. A filter is provided to further smoothen it. It is significant to note that the subcarrier that is generated at the receiving end to affect demodulation must be of correct frequency and phase *i.e.*, it must be the true replica of the carrier used for modulation at the transmitting end.

4.2 QUADRATURE AMPLITUDE MODULATION (Q.A.M.)

In colour television, two colour-difference signals need two carrier frequencies for their transmission but only one carrier (colour subcarrier) is available for this purpose. This problem is solved by creating two carrier frequencies from the same colour subcarrier without any change in its numerical value. Two separate modulators are used, one for the (B – Y) signal and other for the (R – Y) signal. However, the carrier fed to one modulator is given a relative phase shift of 90° with respect to the other. The two subcarriers thus obtained from a common generator are said to be in quadrature and the method of modulation is known as 'Quadrature Modulation'. After modulation, the two output phasors are combined to obtain the resultant chrominance phasor.

Suppressed Carrier Push-Pull Quadrature Modulator

It is common practice in colour television to employ a push-pull quadrature modulator. It not only suppresses the carrier but also eliminates higher harmonic products and enables easy addition of the two quadrature outputs to form the chrominance signal. The carrier, as explained earlier is suppressed to avoid interference with the Y signal.

Fig. 4.3(*a*) shows block diagram of such a modulator where the subcarrier oscillator at 4.43 MHz feeds the two modulators, one with the inphase frequency ($A_c \sin \omega_c t$) and the other with the quadrature phase frequency ($A_c \cos \omega_c t$). The phase shift of 90° is obtained by a suitable network. The two outputs are called *modulation products*. These two phasors are mutually perpendicular to each other (90° phase difference) because of quadrature modulation. Both the phasors have same frequency as that of the subcarrier and are sinewaves for steady colour-difference signals.

4.3 THE CHROMINANCE SIGNAL

The vectorial addition of the two modulation products is shown in Fig. 4.3(*b*). The resultant phasor (C) is called the chrominance signal. The main characteristics of the thus formed chroma signal are as follows:

(1) The two modulation product vectors which are 90° apart revolve at the modulating frequency in opposite direction with respect to subcarrier.

(2) The magnitude and location of these vectors depends on hue and saturation of the colour element. In fact both (B – Y) and (R – Y) vectors can be positive, zero or negative depending on the colour information being transmitted at the given instant.

(3) The vectoral sum of the two modulation products is the resultant chrominance phasor (C) with magnitude

$$C = \sqrt{(B-Y)^2 + (R-Y)^2}$$

and $$\text{phase angle } \phi = \tan^{-1}\frac{(R-Y)}{(B-Y)}.$$

(4) As shown in Fig. 4.3(*b*) amplitude and location of the two phasors determine the amplitude and phase angle (ϕ) of the chrominance phasor. Thus, the location and amplitude of the resultant subcarrier

phasor (C) depends entirely upon the amplitude and polarity of (B – Y) and (R – Y) vectors. In other words from simple amplitude modulation of the two carriers we obtain amplitude and phase modulation of a single carrier *i.e.*, the colour subcarrier. Similarly, magnitude of the resultant phasor C, which depends on the amplitudes of both (B – Y) and (R – Y), determines saturation of the colour. Maximum amplitude therefore, corresponds to greatest possible saturation and the absence of amplitude means no colour saturation *i.e.*, white or a grey shade. This point is stressed in order to avoid any misconception that hue and saturation are contained in the colour-difference signals.

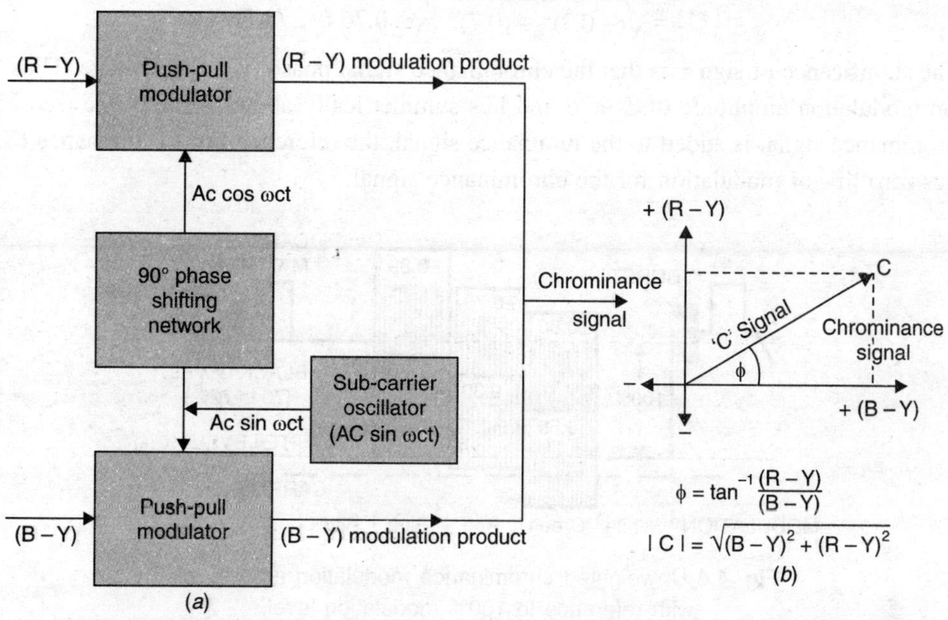

Fig. 4.3 Suppressed carrier double push-pull quadrature modulator (*a*) block diagram (*b*) vectorial representation of the two modulation products.

Zero phase reference. As explained above, phase of the subcarrier is very important because it determines hue of the colour being transmitted. Therefore, it is necessary to standardize it on a reference phase. Thus, the phase of positive direction of (B – Y) axis is called zero phase and all angles are measured anticlockwise from this phase. The carrier f_{sc} switches to a phase + 180° when (B – Y) is negative. The quadrature component of f_{sc} is at +90° when (R – Y) is positive and +270° *i.e.*, – 90° when (R – Y) is negative. Thus, the angular location and magnitude of each carrier phasor gives a direct indication of the polarity and amplitude respectively of the corresponding colour-difference signal. The resultant subcarrier phasor (C) which has an instantaneous amplitude = $V_c \sin(\omega_m t + \phi)$ is also a sinewave and bears evidence of the polarities and amplitudes of the two constituent phasors by its resultant magnitude and angle measured with reference to +(B – Y).

4.4 CHROMA SIGNAL PHASOR LOCATION ON DIFFERENT COLOURS

It is instructive to investigate the amplitude and location of the chrominance signal phasor for primary and complimentary colours. We know that

$$(B - Y) = - 0.3R - 0.59G + 0.89B,$$

$$(R - Y) = 0.7R - 0.59G + 0.11B \text{ and}$$

$$|C| = \sqrt{(B-Y)^2 + (R-Y)^2}$$

Assuming R = G = B = IV *i.e.*, 100% amplitude for pure colour, the values of (B – Y) and (R – Y) phasors can be expressed accordingly. As stated earlier, steady values are considered.

For the red hue (B – Y) = – 0.3 and (R – Y) = 0.7

$$\therefore \quad |C| = \sqrt{(-0.3)^2 + (0.7)^2} = \pm 0.76 \text{ i.e., } (\pm 76\%)$$

The significance of sign ± is that the chrominance signal phasor is an ac voltage. Thus, red has maximum modulation amplitude of ± 0.76 and lies symmetrically about the zero line (see Fig. 4.4). Since chrominance signal is added to the luminance signal, the reference line of luminance (Y) signal constitutes zero line of modulation for the chrominance signal.

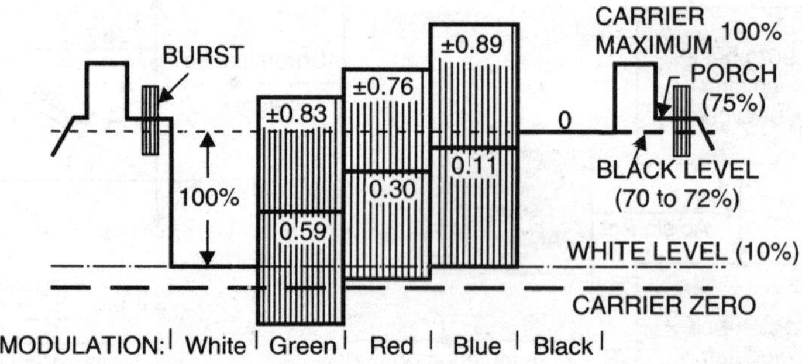

Fig. 4.4 Unweighted chrominance modulation amplitudes with reference to 100% modulation level.

With the red signal, the chrominance amplitude of ± 0.76 can be added to its Y amplitude of 0.3. When put in another way the ac chrominance signal is superimposed on the corresponding Y signal. This addition is clear from Fig. 4.4 where luminance and chrominance signal additions for red, green, blue, white and black are shown.

For green, both the colour-difference signals have the same value of –0.59. Therefore, for a green image, output of the suppressed carrier quadrature modulator referred to as 'C' phasor = $\sqrt{(-0.59)^2 + (-0.59)^2} = \pm 0.83$.

This signal as shown in Fig. 4.4, occupies ± 83% of the available maximum modulation amplitude. Similarly, for blue, (B – Y) = 0.89 and (R – Y) = – 0.11.

$$\therefore \quad |C| = \sqrt{(0.89)^2 + (-0.11)^2} = \pm 0.89$$

This when added to the Y signal amplitude of 0.11 for blue attains the maximum possible total amplitude as shown in the figure.

A similar exercise will enable evaluations of amplitudes for the complementary colours. But it is not necessary to do so since amplitude of a complementary is the same as that of its primary. However, the complementary colour phasors occupy a position that is 180° out of phase with respect to the corresponding primary colour phasors. Fig. 4.5 (page 60) shows the formation of chrominance signal phasor for all the primary and complementary colours.

4.5 WEIGHTING FACTORS

As explained above the reference line of Y signal becomes the zero for addition of chrominance signal to it. However, observe in Fig. 4.4 that it is not practicable to transmit the chroma signal waveform because its peaks would exceed the limits of 100 percent modulation. This means that on modulation with the picture carrier some of the colour signal amplitudes would exceed the limits of maximum sync tips on one side and white level on the other. For example, notice that in the case of red signal the chrominance value of ± 0.76 when added to the luminance amplitude of 0.30 exceeds the limit of 100% of both white and black levels. Similarly, blue signal amplitude greatly exceeds the black level and will cause a high degree of overmodulation. If overmodulation is permitted, the reproduced colours will get objectionally distorted. Therefore, to avoid overmodulation on 100% saturated colour values, it is necessary to reduce the amplitude of colour video signals before modulating them with the colour subcarrier. This reduction in amplitude of (B – Y) and (R – Y) is called *application of weighting factors*.

However, this reduction should not be such that an unfavourable signal-to-noise ratio is obtained with an adverse effect on transmission of the chrominance signal. It was found that the most favourable value of maximum possible overmodulation is 33% above the black level. This is permitted because in practice, the saturation of hue in natural and stage scenes seldom exceeds 75 percent of the maximum possible value. Since the amplitude of chrominance signal is proportional to the saturation of any hue, maximum chroma amplitudes are seldom encountered in practice.

Thus, on the basis of 33% overmodulation, the values of weighting factors must be found out. This is easily done with the help of signal amplitudes shown in Fig. 4.6 (page 61). Notice that with the upper limit set for 33% overmodulation, the chrominance signal for red must have an amplitude not exceeding ± 0.63. This is obtained by adding the Y signal amplitude of 0.3 (which lies below the zero line) and 0.33 because of 33% allowed overmodulation (0.3 + 0.33 = 0.63). Similarly, for blue the amplitude must be limited to ± 0.44 (0.11 + 0.33 = 0.44) as shown in the figure. The maximum allowed value for green has also an excursion above the 100% modulation limits at both ends.

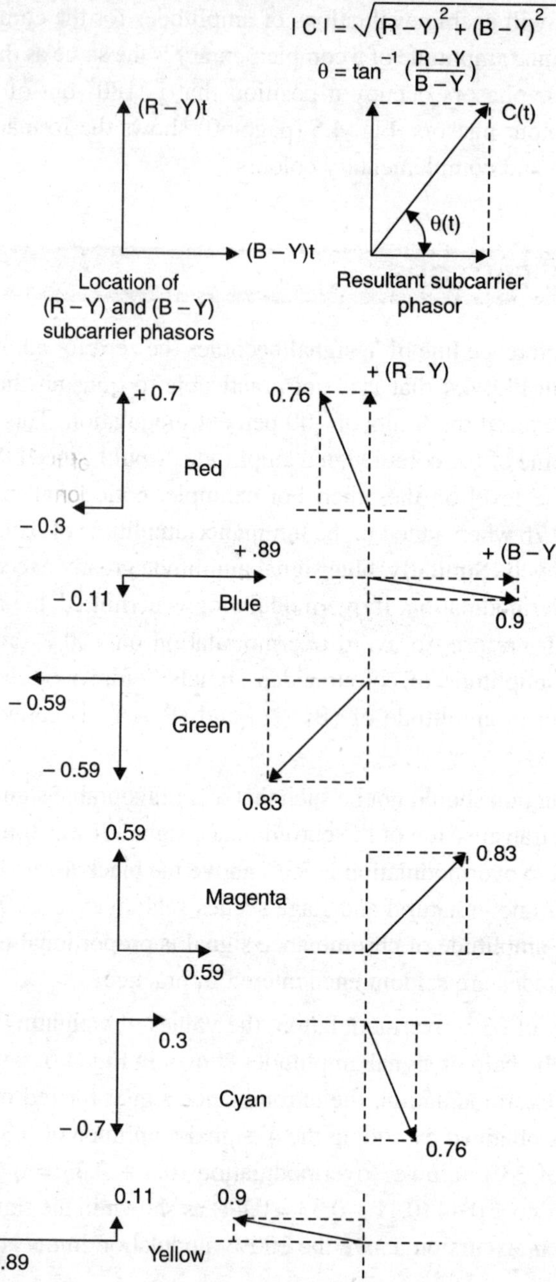

Fig. 4.5 Quadrature amplitude modulated colour-difference signals and carrier phasors for the primary and complementary colours. Note that the magnitudes shown correspond to unweighted values of colour-difference signals.

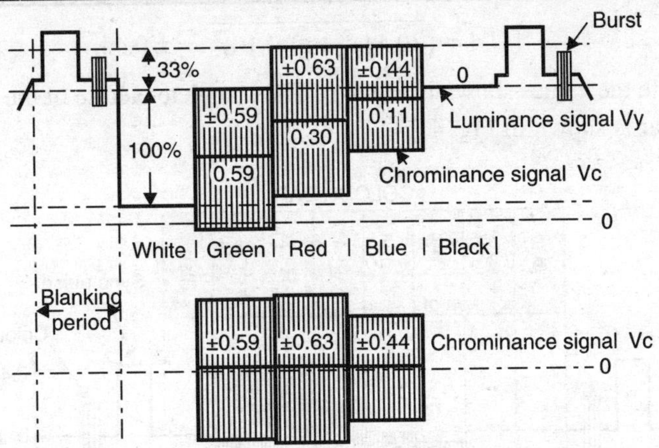

Fig. 4.6 Weighted chrominance modulation amplitudes of the primary colours for 100% amplitude and 100% saturation.

Magnitudes of Weighting Factors

Let 'a' be the factor by which (B – Y) be reduced and 'b' the factor by which (R – Y) be multiplied (*i.e.,* reduced) to keep overmodulation within 33%. This can be expressed as:

$$(B - Y).a = (B - Y)'$$

and

$$(R - Y).b = (R - Y)'$$

Note the prime (′) is to indicate that the colour-difference signals are weighted *i.e.,* reduced in value. Thus, for red the chrominance signal phasor amplitude with weighted values of colour-difference signal can be written as:-

$$|C'| = \pm 0.63 = \sqrt{(-0.3a)^2 + (0.7b)^2}$$

Similarly, for the blue-colour image

$$|C'| = \pm 0.44 = \sqrt{(0.89a)^2 + (-0.11b)^2}$$

the above equations on solution yield the value of $a = 0.493 \approx 0.49$ and $b = 0.877 \approx 0.88$

Now we can write $(B - Y)' = 0.493\,(B - Y)$

and $(R - Y)' = 0.877\,(R - Y)$

The reduced or weighted values of colour-difference signals can also be expressed in terms of R, G and B as under:-

$$(B - Y)' = -0.15R - 0.29G + 0.44B$$

and

$$(R - Y)' = 0.62R - 0.52G + 0.10B$$

Notice that the weighted colour-difference signal values also become zero for whites like the unweighted values.

Since for red $(B - Y)' = -0.15$ and $(R - Y)' = 0.62$ we get

$$|C'| = \sqrt{(-0.15)^2 + (0.62)^2} = \pm 0.63$$

Similarly, with green $|C'| = \sqrt{(-0.29)^2 + (-0.52)^2} = \pm 0.59$ and for blue

$$|C'| = \sqrt{(0.44)^2 + (-0.10)^2} = \pm 0.44.$$

This checks with the values shown in Fig. 4.6. The complete picture of the levels occupied by all the hues in a colour bar is shown in Fig. 4.7.

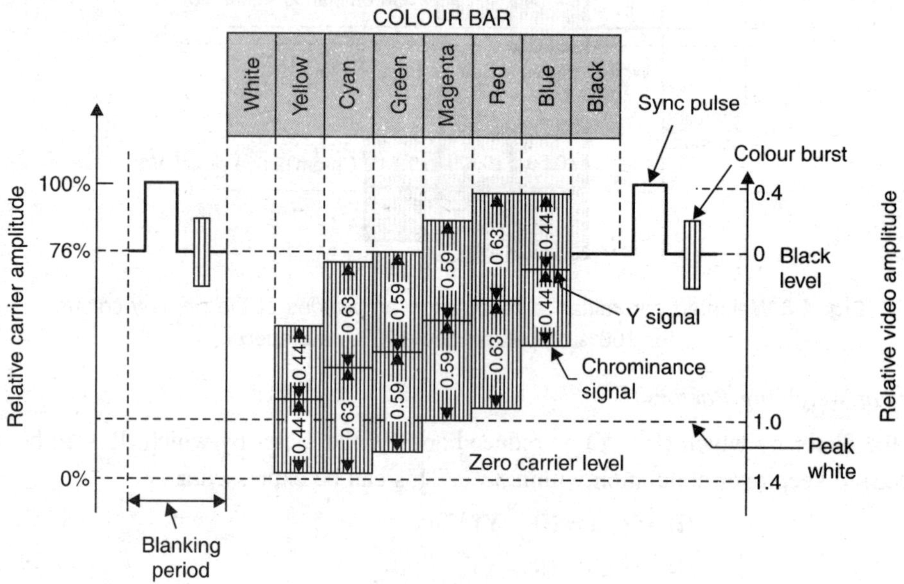

Fig. 4.7 100% saturated, 100% amplitude colour-bar signal in which the colour-difference signals are reduced by weighting factors to restrict the chrominance signal excursions to 33% beyond black and peak white levels.

4.6 SIGNAL PHASORS WITH WEIGHTED VALUES

The amplitude and phase angles of primary colours can be easily found out from weighted values of (B – Y) and (R – Y). For example, when a red image is televised (B – Y)′ = – 0.15 and (R – Y)′ = 0.62 from the expressions given in the previous section. Fig. 4.8 (*a*) shows how these values which are at right angles to each other yield on vectorial addition the red chrominance signal of amplitude 0.63 with a phase angle of 103.5°. As stated earlier all angles are measured with reference to the + (B – Y)′ axis.

For green (B – Y)′ = – 0.29 and (R – Y)′ = – 0.52.

These give a chrominance value of 0.59 with a phase angle of 241°. Similarly, for blue we find that (B – Y) = 0.44 and (R – Y) = – 0.1. This yields the chrominance amplitude of ≈ 0.44 having a phase angle of 347°. The locations of all three colour phasors are marked in Fig. 4.8 (*a*).

Location of (G – Y) After knowing the electrical representation of (B – Y) and (R – Y)′ and the location of three primary colours, it is necessary to find out the value of (G – Y)′ for completion of such a representation.

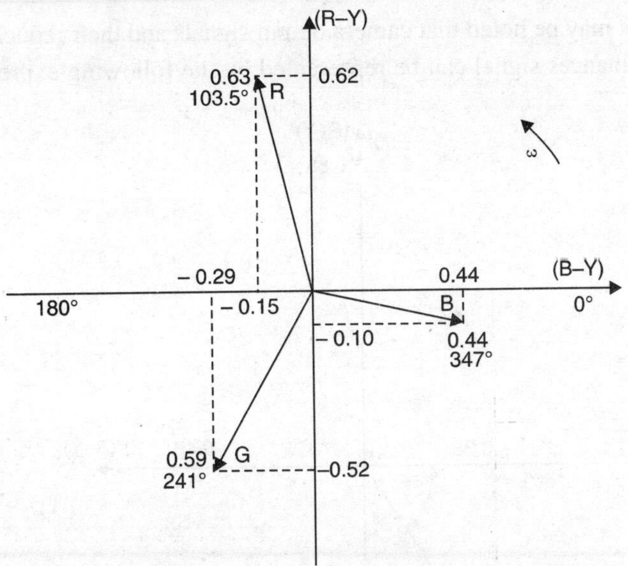

Fig. 4.8 (*a*) Amplitudes and phase angles of primary colours.

The signal (G – Y) is not transmitted separately as it is contained in the other two colour-difference signals. This can be shown easily (see also section 2 – 11) by writing Y = 0.3R + 0.59G + 0.11B.

Y can also be expressed as 0.3Y + 0.59Y + 0.11Y. On combining the above two equations we get 0 = 0.3 (R – Y) + 0.59 (G – Y) + 0.11 (B – Y)

$$\therefore \quad (G - Y) = -\frac{0.3}{0.59}(R - Y) - \frac{0.11}{0.59}(B - Y)$$

Substituting $\frac{(R - Y)'}{0.877}$ for (R – Y) and $\frac{(B - Y)'}{0.493}$ for (B – Y).

$$\text{we get } (G - Y) = -\frac{0.3}{0.59 \times 0.877}(R - Y)' - \frac{0.11}{0.59 \times 0.493}(B - Y)'$$

$$= -0.58\,(R - Y)' - 0.38\,(B - Y)'$$

Thus* (G – Y) can be obtained by a suitable matrix which represents the above expression.

Figure 4.8 (*b*) illustrates how the amplitude and phase angle of (G – Y) can be obtained graphically. The amplitude of

$$(G - Y) = \sqrt{(-0.58)^2 + (-0.38)^2} = 0.7$$

This means that $(G - Y)' = \frac{(G - Y)}{0.7}$ *i.e.*, 1.43 (G – Y)

Notice that while the amplitudes of (B – Y) and (R – Y) get reduced on application of weighting factors, (G – Y) on the other hand increases in amplitude. (G – Y) that is obtained from (B – Y) and (R – Y) consists of the following primary colour components *i.e.*, (G – Y) = – 0.43B + 0.59G – 0.16B.

*(G – Y) is not transmitted as such but obtained from (B – Y) and (R – Y).

CHAPTER 4

In conclusion, it may be noted that camera output signals and their reduced amplitude values for modulating the chrominances signal can be represented by the following expressions.

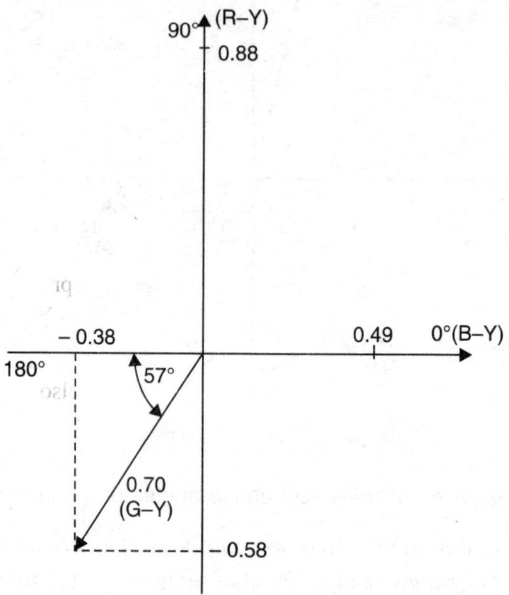

Fig. 4.8 (*b*) Amplitudes and phase angles of (G – Y′).

Camera Signals

$$(V_R - V_Y) = + 0.70\ V_R - 0.59\ V_G - 0.11\ V_B$$
$$(V_B - V_Y) = - 0.30\ V_R - 0.59\ V_G + 0.89\ V_B$$
$$(V_G - V_Y) = - 0.3\ V_R + 0.41\ V_G - 0.11\ V_B$$

Weighted (Reduced) Modulation Amplitudes

$$(V_R - V_Y') = + 0.62\ V_R - 0.52\ V_G - 0.10\ V_B$$
$$(V_B - V_Y') = - 0.15\ V_R - 0.29\ V_G + 0.44\ V_B$$
$$(V_G - V_Y') = - 0.43\ V_R + 0.59\ V_G - 0.16\ V_B$$

On demodulation at the receiver, these signals are deweighted to bring them back to their original values before applying to the picture tube electrodes or feeding the matrixing circuit.

REVIEW QUESTIONS

1. Sketch suitable waveshapes to show how a suppressed carrier amplitude modulated wave looks when modulated by (a) a sinusoidal modulating signal; (b) steady colour-difference signals of opposite polarity.
2. Explain with necessary waveforms how the modulating signal can be recovered from a suppressed carrier A.M. signal by sampling it at the carrier rate. Why is it necessary to generate the carrier at the receiver with exact frequency and phase as employed for modulation at the transmitting end.

3. What do you understand by quadrature amplitude modulation (QAM)? What is its significance in the formation of chrominance signal?
4. Enumerate main characteristics of the modulation products and resultant phasor as applied to suppressed carrier push-pull quadrature modulation.
5. Determine amplitude and location with reference to +(B – Y), of the chrominance signal for all the primary and complementary colours.
6. Show that combined chrominance and Y signals will cause overmodulation of the picture carrier wave on amplitude modulation.
7. What are weighting factors and how are their values found out for the two colour-difference signals that are transmitted. Why a 33% over-modulation is allowed?
8. Determine the values of weighted chrominance phasers for the three primary colours and show by drawing modulation waveforms that with weighted values of (B – Y) and (R – Y) overmodulation does not exceed 33 percent.
9. Determine the value of (G – Y) in terms of (B – Y) and (R – Y) and also in terms of B, R and G. Show that (G – Y) ′ = – 0.43R + 0.59G – 0.16B.
10. Show how by graphical construction the value of R, G and B can be determined from the expressions for (B – Y) ′, and (R – Y) ′. Also show how (G – Y) can be obtained from (B – Y) ′ and (R – Y)′.

5 COMPLETE TRANSMITTER SIGNAL

INTRODUCTION

In PAL colour system, the technique of producing chrominance signal is a bit complex. The method described in the pervious chapter applies to the NTSC system. The complexity was deliberately not introduced because the stress there was on explaining quadrature modulation, need for suppressing the subcarrier and the use of weighting factors to reduce amplitudes of colour-difference signals.

In the PAL system weighted colour-difference signals *i.e.*, (B – Y), and (R – Y)′ are modulated by Quadrature Amplitude Modulation (QAM) as in the NTSC but the phase of subcarrier to the (R – Y) modulator is reversed from + 90° to – 90° at the line frequency rate. In fact the system derives its name 'Phase Alteration by line' (PAL) from this mode of operation. Such a sequence of modulation cancels hue errors which result from unequal phase shifts in the transmitted signal.

The chrominance signal is accommodated in the same video bandwidth by an ingenious method called 'frequency interleaving'. The colour scubcarrier for quadrature modulation is so chosen that its sidebands lie at the higher end of video bandwidth and fall into gaps of the Y signal sideband energy clusters. In order to achieve perfect frequency interleaving, the subcarrier is maintained at an accurate value of 4.43361875 MHz. For brevity it is written as 4.43 MHz.

The colour subcarrier is suppressed during modulation and thus not transmitted. It is, instead regenerated at the receiver. Since its frequency and phase are linked with the reproduction of hues, a signal called 'Colour Burst' is transmitted along with other sync pulses to ensure perfect synch.onization between the subcarrier generated at the transmitter and that at the receiver.

The line and field sync pulses are added to the Y signal before combining it with the chroma signal. The composite signal thus formed is amplitude modulated with the station channel picture carrier and given due amplification. The sound signal on frequency modulation (FM) with the channel sound carrier is then added to it to form the complete transmitter signal. This RF signal is fed via a feeder line to antenna for radiation.

5.1 FREQUENCY INTERLEAVING

Frequency interleaving in television transmission is possible because of the relationship of video signal to the scanning frequencies which are used to develop it. It has been determined that energy content of the video signal is contained in individual energy 'bundles' which occur at harmonics of the line frequency (15.625, 31.250 ... KHz) the components of each bundle being separated by a multiplier of the field frequency (25, 50, 100 ... Hz). The shape of each energy bundle shows a peak at all harmonics of the horizontal scanning frequency. This is illustrated in Fig. 5.1. As shown there, the lower amplitude excursions that occur on either side of the peaks are spaced at 25 Hz intervals and represent harmonics of the vertical scanning rate. The vertical sidebands contain less energy than the horizontal because of the lower rate of vertical scanning. Note that the energy content progressively decreases with increase in the order of harmonics and is very small beyond 3.5 MHz from the picture carrier.

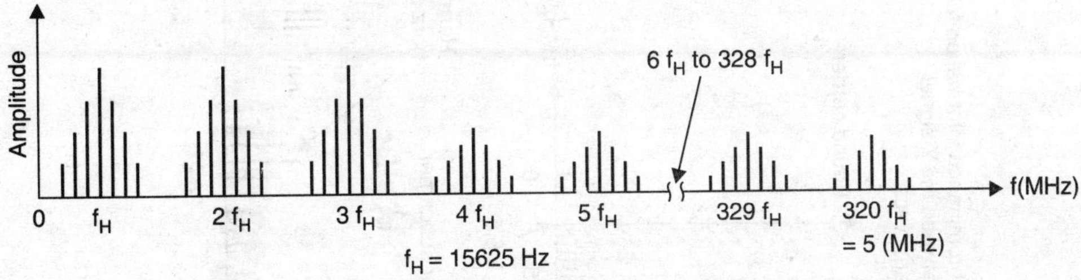

Fig. 5.1 Composition of video information at multiples of line frequency.

It can also be shown that even for the composite video signal the overall spectra still remains 'bundled' around the harmonics of line frequency. Therefore, a part of the bandwidth in the monochrome television signal goes unused because of spacing between the bundles. This suggests that the available space could be occupied by another signal. It is here where the colour information is located by modulating the colour-difference signals on a carrier frequency called 'colour subcarrier'. The subcarrier frequency is so chosen that its sideband frequencies fall exactly mid-way between the harmonics of the line frequency. This requires that frequency of the subcarrier must be an odd multiple of half-the-line frequency. The resultant energy clusters that contain colour information are shown in Fig. 5.2 by dotted chain lines along with the Y signal energy bundles. In order to avoid crosstalk with the luminance signal, the frequency of the subcarrier is chosen rather on the high side of channel bandwidth. It is close to 567 times one-half the line frequency in the PAL system and comes to nearly 4.43 MHz. Note that in the American 525 line system, owing to smaller bandwidth of the channel, the subcarrier employed is 455 times one-half the line frequency *i.e.*, $(2 \times 227 + 1) \times 15750/2$ and is approximately equal to 3.58 MHz.

CHAPTER 5

5.2 CHOICE OF SUBCARRIER FREQUENCY

As already explained, colour subcarrier (f_{sc}) should be so chosen that its sideband clusters of energy occupy gaps occurring in the energy distribution of Y signal. However, its exact value is influenced by the following factors:

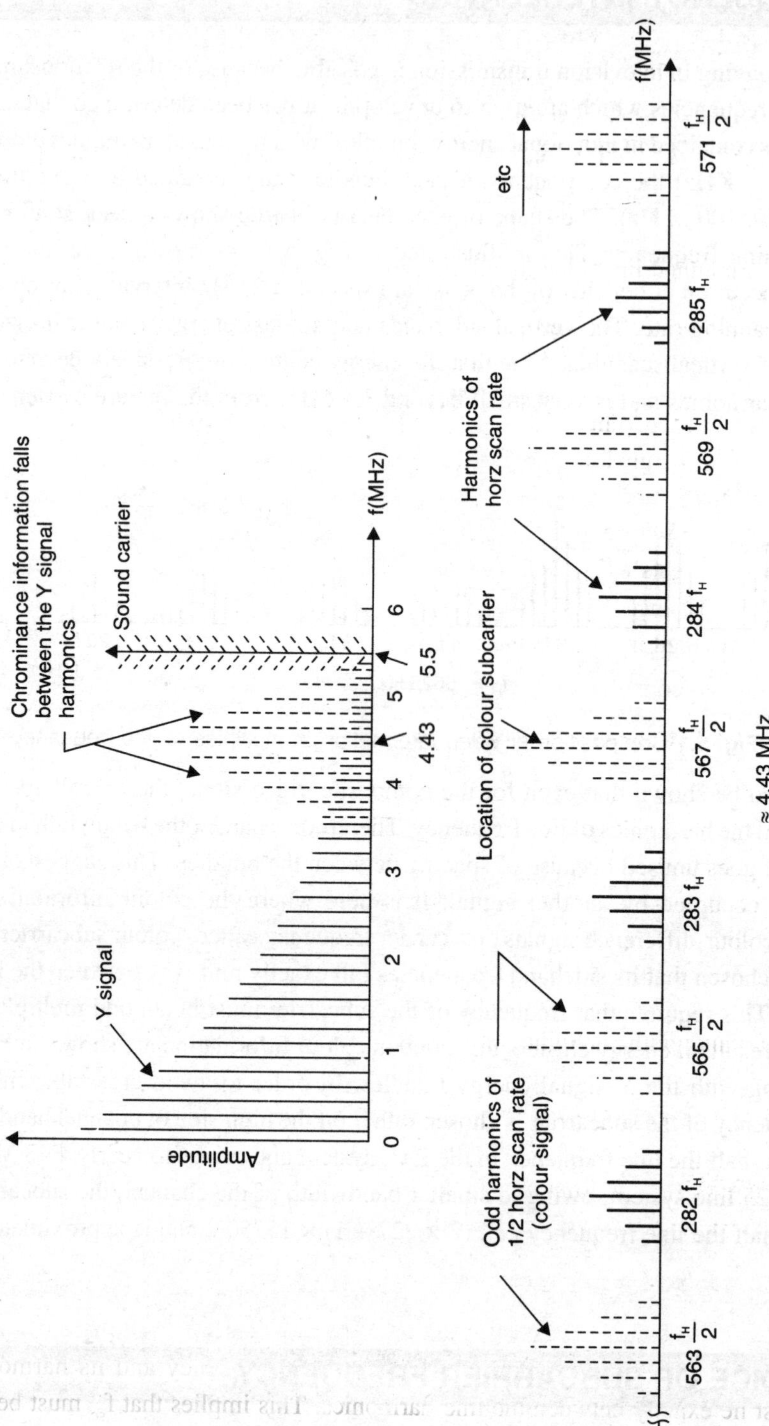

Fig. 5.2 Interleaving of the colour signal.

Cross-talk between luminance and chrominance signals. The picture carrier and colour subcarrier should be located quite apart from each other to avoid any beat interference between the two signals due to some overlaps and imperfect frequency interleaving. It can be minimized by placing the chroma signal near high frequency end of the Y signal spectrum where its energy content is very small. Thus, it is desirable to assign the colour subcarrier a frequency as high as possible.

Dot pattern visibility on a monochrome receiver. The chroma signal lies in the pass-band of luminance signal and can thus reach the picture tube input of a B & W receiver. The tube's Ib/Vg characteristics being non-linear can cause amplitude distortion of the applied signal. Thus, when a colour transmission is tuned in, f_{sc} and its harmonics cause dot patterning on the receiver screen. The visibility of such an interference can be minimized by making the dots very small. This is possible by choosing f_{sc} to be as high as possible.

Dot pattern on a colour receiver. In a colour receiver, a notch (trap) or comb filter is provided in the luminance channel to prevent passage of chrominance signal. Since it is not desirable to remove any part of the luminance signal, the notch filter design is a compromise between total removal of the chrominance signal and any undesirable deletion of the luminance information. Thus, some dot patterning may appear on the colour picture due to passage of some sideband components of the colour subcarrier alongwith the Y signal. This effect can again be reduced by choosing a higher subcarrier frequency. Since there is little luminance signal energy close to the end of its spectrum, a relatively wide-band notch filter can be provided in the luminance channel if the colour signal is interleaved in this region.

While the above considerations indicate that higher the subcarrier lesser will be the mutual interference between signals, too high a value of f_{sc} cannot be chosen for the following reasons:-

(*i*) Keeping the subcarrier very high would mean single sideband transmission of the chroma signal with the consequent increase in receiver design complexity. The chroma signal requires atleast a bandwidth of 2 MHz centered of the subcarrier. Thus, if both the sidebands are to be fully accommodated, the highest possible value of f_{sc} is around 4 MHz (5 MHz – 1 MHz).

(*ii*) A very high subcarrier will bring it too close to the sound signal spectrum and cause another type of interference due to mutual interference.

(*iii*) It is technically difficult to obtain reasonably linear phase characteristics near the cut-off point of the video bandwidth. A higher values of f_{sc} would place the complex chroma signal in this region causing its distortion. Any phase shift of the chroma signal affects hues and hence too high a value of f_{sc} is not desirable.

5.3 THE PAL SUBCARRIER FREQUENCY

As explained in section 5.1, Y signal energy is distributed throughout the bandwidth in bunches centered on harmonics of the line frequency. It is in the interleaving gaps between the energy clusters where chroma signal needs to be fitted. For this, the subcarrier frequency and its harmonics (sideband frequencies) must lie exactly between the line harmonics. This implies that f_{sc} must be offset by half-the-line frequency from the harmonics. This is known as 'half-line-offset' and involves choosing a subcarrier frequency equal to an odd multiple of half-the-line frequency. This means that f_{sc} must be set equal to $(2n + 1) f_{L/2}$ where 'n' is a suitable integer to keep f_{sc} towards high frequency end of the channel bandwidth and f_L is the line scanning frequency.

Keeping in view the above constraints, 'n' equals 283 for the PAL colour system. Thus

$$f_{sc} = (2 \times 283 + 1) \frac{15625}{2} \approx 4.43 \text{ MHz.}$$

Dot Pattern Cancellation

The number of subcarrier half cycles per line is equal to

$$\frac{f_{sc}}{f_{L/2}} = \frac{(2n+1) f_{L/2}}{f_{L/2}} = 2n + 1$$

Since (2n + 1) is an odd number, there are, therefore, an odd number of half cycles of subcarrier on each line. Keeping this fact in mind, the dot pattern cancellation can be easily explained if the receiver and picture tube characteristics are assumed to be fully linear.

Assume that the interfering colour signal has a sinusoidal variation which rides on the average brightness level of the monochrome signal. This produces white and black dots on the screen. If the colour subcarrier happens to be a multiple of the line frequency ($n \times f_L$) the phase position of the disturbing colour frequency will be same on successive even or odd fields. Thus, black and white dots will be produced at the same spots on the screen and will be seen as a persistent dot pattern interference. However, if a half-line offset is provided by fixing the subcarrier frequency to be an odd multiple of the half-line frequency, the disturbing colour signal frequency will have opposite polarity on successive odd and even fields. Thus, at the same spot on the display screen a bright dot image will follow a dark one alternately. The cumulative effect of this on the eye would get averaged out and the dot pattern will be suppressed.

As an illustration of this phenomenon assume that a simple five-line scanning system is being used. Figure 5.3 shows the effect of sinewave luminance signal that is an odd harmonic of one-half of the scanning frequency. Each negative excursion of the signal at the cathode of the picture tube will produce a unit area of brightness on the screen while the positive going excursion of the signal will cause a unit dark area on the picture. In the illustration under consideration where the sinewave completes 3.5 cycles during one active horizontal line, four dark areas and three areas of brightness will be produced

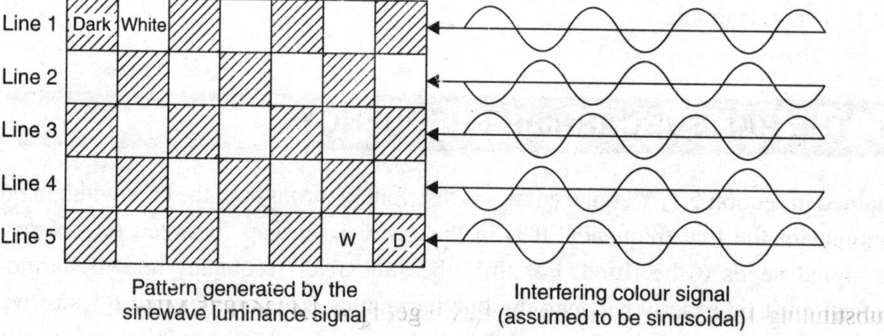

Fig. 5.3 Illustration of the technique used to reduce dot-pattern interference due to the luminance signal.

during the first line scan. Because of the extra half-cycle the next horizontal scan begins with an area of brightness and the entire line contains only three dark areas. The same off-set occurs on each succeeding line, producing a checkerboard pattern on the screen. Since the scanning rate in the example utilizes an odd number of horizontal scans for each complete presentation, the luminance signal will be 180° out of phase with pervious signal as line number one is again scanned. Thus, the pattern obtained on the screen will be the reverse of that which was generated originally. The total effect of the above process on the human eye is one of cancellation.

However, in actual practice due to marked non-linearity of the picture tube characteristics, compensation or cancellation of black and white for the fields in counter-phase is incomplete, especially on dark places of the image where control voltage is to some extent below the cut-off point of the cathode-to-control grid voltage. In that case interference appears as a dot pattern of fine structure. However, its visibility on the screen is considerably reduced because scanning takes place at a very fast rate and the eye due to its persistance of vision blends the dots together. Thus, dot pattern interference goes almost unnoticed on a monochrome receiver screen.

Exact Value of Subcarrier Frequency

In the PAL colour system phase of the subcarrier is switched line by line to overcome another type of distortion called '* Phase Error'. Basically one of the two subcarrier components is inverted on each successive line. This is equivalent to moving the dots produced by this component along by half-a-cycle on alternate lines. This partially negates the simple corrective result which the half-line offset gives. Thus, if the subcarrier is chosen entirely on the half-line offset basis, on certain hues, dots tend to line up vertically to give an annoying 'interference'. To overcome this, 'Quarter line-offset' is used. Here, f_{sc} is made equal to an odd multiple of one quarter of the line frequency. With quarter line offset, four lines are needed to contain an even number of subcarrier cycles and four picture (eight fields) are needed to produce a total number of lines which is divisible by four. The dot pattern therefore follows a four picture or eight-field sequence. Its visibility gets reduced as in the case of half-line offset. However, for optimum results, the value of f_{sc}, as obtained by employing quarter-line offset is slightly modified by adding 25 Hz to it to provide exact reversal on each successive field. Thus, actual relationship between f_{sc}, f_L and f_V can be expressed as:-

$$f_{sc} = \frac{f_L}{4}(2n+1) + \frac{f_V}{2}$$

Choosing n = 567 to keep f_{sc} close to 4.43 MHz we get

$$f_{sc} = \frac{f_L}{4}(2 \times 567 + 1) + \frac{f_V}{2}$$

$$= \frac{f_L}{4}(1135) + \frac{f_V}{2}$$

or

$$= f_L(284 - 1/4) + \frac{f_V}{2}$$

Substituting 15625 for f_L and 50 for f_V we get f_{sc} = 4.43361875 MHz.

This is the exact value of subcarrier used in the PAL colour system.

---

*Phase Error is fully discussed in the next chapter.

CHAPTER 5

Finally, it may be summarized by noting that the chroma signal is filtered out of the luminance channel in a colour receiver. Dot patterning is essentially a compatibiity consideration and is of major concern in B & W receivers when switched to receive colour transmission. As explained in a previous section, the chrominance signal disappears on grey and white details. So, dot pattern interference can appear only on coloured portions of the picture. Since all colours appear as grey on a monochrome rreceiver, and with cancellation of dots by an optimum choice of f_{sc}, any resulting dot pattern is of very low visibility and does not present any problem.

5.4 BASIC PAL CODER

The word coder or encoder as used in computers means to translate or prepare a programme. In colour television technology, it is taken to mean generation of composite video signal from the three camera outputs. Thus, it includes gamma correction, matrixing, weighting, quadrature modulation and combining circuits. The chrominance signal thus formed is combined with the Y signal by interleaving the two signals.

In the PAL colour system the coder has additional complexity of the need to change phase of the subcarrier to one of the modulators from + 90° to – 90° line-by-line. This, as pointed out earlier, cancels hue errors caused by unequal signal delays between the burst and chrominance signal during transmission.

The basic block diagram of the PAL coder is shown in Fig. 5.4(*a*). It is usual in PAL system to call the weighted colour-difference voltages as U and V video signals. Thus,

$$U = (B - Y)' = 0.493\ (B - Y)$$

and

$$V = (R - V)' = 0.877\ (R - Y)$$

These signals on mudulation with the subcarrier become U and V modulation products which are phasors at the subcarrier frequency. As shown in the figure the gamma corrected signals from the camera are combined in the matrix circuit to form luminance and colour-difference signals. Weighting factors are applied in the same matrix network to provide U and V colour-difference video signals. Burst gating pulses are also fed to the matrixing circuit and thus become available along with the U and V signals. The U signal feeds the U balanced modulator through a filter to limit its bandwidth. The modulator also receives subcarrier f_{sc} at a phase angle 0°. Similarly, V signal is fed to the V balanced modulator through a similar band-pass filter. However, the subcarrier to this modulator passes through a phase switching circuit which changes its phase from +90° to –90° on alternate lines. Thus, V modulator receives the subcarrier at f_{sc} + 90° on one line and f_{sc} + 270° on alternate lines. Since one switching cycle takes two lines, the square-wave switching cycle to the phase switch is of half-line frequency. *i.e.*, approximately 7.8 KHz.

The double-side band suppressed carrier modulated signals called U and V modulation products are added in an adder circuit to yield the Quadrature Amplitude Modulated (QAM) chrominance signal. As shown in Fig. 5.4(*a*), this passes through a filter which removes harmonics of the sub carrier frequency.

The Y signal from the matrix circuit is combined with line and field blanking and sync pulses in an adder circuit. Because the colour-difference signals are bandwidth restricted to be in the range of about 0-1.3 MHz, they suffer a small delay relative to the Y signal which has a broader bandwidth of nearly 5 MHz. To bring the Y and chrominance signals into step at the point where they are added, it is necessary to insert a compensating network into the Y path. This is achieved by a delay line as shown in the figure.

The Y and sync signals as obtained from the delay line and 'C' signal as available at the output of harmonic filter are combined in another adder circuit to form a composite video signal. This goes forward to the transmitter via an amplifier to modulate the picture carrier in the normal way.

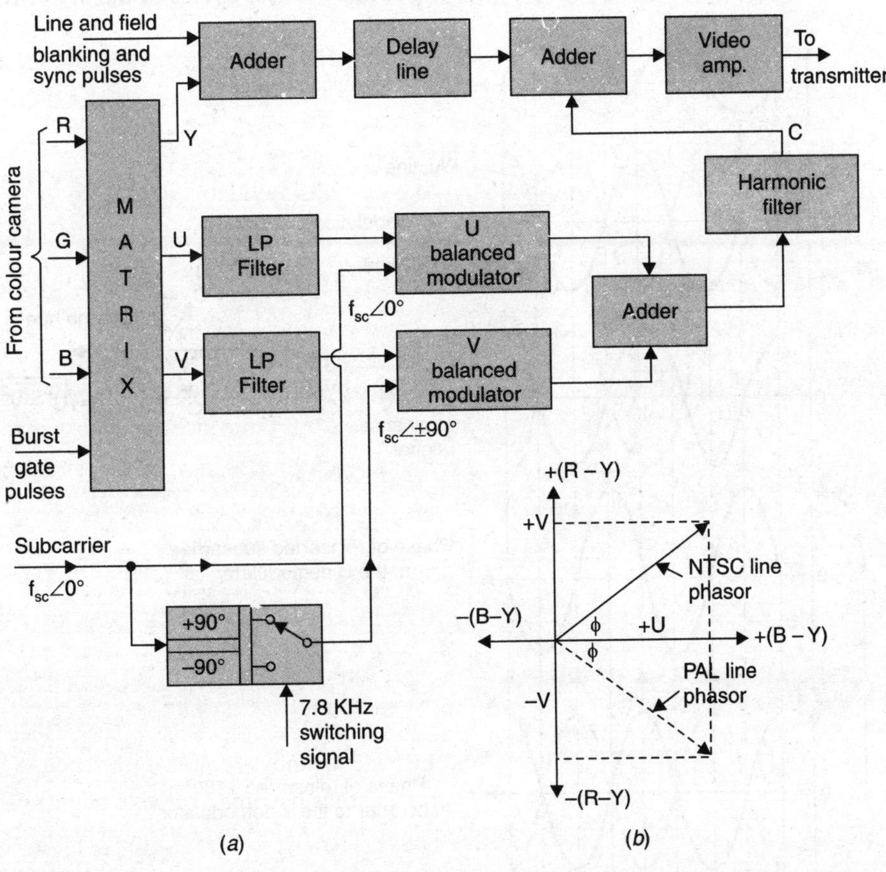

Fig. 5.4. The PAL Coder—(*a*) Basic block diagram, (*b*) Associated phasor diagram.

Phasor diagram. The phasor diagram of Fig. 5.4(*b*) shows the location of phasor 'C' an alternate lines. Since the phasor can be in any one of the four quadrants, the one shown in the figure is for a purple hue where both (B – Y) and (R – Y) are positive. Assume that for line number 'n', the phase angle of the subcarrier to the V modulator is + 90°. The chrominance signal phasor will thus lie in the first quadrant. On the next line *i.e.*, (n +1), when the subcarrier phase angle changes to become –90° (+ 270°), the 'C' phasor shifts to the 4th quadrant in an image symmetry. This is shown by a dotted line phasor. The phasor location corresponding to f_{sc} + 90° is called the NTSC line signal because it is the same as in the NTSC colour system. However, the phasor location corresponding to f_{sc} – 90° is called the PAL line signal because it is peculiar to the PAL colour system.

Modulation Products and Chrominance Signal

To further consolidate the concept of U and V modulation products in the PAL system and exmine how these are demodulated at the receiver; signals for a orange hue bar are considered as shown in Fig. 5.5. The corresponding phasor diagram is also shown alongside.

At (*a*) and (*b*) are shown the U and V suppressed-carrier amplitude modulated signals, referred to in the text as modulation products, whilst (*c*) shows the corresponding QAM signal (Chrominance Signal) formed by the addition of (*a*) and (*b*). These waveforms can be seen on a C.R.O. screen by making the horizontal sweep frequency very high so as to show a few cycles of the 4.43 MHz wave.

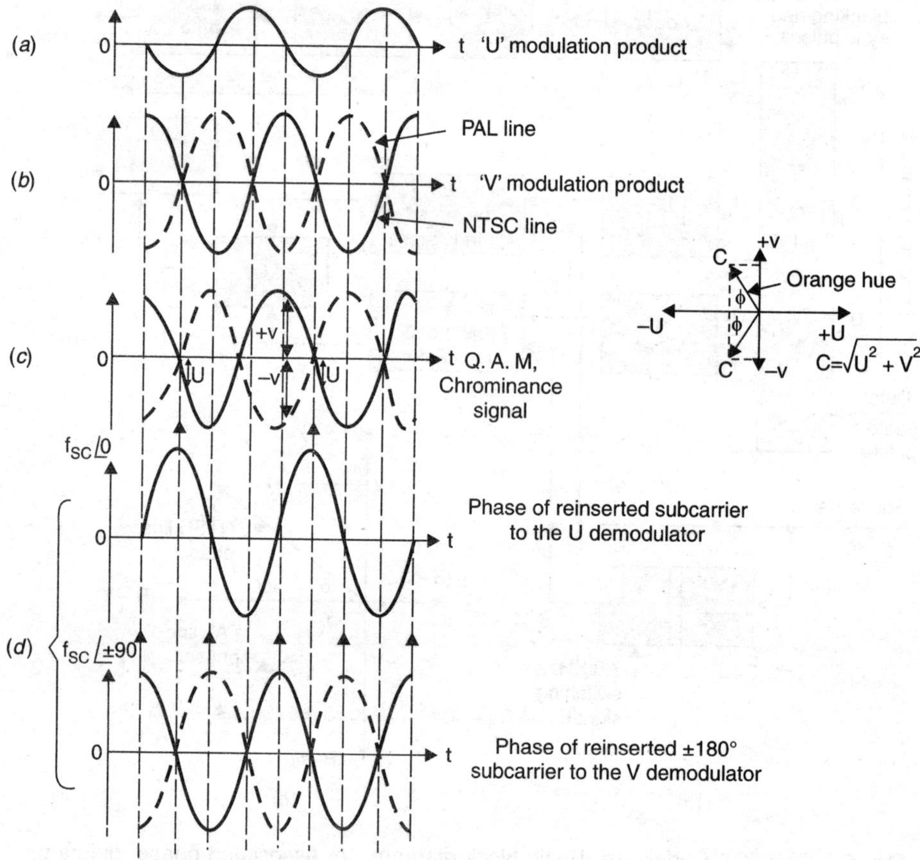

Fig 5.5 U and V modulation products and the combined chrominance signal waveforms for an orange colour-bar.

Note that the 'U' component representing (B – Y) colour-difference signal is negative. The V modulation product representing (R – Y) colour-difference signal is positive. Since the V component has its phase inverted on each line, two sinewaves are shown 180° out of phase with respect to each other in full and dotted lines, respectively. Since there is no inversion of the U component, a single sinewave is only shown in (*a*).

Notice that for an orange bar, (R – Y) colour-difference signal is larger than the (B – Y) and hence, the chrominance signal which is the sum of two, is very much influenced by the phase of

(R – Y). The solid line waveshape in (*c*) represents the NTSC line signal and the dotted line waveshape the PAL line signal.

At the receiver the U demodulator is fed with a subcarrier signal $f_{sc} \angle 0°$ and the V demodulator receives it at $\angle \pm 90°$. Thus, the sampling instants as shown in (*d*) are the correct ones to produce negative (B – Y) and positive (R – Y) video outputs.

5.5 CHROMA SIGNAL PHASOR DIAGRAM

The chrominance signal phasor may rest in any of the four quadrants of a circle because there are four possible combinations of polarity of its two components. Also, when the 'V' component is inverted on alternate lines in the PAL signal, the effect is to cause the chrominance phasor to switch across the 'x' *i.e.*, (B – Y) or U axis to the quadrant adjacent to the one it occupies on the NTSC line phase. These facts are summarized in Table 5.1.

Table 5.1

HUE	*Colour-difference signal polarities*		*Subcarrier phasors*		*Chrominance signal*
	(B – Y)	*(R – Y)*	*U*	*V*	*Quarter in which the phasor lies*
Purples	+	+ (–)	0°	90° (270°)	1st (4th)
Red, orange, yellow	–	+ (–)	180°	90° (270°)	2nd (3rd)
Yellow-greens, greens	–	– (+)	180°	270° (90°)	3rd (2nd)
Blue-greens, blue	+	– (+)	0°	270° (90°)	4th (1st)

Note (*a*) The figures in brackets show the polarities and angles on 'alternate' PAL lines.

(*b*) Purples are often called magentas and blue-greens as cyans.

The information given in the above table is a good guide to the location of chrominance phasor. However, in order to draw a phasor diagram with correct amplitude and exact location on the colour circle, it is necessary to know the weighted values of U and V for the 100% saturated 100% amplitude primary and complementary colours. Such an information is given in Table 5.2.

As obvious, the compensated (weighted) colour-difference signal values result in a change in both amplitudes and phase angles. With the values shown in Table 5.2, the chrominance phasor diagram for primary and secondary colours is shown in Fig. 5.6. Note that each complementary colour is diametrically opposite to its associated primary colour.

Thus, yellow is opposite blue, magenta opposite green and cyan opposite red. The amplitudes of these opposite phasors are equal since both are of the same (100%) saturation. Their algebraic sum is zero since together they make white. However, this does not mean that these phasors are present on white since both U and V are zero on white or grey shades. Put in another way, suppose cyan is present on its own at 100% saturation and if red is added, the cyan desaturates and the chrominance phasor gets shorter. As far as the diagram is concerned, effect on the cyan phasor may be imagined to be caused by the presence of a diametrically opposite red phasor which reduces the net amplitude of the cyan phasor.

CHAPTER 5

In reality, of course only one chrominance phasor can be present at any time. The desaturation affects the colour-difference signals before the chrominance signal is built.

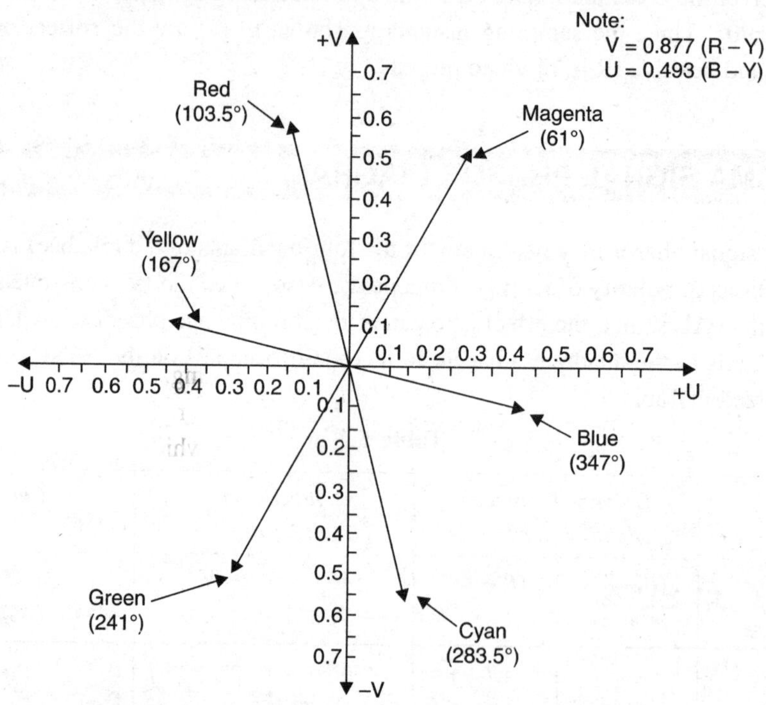

Fig. 5.6 Chrominance signal phasor positions for the primary and complementary colours. The relative amplitudes are those of 100% saturated 100% amplitude signal. All angles are measured with reference to + U axis.

Table 5.2

Colour Bar	*Y*	*Unweighted*		*Weighted*		*Chrominance amplitude* $\mid C \mid = \sqrt{U^2 + V^2}$	*Chrome phase angle NTSC line*
		(B – Y)	*(R – Y)*	*(U = 0.493 × (B – Y)*	*V = 0.877 × (R – Y)*		
White	1	0	0	0	0	0	0
Yellow	0.89	– 0.89	+ 0.11	– 0.4385	+ 0.0965	0.44	167°
Cyan	0.7	+ 0.3	– 0.7	+ 0.148	– 0.614	0.63	283°
Green	0.59	– 0.59	– 0.59	– 0.29	0.5174	0.59	241°
Magenta	0.41	+ 0.59	+ 0.59	+ 0.29	+ 0.5174	0.59	61°
Red	0.3	– 0.3	+ 0.7	+ 0.148	0.614	0.63	103°
Blue	0.11	+ 0.89	0.11	+ 0.4388	– 0.0965	0.44	347°
Black	0	0	0	0	0	0	0

Another point that is commonly misunderstood is the location of colour phasors. Red does not lie on the (R – Y) axis nor does blue lie on the (B – Y) axis. Furthermore, it is a common fallacy to think that on red, (B – Y) must be zero and (R – Y) maximum. What is true is that on reds, (R – Y) is large and (B – Y) small. It is vice-versa on blue because both (B – Y) and (R – Y) combine to give the phasor a particular location which is different for different colours.

The position of the phasor may be found our after knowing the quadrant in which the phasor lies. If it is in the first quadrant, $\phi = \tan^{-1}\dfrac{V}{U}$, for the second quadrant $\phi = 180° - \tan^{-1}\dfrac{V}{U}$, for the third quadrant $\phi = 180° + \tan^{-1}\dfrac{V}{U}$, and for the fourth quadrant $\phi = 360° - \tan^{-1}\dfrac{V}{U}$.

It is useful to remember chroma phase angles for the three primary colours:- red = 103.5° green = 241° and blue = 347°. The complementary angles are each deduced by adding 180° to the appropriate opposite primary angles. For example, since red = 103.5°, the cyan = 103.5° + 180° = 283.5°. Similarly, magenta = 61° *i.e.*, 241° + 180° = 421° which is the same as 421° – 360° = 61°. The angle for yellow comes to 167°.

5.6 COLOUR BURST

The colour subcarrier f_{sc} is suppressed during modulation at the transmitter to minimize dot pattern effect on the reproduced picture. But since the carrier is necessary for demodulation it is generated in the receiver. This must be done with correct frequency and phase as colour reproduction is linked with relative phase angle of the carrier. In other words, the carrier must have the correct phase to recognise phase modulation on the transmitted modulation products.

For this a signal called 'Colour Burst' is transmitted alongwith sync pulses at the back porch of the line blanking period. This consists of about ten cycles of the subcarrier.

The transmitter provides a reference phase by generating a continuous sinewave of subcarrier frequency at a phase standardized at 180° *i.e.*, along with the—(B – Y) axis. The reason for the choice is that such a colour burst signal will produce a negative voltage at the output of synchronous detector and as such is not capable of causing any beam current because during this period (blanking) the luminance signal is zero and the picture tube is therefore blocked. Thus, there is practically no visibility on the receiver screen due to the colour burst signal. The peak-to-peak amplitude of the colour burst signal is the same as the height of the sync pulse *i.e.*, 0.3 of the peak white transmitted signal.

The colour burst signal is gated out at the receiver and used in conjunction with a phase comparator circuit to lock the local subcarrier oscillator frequency and phase with that at the transmitter.

5.7 BANDWIDTH OF COLOUR DIFFERENCE SIGNALS

The Y signal is transmitted with full frequency bandwidth of 5 MHz for maximum horizontal detail. This implies that very fine details may be described in black and white. However, such a large frequency spectrum is not necessary for colour video signals because the eye does not have perception for such fine details in colour. If coloured stripes that are quite wide are painted side by side on a white board and held some 5 meters away from the eye, the colours are seen clearly. As the size of stripes is decreased in steps, a stage comes when the eye no longer sees the stripes as coloured but gains only a brightness impression of them. Thus, beyond a certain size the eye perceives only the brightness but not the colour. Detailed studies have shown that eye's perception of colours which are produced by combining the three primary colours is limited to objects which have relatively large coloured areas (0.2-cm wide or more on a screen 50 cm wide). On scanning such stripes, video frequencies which are generated do not exceed 0.5 MHz. Further, for medium-size objects or areas which produce a video frequency spectrum between 0.5 and 1.5 MHz, only two primary colours are needed because for finer details the eye fails to distinguish purple (magenta) and green-yellow hues from greys. As the coloured areas become very small in size (width) the red and cyan also become indistinguisable as colours and appear only in grey shades. Thus, for very fine colour details produced by frequencies from 1.5 to 5 MHz all persons with normal vision are colour blind and see only changes in brightness for colour areas. Therefore, for colour details a bandwidth of about one fifth of that allowed for the luminance signal is found to give satisfactory reproduction.

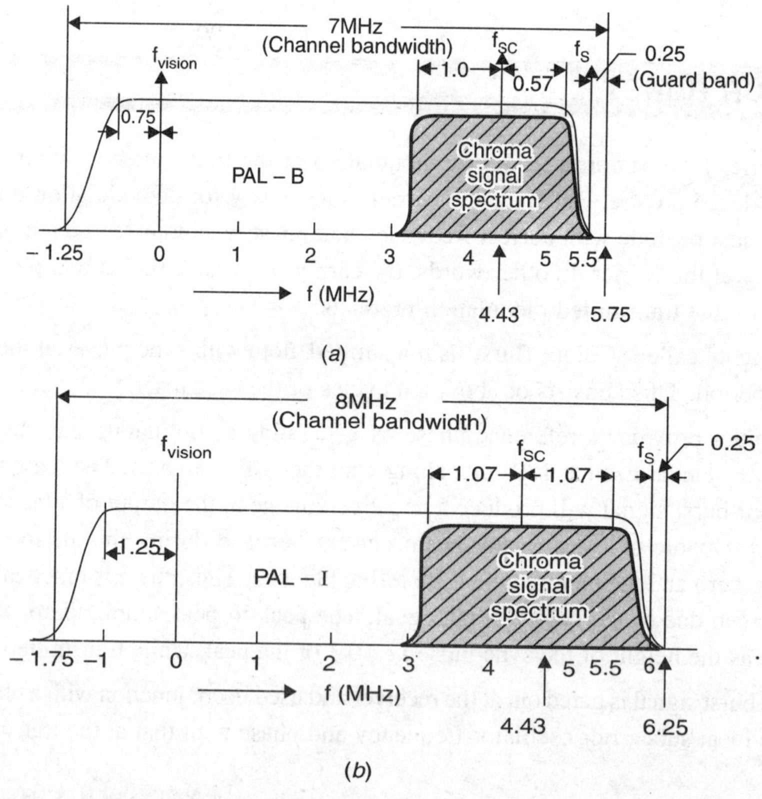

Fig. 5.7 Channel bandwidth details (*a*) PAL 'B' (*b*) PAL 'I'.

In some PAL colour versions a total bandwidth of 2 MHz is allowed for the colour-difference signals. In PAL 'B' system (adopted by India) where video bandwidth = 5 MHz and channel-width = 7 MHz, the colour subcarrier is at 4.43 MHz. Thus, the chrominance signal is of vestigial sideband type.

The upper sideband attenuation slope starts at 0.57 MHz (*i.e.*, 5.0 – 4.43 = 0.57 MHz) but the lower sideband extends to 1 MHz before attenuation begins. (details shown in Fig. 5.7) (*a*)). Note that in the signal that is actually rediated the attenuation slopes of upper and lower sidebands of the chrominance signal are not identical as shown in the figure for simplicity. The upper sideband is dictated by the upper edge of system's picture channel characteristics. The lower sideband slope is made less steep and extends to nearly 0.5 MHz. Corrective measures are taken in the chroma band-pass amplifier to compensate the loss of high frequency signal caused by vestigial sideband transmission to ensure correct saturation of fine coloured details in the picture. In the PAL-1 colour system where video bandwidth extends up to 5.5 MHz with a total channel bandwidth of 8 MHz, the same colour subcarrier of 4.43 MHz is used but the chrominance signal is double sideband up to 1.07 MHz (*i.e.*, 5.5 – 4.43 = 1.07 MHz). However, the attenuation slope at higher freqencies is more rapid at the upper sideband than at the lower. The U.K. PAL (1) channel details are shown in Fig. 5.7 (*b*).

5.8 COMPOSITE COLOUR SIGNAL

After having explored the hows and whys of producing a compatible colour signal the next logical step is to examine the formation of composite video signal which is modulated with the channel picture carrier frequency before transmission. It is called composite signal because it is formed by combining the luminance, chrominance and synchronizing signals. We will first examine step by step formation of the chroma signal and then explore how it combines with the Y signal. The blanking and sync pulses are added to it later to form the composite colour signal.

Ladder diagram. The ladder diagram of Fig. 5.8 is arranged in such a way that each voltage waveform for the colour bar pattern is positioned one above the other so that waveform changes that take place in each circuit till the formation of composite colour signal can be easily followed.

The scene in front of the camera is supposed to consist of vertical stripes of white, red, green, blue and black which will produce corresponding colour signals in the horizontal scanning direction as shown in the figure. From the three gamma corrected signals R, G and B, is derived the Y signal which consists of amplitude values 1.0, 0.3, 0.59, 0.11 and 0. The matrixing circuit develops weighted colour-difference signals from the same camera outputs. The values of (R – Y)′ in the same order as for the Y signal are 0, 0.62, – 0.52, – 0.11 and 0 for the colour bar stripes. Similarly, corresponding values for (B – Y)′ are 0, – 0.15, – 0.29, 0.44 and 0 in the same order of succession. Note that these are rounded off values of those given in Table 5.2.

The varying colour-difference signal (R – Y)′ modulates the colour subcarrier V_c cos ωt (± 90°) in the quadrature modulator. At the same time (B – Y) modulates the subcarrier V_c sin ωt (0°). Note that the waveshapes shown are for a NTSC line because this enables a clearer follow up of the process of forming the composite signal.

The modulation product V *i.e.*, (R – Y)′ cos ωt has amplitude values of 0, ± 0.62, ± 0.52, ± 0.10 and 0 as shown in Fig. 5.8. The opposite order of succession of polarities indicates that the carrier oscillation starts either by +cos ωt or by –cos ωt. In the same manner, the modulation product U *i.e.*, (B – Y)′ has values of 0, ± 0.15, ± 0.29, ± 0.44 and 0. As is clearly evident from the modulation product waveforms, whenever the colour-difference signal changes from positive to negative or vice versa *i.e.*, at zero passage, a phase shift of 180° occurs in the modulation product. The chrominance signal which is actually obtained as the output from the subcarrier quadrature modulator is the sum of V and U and is of the form $V_c \sin(\omega t + \phi)$ where its instantaneous amplitude and phase angle ϕ denote saturation and hue of the colour being scanned. Colour burst signal is added to the chrominance signal in the modulation process during back porch periods through appropriate burst pulses obtained from the line oscillator circuitry.

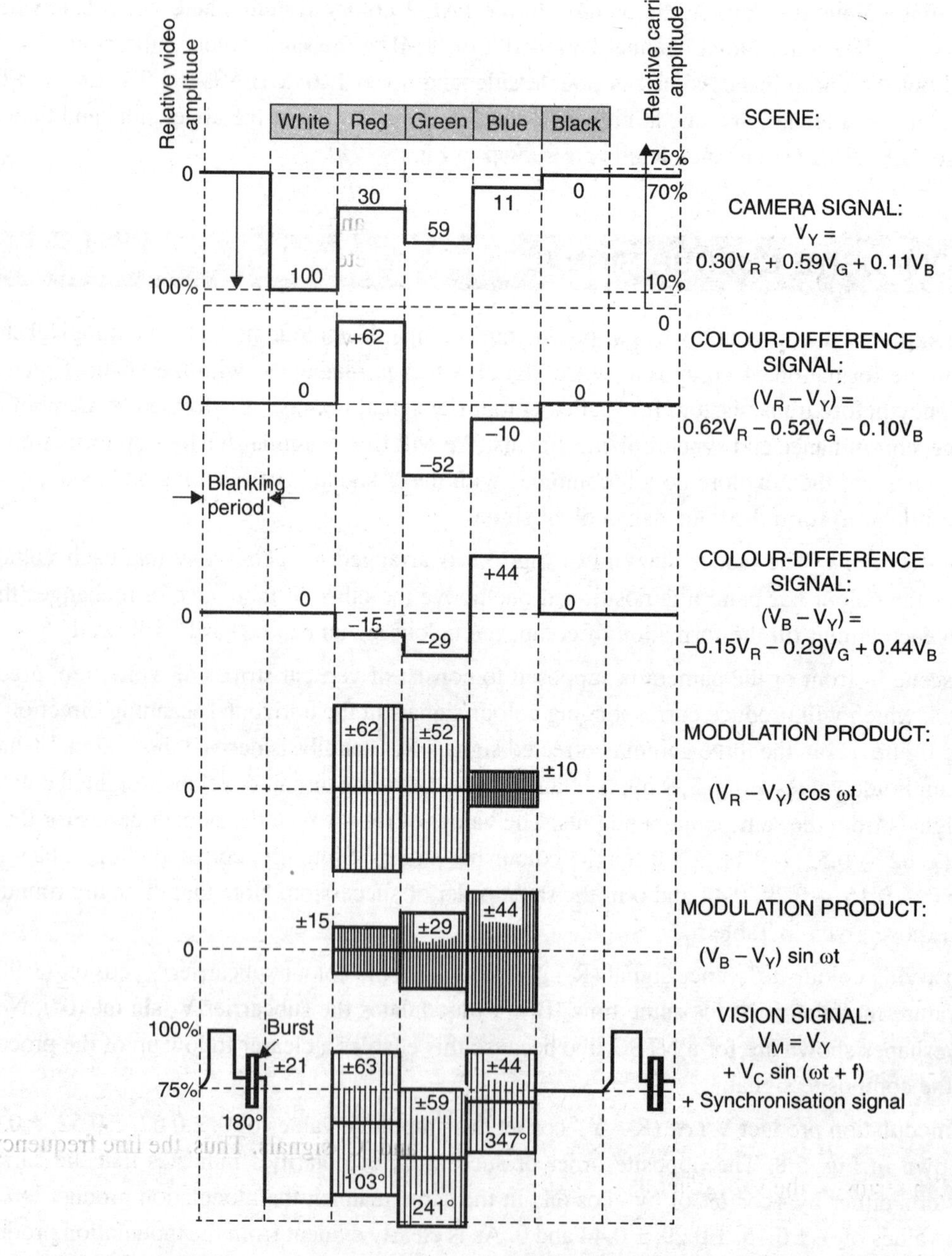

Fig. 5.8 Step by step conversion of camera signals into picture carrier modulation.

Complete Colour Signal

The output from the harmonic filter is combined with the Y signal as shown in the last waveform of Fig. 5.8. The line and field sync pulses are added earlier to the Y signal. As is obvious from the final waveform, the chrominance signal gets superimposed on the luminance signal. The reduction of colour-difference signals on multiplication with weighting factors restrict the mudulation product amplitudes in such a way that the composite signal on modulation with the picture carrier of the CTV transmitter keeps the modulated output within limits of black level at nearly 70% and white level that is close to 10% of the maximum.

5.9 PAL COLOUR TELEVISION TRANSMITTER

A simplified block diagram of a colour television transmitter is shown in Fig. 5.9. It shows various building blocks that are employed to obtain composite colour signal and its modulation with the picture (vision) carrier. In addition, it also shows formation of the complete signal that is transmitted after adding the frequency modulated sound signal to it.

The camera circuitry yields gamma corrected R, G and B video signal. The matrixing and weighting factor multiplication circuit then provides U, V and Y signals. Band-pass filters are used to restrict the bandwidth of colour-difference signal to about 1.5 MHz. Later these are quadrature modulated with the colour subcarrier of 4.43 MHz and added to form the chrominance signal.

Time Coincidence of Luminance and Chrominance Signals

The bandwidth allowed for the Y signal is 5 MHz whereas the colour-difference signals are restricted up to about 1.5 MHz. Since the bandwidths of the Y and C signals are different, a sharp transition in the picture will give rise, in general, to a sharp transition of the Y signal but a lower transition to the colour-difference signals during their passage through bandwidth limited amplifiers. The fastest possible rate of rise, measured from 10% to 90% of the amplitude change, is given by $T = \frac{1}{2}2f$, where f is the bandwidth of the channel through which the signal passes. Thus, the delay of each signal depends upon bandwidth of the channel and the attenuation slope of its cut-off frequency.

For the luminance and chrominance transitions to appear to be coincident in time, centres of these transitions must coincide. This is achieved by delaying the Y signal with respect to chrominance signals. A delay line in the path of Y signal is used for this purpose. This explains the need of the delay line block shown after the matrixing network in Fig. 5.9.

Sync Pulses

The subcarrier frequency has to be exact at a value of 4.43361875 MHz. It is thus generated by a crystal controlled oscillator. The line frequency f_L is locked with f_{sc} by frequency countdown circuits, so that correct numerical relationship between the two is maintained at all times. This is necessary for proper frequency interleaving and least cross-talk between 'Y' and 'C' signals. Thus, the line frequency is often stated in terms of the subcarrier as:-

$$f_L = \frac{f_{sc} - 25}{284 - 1/4} = 15625 \text{ Hz.}$$

CHAPTER 5

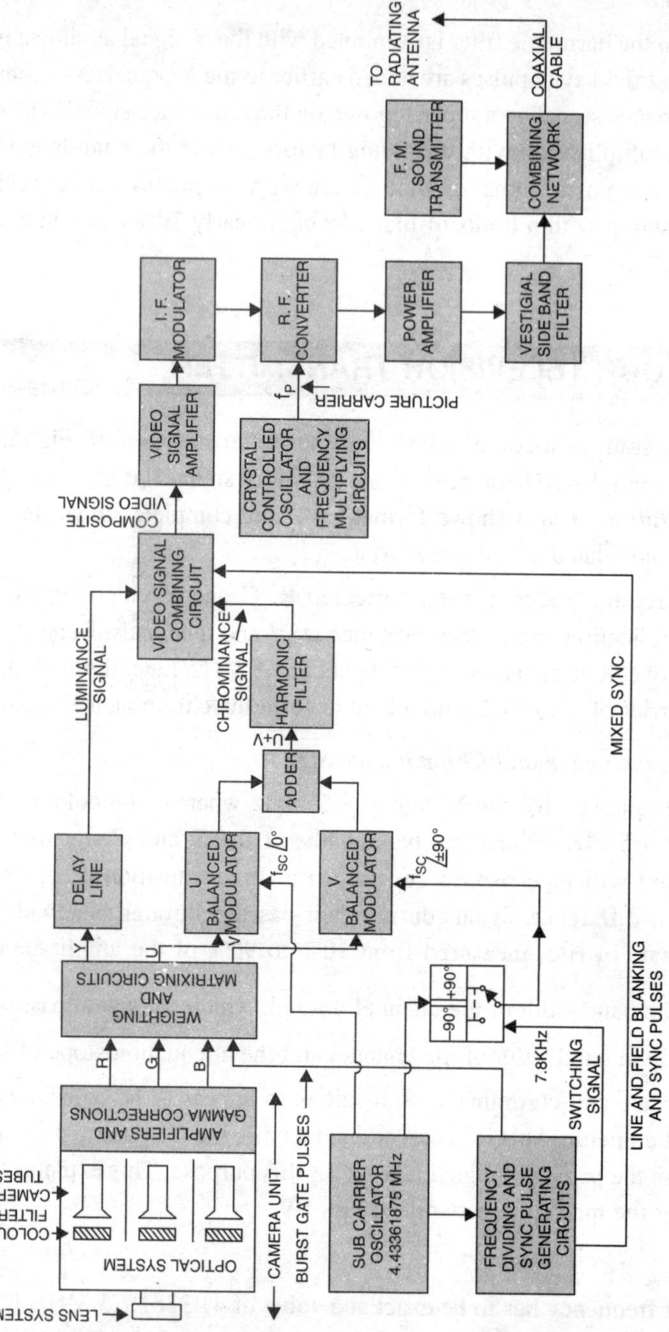

Fig. 5.9 Simplified block diagram of a colour television transmitter.

Although the field frequency f_v is equal to the nominal mains frequency of 50 Hz, it is not locked to the mains. The reason being that mains frequency varies slightly with loading and if f_v is locked with it, and error would develop in the relationship between f_{sc}, f_L and f_v as is obvious from the expression for the line frequency. Thus, in order to avoid this, the field frequency is maintained accurately at 50 Hz independently of the mains frequency. The system is then said to be asynchronous since it is not mains synchronized. As such any mains hum pattern on the screen due to faulty conditions is no longer stationary but drifts at the rate determined by the difference between the field and mains frequencies. The horizontal, vertical and associated sync pulses are derived by suitable circuitry from corresponding deflection frequencies and added to the Y signal in the video signal combining circuit. The 7.8 KHz line switching signal is also derived from the same basic circuitry as shown in Fig. 5.9.

Colour Burst Gating

As already explained a few cycles of subcarrier are carried at the back porch of each line blanking pulse for synchronizing the colour subcarrier generator in the TV receiver. Its location is shown in Fig. 5.10 (*a*). In all about 9 cycles of burst are sent after each active line of scanning.

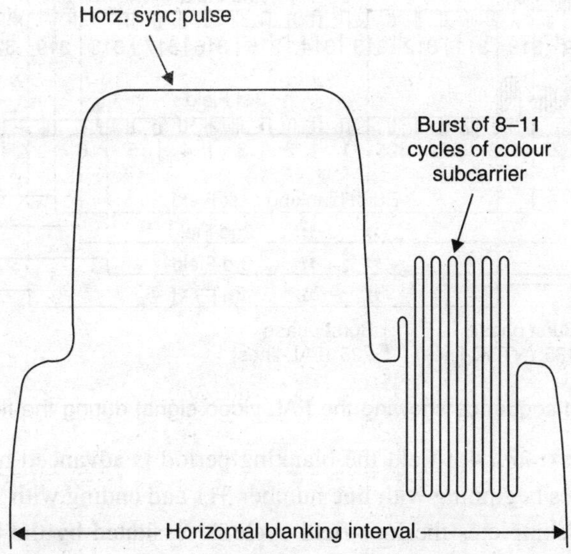

Fig. 5.10 (*a*) Location of colour burst on the back porch of horizontal blanking pulse.

In the PAL signal, in order to make the burst phase start and stop in identical direction at the beginning and end of each field, the burst blanking pulse is shifted as shown in Fig. 5.10 (*b*) so that it follows a four-field cycle. The duration of the burst blanking pulses is about 9 lines period.

If the burst blanking pulse were not staggered in this way but repeated at regular intervals of 1/50 seconds (*i.e.*, at intervals of 312.5 lines precisely), the burst phase at the end of each field and the start of the next would be in one direction for fields 1 and 4 but opposite for fields 2 and 3. This may be seen by applying the same field blanking pulse, say the third, to all the four fields.

For a better understanding of the need for staggering the blanking pulse, consider each field sequence saparately. Note that the arrow-head pointing upwards denotes a burst phase of 135° (NTSC line) while the one pointing downward denotes a phase angle of 225° (PAL line). For field No. 1 the nine burst blanking pulses occupy line numbers 623 to 6. Since line No. 622 is an NTSC line, and alternate lines are NTSC and PAL, line number 7 must be an NTSC line. The upward pointing arrow before line 7 confirms this. For the second field after an exact number of 312.5 lines (7 + 312.5 = 319.5) the end of burst blanking pulse falls in the middle of the line but this is not permitted. Therefore, the field burst blanking pulse is delayed by 1/2 line for the second field so that it ends just after line

CHAPTER 5

number 318. The 319th line is then an NTSC line as desired. Similarly for field 3, the blanking pulse is further shifted by 1/2 line to account for the half-line discrepancy.

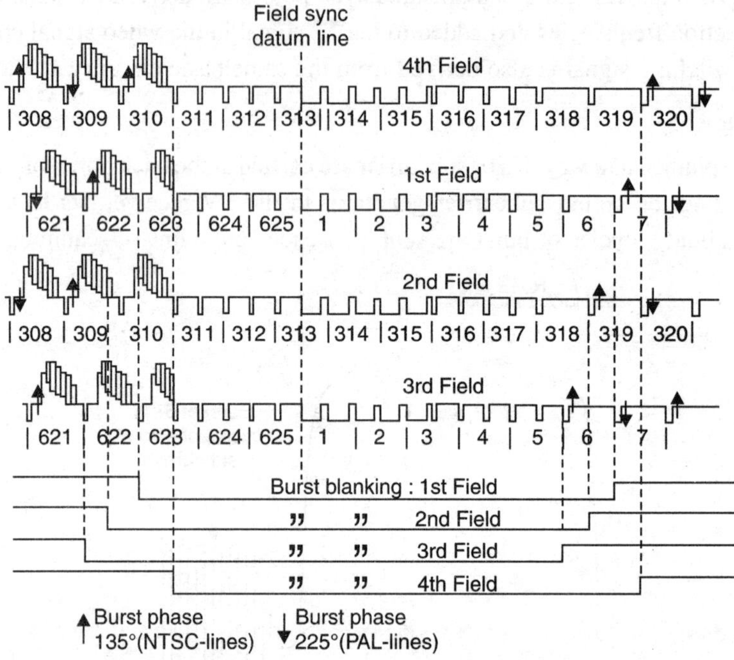

Fig. 5.10 (*b*) Four field sequence showing the PAL video signal during the field blanking period.

However, for the next *i.e.*, 4th field the blanking period is advanced by 1½ lines to provide a blanking period of nine lines beginning with line number 311 and ending with 319. Line number 320 is then the NTSC line. This completes the four-field cycle necessitated by the fact that alternate lines must be NTSC and PAL and each field has a half line because the total number of lines is odd *i.e.*, 625. Note that the burst signal is not sent during the field blanking period.

Channel Carrier Modulation

The composite signal available at the output of combining network is given due amplification before modulating it with the channel picture carrier. The modulation is usually affected in two stages. In the block labelled IF modulator, a carrier frequency of 40 MHz or so is employed. It is a balanced modulator and a band ranging from 35 to 45 MHz is obtained at its output. It is then mixed with an appropriate higher frequency source to get the desired channel carrier frequency and its sidebands as the difference product. The output of this RF converter is amplified in atleast two stages of power amplifiers. The final shaping of the amplitude modulated RF signal is done by passing it through the vestigial sideband filter.

Complete Transmitter Signal

A separate section of the transmitter modulates the sound signal associated with the scene being televised. It is done in the same way as in a monochrome transmitter. An Armstrong frequency modulator is employed with a carrier frequency equal to the channel sound carrier located at 5.5 MHz away from the picture carrier. The two modulated signals are then combined in a network and sent to the antenna array through a coaxial cable. The radiating array is usually located on top of a high tower, hill or a tall building.

5.10 COMPOSITION OF THE TRANSMITTED SIGNAL

In order to fully appreciate the functioning of a colour receiver it is necessary to be fully conversant with the composition of received signal. While processing of each part of the colour television signal has already been covered, various components of the signal transmitted by a CTV station are given below for a quick review of the composition of the signal.

Colour camera output. The R, G and B colour voltages developed by the colour camera represent full information of the brightness and hue of various colours present in the scene being televised. These are gamma corrected to account for non-linear characteristic of the picture tube.

Luminance (Y) signal. This is the output as obtained from the Y matrix and carries brightness information of the picture. It is the same signal as developed by a black and white camera. The allowed bandwidth to the Y signal is nearly 5 MHz and hence enables reproduction of very fine details of the scene.

Colour-difference signals. The colour-difference signals (B – Y) and (R – Y) as formed by the matrix circuit are multiplied by weighting factors to prevent excessive overmodulation of the picture carrier. Their bandwidth is restricted to about 1.5 MHz because of the limitation of the eye to perceive finer details in colour. The weighting and bandwidth limiting is done before modulation with the colour subcarrier. On demodulation at the receiver both (B – Y)′ and (R – Y)′ are deweighted to restore their original amplitude values.

Chrominance signal. The chrominance signal is the colour signal bearing modulation of both U and V video signals. Its bandwidth is also limited by filtering out higher harmonics at the output of quadrature modulator and combining circuits. It is combined with the Y signal by frequency interleaving.

Line sync pulses. These are 4.7 μs pulses that form part of the 12 μs line blanking period. These occur at the frequency of 15625 Hz and are added to the Y signal.

Field sync pulses. After each field of scanning a blanking period of 160 μs follows. It repeats at a frequency of 50 Hz and bears field sync and pre- and post-equalizing pulses. The field pulse train is also added to the Y signal before combining it with the chrominance signal.

Colour burst. At the frequency of subcarrier, 9 to 11 cycles of this signal are suitably gated to be placed at the back porch of line blanking pedestal. Their average reference phase is—(B – Y) axis. Due to the line-by-line reversal of (R – Y) phasor, an identification signal at half the line frequency (≈ 7.8 KHz) is in built in the colour burst-signal. This enables identification of NTSC and PAL lines. The colour burst signal is added to the chrominance signal in correct phase and sequence and thus becomes a part of the chrominance signal.

Sound signal. This is the output signal from microphones used for picking up sound information associated with the scene. Its bandwidth is up to 15 KHz. It is amplified and frequency modulated on the channel sound carrier. Pre-emphasis is a normal feature to improve signal to noise ratio. A similar de-emphasis circuit is used at the receiver to restore the audio signal to its original value.

Picture carrier. This is a continuous wave (C.W.) sinusoidal signal in the VHF/UHF range depending on the channel number. For example, it has a frequency of 62.25 MHz of channel No. 4 and 189.25 MHz for channel No. 7. The composite signal that is formed by combining Y, C, line sync, field sync and colour burst signals is modulated on this picture carrier before transmission. This gives identity to the station and enables its selection at the receiver.

Sound carrier. The sound signal associated with the scene is frequency modulated with the channel sound carrier for purposes of transmission. Its frequency is 5.5 MHz higher than the corresponding picture carrier. The FM sound carrier is combined with the AM picture carrier in a suitable network.

The complete transmitter RF signal thus formed is radiated by the antenna system of the colour television station.

REVIEW QUESTIONS

1. Explain how by frequency interleaving, colour information is accommodated within the same channel bandwidth.
2. Discuss the factors which influence the choices of subcarrier frequency. Why it must be kept as high as possible?
3. What do you understand by half-line and quarter-line offsets as used for choosing the value of subcarrier frequency? Explain why f_{sc} = 4.43361875 MHz is chosen in the PAL colour system.
4. Draw block diagram of a PAL Encoder and explain how the composite colour signal is formed. Draw a phasor diagram to show that amplitude of the resultant chrominance signal phasor determines saturation and its angle tells hue of the colour being transmitted.
5. Justify that the chrominance signal phasor may rest in any of the four quadrants of the colour circle. Sketch phasor locations for all the primary and complementary colours.
6. Explain why a colour burst signal must accompany the chrominance signal and follow a four-field cycle to ensure that it starts and stops in identical directions at the beginning and end of each field.
7. What determines the bandwidth necessary for transmission of colour-difference signal? Why is it much less than the bandwidth allotted for the Y signal? What is the function of a delay line that is introduced in the path of Y signal?
8. Draw a ladder diagram to show step by step formation of the chrominance signal and its modulation on the channel picture carrier. Assume that a colour bar having white, yellow, magenta, cyan and black stripes are scanned in the horizontal direction.
9. Draw a simple block diagram of a colour television transmitter and explain how the camera and microphone output voltage are processed to form the complete transmitter signal.
10. The RF signal transmitted by a colour transmitter is quite complex. Give its composition and describe briefly the function of each constituent that forms part of this signal.

6

COLOUR TELEVISION SYSTEMS–NTSC, PAL AND SECAM

INTRODUCTION

The three different standards of black and white television resulted in the development of three colour systems each compatible with the corresponding monochrome system. These are :— NTSC, PAL and SECAM. The NTSC was the first to be developed and adopted by U.S.A, Japan and many other countries. The PAL, though similar to NTSC is compatible with the European 625 line monochrome system. It was adopted by the Federal Republic of Germany and U.K. in 1967. Subsequently, many Middle East and Asian countries including India also opted for it. The third colour system in use is the SECAM which was developed in France and adopted by them in 1967. Later a modified version of SECAM came into use in Russia and many East European countries. In any case, the adoption of any one colour system by different countries was largely influenced by the monochrome system already in use.

It is necessary to gain familiarity with the technique of encoding and decoding of colour information in each system. This chapter is, therefore, devoted to basic description of the three colour systems together with their relative merits and demerits.

As stated above the three colour television systems in use are :—

(*i*) The American NTSC (National Television Systems Committee) System.

(*ii*) The German PAL (Phase Alteration Line by Line System).

(*iii*) The French SECAM (Sequential Couleures a Memorie) System.

All the three systems describe the picture in terms of luminance and colour-difference signals. It is the way in which colour-difference signals are modulated that makes one system different from the other. In many respects, transmission and reception techniques of NTSC and PAL are similar. These are, therefore, treated first. The SECAM, being very much different from the first two is described in the later part of this Chapter.

6.1 NTSC COLOUR TELEVISION SYSTEM

The NTSC colour system is compatible with the American 525 line black and white system. In this system, phase angles are measured relative to – (B – Y) phasor. This location is designated 0° or the reference phase position as shown in Fig. 6.1. This is also the phase of colour burst that is transmitted on the black porch of each horizontal blanking pulse. As shown in the figure, the weighted (or compensated) phasor for the colour magenta has a magnitude of 0.59 and is located at an angle of 119° with respect to – (B – Y). Similarly, magnitudes and phase angles for other colour signals are also shown in Fig. 6.1. Note that primary colours, as explained in a previous chapter, are 120° apart and complementary colours differ in phase by 180° from corresponding primary colours.

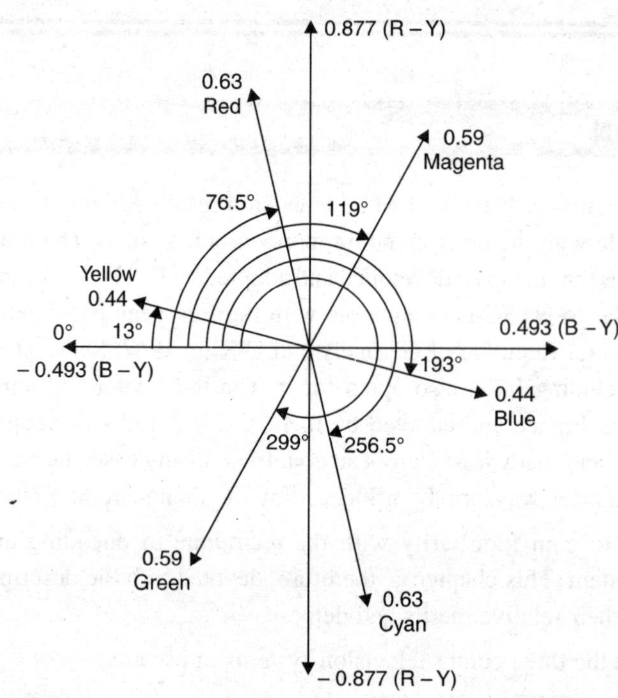

Fig. 6.1 Magnitude and phase relationships of compensated chrominance signals for the primary and complementary colours.

Bandwidth Reduction

In order to save bandwidth, advantage has been taken of the fact that eye's resolution of colours along the reddish blue-yellowish green axis on the colour circle is much less than for colours which lie around the yellowish red-greenish blue axis. Therefore, two new colour video signals, which correspond to these colour regions, are generated. These are designated as I and Q signals. The I signal lies in a region 33° counter clockwise to + (R – Y) where the eye has maximum colour resolution. *It is derived from the (R – Y) and (B – Y) signals and equals 0.60R – 0.28G – 0.32B. As shown in Fig. 6.2(*a*) it is located at an angle of 57° with respect to colour burst. Similarly, the Q** signal is derived from colour

*I = 0.74(R – Y) – 0.27(B – Y).

**Q = 0.48(R – Y) + 0.41(B – Y).

difference signals by a suitable matrix and equals 0.21R – 0.52G + 0.31B. It is located 33° counter clockwise to the + (B – Y) signal and is thus in quadrature with the I signal. As illustrated in Fig. 6.2(*a*) the Q signal covers regions around magenta (reddish-blue) and yellow-green shades. Similarly, orange hues correspond to phase angles centred around + I and the complementary blue-green (cyan) hues are located around the diametrically opposite – I signal. Since the eye is capable of resolving details in these regions, I signal is allowed to have bandwidth up to 1.5 MHz. However, the eye is least sensitive to colours that lie around the ± Q signal axis, and therefore Q is allowed a bandwidth of only ± 0.5 MHz around the colour subcarrier.

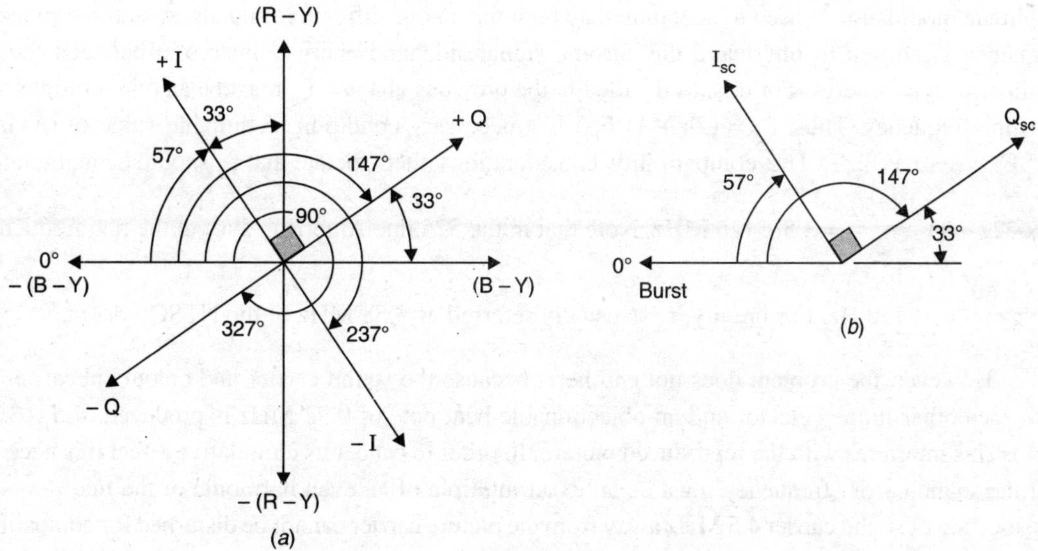

Fig. 6.2 Phasor diagram of the I and Q signals in the NTSC system.

It may be noted that both I and Q signals are active up to 0.5 MHz and being at right angles to each other, combine to produce all the colours contained in the chrominance signal. However, Q signal drops out after 0.5 MHz and only I signal remains between 0.5 and 1.5 MHz to produce colours, the finer details of which the eyes can easily perceive.

To help understand this fact it may be recalled that only one colour-difference signal is needed for producing colours which are a mixture of only two colours. Thus, Q signal is not necessary for producing colours lying in the region of orange (red + green) and cyan (green + blue) hues. Hence, at any instant when Q = 0 and only I signal is active, the colours produced on the screen will run the gamut from reddish orange to bluish green.

Double sideband transmission is allowed for the Q signal and it occupies a channel bandwidth of 1 MHz (± 0.5 MHz). However, for the I signal, upper sideband is restricted to a maximum of 0.5 MHz while the lower sideband is allowed to extend up to 1.5 MHz. As such, it is a form of vestigial sideband transmission. Thus in all, a bandwidth of 2 MHz is necessary for colour signal transmission. This is a saving of 1 MHz as compared to a bandpass requirement of 3 MHz if (B – Y) and (R – Y) are directly transmitted.

It is now obvious that in the NTSC system, advantage in taken of the limitations of human eye to restrict the colour signal bandwidth, which in turn results in reduced interference with the sound and picture signal sidebands. The reduction in colour signal sidebands is also dictated by the relatively narrow channel bandwidth of 6 MHz in the American TV system.

6.2 EXACT COLOUR SUBCARRIER FREQUENCY

The method of transmitting colour signal information is nearly the same in NTSC and PAL. Quadrature amplitude modulation is used to accommodate both the colour-difference signals. A suitable subcarrier frequency is chosen to interleave the chroma signal sideband energy clusters in-between those of luminance signal energy. For this, as detailed in the previous chapter, f_{sc} must be an odd multiple of the half-line frequency. Thus, $f_{sc} = (2n + 1)\ F_{L/2}$ is a necessary condition. A suitable value of 'n' in the NTSC system is 227. The compatibility considerations then dictate that f_{sc} should be maintained at $(2 \times 227 + 1)\ \dfrac{15750}{2} = 3.583125$ MHz. Note that in the 525 line American system the line frequency = $\dfrac{525 \times 60}{2} = 15750$ Hz. For brevity, f_{sc} is usually referred as 3.58 MHz in the NTSC system.

However, the problem does not end here, because the sound carrier and colour subcarrier beat with each other in the detector and an objectionable beat note of 0.92 MHz is produced (4.5 – 3.58 = 0.92). This interferes with the reproduced picture. In order to cancel its cumulative effect it is necessary that the sound carrier frequency must be an exact multiple of an even harmonic of the line frequency. The location of sound carrier 4.5 MHz away from the picture carrier cannot be disturbed for compatibility reasons and so 4.5 MHz is made to be the 286th harmonic of the horizontal deflection frequency. Therefore, $f_L = 4.5$ MHz/286 = 15734.26 Hz is chosen. Note that this is closest to the value of 15750 Hz used for horizontal scanning in monochrome transmission. A change in the line frequency necessitates a change in the field frequency since 262.5 lines must be scanned per field. Therefore, the field frequency (f_v) is changed to be 15734.26/262.5 = 59.94 Hz in place of 60 Hz.

The slight difference of 15.74 Hz in the line frequency (15750 – 15734.26 = 15.74 Hz) and of 0.06 Hz in the field frequency (60 – 59.94 = 0.06 Hz) has practically no effect on the deflection oscillators because an oscillator that can be triggered by 60 Hz pulses can also by synchronized to produce 59.94 Hz output. Similarly, the AFC circuit can easily adjust the line frequency to a slightly different value while receiving colour transmission.

As explained earlier, the colour subcarrier frequency must be an odd multiple of half-the-line frequency to suppress do pattern interference. Therefore, f_{sc} is fixed at $(2n + 1)\ f_{L/2}$, *i.e.*, 455 × 15734.26/2 = 3.579545 MHz. To obtain this exact colour subcarrier frequency, a crystal controlled oscillator is provided.

6.3 NTSC CODER

Figure 6.3 illustrates the encoding process of colour signals at the NTSC transmitter. A suitable matrix is used to get both I and Q signals directly from the three camera outputs. Since I = 0.60R – 0.28G – 0.32B, the green and blue camera outputs are inverted before feeding them to the appropriate matrix. Similarly, for Q = 0.21R – 0.52G + 0.31B, an invertor is placed at the output of green camera before mixing it with the other two camera outputs.

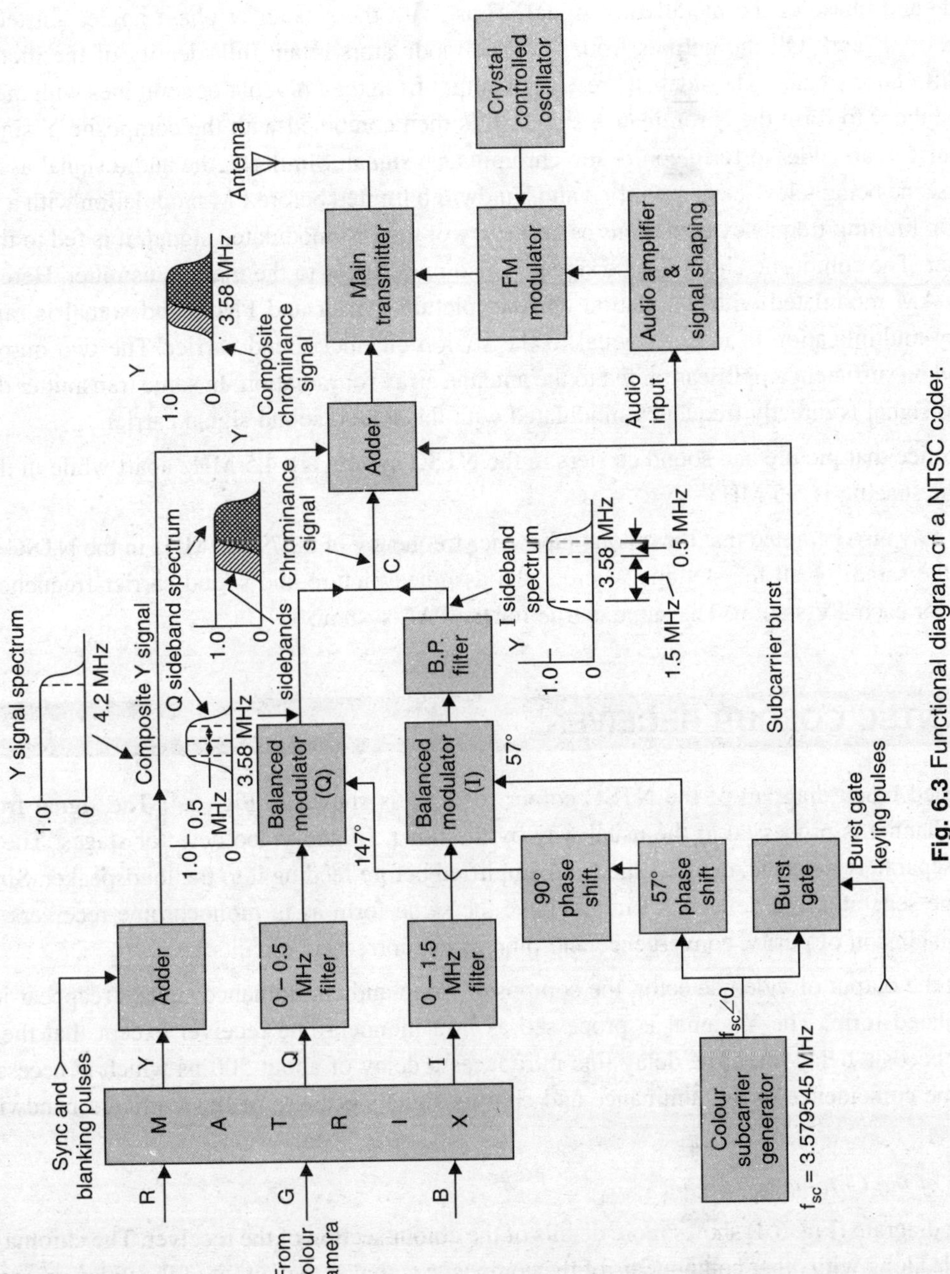

Fig. 6.3 Functional diagram cf a NTSC coder.

The bandwidths of both I and Q are restricted before feeding them to corresponding balanced modulators. The subcarrier to I modulator is phase shifted 57° clockwise with respect to the colour burst. The carrier is shifted by another 90° before applying it to the Q modulator. Thus, relative phase shift of 90° between the two subcarriers is maintained for quadrature amplitude modulation [see Fig. 6.2(*b*)].

It is the characteristic of a balanced modulator that while it suppresses the carrier and provides frequency translation, both amplitude and phase of its output are directly related to the instantaneous amplitude and phase of the modulating signal. Thus, with the subcarrier phase angles shifted to the locations of 'I' and 'Q', the outputs from both the modulators retain full identity of the modulating colour-difference signals. The sideband restricted output from the I modulator combines with the output of Q modulator to form the chrominance signal. It is then combined with the composite Y signal and colour burst in an adder to form composite chrominance signal. Similarly, the audio signal associated with the scene being televised is amplified and bandwidth limited before FM modulation with a low RF carrier. On limiting frequency deviations as necessary of the FM modulated signal, it is fed to the main transmitter. The composite chrominance signal is also made input to the main transmitter. Here the '*c*' signal is AM modulated with the station channel picture carrier and FM sound signal is raised by frequency multiplication to make it equal to the station channel sound carrier. The two outputs are added and on sufficient amplification fed to the antenna array for radiation. In some transmitter designs, the sound signal is directly frequency modulated with the station sound signal carrier.

Notice that picture and sound carriers in the NTSC system are 4.5 MHz apart while in the PAL system the spacing is 5.5 MHz.

It may also be noted that the colour subcarrier frequency of 3.579545 MHz in the NTSC system remains the same for all the stations whereas the assigned picture and sound carrier frequencies are different for each TV station. The same is true for the PAL system.

6.4 NTSC COLOUR RECEIVER

A simplified block diagram of the NTSC colour receiver is shown in Fig. 6.4. The signal from the selected channel is processed in the usual way by the tuner, IF and video detector stages. The sound signal is separately detected, demodulated and amplified before feeding it to the loudspeaker. Similarly AGC, sync separator and deflection circuits have the same form as in monochrome receivers except from the inclusion of purity, convergence and pincushion correction circuits.

At the output of video detector, the composite video and chrominance signals reappear in their pre-modulated form. The Y signal is processed as in a monochrome receiver except that the video amplifier needs a delay line. The delay line introduces a delay of about 500 ns which is necessary to ensure time coincidence of the luminance and chroma signals because of the restricted bandwidth of the latter.

Decoding of the Chroma (C) Signal

The block diagram (Fig. 6.4) shows more details of the colour section of the receiver. The chroma signal is available along with other components of the composite signal at the output of the video preamplifier (1st video amp). It should be noted that the chrominance signal has colour information during active trace time of the picture and the burst occurs during blanking time when there is no picture. Thus, although the 'C' signal and burst are both at 3.58 MHz, they are not present at the same time.

Chrominance Bandpass Amplifier

The purpose of the bandpass amplifier is to separate the chrominance signal from the composite video signal, amplify it and then pass it on to the synchronous demodulators. The amplifier has fixed tuning with a bandpass wide enough (≈ 2. MHz) to pass the chroma signal. The colour burst is prevented from appearing at its output by horizontal blanking pulses which disable the bandpass amplifier during horizontal blanking intervals. The blanking pulses are generally applied to the colour killer circuit which in turn biases-off the chrominance amplifier during these periods.

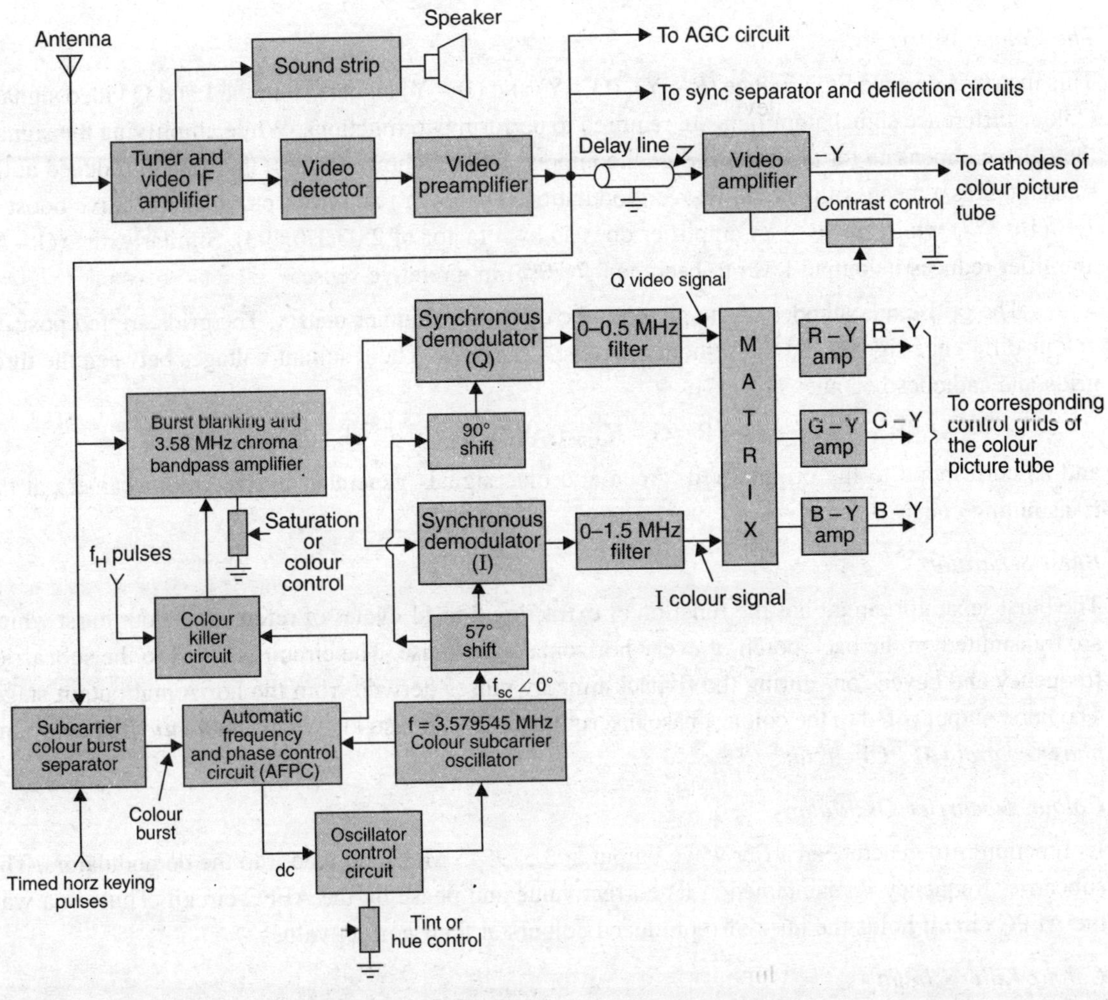

Fig. 6.4 Simplified block diagram of a NTSC receiver.

Colour Demodulators

Synchronous demodulators are used to detect the modulating signal. Such a demodulator may be thought of as a combination of phase and amplitude detectors because the output is dependent on both phase and amplitude of the chroma signal. As shown in the block diagram (Fig. 6.4) each demodulator has two input signals, the chroma which is to be demodulated and a constant amplitude output from the local subcarrier oscillator. The oscillator output is coupled to the demodulators by phase-shifting networks. The 'I' demodulator oscillator voltage has a phase of 57° with respect to the burst phase (f_{sc} 0°) and so has the correct delay to detect the 'I' colour difference signal. Similarly, the oscillator voltage to the 'Q' demodulator is delayed by 147° (57° + 90°) for detecting the Q colour-difference signal. Thus, I and Q synchronous demodulators convert the chroma signal (a vector quantity) into its right-angle components (polar to rectangular conversion).

The Colour Matrix

This matrix is designed to produce (R – Y), (G – Y) and (B – Y) signals from the I and Q video signals. Colour-difference signal amplifiers are required to perform two functions. While amplifying the signals they also compensate for the chroma signal compression (weighting factors) that was introduced at the transmitter as a means of preventing overmodulation. The (R–Y) amplifier provides a relative boost of 1.14(1/0.877) while the (B – Y) amplifier does so by a factor of 2.03(1/0.493). Similarly, the (G – Y) amplifier reduces its output level to become 0.7(70%) in a relative sense.

The grids and cathodes of the picture tube constitute another matrix. The grids are fed positive colour-difference signals and the cathodes receive – Y signal. The resultant voltages between the three grids and cathodes become:

$$(R - Y) - (-Y) = R,\ (G - Y) - (-Y) = G \quad \text{and} \quad (B - Y) - (-Y) = B$$

and so correspond to the original red, green and blue signals generated by the colour camera at the transmitting end.

Burst Separator

The burst separator circuit has the function of extracting 8 to 11 cycles of reference colour burst which are transmitted on the back porch of every horizontal sync pulse. The circuit is tuned to the subcarrier frequency and keyed 'on' during the flyback time by pulses derived from the horizontal output stage. The burst output is fed to the colour phase discriminator circuit also known as *automatic frequency* and *phase control (AFPC) circuit.*

Colour subcarrier Oscillator

Its function is to generate a carrier wave output at 3.579545 MHz and feed it to the demodulators. The subcarrier frequency is maintained at its correct value and phase by the AFPC circuit. Thus, in a way the AFPC circuit holds the hues of reproduced colours at their correct values.

Colour Killer Circuit

As the name suggests this circuit becomes 'on' and disables the chroma bandpass amplifier during monochrome reception. It thus prevents any spurious signals which happen to fall within the bandpass of chroma amplifier from getting through the demodulators and causing coloured interference on the screen. This colour noise is called 'confetti' and looks like snow but with large spots in colour. The receiver thus automatically recognizes a colour or monochrome signal by the presence or absence of colour sync burst. This voltage is processed in the AFPC circuit to provide a dc bias that cuts-off the colour killer circuit. Therefore, when the colour killer circuit is 'off', the chroma bandpass amplifier is 'on' to pass colour information. In some receiver designs the colour demodulators are disabled instead of the chroma bandpass amplifier during monochrome reception.

Manual Colour Controls

The two additional operating controls necessary in the NTSC colour receivers are colour (saturation) level control and tint (hue) control. These are provided on the front panel of colour receiver. The colour control changes gain of the chrominance bandpass amplifier and thus controls the internsity or amount of colour in the picture. The tint control varies phase of the 3.58 MHz oscillator with respect to the colour sync burst. This circuit can be either in the oscillator control or AFPC circuit.

6.5 LIMITATIONS OF THE NTSC SYSTEM

The NTSC system is sensitive to transmission path differences which introduce phase errors that result in colour changes in the picture. At the transmitter, phase changes in the chroma signal take place when changeover between programmes of local and television network systems takes place and when video tape recorders are switched on. The chroma phase angle is also effected by the level of signal while passing through various circuits. In addition, crosstalk between demodulator outputs at the receiver causes colour distortion. All this requires the use of an automatic tint control (ATC) circuit with provision of a manually-operated tint control.

6.6 PAL COLOUR SYSTEM

The PAL system which is a variant of the NTSC system, was developed at the Telefunken Laboratories in the Federal Republic of Germany. In this system, the phase error susceptibility of the NTSC system has been largely eliminated. The main features of the PAL system are :—

(*i*) The weighted (B – Y) and (R – Y) signals are modulated without being given a phase shift of 33° as is done in the NTSC system.

(*ii*) On modulation, both the colour-difference quadrature signals are allowed the same bandwidth of about 1.3 MHz. This results in better colour reproduction. However, the chroma signal transmission is of vestigial sideband type. The upper sideband attenuation slope starts at 0.57 MHz, *i.e.*, (5 – 4.3 = 0.57 MHz), but the lower sideband extends to 1.3 MHz before attenuation begins.

(*iii*) The colour subcarrier frequency is chosen to be 4.43361875 MHz. It is an odd multiple of one-quarter of the line frequency instead of the half-line off-set as used in the NTSC system. This results in somewhat better cancellation of the dot pattern interference.

(*iv*) The weighted (B – Y) and (R – Y) signals are modulated with the subcarrier in the same way as in the NTSC system (QAM) but with the difference, that phase of the subcarrier to one of the modulators (V modulator) is reversed from +90° to – 90° at the line frequency. In fact the system derives its name, phase alteration by line (*i.e.*, PAL), from this mode of modulation. This technique of modulation cancels hue errors which result from unequal phase shifts in the transmitted signal. The switching action naturally occurs during the line blanking interval to avoid any visible disturbance.

As explained in the previous chapter, the colour-difference signals are scaled down with weighting factors before feeding to the quadrature modulator. These on modulation become U and V modulation products of the chrominance signal. Thus, as illustrated in Fig. 6.5 (*a*),

$$C_{PAL} = U \sin \omega_{sc}t \pm V \cos \omega_{sc}t$$

$$= \sqrt{U^2 + V^2}\ \sin (\omega_{sc}t \pm \theta) \text{ where } \theta = \tan^{-1} \frac{V}{U}$$

6.7 THE PAL BURST

If the signal were applied to an NTSC type decoder, the (B–Y) output would be U as required but the (R–Y) output would alternate as +V and –V from line to line. Therefore, the V demodulator must be switched at half the horizontal (line) frequency rate to give '+ V' only on all successive lines. Clearly the PAL receiver must be told how to achieve the correct switching mode. A colour burst (10 cycles at 4.43 MHz) is sent out at the start of each line. Its function is to synchronize the receiver colour oscillator for reinsertion of correct carrier into the U and V demodulators.

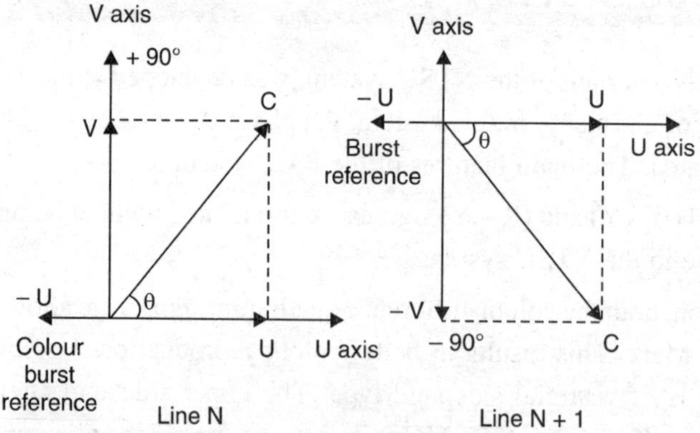

Fig. 6.5 (*a*) Sequence of modulation *i.e.*, phase change of 'V' signal on alternate lines in the PAL colour system.

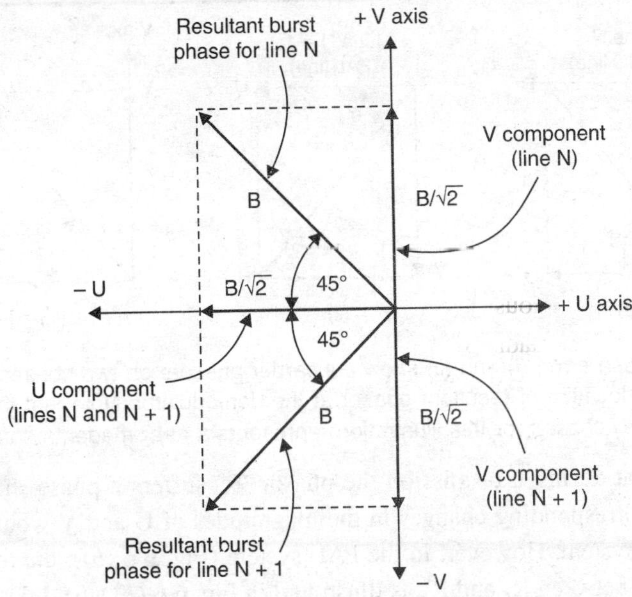

Fig. 6.5 (*b*) Illustration of PAL colour burst swing.

While in NTSC the burst has the phase of – (B – Y) and a peak-to-peak amplitude equal to that of the sync, in PAL, the burst is made up of two components (see Fig. 6.5 (*b*)), a – (B – Y) component as in NTSC but with only $1/\sqrt{2}$ of the NTSC amplitude and an (R – Y) component which like the (R – Y) signal is reversed in phase from line to line. This ± (R – Y) burst signal has an amplitude equal to that of the – (B – Y) burst signal, so that the resultant burst amplitude is same as in NTSC. Note that the burst phase actually swings ± 45° (see Fig. 6.5 (*b*)) about the – (B – Y) axis from line to line. However, the sign of (R – Y) burst component indicates the same sign as that of the (R – Y) picture signal. Thus, the necessary switching mode information is always available. Since the colour burst shifts on alternate lines by ± 45° about the zero reference phase it is often called the swinging burst.

6.8 CANCELLATION OF PHASE ERRORS

As already pointed out the chroma signal is suceptible to phase-shift errors both at the transmitter and in the transmission path. This effect is sometimes called 'differential phase error' and its presence results in changes of hue in the reproduced picture. This actually results from a phase shift of the colour sideband frequencies with respect to the colour burst phase. The PAL system has a built-in protection against such errors provided the picture content remains almost the same from any one line to the next. This is illustrated by phasor diagrams. Figure 6.6 (*a*) shows phasors representing particular U and V chroma amplitudes for two consecutive lines of a field. Since there is no phase error the resultant phasor (R) has the same amplitude on both the lines. Detection along the U axis in one synchronous detector and along the V axis in another accompanied by sign switching in the latter case yields required U and V colour signals. Thus, correct hues are produced in the picture.

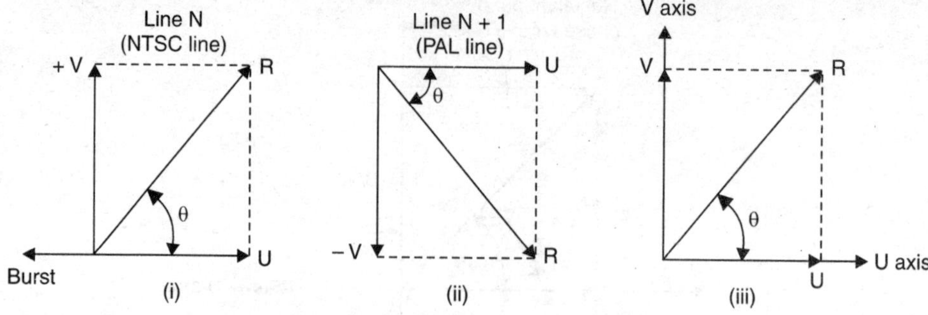

Fig. 6.6 (*a*) No phase error, (i) and (ii) show subcarrier phasors on two consecutive lines while (iii) depicts location of resultant phasor at the demodulator. Note that the phasor (U ± jV) chosen for this illustration represents a near magenta shade.

Now suppose that during transmission the phasor 'R' suffers a phase shift by an angle δ. As shown in Fig. 6.6 (*b*) corresponding changes in the magnitudes of U and V would mean a permanent hue error in the NTSC system. However, in the PAL system (Fig. 6.6 (*b*)), the resultant phasor at the demodulator will swing between R_1 and R_2 as illustrated in Fig. 6.6 (*b*) (*iii*). It is now obvious that the phase error would cancel out if the two lines are displayed at the same time. In actual practice however, the lines are scanned in sequence and not simultaneously. The colours produced by two successive lines, therefore, will be slightly on either side of the acutal hue.

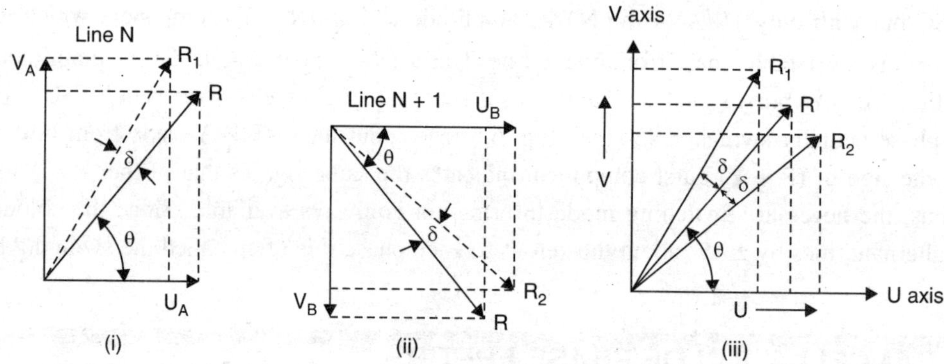

Fig. 6.6 (*b*) Cancellation of phase error, (δ) (i) and (ii) show subcarrier phasors on two consecutive lines when a phase shift occurs (iii) location of resultant phasor at the demodulator.

Since the lines are scanned at a very fast rate the eye due to persistence of vision will perceive a colour that lies between the two produced by R_1 and R_2, respectively. Thus, the colour seen would more or less be the actual colour. It is here, where PAL system claims superiority over the NTSC system.

6.9 PAL-D (PAL STANDARD) COLOUR SYSTEM

The use of eye as the averaging mechanism for the correct hue is the basis of 'simple PAL' colour system. However, beyond a certain limit the eye does see the effect of colour changes on alternate lines, and so the system needs modifications. Remarkable improvement occurs in the system if a delay line is employed to do the averaging first and then present the colour to the eye. This is known as PAL-D or Delay Line PAL, or (PAL standard) method and is most commonly used in PAL colour receivers. As an illustration of PAL-D averaging technique, Fig. 6.7 shows the basic circuit used for separating individual U and V products from the chrominance signal. For convenience, both U and V have been assumed to be positive, that is, they correspond to some shade of magenta (purple). Thus, for the first line when the V modulator product is at + 90° to the + U axis, the phasor can be expressed as (U + jV). This is called the *NTSC line*. But on the alternate (next) line when the V phase is switched to – 90°, the phasor becomes (U – jV) and the corresponding line is then called the *PAL line*.

As shown in Fig. 6.7 (*a*), delay line together with adding and subtracting circuits are interposed between the chrominance amplifier and demodulators. The object of delay line is to delay the chrominance signal by almost exactly one line period of 64 μs. The chrominance amplifier feeds the chrominance signal to the adder, the subtractor and the delay line. The delay line in turn feeds its output to both the adder and subtractor circuits. The adder and subtractor circuits, therefore, receive the two signals simultaneously. These may be referred to at any given time as the direct line and delay line signals. For the chosen hue, if the present incoming line is an NTSC line, the signal entering the delay-line and also reaching the adder and subtractor is (U + jV). But then the previous line must have been the PAL line

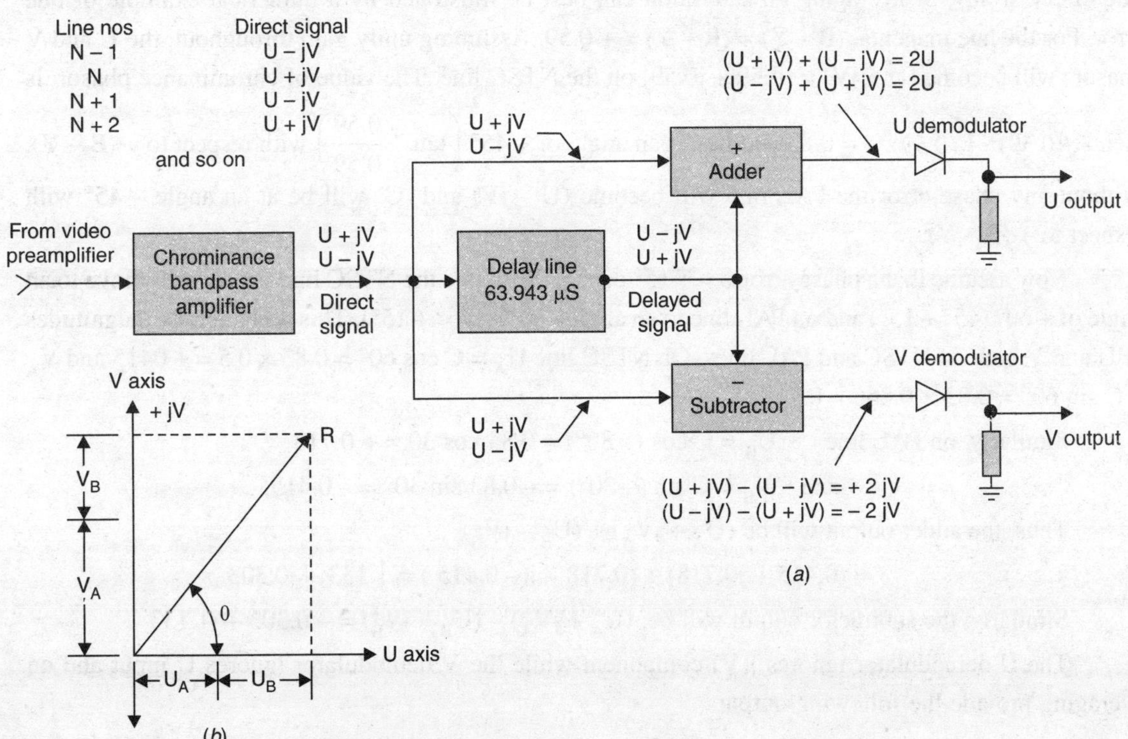

Fig. 6.7 Separation of U and V modulation products (*a*) block diagram, (*b*) summation of consecutive line phasors in the PAL-D receiver.

i.e., (U – jV) and this signal is simultaneously available from the delay line. The result is (see Fig. 6.7 (*a*)) that the signal information of two picture lines, though transmitted in sequence, are presented to the adder and subtractor circuits simultaneously. The adder yields a signal consisting of U information only but with twice the amplitude (2U). Similarly, the subtraction circuit produces a signal consisting only of V information, with an amplitude twice that of the 'V' modulation product.

To permit precise addition and subtraction of direct and delayed line signals, the delay line must introduce a delay which is equivalent to the duration of an exact number of half-cycles of the chrominance signal. This requirement would not be met if the line introduced a delay of exactly 64 μs. At a frequency of 4.43361875 MHz the number of cycles which take place in 64 μs are $4.43361875 \times 10^{+6} \times 64 \times 10^{-6}$ = 283.7485998 Hz, ≈ 283.75 Hz. A delay line which introduces a delay equal to the duration of 283.5 sub carrier cycles is therefore suitable. This is equal to a time delay of 63.943 μs ($1/f_{sc} \times 283.5$).

The addition and substraction of consecutive line phasors can also be illustrated vectorially by phasor diagrams. Figure 6.7 (*b*) is such a phasor diagram pertaining to phase error illustration of Fig. 6.6 (*b*). The U and V signals thus obtained are fed to their respective synchronous demodulators for recovery of colour-difference signals.

Effect of Average on Saturation

The effect, if any, of averaging on saturation can best be illustrated by a numerical example of hue error. For the hue magenta, (B – Y) = (R – Y) = + 0.59. Assuming unity gain throughout, the U and V phasors will become (U + jV) = (0.59 + j0.59) on the NTSC line. The value of chrominance phasor is then $\left(\sqrt{(0.59)^2 + (0.59)^2}\right) = 0.83$ and lies at an angle of + 45° $\left(\tan^{-1} \dfrac{0.59}{0.59}\right)$ with respect to + (B – Y). Without any phase error the PAL line will become (U – jV) and 'C' will be at an angle – 45° with respect to + (B – Y).

Now assume that a phase error δ = + 15° occurs. With this the NTSC line phasor will move to an angle of + 60° (45° + 15°) and on PAL line to an angle –30° (– 45° + 15°). This will alter the magnitudes of U and V both on NTSC and PAL lines. On NTSC line $U_A = C \cos 60° = 0.83 \times 0.5 = +0415$ and $V_A = C \sin 60° = 0.83 \times 0.866 = 0.718$.

Similarly, on PAL line $U_B = C \cos(-30°) = 0.83 \cos 30 = +0.718$

and $V_B = C \sin(-30°) = -0.83 \sin 30° = -0.415$.

Thus, the adder output will be $(U_A + jV_A) + (U_B - jV_B)$

$$= (0.415 + j0.718) + (0.718 - j(-0.415)) = 1.133 + j0.303$$

Similarly, the subtractor output will be: $(U_A + jV_A) - (U_B - jV_B) = -0.303 + j1.133$

The U demodulator ignores j(V) component while the V demodulator ignores U input and on averaging provide the following outputs:

$$\frac{U_A + V_B}{2} = +\frac{1.133}{2} = +0.567$$

and $$\frac{U_A - V_B}{2} = +\frac{1.133}{2} = +0.567$$

The magnitude of C becomes $\sqrt{(0.567)^2 + (0.567)^2} = 0.8$. Notice that the chrominance signal has returned to its original position with the same angle of + 45° $\left(\tan^{-1}\frac{0.567}{0.567}\right)$ with reference to + (B – Y) axis and hue distortion has disappeared. However, there is a small reduction in saturation equal to $\frac{(0.83 - 0.8)}{0.83} \times 100 = 3.6\%$. In practice, the phase errors are usually smaller than assumed in the illustration and there is no significant change of saturation in the PAL system.

6.10 COLOUR SUBCARRIER GENERATION

The colour subcarrier frequency of 4.43361875 MHz is generated with a crystal controlled oscillator. In order to accomplish minimum raster disturbance through the colour subcarrier it is necessary to maintain correct frequency relationships between the scanning frequencies and subcarrier frequency. It is therefore usual to count down from the subcarrier frequency to obtain twice the line frequency ($2f_L$) pulses to be fed to sync pulse generators. There are several ways in which frequency division can be accomplished. In the early days of colour television it was necessary to choose frequencies which had low-order factors so that division could take place in easy stages, but such constraints are no longer necessary. The PAL subcarrier frequency is first generated directly. Then the 25 Hz output obtained from halving the field frequency is subtracted from the subcarrier frequency. The frequency thus obtained is first divided by 5 and then by 227. Such a large division is practicable by using a chain of binary counters and feeding back certain of the counted down pulses to earlier points of the chain so that count down is changed from a division of 2^8 to a division of 227.

6.11 THE PAL CODER

Figure 6.8 is the functional diagram of a PAL coder. The gamma corrected R, G and B signals are matrixed to form Y and weighted colour-difference signals. The bandwidth of both (B – Y) and (R – Y) video signals is restricted to about 1.3 MHz by appropriate low-pass filters. In this process, these signals suffer a small delay relative to the Y signal. In order to compensate for this delay, a delay line is inserted in the parth of Y signal.

The weighted colour-difference video signals from the filters are fed to corresponding balanced modulators. The sinusoidal subcarrier is fed directly to the U modulator but passes through a $\angle\pm 90°$ phase switching circuit on alternate lines before entering the V modulator. Since one switching cycle takes two lines, the squarewave switching signal from the multivibrator to the electronic phase switch is of half-line frequency *i.e.*, approximately 7.8 KHz. The double sideband suppressed carrier signals from the modulators are added to yield the quadrature amplitude modulated (QAM) chrominance (C) signal. This passes through a filter which removes harmonics of the subcarrier frequency and restricts the upper and lower sidebands to appropriate values. The output of filter feeds into an adder circuit where it is combined with the luminance and sync signals to form composite colour video signal. The bandwidth and location of the composite colour signal (U and V) is shown along with the Y signal in Fig. 6.8.

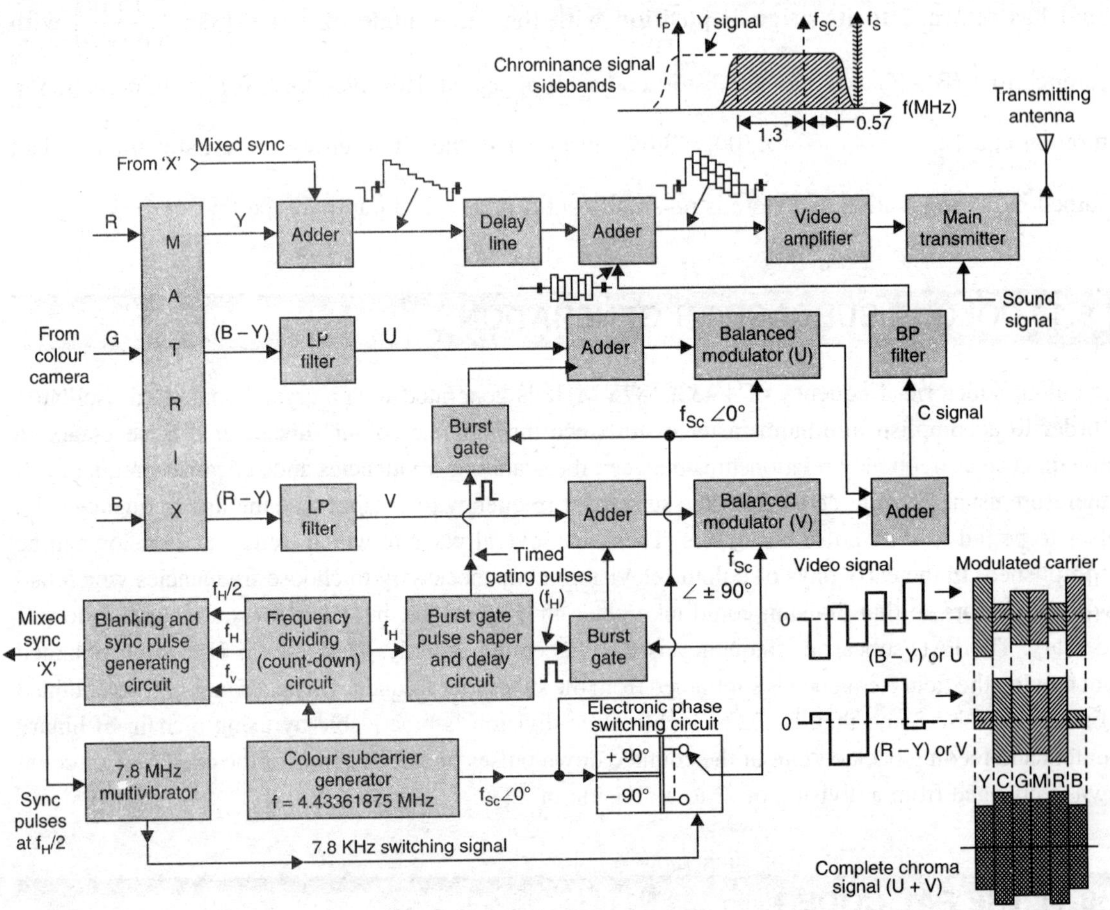

Fig. 6.8 Basic organisation of the PAL coder.

Notice that the colour burst signal is also fed to the modulators along with the U and V signals through the adders. The burst signals are obtained from circuits that feed colour subcarrier signal to the two modulators. However, before feeding the burst signals to U and V adders these are passed through separate burst gates. Each burst gate is controlled by delayed pulses at f_L rate obtained from the frequency dividing circuit. The gating pulses appear during the back porch period. Thus, during these intervals the (B – Y) *i.e.*, U modulator yields a subcarrier burst along – U while the (R – Y) *i.e.*, V modulator gives a burst of the same amplitude but having a phase of ± 90° on alternate lines relatives to – U phasor. At the outputs of two modulators, the two burst components combine in the adder to yield an output which is the vector sum of the two burst inputs. This is a subcarrier sinewave (= 10 cycles) at + 45° on one line and – 45° on the next line with reference to – U phasor (see Fig. 6.5(*b*)).

The colourplexed composite signal thus formed is fed to the main transmitter to modulate the station channel picture carrier in the normal way. The sound signal after due processing is frequency modulation with the channel sound carrier frequency also forms part of the RF signal that is finally radiated through the transmitter antenna system.

6.12 PAL DECODER

The general pattern of signal flow in PAL and NTSC receivers is nearly the same. In fact the tuner, sound section, AGC, sync-separator, deflection circuits and luminance channel function in a similar manner and are almost the same as in the compatible monochrome receiver. The main difference lies in the chrome section of two receivers. A simple functional block diagram of the PAL-D receiver is shown in Fig. 6.9 where all the blocks of chroma section are shown separately.

Chroma signal path. As in any receiver, output from the IF section is detected at the video detector. The output is the Y signal mixed with chrominance signal. It is amplified by the 1st video amplifier before feeding to different sections of the receiver. As shown in Fig. 6.9 output from the 1st video amplifier feeds both a 4.43 MHz bandpass amplifier and a gatedburst amplifier located in the chroma section of the receiver. The two paths are described separately.

Chroma Band Pass Amplifier

This band-pass amplifier is tuned to accept only the chrominance signal out of the composite video signal. It is thus designed to have a centre frequency of 4.43 MHz (f_{sc}) with a bandwidth of nearly 2 MHz (± 2MHz). The colour burst is not allowed to pass through this amplifier. For this, input of the chroma amplifier is shut down by a burst blanking circuit at the end of each active line thus preventing the burst to pass through. The burst blanking circuit is driven by timed pulses obtained from the line time-base circuit. Each pulse is suitably timed to open circuit the chrominance signal path during backporch periods when the burst is present.

Colour Killer Circuit

The purpose of colour killer circuit is to make the chrominance bandpass amplifier inoperative when the receiver is tuned to receive a black and white programme. It thus prevents any luminance signal which happens to fall within the bandwidth of chrominance amplifier from getting through to the demodulators. The passage of such random signals, on demodulation and amplification by colour-difference amplifiers, cause annoying colour interference on the reproduced picture. Thus, in effect, the colour killer circuit acts as an electronic switch in the path of chroma signal. It closes during reception of colour transmission and opens for monochrome programmes. The killer circuit is actuated by another circuit block (7.8 KHz-tuned amplifier), the operation of which depends on the presence or absence of colour burst.

Separation of U and V Modulation Products

As explained in section 6.9, delay line technique is used to separate U and V modulation products. The output from the band-pass amplifier becomes input to the adder, subtractor and delay line circuits. The U modulated colour-difference signal that becomes available at the output adder circuit, feeds into input of the U demodulator. Similarly, the V modulated signal from the subtractor circuit becomes input to the V synchronous demodulator.

Synchronous Demodulators

In a synchronous demodulator or detector, the incoming signal and the locally generated sinusoidal output at the subcarrier frequency are multiplied together. Each input has a frequency of approximately

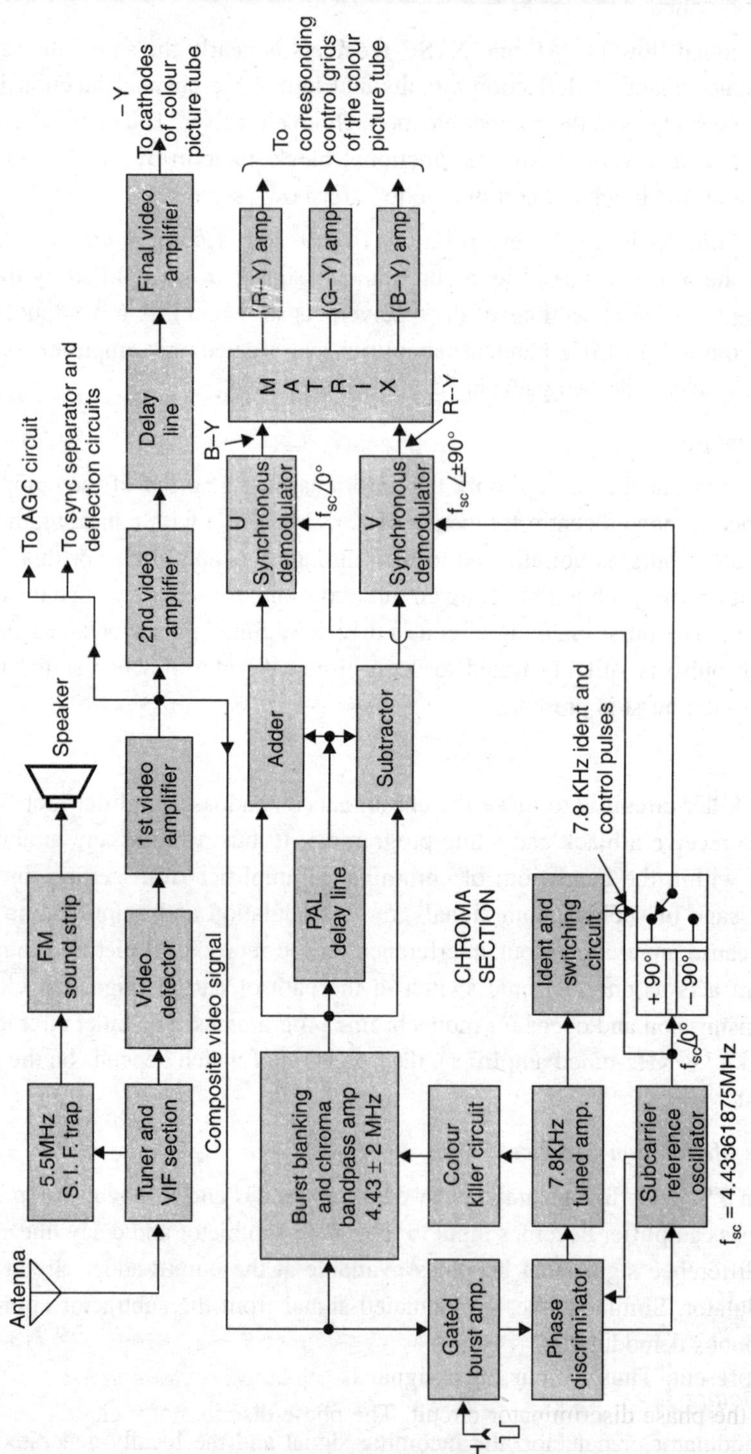

Fig. 6.9 Functional block diagram of a PAL-D colour receiver.

4.43 MHz and the required output is left as the beat between them after the high frequency terms have been filtered out. Practical circuits may be represented by switches in the signal path which are opened and closed by the local subcarrier frequency waveform, the sampling occurring once each cycle at the desired instant. Thus, as shown in the block diagram, the U modulation product feeds the U demodulator along with $f_{sc} \angle 0°$ from reference oscillator circuit and the output is (B – Y) signal. Similarly, the V modulation product and $f_{sc} \angle \pm 90°$ are the inputs to the V demodulator which yields (R – Y) as its output.

Colour-Difference Signal Matrixing & Amplifiers

The (R – Y) and (B – Y) colour-difference signals as obtained from the two demodulators are fed to the matrixing circuit which not only generates the (G – Y) signal but also provides necessary amplitude correction to the three colour-difference signals that is necessary because of the weighting factors applied at the transmitting end. *The three signals thus obtained are fed to corresponding control grids of the picture tube after due amplification. The –Y signal as obtained from the luminance channel is fed at the cathodes of three guns. The colour-difference voltages at the control grids and –Y at the common cathode terminal amounts to matrixing where the net voltages to drive the guns are V_R, V_G and V_B. Thus the three beam currents vary in accordance with the colour voltages which are proportional to those developed at the camera. The beam currents are converged and focussed in such a way that on striking the inner side of screen they hit corresponding colour phosphor dots/stripes in each triade/stripe set and cause radiation of colour lights. The varying intensities of red, green and blue lights when viewed together by the eye are perceived as one resultant colour. Thus the varying beam currents cause colour reproduction as present in the original scene.

Subcarrier Generation and Control

This section is designed to generate a sinusoidal carrier wave of correct frequency and phase to replace the subcarrier (f_{sc}) that is suppressed while modulating the colour-difference signals. The subcarrier thus generated should have exactly the same frequency and phase as that fed to the modulators at the transmitter. For this, a crystal controlled oscillator is built in chroma section of the receiver. It is forced to work at the correct frequency and phase by the action of an automatic frequency and phase control circuit. It is called 'phase discriminator' as labelled in Fig. 6.9. It compares the burst signal as obtained from the gated burst amplifier (see figure) and locally generated reference subcarrier to develop a control voltage which in turn forces the local oscillator tuned circuit to maintain correct frequency and phase.

7.8 KHz-Tuned Amplifier

The PAL swinging burst as introduced at the transmitter has swing of ± 45° about the – (B – Y) axis from line to line. Thus, it swings at half-the-line frequency rate *i.e.*, 15.625/2 KHz ≈ 7.8 KHz. The gated burst amplifier is fed with timed pulses from the line time-base circuit. These pulses which are at the line frequency rate, permit operation of the normally cut-off amplifier only during the period, the colour burst is present. Thus, colour burst signal is separated by the gated burst amplifier and fed continuously to the phase discriminator circuit. The phase discriminator circuit, besides developing a

*In later receive designs – V_R, – V_G and – V_B are applied to corresponding cathodes and the control grids are grounded.

control voltage for the oscillator also senses the swinging rate of the burst and develops a small voltage across a tuned circuit that varies at 7.8 KHz. It is picked up by the 7.8 KHz tuned amplifier and given necessary amplification. The output of this amplifier (see Fig. 6.9) feeds both the 'Colour killer' and 'Ident' circuits. The colour-killer circuit is thus activated only when burst is present which in turn keeps the band-pass amplifier 'on' during reception of colour transmission. Since there is no colour burst signal on monochrome transmission, the 7.8 KHz amplifier has no output and hence no voltage to feed the colour-killer circuit. Thus, the band-pass amplifier remains cut-off during reception of black and white programme.

Ident & Phase Shifting Circuits

In Fig. 6.9, a single pole two-way switch is shown to suggest alternate reversal of the subcarrier phase before applying it to the 'V' demodulator. The switch in the receiver circuitry is actually a bistable multivibrator triggered by line rate pulses obtained from the flyback (line-output) transformer. These pulses switch the multivibrator at the required rate of $f_{L/2} \approx 7.8$ KHz. However, as stated earlier, it is necessary that both the sequence and instant of switching remain synchronized with the swinging burst. For this an 'IDENT' or identification signal is developed from the output of 7.8 KHz tuned amplifier. The 'Ident' circuit is essentially a discriminator circuit which develops a dc voltage of a particular polarity. This, as shown in the block schematic circuit triggers the 7.8 KHz multivibrator (electronic switch) in such a way that the subcarrier oscillator switches at correct instant and phase sequence. This ensures that the synchronous demodulators are fed with correct frequency and phase of the subcarrier necessary for proper detection of U and V signals.

6.13 MERITS AND DEMERITS OF THE PAL SYSTEM

The problem of differential phase errors has been successfully overcome in the PAL system. This is its main merit. In addition the use of PAL-D technique in receivers for electrically accumulating adjacent line colour signals further reduces hue error effects. Thus, a manual hue control becomes unnecessary. However, the delay line technique of reception involves a reduction in the vertical resolution of the chrominance signal but the effect is less pronounced because the two chrominance signals are radiated continuously and the receiver interpolates between the signals of two consecutive lines.

The use of phase alternation-by-line technique and associated control circuitry together with the need of a delay line in the receiver makes the PAL system more complicated and the receiver cost is higher. In addition, the PAL system presents problems in magnetic recording since a complete colour coding sequence requires eight fields instead of four necessary in the NTSC system.

6.14 SECAM COLOUR SYSTEMS

The SECAM system was developed in France. The fundamental difference between the SECAM system on the one hand and NTSC and PAL on the other is that the latter transmit and receive two chrominance signals simultaneously while the SECAM system is "sequential a memoire" *i.e.*, only one of the two colour-difference signals is transmitted at a time. The subcarrier is frequency modulated by the colour-difference signals before transmission. The magnitude of frequency deviation represents saturation of colours and rate of deviation their fineness.

If the red colour-difference signal is transmitted on one line then the blue colour-difference signal is transmitted on the following line. This sequence is repeated for the remaining lines of the raster. Because of the odd number of lines per picture, if nth line carries (R–Y) signal during one picture, it will carry (B – Y) signal during scanning of the following picture. At the receiver an ultrasonic delay linc of 64 μs is used as a one-line memory device to produce decoded output of both the colour-difference signals simultaneously. The modulated signals are routed to their correct demodulators by an electronic switch operating at the rate of line frequency. The switch is driven by a bistable multivibrator triggered from the receiver's horizontal deflection circuitry. The determination of proper sequence of colour lines in each field is accomplished by identification (Ident) pulses which are generated and transmitted during vertical blanking intervals.

SECAM III

During course of development, the SECAM system has passed through several stages and the commonly used system is known as SECAM III. It is a 625 line 50 field system with a channel bandwidth of 8 MHz. The sound carrier is +5.5 MHz relative to the picture carrier. The nominal colour subcarrier frequency is 4.4375 MHz. As explained later, actually two subcarrier frequencies are used. The Y signal is obtained from the camera outputs in the same way as in the NTSC and PAL systems. However, different weighting factors are used and the weighted colour-difference signals are termed D_R and D_B where $D_R = 1.9$ (R – Y) and $D_B = 1.5$ (B – Y).

Modulation of the Subcarrier

The use of Frequency Modulation for the subcarrier means that phase distortion in the transmission path will not change hue of the picture area. Limiters are used in the receiver to remove amplitude variations in the subcarrier. The location of subcarrier, 4.4375 MHz away from the picture carrier reduces interference and improves resolution. In order to keep the most common large deviations away from the upper end of the video band, a positive frequency deviation of the subcarrier is allowed for a negative value of (R – Y). Similarly, for the blue colour-difference signals a positive deviation of the subcarrier frequency indicates a positive (B – Y) value. Therefore, the weighted colour signals are : $D_R = -1.9$ (R – Y) and $D_B = 1.5$ (B – Y). The minus sign for D_R indicates that negative values of (R – Y) are required to give rise to positive frequency deviations when the subcarrier is modulated.

In order to suppress visibility of a dot pattern on monochrome reception, two different subcarriers are used. For the red difference signal it is 282 f_L = 4.40625 MHz and for the blue difference signal it is 272 f_L = 4.250 MHz.

Pre-Emphasis

The colour-difference signals are bandwidth limited to 1.5 MHz. As is usual with frequency modulated signals, the SECAM chrominance signals are pre-emphasized before transmission. On modulation the subcarrier is allowed a linear deviation = 280 D_R KHz for the red difference signals and 230 D_B for the blue difference signals. The maximum deviation allowed is 500 KHz in one direction and 350 KHz in the other for each signal although the limits are in opposite direction for the two chroma signals.

After modulating the carrier with the pre-emphasised and weighted colour-difference signals (D_R and D_B), another form of pre-emphasis is carried out on the signals. This takes the form of increasing amplitude of the subcarrier as its deviation increases. Such a pre-emphasis is called *high-frequency pre-emphasis*. It further improves signal to noise ratio and interference is thus very much reduced.

Line Identification Signal

The switching of D_R and D_B signals line-by-line takes place during the line sync pulse period. The sequence of switching continues without interruption from one field to the next and is maintained through the field blanking interval. However, it is necessary for the receiver to be able to deduce as to which line is being transmitted. Such an indentification of the proper sequence of colour lines in each field is accomplished by identification pulses that are generated during vertical blanking periods. The signal consists of a sawtooth modulated subcarrier (see Fig. 6.10) which is positive going for a red colour-difference signal and negative going for the blue colour-difference signal. At the receiver the Ident pulses generate positive and negative control signals for regulating the instant and sequence of switching.

SECAM Coder

Figure 6.10 is a simplified functional diagram of a SECAM III coder. The colour camera signals are fed into a matrix where they are combined to form the luminance ($Y = 0.3R + 0.59G + 0.11B$) and colour-difference signals. The SECAM weighting and sign factors are applied to the colour-difference signals so that the same subcarrier modulator can be used for both the chrominance (D_R and D_B) signals. The Ident signal is also added in the same matrix.

An electronic switch which changes its mode during every line blanking interval directs D_R and D_B signals to the frequency modulator in a sequential manner *i.e.*, when D_R is being transmitted on the line, then D_B is not used and vice versa.

Sync Pulse Generation and Control

The line frequency pulses from the sync pulse generator are passed through selective filters which pick out the 272nd and 282nd harmonics of f_L. These harmonics are amplified and used as the two subcarrier references. The sync pulse generator also synchronizes the switching control unit which in turn supplies operating pulses to the electronic switch for choosing between D_R and D_B signals. The switching control also operates the circuit which produces modulated waveforms of the Ident signal. These are added to the chrominance signals during field blanking period and before they are processed for modulation.

The output from the electronic switch passes through a low-pass filter which limits the bandwidth to 1.5 MHz. The bandwidth limited signals are pre-emphasized and then used to frequency modulate the subcarrier. The modulator output passes through a high frequency pre-emphasis filter having a bell-shaped response before being added to the Y signal. The sync and blanking pulses are also fed to the same adder. The adder output yields composite chrominance signal which is passed on to the main transmitter.

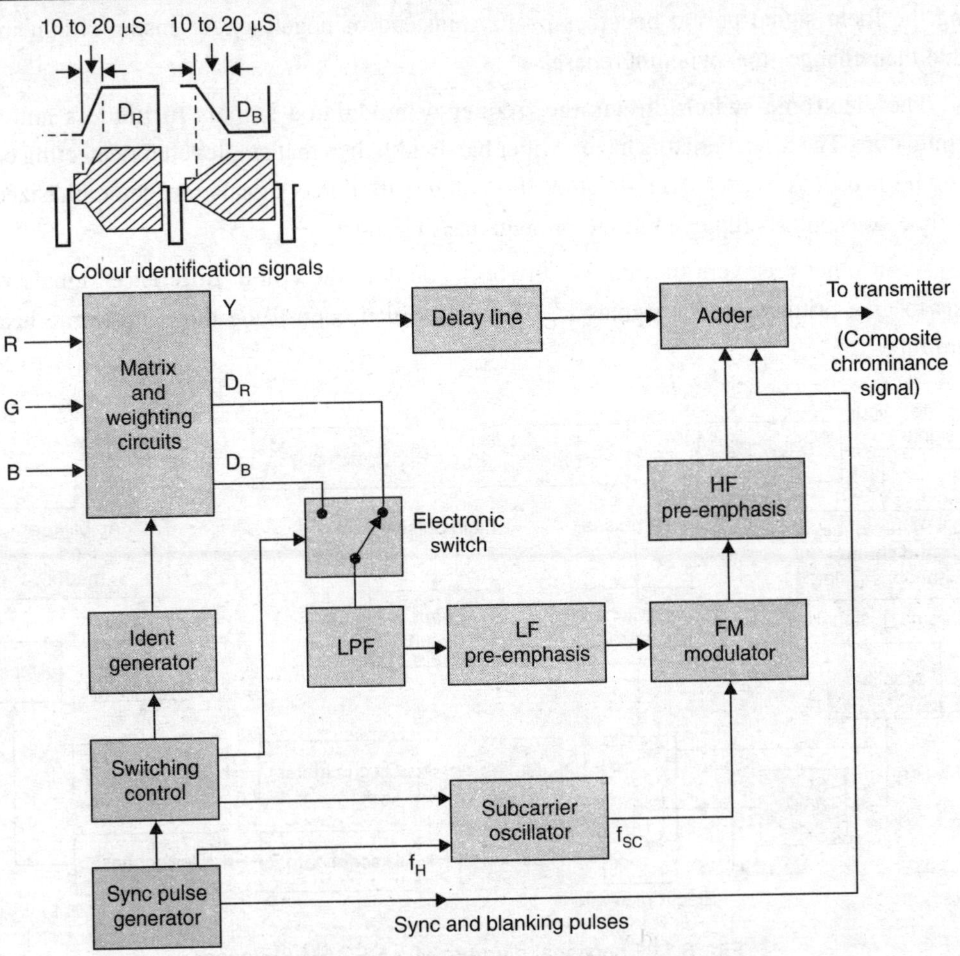

Fig. 6.10 Functional diagram of a SECAM III coder.

SECAM Decoder

SECAM receivers are similar in most respects to the NTSC and PAL colour receivers and employ the same type of colour picture tubes. The functional diagram of a SECAM III decoder is shown in Fig. 6.11. The chroma signal is first filtered from the composite colour signal. The bandpass filter, besides rejecting unwanted low frequency luminance components, has inverse characteristics to that of the bell shaped high frequency pre-emphasis filter used in the coder. The output from the bandpass filter is amplified and fed to the electronic line-by-line switch via two parallel paths. The 64 µs delay line ensures that each transmitted signal is used twice, once on the line on which it is transmitted and a second time on the succeeding line of that field. The electronic switch ensures that D_R signals, whether coming by the direct path or the delayed path, always go to the D_R demodulator. Similarly, D_B signals are routed only to the D_B demodulator.

The switch is operated by line frequency pulses. In case phasing of the switch turns out to be wrong, *i.e.*, it is directing D_R and D_B signals to the wrong demodulators, the output of each demodulator during the Ident signal period becomes positive instead of negative. A sensing circuit in the Ident module then changes the switching phase.

The electronic switch directs the frequency modulated signals to limiters and frequency discriminators. The discriminators have a wider bandwidth than that employed for detecting commercial FM sound broadcast. After demodulation the colour-difference signals are de-emphasized with the same time-constant as employed in the pre-emphasis circuit.

As in other receivers the matrix networks combine the colour-difference signals with the Y signals to give primary colour signals R, G, and B which control the three electronic beams of the picture tube.

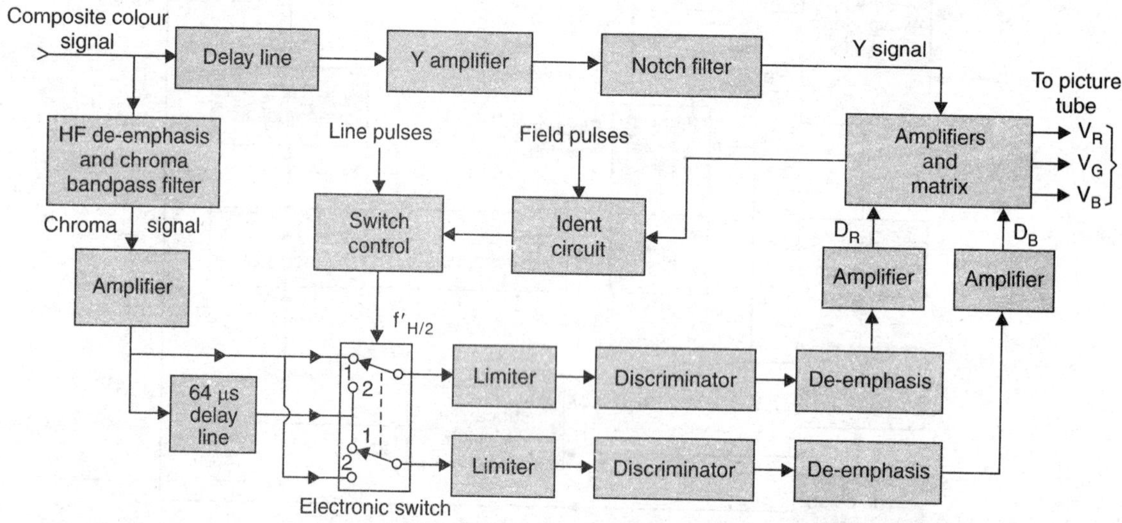

Fig. 6.11 Functional diagram of a SECAM III decoder.

It may be noted that a SECAM colour receiver requires the same two controls—brightness and contrast, as needed on the compatible monochrome receiver. The saturation and hue controls are not needed because the system is immune to these distortions. This is so because colour signals are constant amplitude, frequency-modulated signals and the frequency deviations which carry colour information are not effected during transmission.

**NIR SECAM*

The advent of commercially available delay lines primarily developed for SECAM and PAL systems led to the development of other SECAM systems. In the later versions known as SECAM IV and SECAM V, quadrature amplitude modulation (as in the NTSC system) and synchronous detectors are used instead of the frequency modulators and discriminators. These were developed at the Russian National Institute for Research (NIR) and are sometimes referred to an NIR-SECAM systems. Here the transmitted chrominance signal amplitude is not proportional to the vector sum of the two chrominance components but is proportional to the square root of the vector sum. The signals are, however, transmitted

*NIR SECAM is not used for commercial broadcasting.

on a sequential basis which is the distinguishing feature of all SECAM systems. At the receiver, one line time delay is used in the decoder so that both types of transmitted signals are available during all the lines of the picture. A change-over switch is required to route the constant one phase signal over one path and the varying or other phase signal over another path. The rest of the receiver circuit is similar and produces R, B and G signal voltages at the electrodes of the picture tube to reproduce all colours on the screen.

6.15 MERITS AND DEMERITS OF SECAM SYSTEMS

Several advantages accrue because of frequency modulation of the subcarrier and transmission of one line signal at a time. Because of FM, SECAM receivers are immune to phase distortion. Since both the chrominance signals are not present at the same time, there is no possibility of cross-talk between the colour-difference signals. There is no need for the use of QAM at the transmitter and synchronous detectors at the receiver. The subcarrier enjoys all the advantages of FM. The receiver does not need ATC and ACC circuits. A separate manual saturation control and a hue control are not necessary. The contrast control also serves as the saturation control. All this makes the SECAM receiver simple and cheaper as compared to NTSC and PAL receivers.

It may be argued that vertical resolution of the SECAM system is inferior since one line signal combines with that of the previous to produce colours. However, subjective tests do not bring out this deficiency since visual perception for colours is rather poor.

In addition, while SECAM is a relatively easy signal to record, there is one serious drawback in this system. Here luminance is represented by the amplitude of a voltage but hue and saturation are represented by deviation of the subcarrier. When a composite signal involving luminance and chrominance is faded out in studio operation, it is the luminance signal that is readily attenuated and not the chrominance. This makes the colour more saturated during fade to black. Thus, a pink colour will change to red during fade-out. This is not the case in NTSC or PAL systems. Mixing and lap dissolve also present similar problems.

6.16 MAIN CHARACTERISTICS OF THE THREE COLOUR TV SYSTEMS

After learning essential features of NTSC, PAL and SECAM, it is desirable to compare distinguishing features of the three-colour systems. A comparative chart is an easy way to do so and is given on the following pages. It may be noted that PAL-D is essentially the same as PAL-B. The only difference is that in PAL-D, the averaging of the two consecutive phasors is done electronically by using delay-line technique as explained in section 6.9.

Sr. no.	Characteristics	525 Line NTSC	625 Line PAL-B	625 Line SECAM
1.	Channel bandwidth	6.0 MHz	*7.0 MHz	8.0 MHz
2.	Sound carrier relative to picture carrier	4.5 MHz	*5.5 MHz	6.5 MHz
3.	Luminance signal (Y)	Y = 0.3R + 0.59G + 0.11B	Y = 0.3R + 0.59G + 0.11B	Y = 0.3R + 0.59C + 0.11B
4.	'Y' Signal bandwidth (highest modulation frequency)	4 MHz	**5 MHz	6 MHz
5.	Luminance signal modulation	AM (Negative)	AM (Negative)	FM
6.	Sound signal modulation	F.M.	F.M.	A.M.
7.	Chrominance signal (C)	I = 0.74 (R – Y) – 0.27 (B – Y) Q = 0.48 (R – Y) + 0.41 (B – Y)	U = 0.493 (B – Y) V = 0.877 (R – Y)	D_R = – 1.9 (R – Y) D_B = 1.5 (B – Y)
8.	Colour subcarrier-frequency (f_{sc})	3.579545 MHz	4.43361875 MHz	4.43375 MHz
9.	Chrominance signal	Q.A.M (Suppressed Carrier)	Q.A.M (Suppressed Carrier)	F.M
10.	Chrominance signal-modulation sequence	I & Q signals on each line	U & V signals on each line but f_{sc} to the V modulator is reversed from + 90° to – 90° at the line frequency.	D_R and D_B signal alternatively on successive lines
11.	Chrominance signal bandwidth on modulation	I = + 0.57 MHz and – 1.30 MHz Q = ± 0.5 MHz	*U = + 0.57 MHz and – 1.30 MHz *V = + 0.57 MHz and – 1.3 MHz (vestigial)	f_{sc} D_R = 4.40625 MHz D_R (max deviations – 500 KHz and + 350KHz $f_{sc}D_B$ = 4.250 MHz D_B (max deviations) + 500 KHz and – 350KHz Max permitted bandwidth = 1.5 MHz
12.	Subcarrier reference phase	– (B – Y) axis	– (B – Y) axis	Not necessary

* In the British PAL-I system the channel bandwidth is 8 MHz and sound carrier is located 6 MHz away from the picture carrier.

** 5.5 MHz in the British system.

*** In the British PAL-I system, both U and V have the same bandwidth = ± 1.3 MHz.

Sr. no.	Characteristics	525 Line NTSC	625 Line PAL-B	625 Line SECAM
13.	Colour sync signal	10 ± 2 cycles of the subcarrier on the backporch of line blanking pulses.	± 45° swinging burst having 10 ± 2 cycles of f_{sc} carried on the back porch of line blanking pulses.	Pulses of f_{sc} carried on back porch of the blanking pedestal but phase reverse at $f_{L/2}$ rate.
14.	Identification signal	(not needed)	± 45° swinging burst at ≈ 7.8 KHz ($f_{L/2}$)	Modulated subcarrier with positive amplitude sawtooth slope for D_R and negative amplitude sawtooth slopes for D_B.
15.	Colour signal processing on demodulation	I & Q signals on each line	Two successive line signals (U&V) added and subtracted with delay line to obtain U&V modulated signals separately.	One line signal is delayed by one line time and then added.
16.	Colour resolution (vertical)	FULL	Half—but colour registration not much impared.	Though half but colour registration is better than PAL-D.
17.	Phase distortion on transmission	Differential phase error occurs.	Differential phase error mostly eliminated.	The system is immune to phase error because of FM (no ATC or ACC circuits are necessary).
18.	Noise immunity	High	High	Very high
19.	Studio operations (mixing)	Easy	Easy	Difficult because of FM
20.	Colour reproduction	Good	Very good	Excellent
21.	Receiver controls	(i) Brightness (ii) Contrast (iii) Saturation (Colour) (iv) Tint (Hue)	(i) Brightness (ii) Contrast (iii) Saturation (colour)	(i) Brightness (ii) Contrast
22.	Cost	Less than PAL	Somewhat high	Least of the three

REVIEW QUESTIONS

1. Describe main features of the NTSC colour TV system and in particular explain why the modulated subcarrier phasors are shifted by 33° to constitute Q and I signals. Why different bandwidths are assigned to Q and I signals?

2. Draw block diagram of a NTSC coder and label it fully. Justify the choice of 3.579545 MHz as the colour subcarrier in this system. How does it affect the line and field frequencies?
3. Draw a fully labelled block diagram of a NTSC decoder and explain briefly how chroma signal is processed to obtain the original, R, G and B colour signals. Enumerate limitations of the NTSC system which lead to the development of other colour systems.
4. Explain how the differential phase error is continuously corrected in the PAL system while affecting QAM of the colour-difference signals. Establish the value of f_{sc} as used in the PAL system. How does it help to suppress dot patterning on the receiver screen?
5. Enumerate main features of the PAL colour system. Explain with the help of a suitable block diagram the method of encoding colour signals in this system. Why is the colour burst transmitted after each scanning line?
6. Explain the delay line method of separating U and V signals in a PAL-D receiver. Justify the use of a 63.943 ns delay line for this purpose. How is it superior as compared to PAL simple method of detection?
7. Describe how a PAL decoder functions to recover colour-difference signals, and explain fully how matrixing is done to obtain original colour signals. What is the function of saturation control in a colour receiver and where is it located?
8. Describe merits and demerits of the PAL colour system. Justify the need of Ident signal for correct demodulation of (R – Y) colour-difference signal.
9. Describe briefly the basic difference between SECAM and other colour systems. Draw block diagrams of the SECAM coder and decoder and explain how colour signal is processed in these blocks.
10. Compare the three colour systems (525 line NTSC, 625 line PAL and 625 line SECAM) in all respects by listing their main characteristics in a tabular form. What is the main difference between PAL-B and PAL-G. What are the distinguishing features of PAL-I system adopted by U.K.

7

PAL COLOUR RECEIVER

INTRODUCTION

In U.S.A., France, West Germany and many other countries where colour transmission started soon after CTV was developed, receiver manufacture passed through vacuum tube, hybrid and solid state versions to the present state of the art where integrated circuits are used in nearly all sections of the receiver. In India, where CTV was introduced quite late, receiver manufacturers looked for what was latest in TV technology and opted for designs, which though complex, are very efficient and employ dedicated ICs in almost all sections of the receiver. Therefore, an attempt must be made to learn all that is latest in receiver design without having to master earlier techniques and circuits. For this, it is easier to first grasp functions of various blocks of a modern colour receiver and then turn to its circuitry. Accordingly, this chapter is designed to acquaint the reader with essentials of each section of the receiver and the manner in which modulated video and sound signals are processed from antenna to picture tube and loudspeaker. As necessary, more attention has been paid to the chroma section which makes a colour receiver different from a black and white set.

A *detailed block diagram of a PAL colour receiver is given on the fold-out page (Fig. 7.1) and is referred to throughout this chapter. Since many of the monochrome circuits in a colour receiver are slightly different from those of black and white sets, the entire receiver is briefly described.

7.1 RF TUNER

The RF requirements of both monochrome and colour receivers are nearly same. However, for colour reception frequency response and tuning of various resonant circuits over the channel bandwidth is more critical. For example, a dip of about 20 percent in the central portion of the channel response curve can be tolerated in monochrome receivers but a nearly flat response is necessary for colour

---

*The colour receiver chosen does not represent a circuit where ICs are used in most sections. This is deliberately done with an aim to describe TV signal processing in sufficient depth.

reception so that there is no degradation of picture quality due to unequal amplification of chroma sideband frequencies. Thus, tuners of colour receivers should be of good design and tuned critically to ensure adequate amplification of all signals carried by each channel.

Tuner Types

All colour receivers have both, a VHF and a UHF tuner. Earlier VHF tuners were mostly of turret type where coils are switched into and out of the tuned circuits as the selector switch is moved from channel to channel. Similarly, UHF tuners were of continuous tuning type and veiwers often found it difficult to select the desired channel. Modern receivers employ electronic tuners where band selection is done by switching diodes and tuning by varactor diodes.

Varactor Tuning

The varactor diode or voltage variable capacitor (varicap) is a special solid state diode that is used as a capacitor. Its capacitance varies inversely with the amount of reverse bias voltage applied across it. It is so, because, with increase in reverse bias the depletion region on either side of p-n junction widens and density of carriers decreases. As obvious, all varactor diodes operate only with reverse bias. In tuned circuits that use them, resonant frequency is changed simply by changing magnitude of reverse bias voltage across the diode. Greater the reverse bias voltage, lower the varactor capacitance and hence, higher the tuned-circuit resonant frequency.

Varactor Circuit

As an illustration of varactor tuning Fig. 7.2 shows a method of channel selection by varactor diodes. The depression of any one of the selector buttons connects reverse bias voltage to the four varactor diodes. The capacitors C_1 through C_4 are large enough to provide a short for RF signals. Thus the varactors are effectively in parallel with the inductors. For tuning any channel, the potentiometer associated with the chosen selector button is varied to change reverse bias voltage across the varactor diodes. The consequent change in junction capacitance of these diodes enables tuning-in of the desired channel. Only three selector switches and associated potentiometers are shown in Fig. 7.2 but actually 8 to 12 such selectors are provided. Thus, all available channels can be pretuned and later switched-in just on the depression of corresponding selector buttons. It may be noted that selector switching mechanism is so designed that when a selector button is depressed, the one already closed pops up to cause open circuit. The location of varactors in VHF and UHF tuners is also shown in Fig. 7.1.

Band Selection

The lower VHF band (Band—I) covers a frequency range from 41 to 68 MHz. Similarly VHF band III has a frequency range of 174 to 230 MHz. The two together cover too wide a frequency spectrum and the capacitance range of a varactor is not enough to tune all VHF channels with the same inductor. Therefore, band switching becomes necessary in VHF tuners. This, as shown in Fig. 7.3 is accomplished by switching diodes instead of mechanical switch contacts. For band I channels (see Fig. 7.3 (*a*)) diode D_1 is kept reverse biased and the entire coil remains in parallel with the varactor.

For higher VHF channels (Band III) the diode is forward biased (see Fig. 7.3(*b*)) and this amounts to bypassing part of coil inductance through C_1 leaving lesser inductance for tuning higher VHF channels. Since there are four tuned circuits in the tuner, four switching diodes are necessary, one across each coil. This is shown in Fig. 7.2 where connecting points for switching diodes on all the four coils are suitably labelled.

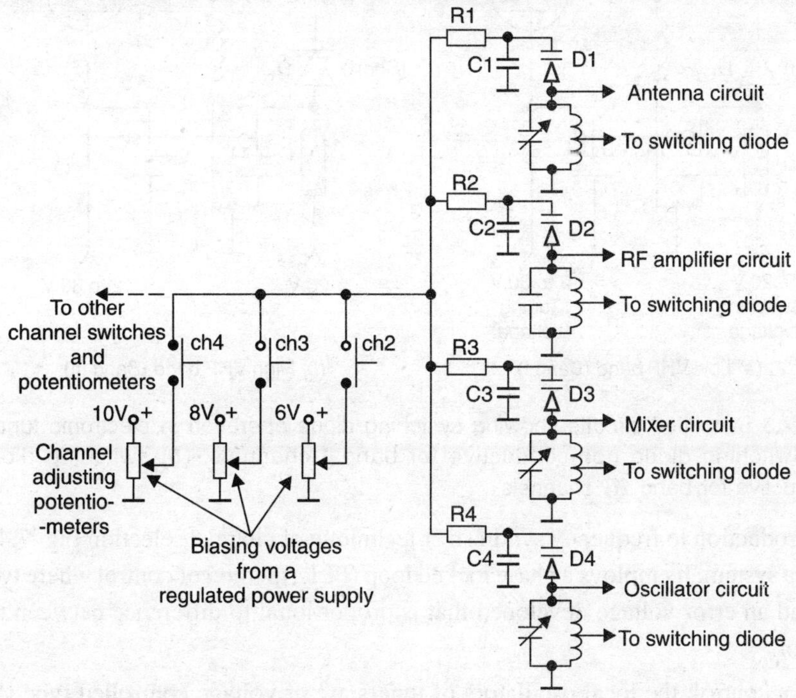

Fig. 7.2 Tuner circuit showing channel tuning by varactor diodes.

All tuners that employ this type of band switching and channel selection have a three position band switch. This enables separate selection of band-I, band III and UHF band channels. In the block diagram of Fig. 7.1 such details are not shown but are fully explained in the following two chapters.

7.2 ALL ELECTRONIC TUNING

The latest in tuner design is the use of digital means for channel tuning. This system does not employ any moving parts like band-change switches and potentiometers. The random access digital method makes it possible to switch from one channel to another in a fraction of a second. Digital systems employ microcomputers (μC) for automatic channel tuning. The μC provides correct dc voltage to be applied to the varactor diodes for any given channel and keeps it locked by a phase lock loop method of control.

Frequency Synthesis Method of Tuning & Control

This is a microcomputer controlled digital system of channel selection and tuning which does not need any fine tuning. It is very accurate and enables selection of all available channels in a numerical sequence just on the depression of a single button. Manual selection in any desired sequence is also possible. A cordless remote control unit can select channels from a distance and perform many other receiver functions.

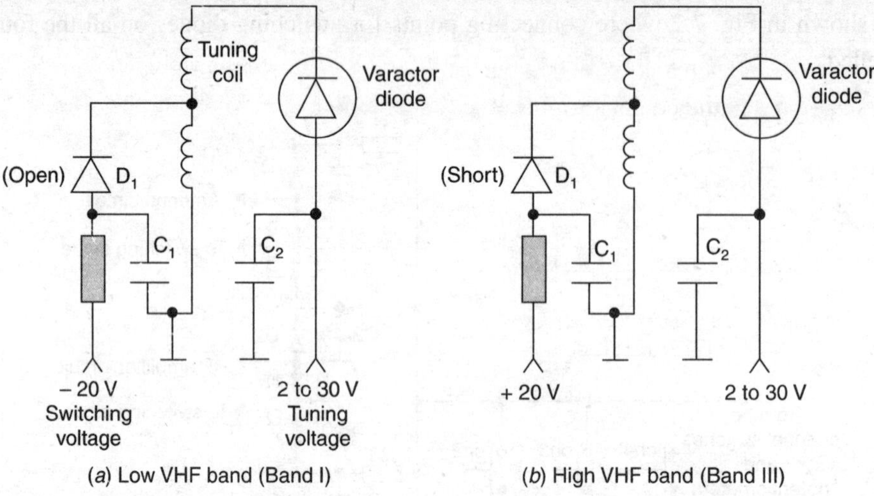

Fig. 7.3 Simplified circuits showing switching diode operation in electronic tuners (*a*) Switching diode non-conductive for band I channels, (*b*) switching diode conductive for band III channels.

As an introduction to frequency synthesizer technique of channel selection Fig. 7.4. shows block diagram of such a system. It employs a phase locked loop (PLL) method of control where two frequencies are compared and an error voltage developed that is proportional to difference between the two inputs to the comparator.

For such a control, the local oscillators of tuners are of voltage controlled type (VCO) and the error voltage that is fed to the varactors is such that the scaled down oscillator frequency becomes equal to the reference frequency.

As shown in the figure, the reference oscillator is crystal controlled. Its frequency is reduced through a chain of dividers to a very low frequency, say 1 KHz. This serves as reference input to the comparator. Similarly the local oscillator frequency is first reduced to a lower frequency by a fixed ratio frequency scaler. The aim being to divide down the high RF frequency to a range that can be processed and counted by standard logic circuits. The output of the prescaler feeds into a programmable divider, the divide ratio of which is determined by the microcomputer (μC). The memory of this μC is loaded with the divide ratio necessary for each channel. On selection of any channel, corresponding divide ratio is retrieved in binary form and fed to the programmable divider. The aim in each case is to scale down local oscillator frequencies of all channels to the same frequency as that obtained from the reference oscillator by successive divisions. Thus the two inputs to the comparator are of the same frequency (assumed 1 KHz) if the local oscillator frequency is at its correct value. For this, the dc error voltage developed by the PLL filter is zero and no correction gets applied.

As shown in Fig. 7.4, the tuning voltage is obtained from 115V dc source through a zener regulator. The μC memory also stores for each channel the magnitude of tuning voltage that must be fed to the varactor diodes. On channel selection it is retrieved and fed to the VCO in the tuner.

If the local oscillator frequency is not correct, its reduced input to the comparator will not be 1 KHz and can be less or more depending on whether the local oscillator frequency is more or less. The comparator senses the difference and develops a proportionate error voltage. It gets added to the applied tuning voltage and is of such a polarity that the oscillator is forced to return to the correct frequency. In turn the frequency fed to the comparator becomes equal to that obtained from the reference source and control voltage returns to zero. The PLL circuit continuously monitors the local oscillator frequency and applies necessary correction.

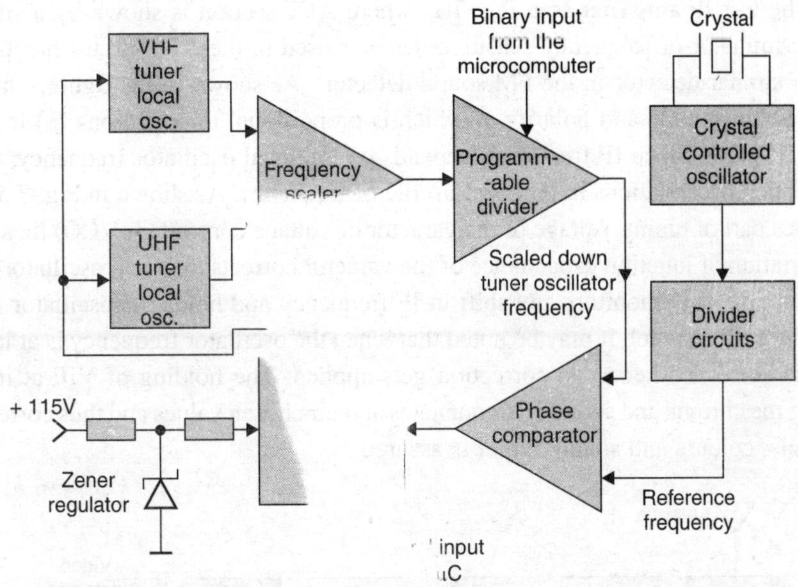

Fig. 7.4 Simplified block schematic of a frequency synthesized PLL controlled system of channel tuning.

Such a control has the precision of a crystal and is so accurate that no fine tuning is necessary. The selection of any VHF or UHF channel is thus instantaneous. In addition to such a control the μC has necessary interface circuitry to generate video signals for displaying the tuned channel number for a short duration on the picture tube screen. In some designs the time of the day is also displayed along with the channel number. In another control system the channel number is continuously displayed on a NIXIE tube located in a window on the front panel of the receiver.

7.3 AUTOMATIC FINE TUNING

Automatic fine tuning control (AFT) is actually automatic frequency control (AFC). It is designed to maintain vision (picture) IF at its correct value of 38.9 MHz. This is possible only if local oscillator frequency of the tuner is maintained at a value which is exactly 38.9 MHz higher than the incoming channel carrier frequency.

In monochrome receivers some deviation of the local oscillator frequency and hence that of IF can be tolerated but in colour receivers even a small shift from the correct value can result in serious deterioration or complete loss of colour. Receivers employing frequency synthesis method of channel selection and tuning have a built-in control of the local oscillator frequency by a phase locked loop

(PLL) control method and may not employ any AFT circuitry. However, such a control is provided to correct channel offsets and inaccurate frequencies of a cable TV system. Thus all colour receivers employ AFT control to maintain local oscillator frequency at its correct value.

AFT Operation

The block diagram of a AFT control system is shown in Fig. 7.5. The input to the circuit is usually from the last IF amplifier (see Fig. 7.1) where AFT section is shown by a single block. The frequency discriminator or frequency sensitive detector used in the AFT circuit functions exactly the same way as a similar detector in the FM sound detector. As shown in the figure, the output is a dc error voltage the magnitude and polarity of which is proportional to deviations (±) in the value of IF frequency (38.9 MHz). Since IF frequency depends on the local oscillator frequency, the AFT circuit output is a function of deviations in the local oscillator frequency. As shown in Fig. 7.5, the dc control voltage becomes part of tuning voltage of the varactor of voltage controlled (VCO) local oscillator. The consequent variation of junction capacitance of the varactor corrects the local oscillator frequency. The AFT control continuously monitors any shift in IF frequency and holds the oscillator frequency at its correct value for each channel. It may be noted that when the oscillator frequency is at its correct value, error voltage is zero and hence no correction gets applied. The holding of VIF at its correct value means keeping the chroma and sound IF frequencies at their chosen values and thus correct reproduction of picture details, colours and sound output is assured.

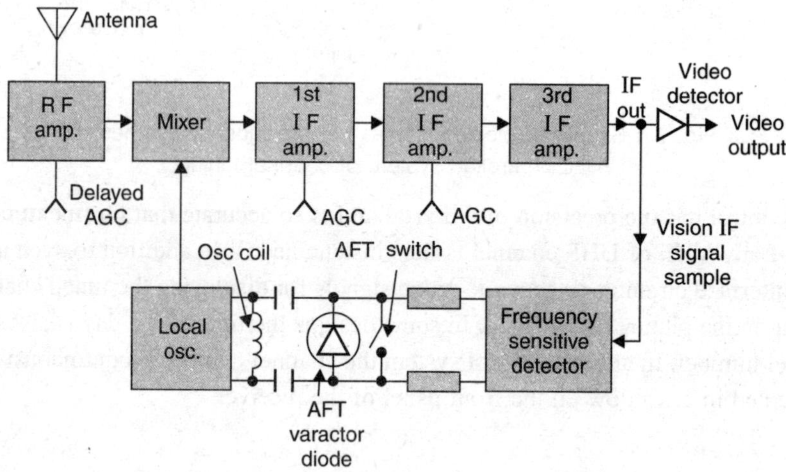

Fig. 7.5 Simplified block diagram of the AFT system of a TV receiver.

AFT Defeat Switch

A push button ON-OFF switch is often connected across output terminals of the discriminator (see Fig. 7.5) and is located on the front panel along with other switches and controls. The purpose of this switch, called the AFT switch, is to disconnect AFT control when desired. The switch is open when depressed and the AFT is in operation. Thus, when not depressed it stays closed and no AFT is applied to the local oscillator circuit. In some receivers a toggle switch is provided for this purpose on the rear of main chassis.

7.4 TUNER OPERATIONS

As shown in Fig. 7.1 both tuner circuits are fed from a common antenna through impedance matching and other necessary networks.

UHF Tuner

The input to the RF amplifier is through an impedance matching network and high-pass filter. The amplifier either employs a transistor as a grounded base amplifier or a MOSFET configuration. The FET has the added advantage that AGC can be easily fed at the second gate. Two varactor diodes, one at the input tuned circuit and another across the output tuned circuit enable proper selection of any channel. A dc voltage variation from 1.5V to 30V across the varactor diodes is enough to select all the UHF channels. The amplified signal feeds into the mixer which is a diode. A separate transistor operates as local oscillator. In another arrangement, the same transistor is used as local oscillator and mixer. Another varactor is provided in the tuned circuit of the oscillator for changing local oscillator frequency when different channels are selected. It also receives AFC voltage from the AFT circuit. The IF output from the mixer is selected by a tuned circuit and fed via a diode (D_1) to the mixer of VHF tuner. The diode is forward biased only when a UHF channel is selected. In addition, for all UHF channels VHF tuner is made inoperative and its mixer connected as IF amplifier. The idea is to raise the weak IF output of the UHF tuner to a reasonable level before feeding it to the IF section.

VHF Tuner

The VHF tuner has a balun at its input (see Fig. 7.1) to match the 300 ohm antenna impedance to 75 ohm input of the RF amplifier. Several IF rejection circuits are also included at the input of this amplifier. The rest of the circuit is similar to the UHF tuner but for another varactor at the input tuned circuit of the mixer. In modern tuner circuits a MOSFET is used as RF amplifier where channel signal feeds at one gate and AGC voltage at the other. The band selection that is necessary for lower and higher VHF channels is obtained through switching diodes. The mixer is also a MOSFET where one gate receives the amplified RF signal and the other output of the local oscillator. When a UHF channel is tuned, the VHF tuner oscillator is disabled and instead IF output from the UHF tuner feeds at the second gate through a diode. As stated earlier, the diode is forward biased only when the UHF tuner is operative.

7.5 IF SECTION

As in the case of tuners, IF section of a colour receiver is essentially the same as in a monochrome receiver. The main requirements are: (*i*) very large gain, (*ii*) high selectivity (*iii*) desired waveshaping (*iv*) exact location of picture (vision), chroma and sound IF frequencies on the response curve, (*v*) vestigial sideband correction and (*vi*) high rejection of adjacent channel signals. These are usually met by employing three or four tuned IF amplifier stages in cascade and selective networks at the input of IF section. The first 2 or 3 stages are AGC controlled to ensure a nearly constant amplitude input to the video-detector while receiving any one of the available stations. Discrete circuitry is no longer used and amplifiers form part of a specially designed IC. The coils for tuning and trap circuits for waveshaping are connected, externally to the integrated circuit. However, the latest in IF section design is the use of Surface Acoustic Wave Filters (SAWF) at the input of IF sub-system. The input transducer of such a filter is so designed that near perfect waveshaping and selectivity is obtained. This makes the design of IF section much simpler because a few external L-C tuned circuits are necessary and very little tuning effort is required.

As shown in the block diagram of Fig. 7.1 a pre-IF amplifier precedes the SAW filter. This is necessary for compensating the attenuation suffered by IF signal in the filter configuration. The input of this amplifier is designed to match the output circuit of mixer. Similarly, an impedance matching transformer is used between the output of SAW filter and input of IC containing IF amplifiers and other allied circuits.

7.6 SAW FILTER

A surface acoustic wave is a non-electromagnetic wave that travels along the surface of a piezo-electric substrate. A piezo-electric substrate undergoes physical stress when an electric potential is applied across it. Similarly when such a material is subjected to any physical stress a proportionate electrical potential is developed. Transducers are provided at the two ends of the piezo-electric wafer or substrate to convert electric energy to acoustic wave energy and then back to electrical energy. The commonly used materials for SAW devices are Quartz and Lithium nibotate.

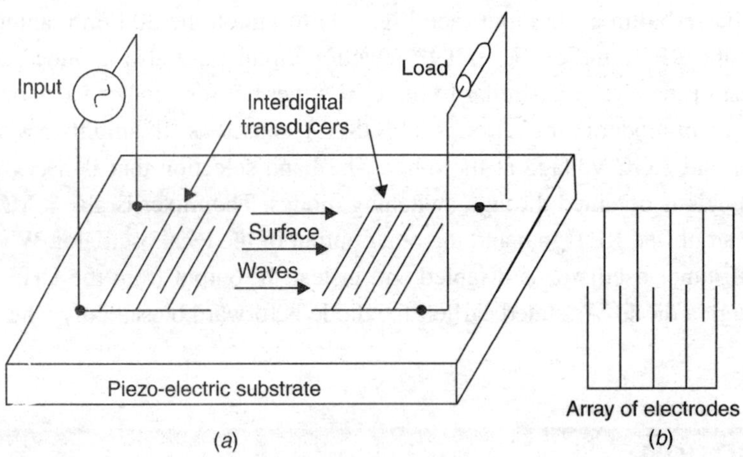

Fig. 7.6 SAW FILTER (*a*) Basic configuration with uniform electrodes for broad band response (*b*) an equally spaced array of electrodes.

The frequency of input signal to the filter can be as high as several gigahertz (10^9) but on conversion to surface acoustic waves (SAW) the signal travels at the speed of sound. However, the SAW waves retain the frequency of their source with possibility of affecting different attenuation to different bands of frequencies.

Wave Shaping

The bandwidth and frequency characteristics of a SAW filter depend on the geometric structure of the transducer array. In general low frequency SAW waves are generated (detected) at that portion of the transducer array where its electrodes (or fingers) are relatively far-apart. Similarly, high frequency SAW waves are generated (detected) where the electrodes are relatively closer. An array of electrodes is called interdigital because of their interlocking nature as shown in Fig. 7.6 (*b*).

The centre frequency f_o of the pass-band of a SAW filter is equal to 'VS' where Vs is the speed of SAW waves on the piezoelectric wafer and "λ_o" the distance between electrodes of the transducer array. As natural, the amplitude of propagating wavc is maximum when wavelength of the SAW or electrical wave is the same as distance between the electrodes. Thus, such frequencies are attenuated least while passing through the filter.

The bandwidth 'B' of the filter configuration is equal to 1/N, where N stands for the number of pairs of opposite interdigital electrodes. Thus, a large number of uniform electrodes cause restricted bandwidth operation and uniform array with only a few fingers has a broad frequency response. For a particular frequeny response pattern as needed in the IF section of a TV receiver, electrodes of the array are given a certain pattern. This is illustrated in Fig. 7.6(*b*) where electrodes overlap either more or less in different sections of the transducer array. Such an arrangement is called amplitude weighting and is designed to obtain desired frequency response.

The SAW filter can thus be freely designed to have any group delay characteristics by changing distance between the electrodes and varying their overlap. In contrast, it may be noted that an inductance-capacitance filter cannot be freely designed to have such precise phase and amplitude characteristics.

SAW Filter for the IF Section

A typical SAW filter that provides complete IF bandpass response is shown in Fig. 7.7(*a*). It consists of input and output interdigital transducers separated by a multistrip coupler. These units are mounted on a polished piezoelectric substrate made from lithium nibotate. It measures about 1.5 cm × 0.5 cm. The multistrap coupler (nearly 100 lines) prevents distortion of surface wave transmission caused by reflection of energy from bottom to top of the substrate.

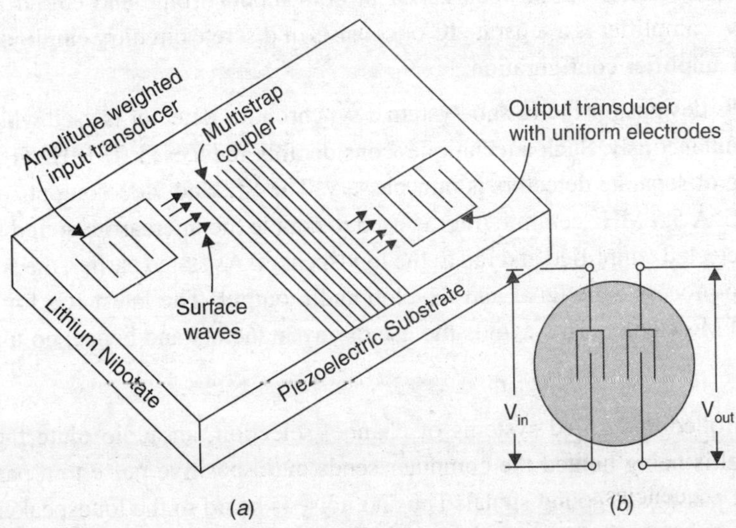

Fig. 7.7 A SAW filter for IF bandpass characteristics of a colour receiver (*a*) simplified diagram of the SAW filter (*b*) symbol and input-output terminals.

As shown in Fig. 7.7(*a*) the input transducer is weighted to obtain necessary selectivity for IF bandpass. The output transducer has only a few uniform electrodes. Thus it has a broad-band response and its major function is to convert the signal from SAW energy back to electrical form. The substrate is mounted on the inside of a ceramic, the bottom of which has a deposit of thin conductive coating connected to the ground of IF circuitry. The package terminals and symbol of a SAW filter are shown in Fig. 7.7(*b*).

Going back to the functioning of IF section and as shown in the block diagram of Fig. 7.1, 1st and 2nd picture (vision) IF (VIF) stages are AGC controlled. In colour receivers, sound and video signals are detected separately. The sound IF detector is fed from the output of 2nd VIF amplifier while output of 3rd VIF stage feeds into the video detector. A 33.4 MHz trap circuit (bypass filter) prevents passage of sound IF and its sidebands to the final IF amplifier and hence, to the video detector. This prevents generation of any strong dot patterning beat rate at 1.07 MHz due to mixing of chroma and sound IF signals (33.47 MHz – 33.4 MHz = 1.07 MHz) in the video detector. As stated earlier, IF amplifiers form part of a specially designed IC which also contains sound and video detectors besides AFC and AGC circuitry.

7.7 SOUND SIGNAL SEPARATION AND PROCESSING

The sound signal at 33.4 MHz is separated at the output of 2nd VIF stage. A parallel L-C tuned circuit coupled to the output circuit of IF amplifier feeds a diode detector where VIF and SIF signals heterodyne to produce intercarrier sound IF signal at 5.5 MHz. As shown in Fig. 7.1, the SIF detector is followed by a SIF trap to permit passage of only the intercarrier sound signal to the SIF amplifier. The amplified SIF signal is processed through a limiter cum FM detector in the same way as in a monochrome receiver. The FM detector is either a differential peak configuration or a phase locked loop type circuit. The use of an IC in the sound section is now universal in both monochrome and colour receivers. In some design, audio power amplifier is a separate IC or consists of discrete circuitry employing complementary transistor type of amplifier configuration.

In recent IC designs for the IF sub-system a synchronous detector is used which detects SIF and video signals simultaneously. Such a technique considerably reduces 1.07 MHz (1070 KHz) beat rate and hence the use of separate detectors is unnecessary. The SIF and video outputs are available at the same pin of the IC. A 5.5 MHz ceramic filter is used to couple the intercarrier sound signal to the sound strip where it is detected, amplified and fed to the loudspeaker. As stated earlier, the sound strip employs one or two ICs to process SIF signal and develop audio output. The latest ICs for the sound section employ a digital FM detector. This avoids the use of any inductors and hence, no tuning is necessary.

Audio Mute

In computer controlled and allied systems of channel selection, an audio mute facility is available. When any channel is being hunted the computer sends out a positive pulse to a particular pin on the sound IC to block passage of sound signal. Thus no noise is heard in the loudspeaker when there is no input signal to the receiver. The same facility is provided on the remote control where on depression of 'audio mute' switch, sound output can be cut-off. Such a provision is very useful while answering a telephone call or responding to the door bell. The picture is continuously displayed on the screen and sound from the loudspeaker is restored on releasing the audio mute button.

7.8 VIDEO DETECTOR AND DISTRIBUTION OF DETECTED SIGNAL

As shown in Fig. 7.1, the last IF stage feeds into AFT and video-detector circuits. The video detector employs a diode as rectifier and a filter network to ground all unwanted high frequency components. The IF signal that becomes available across its load has all the components that constitute the composite video signal. Thus we get back luminance and chrominance signals. All the sync pulses and colour burst are also present. It is necessary to raise the level of detected signal before feeding it to various sections of the receiver. This is done by the 1st video amplifier which operates as a grounded emitter stage. As shown a 5.5 MHz rejection filter is provided at the input of this amplifier to prevent passage of any SIF signal into the Y channel.

The second video amplifier is an emitter follower. The branching of composite video signal is done from its output circuit. This enables distribution of signal from a low impedance source preventing any loss of high frequency content of the video signal through wiring capacitance and shunt paths at the inputs of receiving circuits.

Video Signal Paths

The composite video signal branches-off into the following five paths. (See Fig. 7.1):

(1) Luminance or Y channel

(2) Chroma section

(3) Subcarrier generation and control

(4) AGC circuit

(5) Sync separator

The input circuits of the above listed sections are designed to select necessary component(s) out of the composite video signal. Thus each path uses the received signal in a particular way to perform several functions with the ultimate objective of producing a well defined and synchronized colour picture on the receiver screen.

7.9 THE LUMINANCE CHANNEL

The 3rd, 4th and 5th video amplifiers together with several other blocks constitute the luminance channel. The contrast and brightness controls are provided in this channel. With a view to preventing passage of colour sub-carrier in the Y channel, a 4.43 MHz series resonant (trap) circuit is inserted at the input of luminance channel.

Pedestal Clamp

The signal path from the output of 2nd video amplifier to the input of 3rd is through dc restoration and contrast control networks. The sync pulses in the video signal are not aligned because of ac coupling in the previous stages. Thus, it is necessary to clamp them to the same level if an under or over-bright picture is to be avoided. A separate circuit called 'DC clamp' is used for this purpose. A transistor is used to charge a capacitor in the coupling path to restore dc component. It conducts to charge it for a short period by sync pulses fed at its base and obtained from the sync separator. In-between the arrival of sync pulses, capacitor discharges a bit once it attains a mean value. The discharge path time-constant is quite large and a dc voltage builds across the capacitor. The amount of charge that builds up depends on the nature of video signal. The video signal from a bright scene will cause more dc voltage across the capacitor. Similarly, for a somewhat dark scene the magnitude of condenser voltage will come down. Since dc voltage across the capacitor gets added to the ac component of the video signal, it amounts to dc restoration of the video signal. Thus the condenser voltage continuously varies to clamp sync pulses to the same level and maintain a fixed black level.

Contrast Control

The aim of contrast control is to vary amplitude (peak to peak) of the video signal that feeds into the cathodes of picture tube. This can be done either by controlling gain of one of the video amplifier stages or varying input to any one of them. However, it is necessary to ensure that such a control does not affect bandwidth of the video signal and relative levels are not disturbed. For this a suitable network is inserted between the output of 3rd and input of 4th video amplifiers. In some designs the 4th video stage is also operated as an emitter follower and employs a *p-n-p* transistor if the 3rd video amplifier has a *n-p-n* transistor. This enables near perfect dc clamp, and smooth operation of contrast control without much affecting the bandwidth.

4th and 5th Video Amplifiers

The purpose of last two stages is to:

(i) provide necessary frequency compensation

(ii) incorporate brightness control

(iii) introduce some delay in the path of video signal for correct registration of colour information on the otherwise black and white picture

(iv) raise magnitude of Y signal to the desired level.

The 4th video amplifier is designed to provide enough gain, frequency compensation and brightness control. For this, it is operated as a grounded base amplifier. The video signal from the previous stage is fed at the emitter through a compensating network. The base of this transistor (4th video amplifier) is used for operating the brightness control. The brightness control circuit is essentially a resistive network across a low voltage dc supply. It is so connected that any variation of the brightness control potentiometer alters dc voltage at the base of the transistor. This appears as a large change in quiescent collector voltage. Since dc component forms part of the amplified video signal, the average dc level of the signal changes to cause more or less beam current of the picture tube. Thus any variation of brightness control causes corresponding change in the brightness of the reproduced picture. In some designs, one of the resistors in the brightness circuit is a pre-set. It is adjusted at the factory for optimum operation of the user's brightness control.

The negative going video from the output of 4th video amplifier is fed through a delay line into the base of 5th video amplifier transistor. This is to compensate for the additional delay the colour signal suffers because of limited bandwidth of the chrominance amplifier. The use of a delay line thus ensures time coincidence of chrominance and luminance signals.

Retrace Blanking

Vertical and horizontal blanking pulses obtained from corresponding deflection amplifier outputs are applied at the emitter of 5th video amplifier. The polarity of these pulses is such that the transistor is turned-off during retrace periods. Thus a large dc voltage gets applied at the three cathodes of picture tube and beam currents are cut-off. Therefore, retrace lines are definitely blanked and not seen on the receiver screen. The amplified Y signal and blanking pulses available at the output of last video amplifier are fed simultaneously at the emitter of driver amplifier transistors (see also Fig. 3.14). The three drive *i.e.*, R, G and B amplifiers also receive corresponding colour-difference signals at the base of corresponding transistors. This amounts to matrixing of Y and colour-difference signals. The outputs *i.e.*, R, G and B signals have enough amplitude to drive respective cathodes of the three guns. The resulting beam currents combine to produce coloured pictures on the tube screen. As already stated, the guns are driven to cut-off by blanking pulses to suppress retrace lines.

Modern colour receivers employ a single IC to process both Y and chroma signals. The matrixing also takes place within the IC and the three colour video signals *i.e.*, R, G and B become available at different pins of the package. The driver amplifiers are in any case needed to raise the amplitude of colour signals for fully driving the picture tube. The above discussion pertains to discrete circuitry that is common in earlier colour receivers and is included to give an idea of what needs to be done to the video signal obtained at the output of video detector.

7.10 THE COMB FILTER

It is necessary to remove colour subcarrier and its sidebands from the Y signal if dot patterning is to be avoided. Earlier colour receivers used a specially designed narrow band notch filter (4.43 MHz rejection filter) in the Y channel for this purpose. However, this presented problems in maintaining frequency and phase response of higher video frequencies. On the other hand a relatively wide bandwidth of the notch filter to circumvent this problem meant loss of high frequency content of the Y signal. A solution which became popular is to sacrifice that part of the luminance signal which lies beyond 4 MHz thus making the use of notch filter unnecessary. This is acceptable in colour receivers where colour reproduction dominates the picture quality.

However, in high quality receivers and in particular for those designed for PAL-I colour system, it is not possible to suppress high frequency spectrum of the luminance signal. The answer to separating luminance and chrominance signals without compromising on bandwidth is to use a 'comb filter'. As the name suggests, it selects frequencies that are to be passed without affecting other components in the band. Thus, in colour receivers it combs-off the chroma signal without in any way disturbing the luminance or Y signal.

Elementary Comb Filter

As explained in chapter 5, the luminance signal has the same phase on successive scanning lines and is clustered in energy bundles over the entire video spectrum. The colour subcarrier frequency is so chosen that the chroma signal phase reverses from one scanning line to the next and the colour information energy is interleaved between the luminance energy bundles. The comb filter takes advantage of these factors while separating out the two signals. The basic plan of a comb filter is shown in Fig. 7.8 (*a*) where the composite video signal (v) obtained from video detector is fed simultaneously to a 64 μs delay line and an inverter. Since frequency of horizontal scanning is 15625 Hz (period t = 1/15625 = 64 μs), the delay line retards the Y signal by a full cycle (360°) and there is no phase inversion. However, the inverter reverses phase of all the input signals by 180° *i.e.*, v becomes –v. The delayed signal (v_d) and invertor output (–v) are added in an adder circuit. As obvious the luminance signal components cancel each other being 180° out of phase but the chroma components being in phase add together to yield twice the input value.

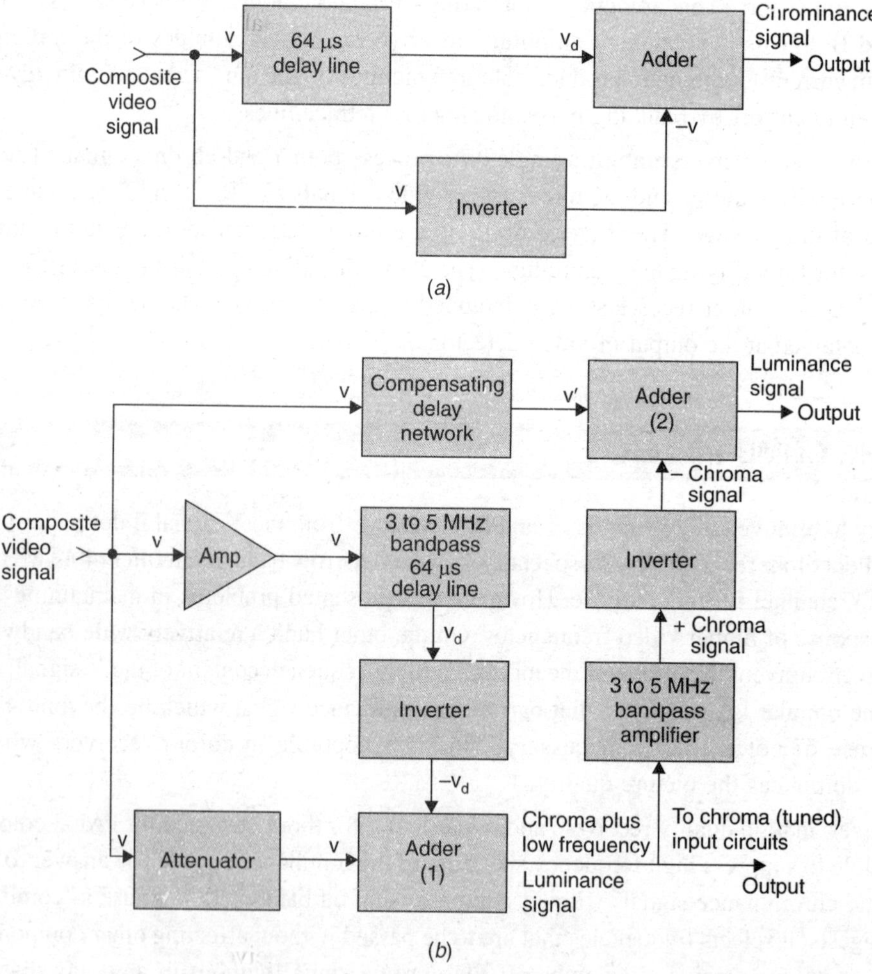

Fig. 7.8 Comb filter (*a*) elementary comb filter representation, (*b*) block diagram of a colour receiver comb filter.

Practical Comb Filter

A detailed block diagram of a colour receiver comb filter is shown in Fig. 7.8 (*b*). The PAL line switching details have been ignored to stress the mode of filter operation. The composite video signal 'v' is fed to the delay line through an amplifier to compensate for the loss that occurs in the line. The bandwidth of the delay line is restricted between 3 to 5 MHz. Therefore, the entire chroma signal and only low amplitude luminance signal present in this range can pass through the delay line. As shown in the figure, the delayed video signal (v_d) is passed through a polarity inverter and becomes – v_d. This and original video signal (v) (after some attenuation) are applied to an adder (1). As explained with reference to Fig. 7.8 (*a*), the luminance signal in the range of 3 to 5 MHz will cancel leaving chroma and lower range (up to 3 MHz) of the Y signal at the output of adder. This is applied to the chroma section input circuits which are tuned to select only chroma and colour burst signals. Thus luminance component present in the input is ignored. It may be noted that luminance signal in the range of 3 to 5 MHz is removed earlier and it is only the chrominance signal that enters chroma section of the receiver.

For obtaining luminance signal for the Y channel, output of adder (1) is first fed to a badpass amplifier designed to have a bandwidth from 3 to 5 MHz. The + chroma signal at its output is phase inverted through an inverter circuit to obtain–chroma signal. This then becomes one of the inputs to the second adder. The second input to this adder is the 'v' signal after passing through a compensating delay network. This delay is introduced to compensate for the delay suffered by the chrominance signal in the bandpass amplifier. The chroma signal contained in the 'v' signal cancels with chroma signal leaving complete luminance signal as output of the adder (2). This feeds into the Y channel and is processed accordingly. Thus this method of separating the two components of composite video signal, though complex, enables to retain full bandwidths of both luminance and chrominance signals.

It is interesting to note that in the American (NTSC) colour system where Y signal does not go beyond 4 MHz and colour subcarrier is fixed at 3.58 MHz, most receiver manufacturers ignore all Y signal information beyond about 3.2 MHz. with a view to avoiding notch filter and keeping low the overall cost of TV receiver. However, now when techniques for digital processing of video signal have become available, comb filters are finding way into such colour receivers to take full advantage of the transmitted signal bandwidth.

7.11 COLOUR DECODER

The function of colour decoder is to recover the three colour video outputs (R, G and B) from the composite signal. For this the decoder has to perform two functions *i.e.*, synchronous detection of the quadrature modulated subcarrier and to obtain R, G, and B signals by matrixing the detected colour difference signals with the Y signal. In addition it is necessary to generate colour subcarrier frequency for feeding the synchronous demodulators. The colour burst in the received signal is used to synchronize the generated subcarrier with that at the transmitter. It is also used to establish correct sequence of phase alteration of the (R–Y) signal.

All modern receivers use ICs for the jobs assigned to the decoder and the approach is somewhat different from that employed in corresponding discrete circuitry. However, with a view to conveying a clear concept of various processes used for decoding colour signals, discrete blocks as shown in Fig. 7.1 are discussed separately.

Chroma Bandpass Amplifier

The chroma bandpass amplifier is normally a three stage configuration. The output from the 2nd video amplifier is fed to the input circuit of the first stage which is tuned to accept the chroma signal. Delayed and suitably shaped line sync pulses are fed at the base of this amplifier transistor which is often connected as an emitter follower. The pulses turn-off the transistor thus blocking passage of colour burst into the decoder circuit. The next stage is a turned amplifier having bandwidth of nearly 2 MHz around 4.43 MHz. This stage also incorporates necessary vestigial side-band correction necessary on account of unequal colour channel bandwidth.

The gain of bandpass amplifier is controlled by the automatic colour control (ACC) circuit to obtain a nearly constant output voltage. The ACC circuit (see Fig. 7.1) is similar to AGC circuit and develops a dc control voltage proportional to the amplitude of colour burst. A manual colour (saturation) control also forms part of the bandpass amplifier. It is designed to bypass part of the chroma signal while feeding it from one stage to the other. It is varied to adjust intensity of colours in the reproduced picture. The last stage is designed to feed chroma signal to the two demodulator circuits.

Colour Killer Circuit

The last chroma bandpass amplifier stage is often called the "delay line driver" because it feeds signal to the demodulators through the delay line network. In addition, it also performs as a switch that can be open circuited or closed by another circuit. The other circuit is called 'colour killer' which in turn is controlled by 7.8 KHz tuned amplifier and IDENT circuits. During reception of any colour transmission, a 7.8 KHz switching rate signal for (R–Y) is available at the automatic phase control (APC) circuit of the reference (subcarrier) oscillator. It is fed to the colour killer circuit through the tuned amplifier where it is rectified to develop a dc voltage. This is used to forward bias the delay line driver amplifier. Thus the switch remains closed during reception of colour programmes. However, when a monochrome program is received there is no 7.8 KHz input to the colour-killer circuit and hence, no forward bias voltage is fed to the driver amplifier. The amplifier thus cuts-off the chroma signal path and any random signal within the chroma bandpass is prevented from reaching the demodulators. Thus there is no colour interference on the black and white picture during monochrome receptions.

7.12 SYNCHRONOUS DEMODULATION

The chrominance signal available at the output of delay line driver amplifier consists of two suppressed carrier amplitude modulated signals called modulation products. These correspond to the two colour-difference signals and bear information about amplitude and polarity of the modulating signals. Because of quadrature modulation, the two modulation product signals have a phase difference of 90° with respect to each other. Thus, when one is passing through its positive or negative peak, the other is passing through zero. The colour demodulators are designed to take advantage of this fact and conduct alternately to sample chrominance signal in step with the injected reference subcarrier signal. The basic block form of the two demodulators is shown in Fig. 7.9 (*a*). The demodulating devices are normally biased to cut-off and conduct only at positive peaks of the reference subcarrier. The (B–Y) demodulator samples the chrominance signal when (R–Y) modulation product is passing through zero. Similarly, the other *i.e.,* (R–Y) demodulator produces output when (B–Y) modulation product is passing through zero. Thus the two modulators produce colour-difference signals without any mutual interference between the two modulation product signals.

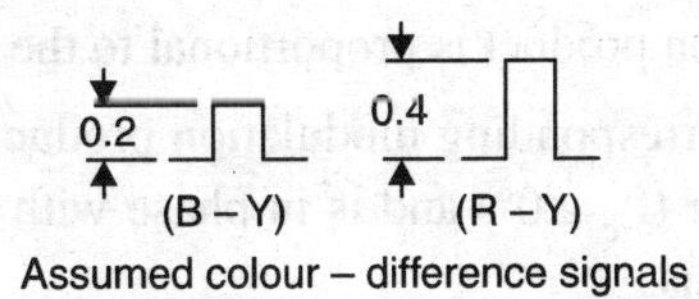

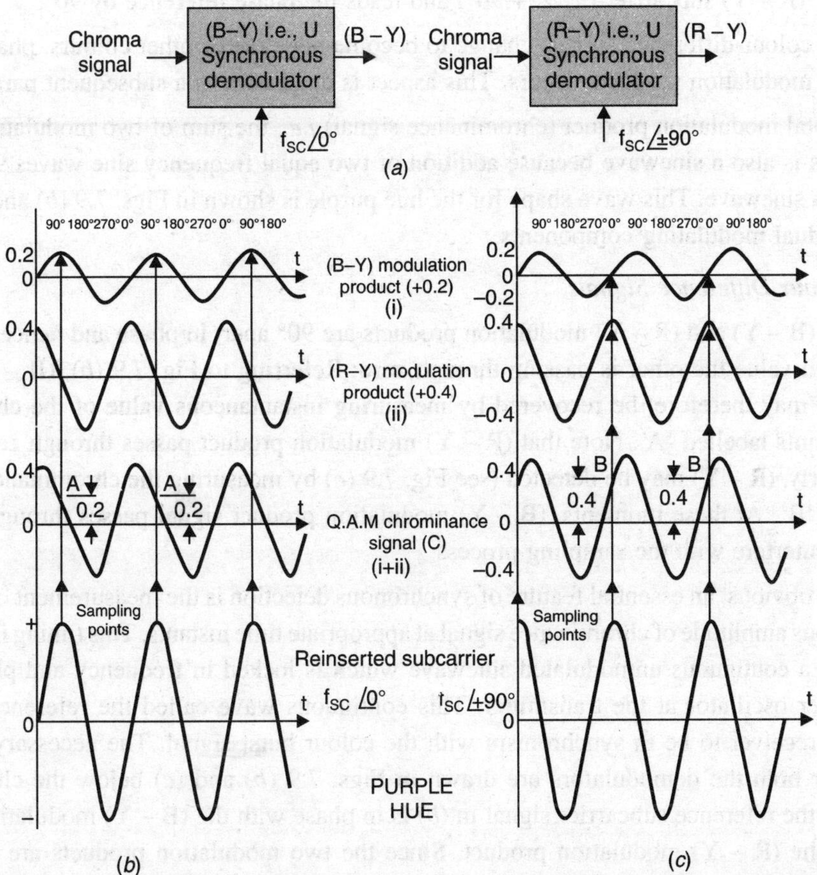

Fig. 7.9 Principle of synchronous demodulation (*a*) basic demodulator representation, (*b*) waveshapes corresponding to (B–Y) demodulator, (*c*) waveshapes corresponding to (R–Y) demodulator.

The process of synchronous demodulation can be best explained by assuming that colour-difference signals are held constant over a period of time. Such signals become available on scanning vertical bar of one particular colour with dark background and are rectangular in shape as shown in Fig. 7.9 (*a*). It has been assumed that (B–Y) = + 0.2 (2 units) and (R–Y) = + 0.4 (4 units). Such colour-difference signals are obtained on scanning a colour bar of purple hue.

Before proceeding further, it is pertinent to note the following with reference to Figs. 7.9 (*b*) and (*c*)

(i) Both the modulation product signals are sinewaves at a frequency of 4.43 MHz (subcarrier).

(ii) Their peak-to-peak values are constant over the period the colour-difference signals remain steady.

(iii) The amplitude of each modulation product is proportional to the value of modulating signal.

(iv) Since (B – Y) is positive, the corresponding modulation product signal has the same phase as that of its reference subcarrier (f_{sc} ∠0°) and is in phase with the system reference phase scale shown above the waveforms.

(v) Similarly, since (R – Y) is positive, its *i.e.,* (R – Y) modulation product has the same phase as that of (R – Y) subcarrier (f_{sc} ∠ + 90°) and leads the phase reference by 90°.

(vi) When colour-difference signals change to become negative or other colours, phase reversal of the modulation products occurs. This aspect is dealt with in a subsequent paragraph.

(vii) The total modulation product (chrominance signal) *i.e.,* the sum of two modulation product signals is also a sinewave because addition of two equal frequency sine waves 90° apart is itself a sinewave. This wave shape for the hue purple is shown in Figs. 7.9 (*b*) and (*c*) below individual modulating components.

Detection of Colour-Difference Signals

As stated above, (B – Y) and (R – Y) modulation products are 90° apart in phase and hence, when one is at its maximum value the other is passing through zero. Referring to Fig. 7.9 (*b*), (B – Y) colour-difference signal may therefore, be recovered by measuring instantaneous value of the chrominance signal at thc instants labelled 'A'. Note that (R – Y) modulation product passes through zero at these moments. Similarly, (R – Y) may be detected (see Fig. 7.9 (*c*) by measuring the chrominance signal at instants labelled 'B'. At these moments, (B – Y) modulation product signal passes through zero and hence, does not interfere with the sampling process.

As is now obvious, an essential feature of synchronous detection is the measurement or sampling of the instantaneous amplitude of chrominance signal at appropriate time instants. This timing information is obtained from a continuous unmodulated sinewave which is locked in frequency and phase to the original subcarrier oscillator at the transmitter. This continuous wave called the reference signal is generated at the receiver to be in synchronism with the colour burst signal. The necessary reference carrier waves for both the demodulators are drawn in Figs. 7.9 (*b*) and (*c*) below the chrominance signal. Note that the reference subcarrier signal in (*b*) is in phase with the (B – Y) modulation product and in (*c*) with the (R – Y) modulation product. Since the two modulation products are 90° out of phase, a phase shift of ∠ + 90° is given to the reference subcarrier (f_{sc} ∠0°) before applying to the (R – Y) demodulator.

While circuit operation of a synchronous demodulator is explained later it is sufficient to note at this stage that positive peaks of the applied reference subcarrier cause conduction of the device employed in the demodulator. Since there is a 90° phase difference between the reference signals fed to them, the two demodulators conduct alternately and the outputs are in the form of short current pulses. However, these occur 4.43 million times ($4.43 \times 10^{+6}$) per second and so the output for all practical purposes can be considered as continuous and uninterrupted. A filter is provided to remove any subcarrier ripple and smoothen the output voltage.

So long as the modulating colour-difference signals remain at one value, as assumed for the present discussion, the current pulses have the same magnitude and a steady output voltage is developed. The two outputs are then exact replicas of the rectangular modulating signals. Any variation of colour-difference signals with changes in hue and/or saturation of the picture elements, affects amplitude and phase angle of the chrominance signals. This in turn changes magnitude and polarity of current pulses produced in the demodulators and hence, the output truly follows changes in colour voltages developed by the camera.

In conclusion it may be stated that the process of synchronous demodulation enables detection of one of the colour-difference signals without any interference from the other. Thus by employing two synchronous demodulators, both the colour-difference signals can be independently recovered. This becomes possible because demodulators are designed to produce an output that is dependent upon both phase and amplitude of the chrominance signal. When put in another way this means that a colour demodulator is a combination of phase and amplitude demodulators and hence, the name 'synchronous demodulator''. It may also be noted that a video detector is only an amplitude detector and the FM discriminator is only a phase detector, while the synchronous detector is both amplitude and phase detector. Also note that while a synchronous demodulator appears like the mixer of a hatrodyne receiver but actually it is not so. The former operates with an external reference signal whose frequency is equal to the carrier frequency and the latter functions on receiving a local oscillator signal of frequency much higher (equal to IF) than the carrier of input RF signal.

7.13 DEMODULATION IN PAL-D RECEIVERS

The demodulation process described above is suitable for NTSC colour receivers. It is somewhat different in PAL-D receivers where a delay line technique is used to separate the modulation products. Each synchronous demodulator receives either U or V modulation product signal. In addition, the reference subcarrier signal to the V demodulator needs to be switched in phase from + 90° to – 90° on alternate lines because of similar switching at the transmitting end.

The manner in which demodulation occurs in a PAL-D receiver is illustrated in Fig. 7.10 where the parameters chosen are same as shown in Fig. 7.9 and pertain to colour-difference signals obtained on scanning a colour bar of purplish hue. The two constituents of chrominance signal are shown separately. The U modulation product on separation is fed to the U demodulator and V to the V demodulator. The reference subcarrier signals that are fed to the two synchronous demodulators are shown below the U and V signal waveforms. The PAL lines (signals) and the associated reference signal are shown in dotted lines for convenience of understanding the demodulation process. The lines with arrows show the moments of conduction. Note that U demodulator passes current pulses when the U modulation product signal is at its positive maximum and the associated V modulation product signal (NTSC or PAL line) is passing through zero. Similarly the V demodulator conducts on positive peaks of the V modulation products signal when U is passing through zero. Note that this is true for both NTSC and PAL lines because the reference subcarrier reverses along with PAL lines thus ensuring that positive peaks of the reference signal coincide with positive peaks of PAL lines. The demodulated V colour-difference signal thus continues to be positive as desired.

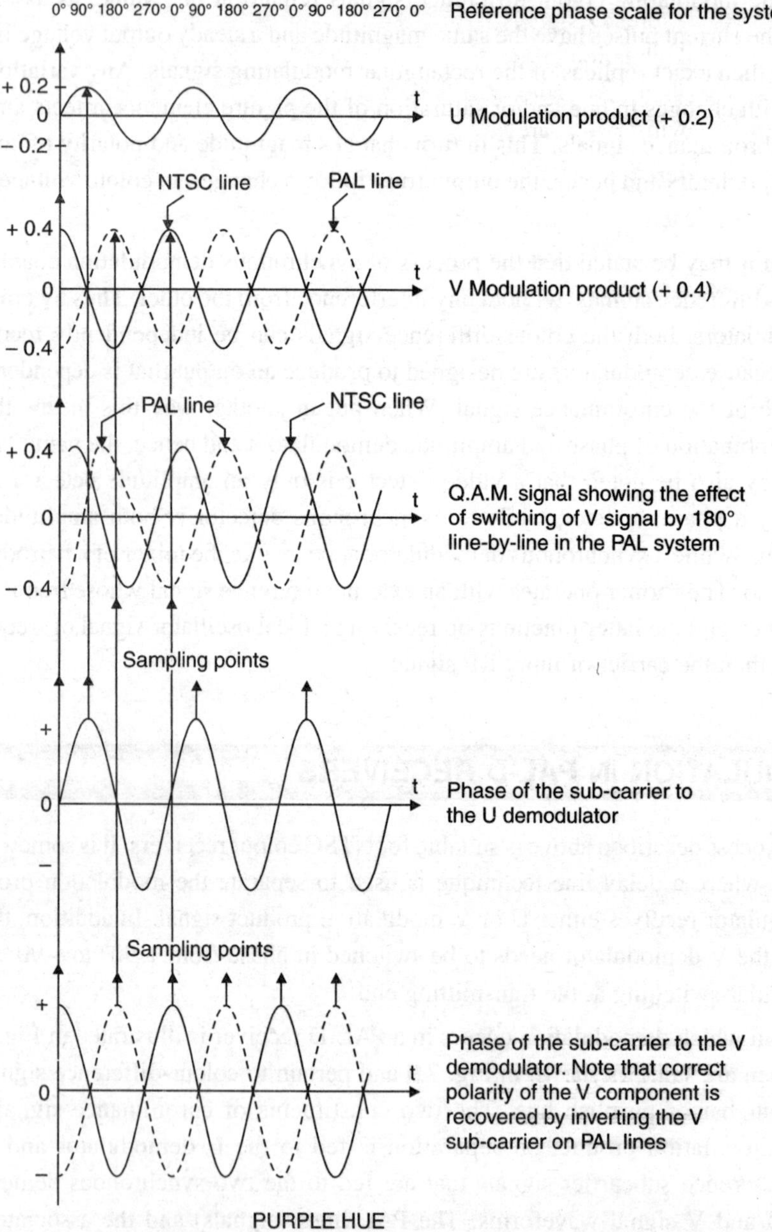

Fig. 7.10 Principle of colour demodulation in a PAL-D receiver.

7.14 DEMODULATION ON DIFFERENT HUES

The polarity of colour-difference signals is different for different hues. This in turn determines the phase angle of U and V modulation products and hence, that of the chrominance signal. This aspect of colour demodulation needs investigation and is illustrated by choosing four different hues. However, before doing so the following points are emphasised:

(a) The colour difference signals may be positive or negative depending on the hue.

(b) Accordingly, each modulation product may have one of the two possible phases (180° apart) depending on polarity of the colour-difference signal.

(c) When (B–Y) is positive the U modulation product has a phase angle of 0° and this changes to 180° when (B–Y) is negative.

(d) When (R–Y) is positive the V modulation product has a phase angle of 90° and changes to 270° when (R–Y) is negative. This is true only for the NTSC lines.

(e) On PAL lines an image inversion of V signal occurs irrespective of its polarity.

(f) The reinserted reference subcarrier signal to the U demodulator must always have a phase lying along the + U axis. It is most important to note that has to be so, whatever the polarity of the U colour-difference signal. When put in another way the reference phase must remain along + U axis irrespective of whether the U modulation product is at 0° or 180°.

(g) Similarly, the reinserted subcarrier to the V demodulator must have a phase angle of + 90° relative to + U axis and this must be true whether the V modulation product has a phase angle of + 90° or + 270°. However, on PAL lines when image inversion of the V modulation product occurs, the reference subcarrier must undergo a similar inversion. This is achieved by shifting phase of the subcarrier from + 90° to – 90° line-by-line by an electronic switch.

Since both (B–Y) and (R–Y) can be positive or negative, four possible combinations of U and V phase angles are considered. For this, four different hues have been chosen. As earlier, the amplitudes of colour-difference signals are held constant on all hues to focus attention on polarities of U and V modulation products and of corresponding demodulated colour-difference signals. Thus amplitudes of U and V modulation products (which are sinewaves when an unchanging colour is being scanned) remain constant in all the four cases. It is only their phase angles that are different depending on the hue they represent. Again, for clarity in drawings (waveshapes) the amplitudes of U and V have been chosen arbitrarily to be in the ratio of 1:2.

Purple Hue

Both the colour-difference signals are positive for this hue. The corresponding wave shapes are shown in Fig. 7.11 (also see Figs. 7.9 and 7.10). The two constituents, U and V of the chrominance signal as separated by the delay line and adder and subtractor circuits are shown in (*a*) and (*b*) of Fig. 7.11. Since (B–Y) is positive, the U modulation product is in phase with the reference subcarrier. Similarly since (R–Y) is positive, the V modulation product has a phase which leads the reference subcarrier by 90°. The dotted line wave-shape is for the PAL lines and is 180° out of phase with the NTSC lines.

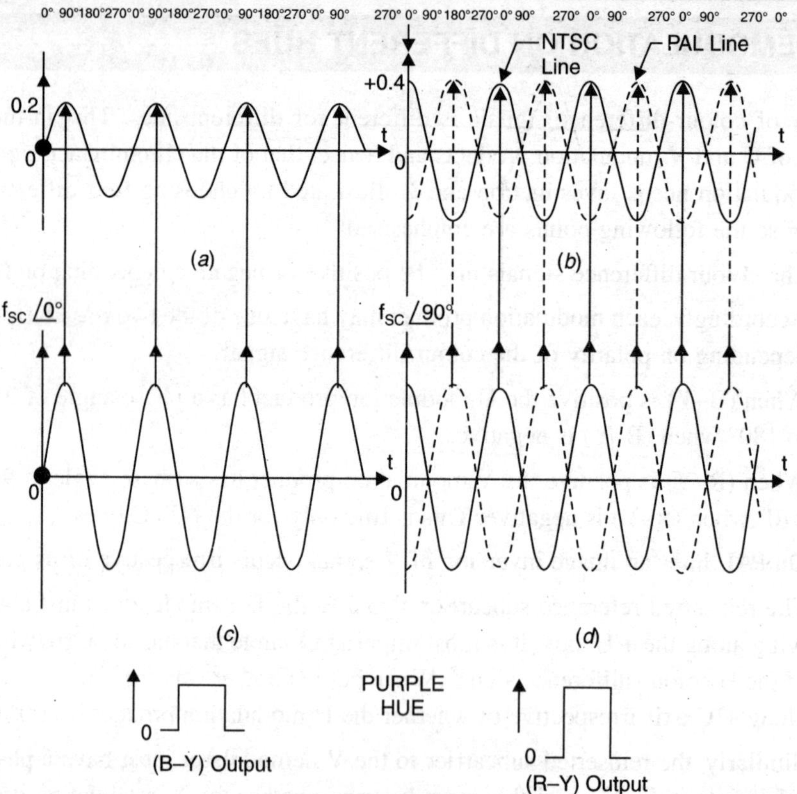

Fig. 7.11 Recovery of (B–Y) and (R–Y) colour-difference signal in a PAL-D system. Note that both the colour-difference signals are positive for a purple hue.

The reference subcarrier or each demodulator is shown below the corresponding modulation product (see (*c*) and (*d*)). The arrows on (*c*) indicate sampling moments for the U demodulator. Looking up from the arrow shows that the device in the demodulator will conduct to pass current pulses during positive peaks of the U modulation product. Thus (B – Y) colour-difference signal available at the output of U demodulator will be of positive polarity as at the modulating end.

Recovery of (R – Y) colour-difference signal is a bit different. For the NTSC lines the reference reinserted carrier has the correct phase ($f_{sc} \angle + 90°$) for sampling positive peaks. However, for PAL lines the phase angle of the subcarrier must be inverted (shifted by 180°) to correspond to the invertion of these lines. It is only then that the positive peaks are sampled on PAL lines to maintain same polarity of the (R – Y) signal on both NTSC and PAL lines. The arrows on the reference signal clearly indicate this. The output as obvious is a positive (R – Y) colour-difference signal.

Orange Hue

An orange hue gives rise to a negative (B – Y) signal and a positive (R – Y) signal. The phase angles of U and V modulation products are, therefore, + 180° and + 90° respectively. All the waveshapes necessary for explaining recovery of colour-difference signal from the two demodulators are shown in Fig. 7.12. The reinserted subcarriers are shown in (*c*) and (*d*). These have the same phase as shown in Fig. 7.11 as necessary for correct demodulation. Looking up at the arrows shows that output from the U demodulator will turn out to be – (B – Y) and that from the V demodulator + (R – Y).

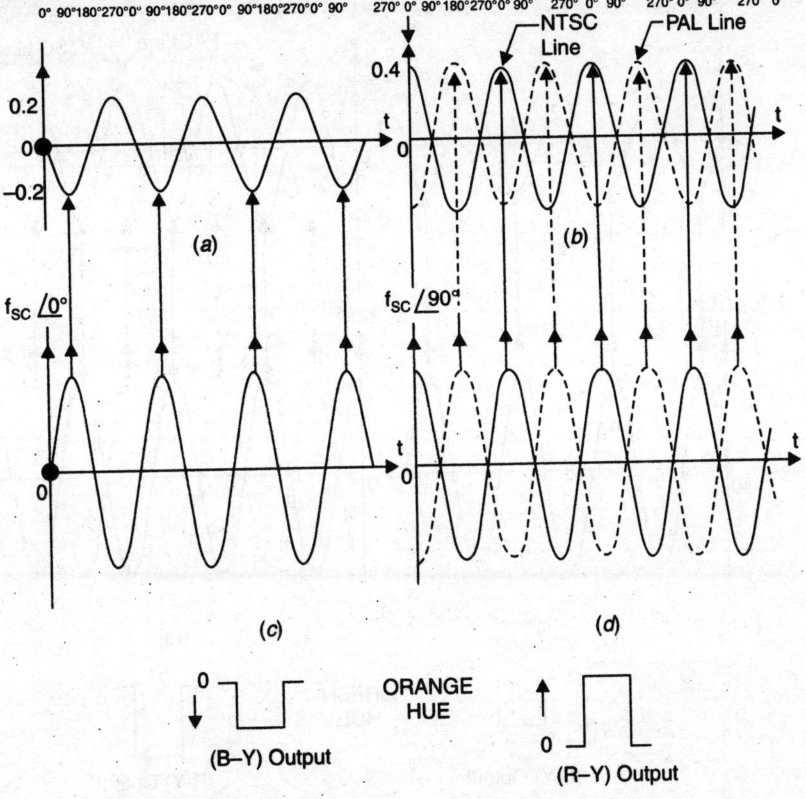

Fig. 7.12 Recovery of colour-difference signal in a PAL-D receiver when (B–Y) is negative and (R–Y) positive, (orange hue).

Green Hue

The green hue represented in Fig. 7.13 gives rise to negative voltages for both (B – Y) and (R – Y) colour difference signals. These in turn cause the U and V modulation products to have phases relative to + U axis of 180° and 270° respectively. The modulation products U and V are shown separately in (*a*) and (*b*). With the same reference reinserted subcarrier phase angles of f_{sc} ∠0° and f_{sc} ∠ ± 90°, it is seen that the U modulation product is sampled during its negative peaks. Similarly, the V modulation signal is sampled at its negative peaks both for NTSC and PAL lines. Thus demodulator outputs have the same polarity as that of modulating colour-difference signals at the transmitting end.

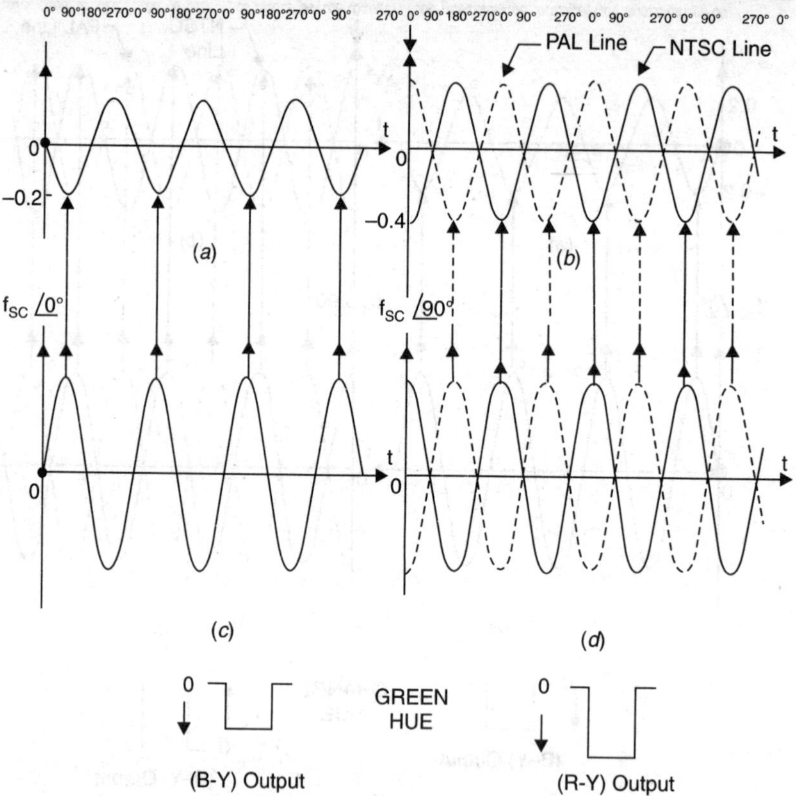

Fig. 7.13 Recovery of colour-difference signal in a PAL-D receiver when (B–Y) and (R–Y) are both negative (green hue).

Bluish-green Hue

The fourth case is shown in Fig. 7.14 where a bluish-green hue gives rise to + (B–Y) and – (R–Y) signals. This is turn causes the U and V modulation products to have phase angles of 0° and 270° respectively (waveshapes (*a*) and (*b*) in the figure).

The U and V reinserted reference subcarrier phases are shown in (*c*) and (*d*). Note that these continue to be the same as in the previous three cases. Looking up at the points (positive peaks) indicated by arrows will show that these correspond to positive peaks of (*a*) and negative peaks of (*b*). Thus, it is again established that correct relative amplitudes and polarities of original colour-difference signals are recovered.

It is obvious from the above discussion that each demodulator does not have to handle QAM (U ± jV) RF signal as is the case in NTSC system. Each demodulator handles either U or Y modulation product signal. Therefore, it is not absolutely necessary to employ synchronous demodulators in PAL-D receivers. However, these are still preferred in practice because they yield an accurate and constant no-colour zero voltage level above and below which (sometimes positive and sometimes negative) the colour-difference signal varies. In addition any inaccuracy in the delay time of the delay line used for separating the two modulation product signals can lead to presence of a small amplitude of U modulation product into the V demodulator and vice-versa. Since a synchronous demodulator while sampling one modulation product completely ignores the other, its use ensures that there will be no mutual interference between the U and V signals.

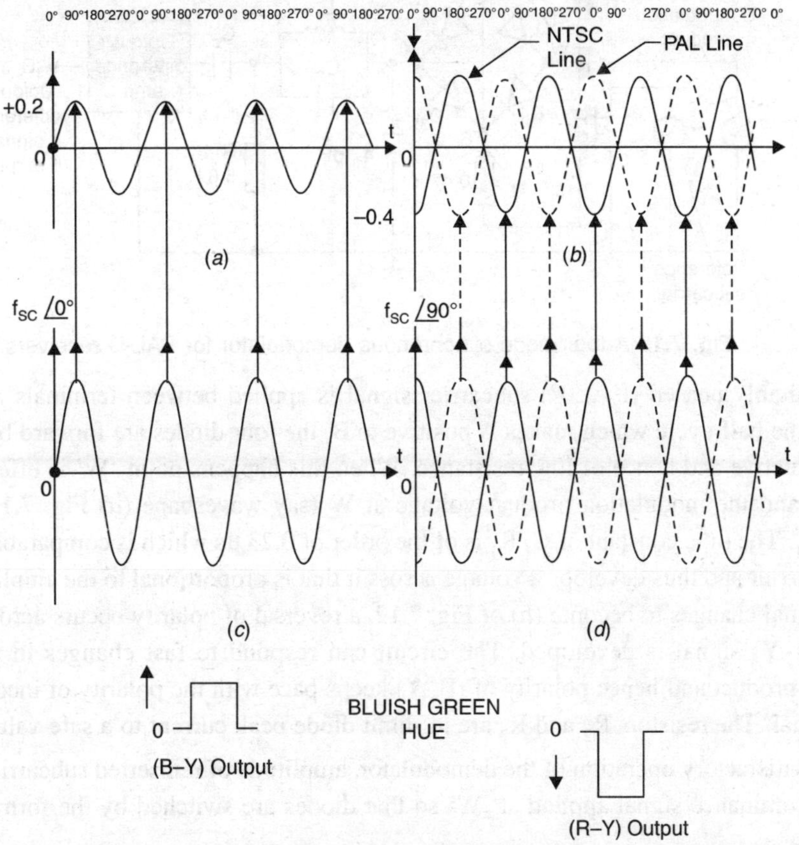

Fig. 7.14 Recovery of colour-difference signals in a PAL-D receiver when (B–Y) is positive and (R–Y) negative (bluish green hue).

7.15 SYNCHRONOUS DEMODULATOR CIRCUIT

It is possible to amplify the chrominance signal so that sufficient colour-difference output is obtained on detection to drive the picture tube directly. This is callled 'high level detection'. However, when chrominance signal is detected with no prior amplification it is called 'low level detection'. In this case colour-difference signals are amplified before feeding them to corresponding picture tube electrodes.

Low level modulation is preferred in practice. It is followed by a filter and amplifier before forming matrix circuits. A simple four diode low level synchronous demodulator is shown in Fig. 7.15. It is often used in PAL-D receivers where only one modulation product is present at the input. As obvious, two such demodulators are used, one for the U modulation product and the other for the V output. The one described is for the U modulation product signal.

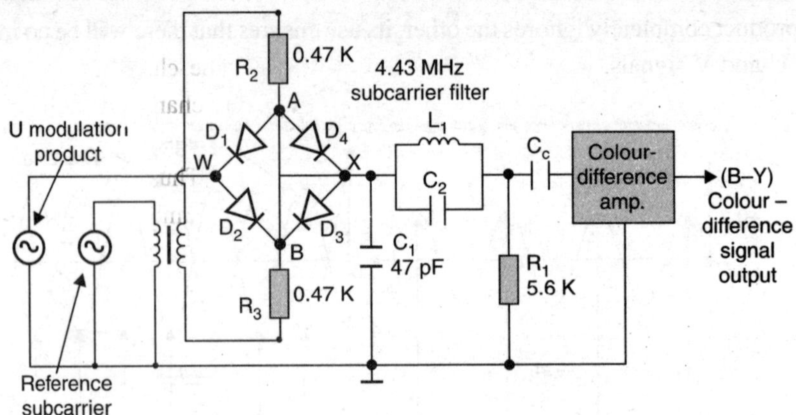

Fig. 7.15 A four diode synchronous demodulator for PAL-D receivers.

A suitably phased (f_{sc} $\angle 0°$) subcarrier signal is applied between terminals A, B of the diode bridge. On the half-cycle which makes A positive to B, the four diodes are forward biased and become highly conductive and hence of low resistance. When this happens point 'W' is effectively connected to point X and the modulation product voltage at W (say waveshape (*a*) Fig. 7.11) appears across capacitor C_1. The time constant of C_1 R_1 is of the order of 0.23 μs which is comparable to periodic time of the subcarrier and thus develops a voltage across it that is proportional to the amplitude of 'U' input. When U signal changes to become (*a*) of Fig. 7.12, a reversal of polarity occurs across C_1 and hence a negative (B–Y) signal is developed. The circuit can respond to fast changes in the polarity of U modulation product and hence polarity of (B–Y) keeps pace with the polarity of incoming modulation product signal. The resistors R_2 and R_3 are to limit diode peak current to a safe value.

For satisfactory operation of the demodulator, amplitude of reinserted subcarrier must be greater than the chrominance signal applied at 'W' so that diodes are switched by the former and not by the latter.

As shown in Fig. 7.15 the demodulator is followed by a filter circuit (L_1-C_2) to remove subcarrier ripple component. The tuned circuit is adjusted to be resonant at f_{sc} and hence, offers high impedance to the unwanted subcarrier signal. The filter output feeds into the colour-difference amplifier.

The other (second) synchronous demodulator is identical to the one described above except for the phase of reference subcarrier to recover other colour-difference signals carried by the V modulation product.

Polarity of Demodulated Colour-Difference Signals

The polarity of demodulated colour-difference signals have so far been shown as positive going *i.e.*, + (B–Y) and + (R–Y). In case it be necessary to develop – (B–Y) and – (R–Y) colour-difference voltages, all that is necessary is to advance the phase of the reference subcarrier fed to the two demodulators by 180°. This would result in shifting the sampling instants by half-cycle and opposite signal peaks (*i.e.*, negative instead of positive and vice-versa) will be sampled thus developing signals 180° out of phase with respect to those obtained without inverting the subcarrier phase. Thus colour-difference signals of opposite polarity can be developed if necessary.

Finally, it may be noted that chrominance signal is usually a complex wave and not sinusoidal as assumed all along while describing demodulators. The reason being that colours in any scene or picture are quite mixed and because of very fast rate of scanning the steady sinusoidal waveshape seldom stays beyond a few micro-seconds. When seen on a oscilloscope, the chrominance signal appears to be complex because of fast changes in its amplitude and polarity with change in colours and their saturation. However, this does not affect the process of demodulation because any complex wave consists of several fundamental sinusoidal components and their harmonics. Thus, the moments of conduction for all constituents of any complex chrominance signal remain the same and correct detection of colour difference signal occurs.

7.16 'V' CHANNEL SWITCHING

The reference subcarrier must be switched in phase on alternate lines for feeding the 'V' demodulator. This operation is carried out during line blanking interval at the end of each active line period. It is usual to employ two semiconductor diodes to carry out the actual switching operation. Each diode has to go 'on' for one line and 'off' for the next. To achieve this a square wave at half the line frequency (15625/2 = 7812.5 Hz) is needed. A component of this frequency is present in the incoming signal because of phase modulation of the 4.43 MHz. subcarrier burst transmitted during the back porch period. Since one complete swing of the burst phase takes place every two lines, cyclic frequency of the 'swinging' burst phase is half the line frequency *i.e.*, 7.8125 KHz. This is usually referred to as 7.8 KHz and called the IDENT Signal.

Fig. 7.16 shows the basic form of such a line switching circuit. The 7.8 KHz square wave switching signal to the diodes is obtained from bistable multivibrator (flip-flop) which is triggered once per line from the line time base and synchronized by the 7.8 KHz IDENT signal. Negative going line sync pulses are fed via steering diodes D_1 and D_2 to the bases of transistors Q_1 and Q_2 respectively. The function of 7.8 KHz 'Ident' signal is simply to block one out of every two of the line pulses to one of the transistors, in this circuit to Q_1. In the absence of 'Ident' signal a continuous succession of line pulses reach both the steering diodes and upon first switching 'on' it is a matter of chance which one of two successive pulses happens to switch 'on' Q_1. When the 'Ident' signal is present, only one out of every two line pulses gets through to trigger Q_1. This puts synchronization of the bistable multivibrator under command of the incoming signal from which the IDENT waveform is derived.

As labelled in Fig. 7.16, the outputs at the collectors of both Q_1 and Q_2 are 7.8 KHz square waves but 180° out of phase with respect to each other. These are applied to the two switching diodes D_3 and D_4. On one line D_3 is 'on' and D_4 'off', thus current from the reference oscillator flows in the upper half of primary of transformer T_1. Similarly, on the next line when D_4 is 'on' and D_3'off' current from the reference source flows into other half of the primary winding. The voltage induced in the secondary is thus 180° out of phase on alternate lines. It is fed to the 'V' synchronous demodulator which as explained in the previous section (for U demodulator) samples V modulation product at desired moments to develop (R–Y) colour-difference signal.

Referring again to Fig. 7.16, the U synchronous demodulator is also shown to draw attention to relative phase of the reference subcarrier input voltages fed to the two demodulators. Inductor Lp introduces a lagging angle to the voltage applied to the primary of T_2 to give its secondary voltage a phase angle of 0° compared with T_1 secondary voltage, which had a phase angle of + 90° on NTSC lines and – 90° on PAL lines. With reference to Fig. 7.1 it may be noted that such details are not shown there. A flip-flop and PAL switch account for the synchronized ± 90° subcarrier supply to the V demodulator while the U demodulator is shown receiving $f_{sc} \angle 0°$ directly from the subcarrier generator.

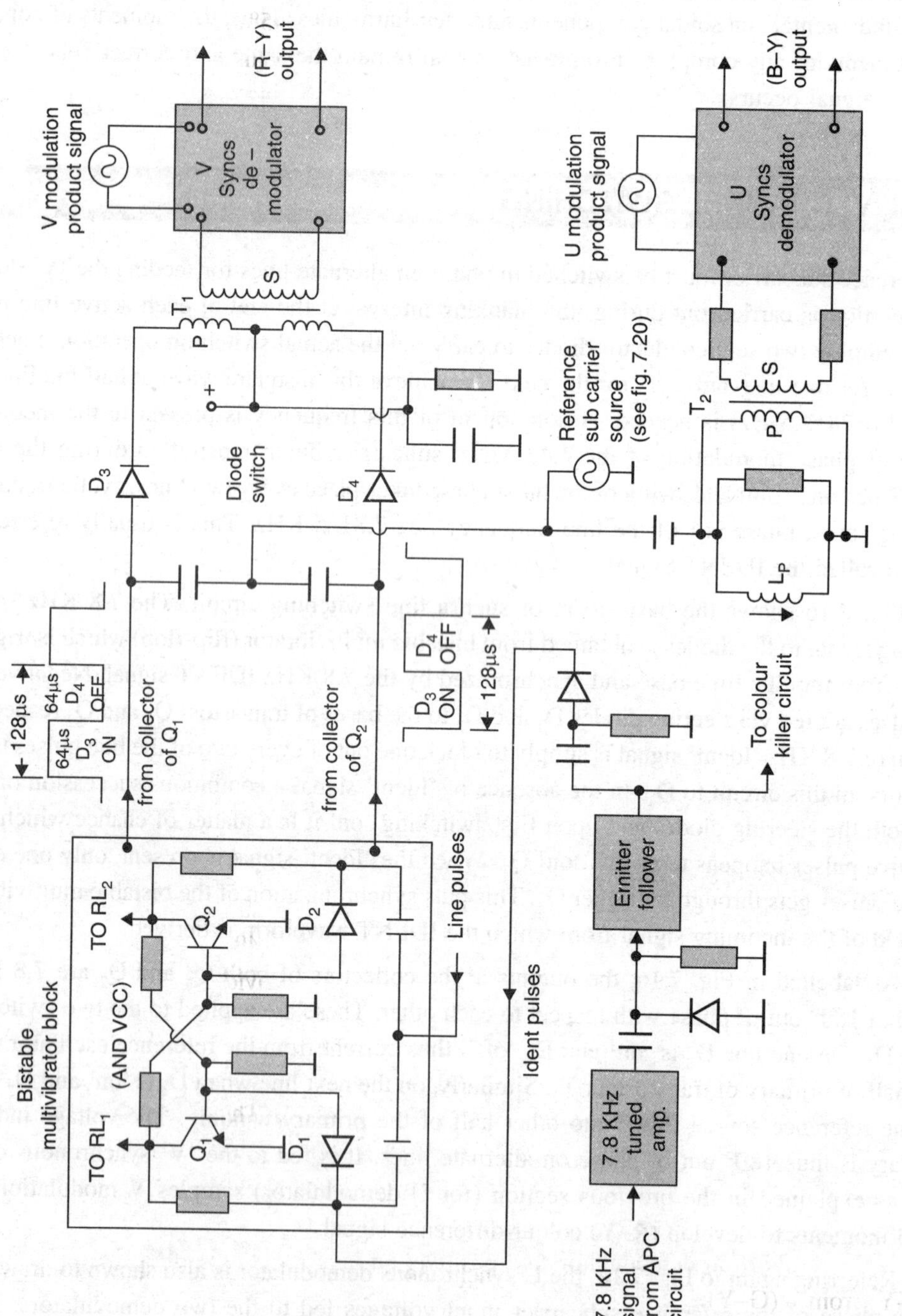

Fig. 7.16 Simplified circuit arrangement of V-channel switching in a PAL-D receiver.

7.17 IDENTIFICATION (IDENT) SIGNAL

As explained in the previous section an Ident signal is used to synchronize the instant and sequence of subcarrier switching to the V demodulator. It is developed from the received burst signal. At the transmitting end the burst phase is switched on successive lines by ± 45° about its mean phase of + 180° to allow the receiver to identify NTSC and PAL lines. For the NTSC lines the burst phase is 135° and for the PAL lines it swings to become 225°. Since one complete swing of the burst phase takes place every two lines, the cyclic frequency of the 'swinging' burst phase is equal to half the line frequency *i.e.,* 7.8 KHz. The identification signal is developed from this in-built information in the colour burst which is available at the automatic phase control (APC) circuit. The 7.8 KHz ripple in APC discriminator (see Fig. 7.16) is picked up and fed to a tuned amplifier. The sinusoidal 7.8 KHz output at the collector of amplifier transistor is clamped by a diode cum capacitor circuit to obtain square wave output. This is referred to as identification or IDENT signal and fed to an emitter follower the output of which connects it to the steering diodes in the bistable multivibrator and colour killer circuits.

7.18 COLOUR SIGNAL MATRIXING

The three electron beams of a colour picture tube must be modulated by gamma corrected R, G and B colour voltages. Therefore, colour-difference signals obtained at the outputs of demodulators are combined properly to reconstruct original colour signals. The circuits used for this purpose are called matrix configurations. In effect, a matrix performs addition and multiplication in order to obtain desired output signals. While different methods have been developed to drive the picture tube but it is first necessary to obtain (G–Y) from the demodulated (B–Y) and (R–Y) signal outputs.

Influence of U and V Weighting Factor on (B–Y) and (R–Y)

A chroma signal compression (weighting factors) is introduced at the transmitter to prevent over-modulation. It is 0.493 for the U and 0.877 for the V components. Thus in order to restore colour-difference signals to their original values, (B–Y) must be raised by a factor of 2.03 (1/0.493) and (R–Y) by 1.14 (1/0.877) besides equal gains to increase their values to be sufficient for driving the picture tube. When put in another way the total (B–Y) gain should be 0.877/0.493 = 1.78 times the (R–Y) gain. This is taken care of while designing colour-difference amplifiers provided after the demodulator (see Fig. 7.15).

(G–Y) Matrix

As deduced in Section 2.8, (G–Y) = – 0.51 (R–Y) – 0.186 (B–Y). Thus (G–Y) can be obtained by choosing above proportions of deweighted (compensated) (R–Y) and (B–Y) signals. Since the magnitudes are less than unity, resistance matrix circuits can be set up to obtain (G–Y). It is necessary to choose correct polarities of (B–Y) and (R–Y). For obtaining + (G–Y), both (B–Y) and (R–Y) should be negative. If outputs from demodulators are positive, an inverter is used after the (G–Y) matrix to obtain (G–Y) from – (G–Y).

Fig. 7.17 (*a*) shows a simple resistance matrix to obtain (G–Y) from deweighted (B–Y) and (R–Y) signals. It is a potential divider arrangement where $\frac{R_3}{R_1 + R_3} = 0.51$ and $\frac{R_3}{R_2 + R_3} = 0.186$. Thus 51% of (R–Y) combines with 18.6% of (B–Y) to develop (G–Y) across R_3. Since output acros R_3 is fed to the (G–Y) amplifier it ie R_3 needs a little adjustment to take into account input resistance of the amplifier. In case (G–Y) matrix is set up using weighted (B–Y) and (R–Y) signals, deweighting is done in the following (G–Y) amplifier. The deweighting can also be obtained by choosing different values of the resistance matrix.

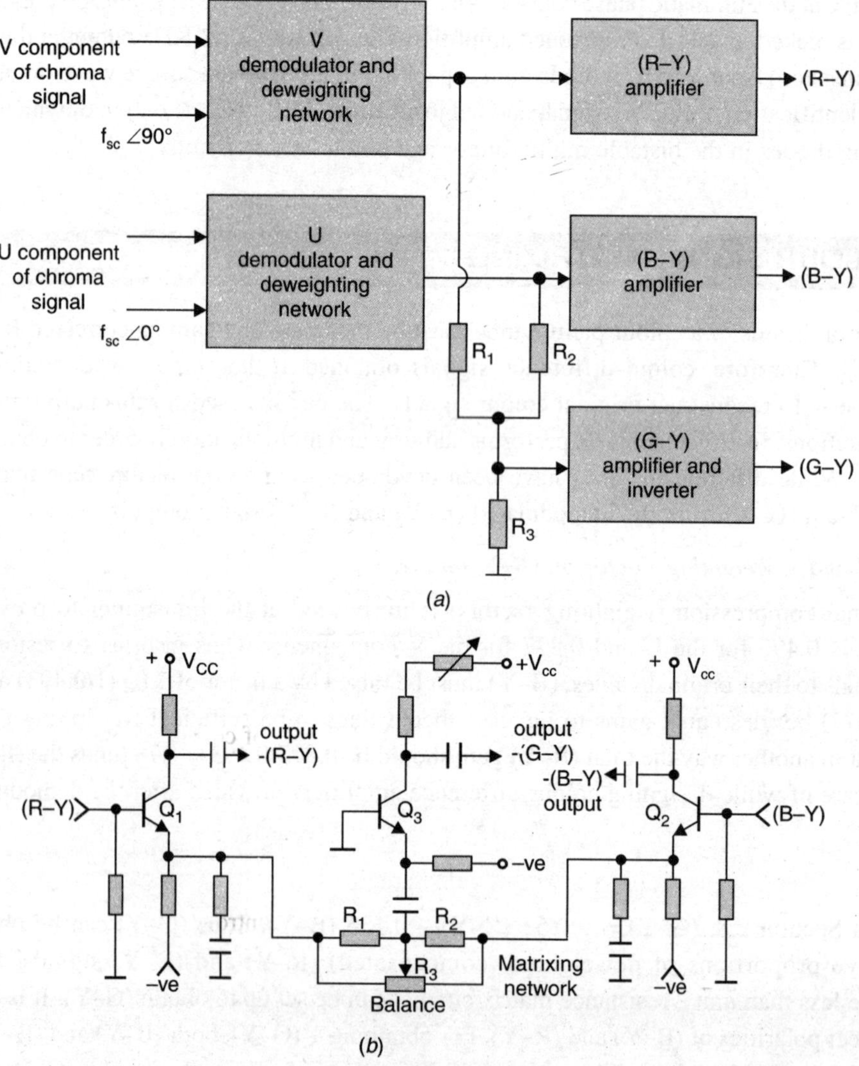

Fig. 7.17 (G–Y) matrix configurations

(*a*) Simple resistance matrix

(*b*) Matrix in-between colour-difference amplifiers.

With weighted values the signal

$$(G-Y) = -0.51 \times 1.14\ (R-Y) - 0.186 \times 2.03\ (B-Y)$$
$$= -0.58\ (R-Y) - 0.377\ (B-Y)$$

Therefore, the ratio of (R–Y) to (B–Y) contribution should be 0.58/0.377 *i.e.* 1.54:1.

Another (G–Y) matrix arrangement is shown in Figure 7.17 (*b*). Here matrixing is carried out within the colour-difference amplifier circuits. The signals obtained at the emitters of (B–Y) and (R–Y) amplifier transistors are combined in the emitter circuit of Q_3 amplifier through R_1, R_2 and R_3 in such a way that the net drive to the amplifier (between base and emitter) is due to (G–Y) current components of two currents and of correct proportion derived from the emitters of Q_1 and Q_2. Thus Q_3 amplifier delivers an output that represents (G–Y) colour-difference signal.

Complete Matrixing Configurations

The two commonly used matrixing configurations are shown in Fig. 7.18. The one used in earlier colour receivers is referred to as a two matrix grid drive configuration (See Fig. 7.18 (*a*)). The first matrix is located after the demodulator to generate (G–Y) by combining suitable proportions of (B–Y) and (R–Y) signals. The second matrix is formed by grids and cathodes of the picture tube. The grids are fed with positive colour-difference signals and the cathodes receive – Y signal. Applying – Y at the cathode is equivalent to applying + Y at the grids so that the tube performs algebraic operations to yield (R–Y) + Y = R, (B–Y) + Y = B and (G–Y) + Y = G. In effect, this means that Y signal produces equal and opposite changes in beam currents so that the net changes are only due to colour voltages. As obvious, the magnitude of colour-difference signals must be greater than the value of Y for obtaining meaningful beam currents.

RGB Matrix

Here, as explained in section 3.14 the colour-difference signals are combined with Y in the driver amplifiers (see Figs. 3.14, 7.1 and 7.18 (*b*)) in such a way that the outputs are R, G and B video signals. These are negative going and of sufficient amplitude to drive cathodes of picture tube directly. In other words the picture tube is not used as the second matrix. In the latest chroma ICs all functions including matrixing and preamplification of R, G and B signals are done within the integrated module. The drive amplifiers simply amplify these signals for feeding to the cathodes of colour guns.

Merits of R, G, B Matrix

Compared to other methods, R, G, B matrixing has the following advantages:

(i) It eliminates the need for high level video amplifiers.

(ii) The picture tube electrode arc protection is better since control grids are at ground or low dc potential.

(iii) Wide bandwidths are available for both colour and luminance signals which makes for crisper pictures.

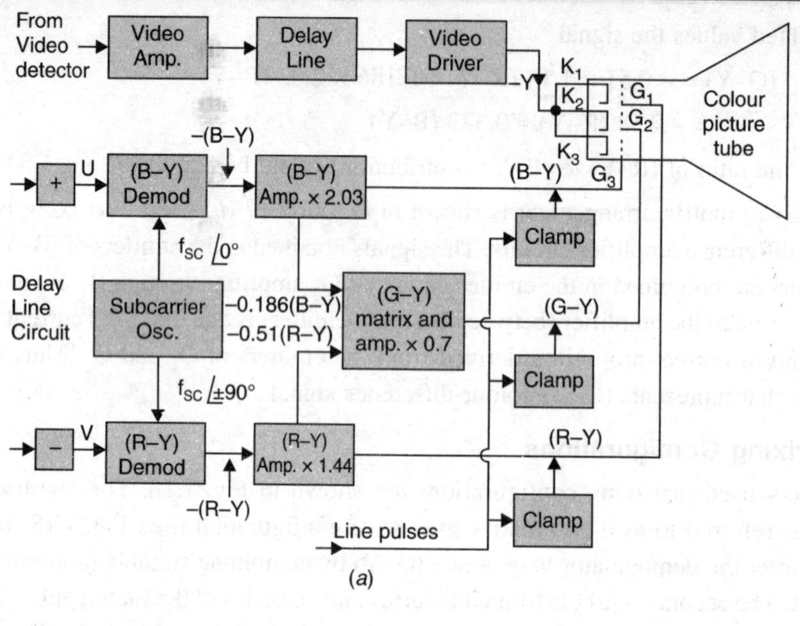

(*a*)

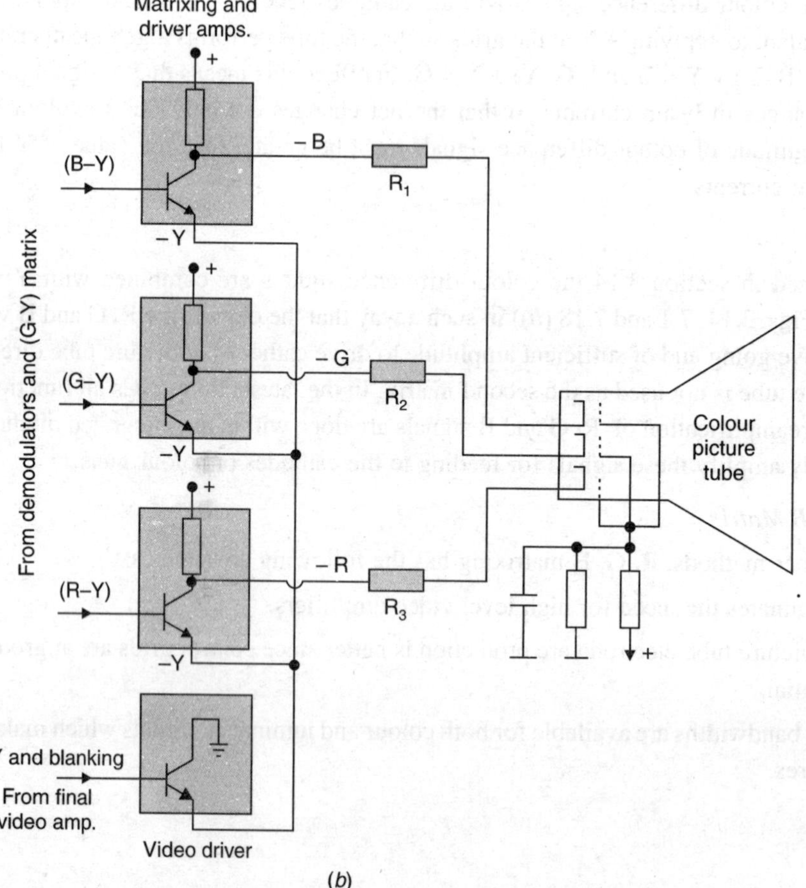

(*b*)

Fig. 7.18 Matrixing circuits (*a*) two matrix circuit for grid drive, (*b*) matrix for cathode drive.

7.19 MATRIXING CIRCUIT REQUIREMENTS

Any matrixing network and associated amplifiers have to meet the following requirements.

Beam Current

It is necessary to remember that a given voltage applied at the cathode(*s*) of a picture tube (for that matter any cathode-ray tube) gives rise to a greater change in beam current than the same input at the control grid(*s*). In fact, cathode drive yields nearly 30% more beam current than grid drive. As an illustration suppose that a 100% amplitude video signal (peak white) is produced by a drive to 100V to the cathode. Since the colour-difference signals are zero on white there is no input voltage at the control grids. To establish the same beam current with grid drive would need nearly 130V (100 × 130/100) input at the control grids. Similarly, for a non-white signal, way 100% saturated, 100% amplitude red colour, the amplitude of (R–Y) if applied at one of the control grids shall have to be increased from 70V to 91V. This is true for any colour and this aspect has to be kept in mind while deciding gain of colour-difference/colour drive amplifiers.

Phosphor Sensitivity

The efficiencies of three phosphors to produce coloured lights (radiation at different wave-lengths) differs and it would be necessary to take this fact into account while making assessment of the maximum voltage to be applied to various electrodes. For example, a given picture tube may require net cathode drives of – 100V, – 95V and – 90V for the red, green and blue guns respectively to yield peak amplitude white. This implies that red phosphor is the least and blue the most efficient. As a correction this would need corresponding adjustments in the magnitudes of colour-difference voltages obtained from respective amplifiers or R, G, B signals obtained from driver amplifiers.

Influence of U and V Weighting Factors

As explained in section 7-18 deweighting of colour-difference signals is necessary before matrixing. Thus (B–Y) must be raised by a factor of 2.03 and (R–Y) by 1.14 to restore them to their original values. This factor is also kept in mind while determining gain of matrixing amplifiers.

Valtage Gain

The need for large signal amplitudes to drive the picture tube makes it necessary to provide atleast one amplifier stage (the driver stage) employing discrete devices and components. High voltage transistors are used in the last amplifier stage. In order to estimate gain, assume that maximum peak-to-peak signal output from the V demodulator to be 600 mV (low level demodulation). This means + 300 m V on red and – 300 m V on cyan. For an estimated 180V on the red gun grid for a 100% saturated, 100% amplitude red, the (R–Y) channel gain must be $\dfrac{180 \times 10^3}{600}$ = 300. Similarly gain requirements of the other two amplifiers can be estimated.

The chroma ICs have in-built amplifiers to boost colour signals to nearly 6V p.p. Thus the discrete stage for red has to provide gain of about 180/6 = 30. A single amplifier stage can provide desired gain. Negative feedback is usually employed to stabilize gain thus making it independent of supply fluctuations and ageing of devices and components.

Bandwidth

The colour-difference amplifiers need a bandwidth from dc to about 1.3 MHz while driver amplifiers must have a wider bandwidth that extends up to about 4.5 MHz. This is provided for with necessary compensation to ward-off shunting effect due to picture tube electrode capacitances.

Coupling

It is necessary to provide dc coupling right from demodulators to drive points. Any loss of dc component in the colour-difference signals produces undesirable hue errors. As an illustration, suppose the whole left-hand side of the picture is a saturated red while the remaining half of it is grey. As shown by the waveshapes in Fig. 7.19, any lack of dc response in the colour-difference amplifiers would tend to give a cyan (complementary of red) cast to the grey half of the picture. In the waveforms a loss of 3 db of the dc component has been assumed.

Notice that although peak-to-peak values of all the signals are unchanged by the loss of dc component, red signal becomes slightly desaturated and grey scene changes to a desaturated cyan.

While the signal assumed for illustrating effect of loss of dc component rarely occurs in practice, it often happens that large areas of the picture are predominantly of one colour so that insufficient dc content would tend to produce a complementary hue spread in the remainder of the picture.

The signal feeding at the cathodes of picture tube has positive going sync pulse. Under these conditions white levels have a magnitude close to zero and its maximum value corresponds to black level. In order to hold the black level constant, it is necessary that dc component of the video signal is retained and hence the need for luminance channel amplifiers to be also dc coupled.

DC Clamps

As explained above, it is necessary to provide dc coupling both in the luminance and chrominance channel amplifiers. This in fact is the practice in modern colour receivers. However, in earlier versions it was common to use ac coupling in post demodulation and Y channel amplifiers then establish no-colour and black levels by employing dc clamps at the last amplifier stage. The clamps (see Fig. 7.18 (a) are usually diodes shunted across the signal path after coupling network of the amplifier. These are driven to full conduction during line blanking periods by pulses derived from the line (horizontal) deflection circuit. The low impedance provided by the forward biased diodes effectively clamps the grids of the guns (cathode in the case of Y amplifier) at a fixed potential set by a potentiometer from a suitable dc course. The discharge time of the ac coupling networks are arranged to be too long for this level to change significantly between the beginning and end of each active line period. This then amounts to dc coupling and no hue errors or black level changes occur despite ac coupling used in the amplifiers.

Blanking'

It must be ensured that retrace lines are not seen on the raster. For this a vertical blanking pulse is obtained from the vertical output stage and a horizontal pulse from a winding on the line output transformer. In the colour-difference amplifier scheme of matrixing, a transistor is used as the blanking driver. The polarity of pulses is such that these saturate the normally 'off' driver transistor. The collector current of drive flows through a common biasing circuit so as to increase conduction of all the three colour-difference amplifiers. The net results is to cause their collect voltages to drop producing very large negative going pulses. These pulses are coupled to the grids of picture tube turning all the three guns'off' during retrace and colour burst intervals.

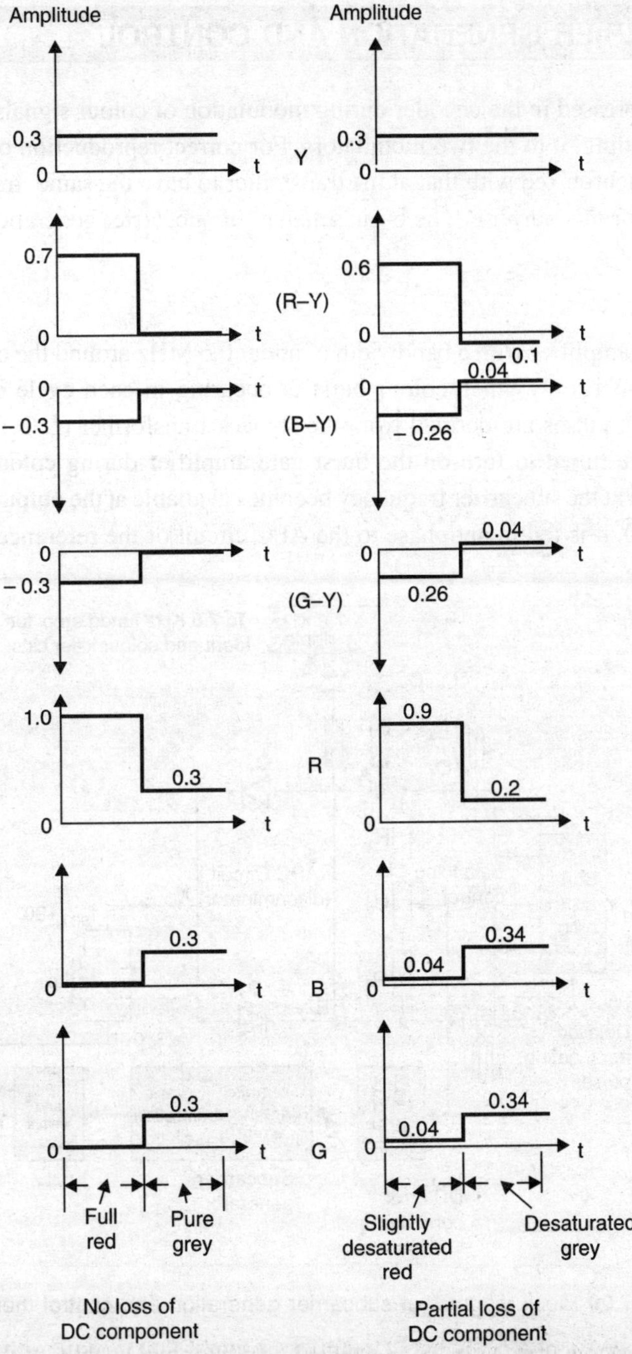

Fig. 7.19 Illustration of hue error resulting from partial loss of dc component of colour-difference signals.

In the other matrixing arrangement (see Fig. 7.1), blanking pulses are combined with the Y signal before it feeds into the emitters of colour driver amplifiers. The net result is the same and picture tube is cut-off during retrace intervals.

7.20 SUBCARRIER GENERATION AND CONTROL

The subcarrier is suppressed in the encoder during modulation of colour signals and is again generated at the receiver for feeding it to the two modulators. For correct reproduction of colours the subcarrier generator must be synchronized with that at the transmitter to have the same frequency and phase. The colour burst is used for this purpose. The basic scheme of subcarrier generation and control is shown in Fig. 7.20.

Burst Gate Amplifier

It is a gated but tuned amplifier with a bandwidth of about 0.6 MHz around the centre frequency of 4.43 MHz and is turned 'ON' only when colour burst is occuring in each cycle of the composite video signal. For this, gating pulses are derived from the flyback transformer (L.O.T.) and suitably delayed and shaped. These are timed to turn-on the burst gate amplifier during colour burst intervals. Thus sinusoidal burst signal at the subcarrier frequency becomes available at the output of burst gate amplifier. As shown in Fig. 7.20, it is fed in antiphase to the APC circuit of the reference oscillator.

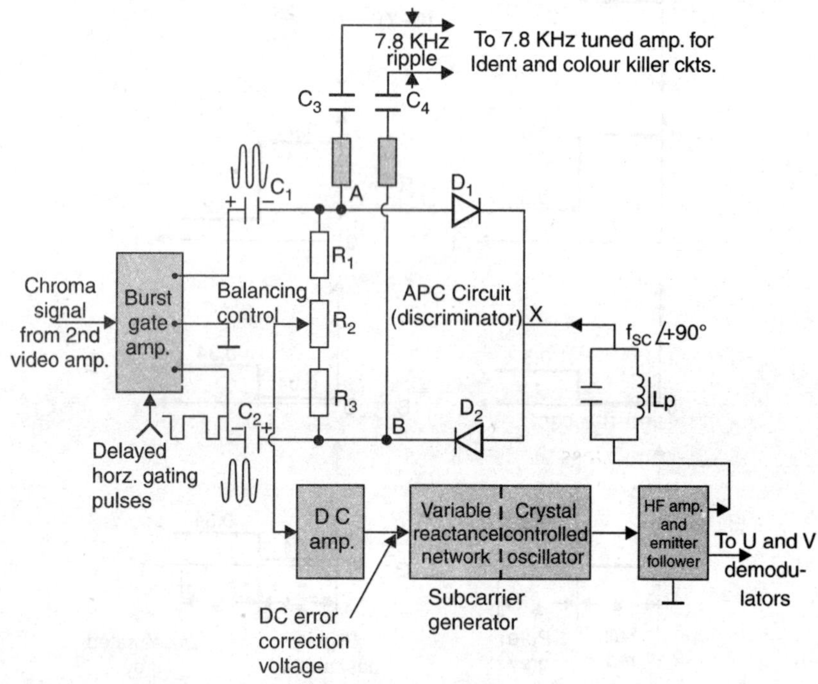

Fig. 7.20 Block diagram of subcarrier generation and control methods.

Reference Oscillator: It is a crystal controlled oscillator and usually of the inverted Colpit type. The variable reactance network has a varactor diode in parallel with the tuned circuit formed by the crystal at the input of transistor in the oscillator. As shown in the figure, the varactor receives an amplified dc control voltage from the APC circuit. Thus any shift in the frequency or phase of the oscillator is automatically and continuously corrected.

The oscillator output at 4.43361875 MHz is ac coupled to a high frequency amplifier to raise its amplitude. The output of this amplifier feeds both U and V demodulators as illustrated in Fig. 7.16. The output is also fed to the APC circuit via a phase shifting network to obtain 90° phase shift with reference to phase of the received subcarrier through colour burst signal.

Automatic Phase Control (*APC*) *Circuit:* As shown in Fig. 7.20, the APC circuit receives two inputs, the locally generated reference subcarrier and the transmitted burst. In the absence of any input from the reference generator, the two diodes, D_1 and D_2 conduct equally to charge C_1 and C_2 with equal voltages but of opposite polarity. If the circuit is perfectly balanced no control voltage is developed. With the reference subcarrier present and applied at point 'X' the circuit behaves in the following manner.

As explained earlier, the subcarrier fed at 'X' is phase shifted to become $f_{sc} \angle +90°$ with reference to colour burst. When subcarrier frequency is of correct value, the two inputs to the APC circuit are in quadrature. The charges on the two capacitors will still remain equal since the subcarrier signal passes through zero potential when diodes are pushed into conduction by peaks of the burst signal. The line-by-line phase alternation of ± 45° about the 180° axis of the burst signal does not affect the circuit balance because mean phase is still 180°. In fact, the burst angle shift of ± 45° favours one diode on one line but the other on the next line. Thus the mean output voltage remains zero.

If the oscillator frequency tends to increase, the reference signal to common connection (X) now passes through zero ahead of the centre of the time interval when the burst pulses the two diodes. The net input to D_2 becomes greater than D_1 and charge on C_2 exceeds that on C_1. This results in a net positive control voltage. Similarly, if the oscillator frequency tends to decrease analogous arguments will establish that net control voltage will become negative.

The control voltage is amplified by a dc amplifier (see Fig. 7.20) before applying to the varactor diode in the input tuned circuit of the oscillator. The amplifier reverses polarity of control voltage and hence, the varactor receives a negative voltage when subcarrier frequency tend to become more and a positive voltage (less negative) when it tends to become less. Any increases in reverse bias decreases capacitance of the varactor and the oscillator frequency decreases to become correct. The opposite is true when the frequency becomes less than the exact value. Thus the APC circuit continuously monitors the reference oscillator frequency and phase to keep it synchronized with the corresponding generator at the transmitter. The potentiometer R_2 is adjusted if necessary to ensure that error voltage is zero when subcarrier generators are fully synchronized.

7.8 KHz Signal

While the mean voltage across A and B in the discriminator circuit of Fig. 7.20 remains steady, a 7.8 KHz ripple is always present due to line-by-line phase alteration of the burst signal. This is picked up via capacitors C_3 and C_4 and fed to the Ident Circuit (see Fig. 7.16) for 'V' channel switching and operation of colour killer circuit.

7.21 AGC AND SYNC SEPARATION

The AGC and sync separator circuits of a colour receiver are the same as in a monochrome receiver. As shown in the block diagram, both are fed with compositive video signal from the 2nd video amplifier. Noise suppression gates (not shown) are used at the input and the outputs are duly amplified.

The AGC circuit is of keyed type and necessary line pulses are obtained from the line output transformer. A suitable delay circuit is provided in the path of AGC to the RF amplifier. In some receiver design very elaborate AGC circuits are used and the control voltage is made less or more dependent on the strength of incoming signal. This enables distortion free amplification of the RF and IF signal with high signal to noise ratio.

The sync separator is of conventional type and the sliced pulses are amplified before feeding them to the integrating circuit in the vertical deflection section and AFC circuit in the horizontal deflection section.

7.22 VERTICAL DEFLECTION CIRCUIT

The vertical oscillator can be a multivibrator, blocking oscillator or a Miller sweep generator. In recent designs where vertical oscillator forms part of an integrated circuit, a VCO (clock) is operated at twice the horizontal frequency (31.250 KHz) and then divided by a chain of binary dividers to obtain 50 Hz. A clock at 15.625 KHz is not used because the then needed divide ratio of 312.5 is not practicable because todays' digital dividers divide only by whole numbers. The VCO output is synchronized by the usual AFC circuit and the derived 50 Hz square wave output is waveshaped before feeding it to the driver circuit. The driver has linearization and vertical height circuitry. With conventional oscillator circuits a frequency control is also provided while the above described IC version does not need any vertical hold control. The output stage is a part of the IC or a discrete class AB operated amplifier and its output transformer provides impedance match to the vertical deflecting coils for obtaining linear deflection of the beam from top to bottom of the screen. The vertical output stage also feeds blanking pulses to the last (5th) video amplifier transistor and also to the E-W dynamic pincushion distortion correction circuit. For smaller screen picture tubes the vertical output amplifier also forms part of the IC and hence no separate transistor amplifier stage is necessary.

7.23 HORIZONTAL DEFLECTION CIRCUIT AND HIGH VOLTAGE GENERATION

The horizontal or line deflection circuit performs a similar function of deflecting the beam from left to right across the screen. However, when compared with vertical section its design and mode of operation is quite different. The reaction scanning technique enables high efficiency, fast retrace and generation of EHT for the picture tube. To be specific the horizontal deflection stage is designed to provide the following:—

(i) generates deflection current for feeding the yoke coils.

(ii) generates high voltage (20 to 25 KV) for feeding final anode of the picture tube.

(iii) generates a high voltage for the focus and screen grid of the picture.

(iv) provides necessary signal at the line frequency to incorporate E-W pincushion distortion correction.

(v) produces low voltage (210V., 30V etc.) dc supplies for feeding various receiver circuit modules.

(vi) provides automatic voltage control against wide supply voltage fluctuations. This is most effective with the switch mode power supply (SMPS).

(vii) contains automatic control to prevent generation of excessive EHT voltage to prevent X-radiation.

(viii) develops L.V. ac supply for the picture tube filaments.

(ix) feeds suitably shaped and timed pulses to AGC, AFC, burst blanking, burst gate and flip-flop in the PAL switch.

(x) feeds horizontal blanking pulses to picture tube drive circuitry.

In discrete versions the oscillator circuits are similar to those employed in the vertical deflection section. However, the drive and output circuits are designed differently to perform the functions listed above. In earlier integrated versions the AFC and oscillator circuits are of conventional type and form part of the same IC that has corresponding vertical deflection circuits. In recent ICs for the deflection section the horizontal (line) oscillator is of the VCO type as explained in the previous section. In fact the same clock oscillator operating at 31.250 KHz provides line frequency of 15625 Hz after a division by 2. Since horizontal and vertical scanning frequencies are derived from the same source which is locked with the incoming sync pulses by the AFC circuit, no horizontal or vertical hold controls are necessary. However, in both discrete and IC versions the horizontal driver and output stage employ high voltage, high wattage transistors because most of the power consumed by the receiver is spent in the horizontal cum power supply section.

7.24 LOW VOLTAGE POWER SUPPLY

All colour receivers have regulated power supplies because voltage requirements in most stages are quite stringent. Most of the power obtained is of the derived type where it is obtained by rectifying several outputs of the horizontal output transformer. The basic source is the mains (50 Hz) supply which is rectified and regulated before feeding to the horizontal output and other sections of the receiver. The two most common types of regulated power supply are (i) series voltage regulated and (ii) switched mode power supply (SMPS).

Series-pass Regulator Power Supply

The block diagram of this type of power supply is given in Fig. 7.21. It includes protection for the series-pass transistor Q_1 against excessive load current and damage to the load in the event of short in Q_1. The transistor Q_2 is the error amplifier, and zener diode provides necessary fixed voltage reference. The monitoring network consists of three resistors in series where the middle one is variable. It is used to adjust the output dc voltage and also provides one of the inputs to the error amplifier. The other input to it is the zener reference voltage. The error amplifier output that is proportional to the difference between the two inputs feeds into the base of Q_1. Thus the collector current of series regulator transistor (Q_1) varies to change voltage drop across its emitter and collector to maintain a steady dc voltage across the load.

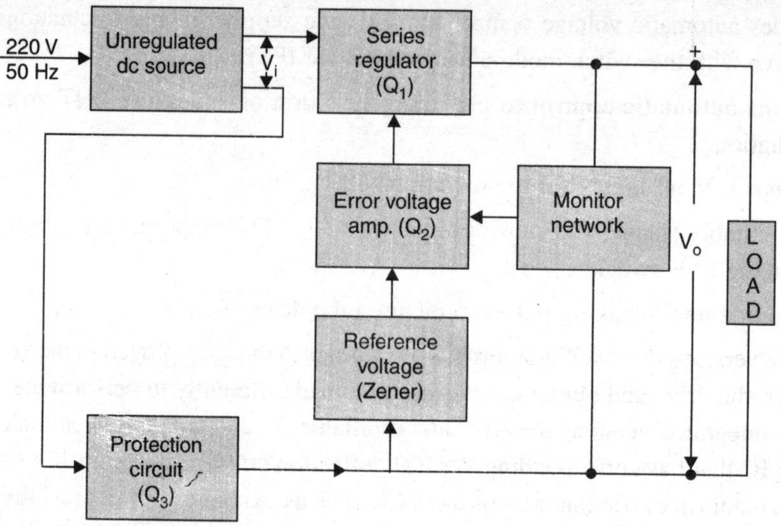

Fig. 7.21 Block diagram of series-pass regulator type LV power supply

The protection circuit (Q_3) is in the negative lead of the unregulated dc source. In the event of either excessive load current or short in Q_1, the tendency is for the output voltage and current to rise. When this happens, the protection circuit increases effective series resistance of Q_3. This reduces unregulated power supply output voltage and current to a safe value. During normal operation, protection transistor Q_3 is saturated and hence does not affect regulation of the power supply.

Switched Mode Regulated Power Supply

The usefulness of series regulated power supplies lead to the development of integrated three terminal regulator modules. However, modern colour receivers prefer switched mode regulated power supply because it is more efficient and provides constant output despite very wide mains voltage fluctuations. Its regulator transistor is switched ON and OFF periodically. Typical switching frequencies are between 10 and 20 KHz. Unlike the series regulator transistor, the switching transistor is always fully ON or fully OFF. Thus much less heat is produced than in a linear regulator.

The basic circuit of a switching regulator is shown in Fig. 7.22. The series pass transistor is Q_1 and the error voltage amlifier is an operational amplifier (OPAMP) which receives reference voltage and a sample of the output voltage as its two inputs. The pulse width oscillator is the heart of SMPS. The oscillator output switches the transistor Q_1, ON and OFF periodically at a frequency as high as 15 KHz. The ON and OFF periods can be varied by changing the pulse width. This is affected by the output of OPAMP which in turn depends on the output dc voltage. Thus the width of output pulses at the emitter of Q_1 varies to affect necessary changes in the output dc voltage. The feedback loop monitors supply variations and hence acts to maintain a near constant output voltage despite input ac fluctuations and load current variations. Because of the high ripple frequency L_1 and C_1 can have relatively small values.

In practical circuits the pulse width source is a blocking oscillator and its pulse width is made dependent on the dc voltage developed in the horizontal output circuit. Several provisions like overload protection, electronic fuse and X-radiation prevention form part of a switched mode power supply.

These details are described in chapter 17 that is devoted to switched mode power supplies.

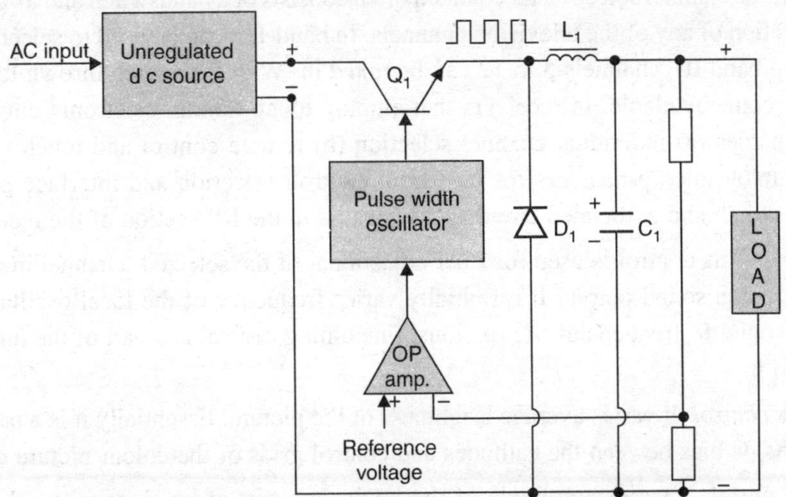

Fig. 7.22 Basic circuit of a switched mode regulated power supply.

7.25 COLOUR RECEIVER CONTROLS

Colour television receivers have a large number of controls which may be classified as follows:-

Operator's or User's Controls

These controls are normally located on the front panel of the receiver. The user or viewer can operate them to select the programme of his or her choice and set the quality of picture and volume of sound to his or her liking.

Service Controls

Such controls are mounted on the chassis and sub-assemblies (PCB's) of the set. The service engineer can manipulate them while repairing the receiver to remove faults and get optimum performance.

Pre-Adjusted Controls

In electronic circuitry it is difficult to select exact values of components for a desired response. The widespread in parameters of active devices and stray capacitances make it necessary to incorporate some variable elements like resistors, capacitors and inductors which can be changed to obtain optimum performance. These are set at the factory during manufacturing process.

While feedback techniques are used to ensure that changing of an active device does not shift its operating point and disturb performance of the circuit yet in cases where a major repair is undertaken; re-adjustment of pre-set controls becomes necessary. For television receivers, special and sophisticated equipment is used to adjust such pre-sets in the receiver. Therefore, these should not be disturbed while repairing a set unless it becomes absolutely necessary to do so. It is thus essential to know, not only the location of all types of controls but also fully understand their functions.

Common Front Panel Controls

Channel selector. The channel selection mechanism often consists of a bandswitch and a variable control. This enables selection of any of the television channels. In band-I, it can be used to select channels 2, 3, or 4. When set on band III, channels 5 to 12 can be tuned in. With the switch thrown to band IV (U), UHF channels become available. In receivers that employ touch tuning, electronic circuitry is used. Such a control enables (a) individual channel selection (b) remote control and touch type operation. Some designs employ microprocessors for initiation, control, selection and interface pursposes. The channel selector switch and associated circuitry are located in the RF section of the receiver.

Fine tuning. This control is used for finer adjustment of the selected channel in any band. It is set for best picture and sound output. It essentially varies frequency of the local oscillator to provide exact vision and sound IF frequencies. As obvious, fine tuning control is a part of the tuner circuit and is located close to it.

Brightness control. It varies average brightness of the picture. Essentially it is a part of the drive circuitry and shifts dc bias beween the cathodes and control grids of the colour picture tube.

Contrast control. It varies amplitude of the luminance part of he composite colour signal and thus is the gain control of video amplifier in the 'Y' channel of the receiver.

Volume control. It varies sound output by changing gain of the audio amplifier. Essentially then, it is a part of the audio strip of the receiver.

Tone control. It controls frequency response of the audio amplifier in the sound section of the receiver. Boosting high frequency response improves treble output while increasing gain for lower audio frequencies varies bass response. Some receivers provide two separate controls, one for treble and another for bass output. The tone control circuitry forms part of the audio amplifier.

Colour or saturation control. This front panel control is a part of the chroma band-pass amplifier prior to demodulation in the decoder section of the receiver. It varies gain of this amplifier thereby changing amplitude of the colour (or colour-difference) signals.

Vertical hold. It forms part of the vertical sweep oscillator and varies its frequency. It is adjusted to prevent any rolling of the picture. In many modern colour receivers special control circuitry is employed to obtain exact vertical deflection frequency and vertical hold control is not provided on the front panel.

Controls Located Inside the Receiver

AFT control. The AFT circuit is actually automatic frequency control of the local oscillator in the tuner. This control aims at obtaining a picture IF frequency of exactly 38.9 MHz at the converter output. To achieve this, IF frequency as obtained from the IF amplifier section (VIF) is measured by a discriminator circuit that forms part of the AFT circuit. A proportionate dc control voltage is developed to be applied to the local oscillator tuning circuit. The AFT control voltage should be zero when the picture IF is exactly 38.9 MHz. The AFT control which is located in the VIF panel is set at the factory for obtaining the above condition. The setting is checked for a steady and clear colour picture with the AFT switch put to both ON and OFF positions.

APC (Automatic Phase Control). It is a part of automatic phase control circuit. This circuit receives two inputs, the locally generated reference subcarrier and the transmitted burst. The comparison circuit develops a control voltage to keep the frequency and phase of the subcarrier oscillator at the correct value. This effectively adjusts correct HUE of the colour. It forms part of the Chroma Strip and is located close to the reference oscillator circuit. It is adjusted at the alignment stage in the manufacturing process.

ACC (Automatic Colour Control). The ACC circuit is similar to the AGC circuit used for automatic gain control of RF and IF sections of the receiver. It develops a control voltage proportional to the amplitude of colour burst. This voltage when fed to the transistors used in the chroma amplifier shifts their operating point to change the stage gain. Thus overall colour signal output from the chroma amplifier remains constant. The ACC control is located in the chroma section. Functionally it has the same effect as the colour or saturation control provided on the front panel. It is adjusted at the factory but might need a touch-up during servicing of chroma section of the receiver.

Colour sync. It is also located in the chroma section of the receiver and is pre-set to prevent any drift of the reference oscillator frequency. Electrically it helps to lock 'f_{sc}' to the transmitted burst signal. It is adjusted to set colour burst output at the correct level.

Drive controls. These form part of the colour driver amplifiers and are adjusted to provide necessary video signal amplitude to the cathodes of picture tube for obtaining full brightness range from black through grey shades to full bright. These per-set type of gain controls are located on the drive amlifier board located near the base socket of picture tube.

Low light adjustment controls. The small inherent difference in the cut-off characteristics of a picture tube results in appearance of a coloured tint. To correct this, 1st anode voltages of the picture tube can be varied. In earlier picture tube circuitry such controls form part of the dc supply to the picture tube anodes and are located on the power supply PCB mounted close to the picture tube. In recent picture tube designs where screen grids have a common external connection, low level adjustment is made by varying bias voltages at the three emitters of drive amplifier transistors. Three such controls are provided on the drive amplifier PCB located close to the picture tube.

High light adjustment controls. To ensure that all other levels of white are also correctly reproduced, it becomes necessary to vary amplitude of colour voltages to the three guns. For this, gain of colour difference amplifiers is varied in circuit where picture tube electrodes are used for 2nd matrixing. However, in modern receivers where R, G and B signals are directly applied at the cathodes, the drive (input) controls are slightly adjusted for optimum high light reproduction.

Focus control. It is located in the high voltage section of the picture tube power supply and feeds voltage to the focus anode of the picture tube. It is varied to obtain sharpness of the beam spot on the screen and set at the factory during final testing.

Purity control. It is a set of two pole magnetic rings located behind the yoke assembly of the picture tube. These are adjusted to prevent patches of colour affecting purity of the raster. Electrically the magnetic rings produce a magnetic field across the neck of the picture tube. The direction and strength of the magnetic field is varied to change orientation of all the three beams so as to converge them at the same hole and make them fall at corresponding dots on the phosphor triads located on the screen. The adjustment is done during testing and alignment and can be varied if necessary during servicing. In modern receivers it is seldom necessary to do so.

Four pole static convergence control. A pair of four pole magnets is located behind the yoke assembly. The resulting magnetic field enables shifting blue and red beams in opposite directions to each other. These are rotated and set for correct reproduction of magenta shades. This is also an intial adjustment and rarely needs any readjustment.

Six pole static convergence control. This set of six pole magnetic rings is located close to the four pole magnetic rings. These enable deviation of red and blue beams in such a way that they converge to combine with the green beam to produce white on the screen. All the magnetic adjustments are made one after the other according to convergence procedure given by receiver manufactures in their service manuals. Since these are factory set, normally there should be no need to readjust them while servicing the colour receiver.

A.G.C. control. The AGC circuit develops a voltage a proportional to the strength of incoming Y signal which in turn controls gain of the RF and IF sections of the receiver to maintain a steady input to the video detector. The AGC control enables variation of this voltage depending on the radiated signal strength in any location. This, then prevents occurrence of a weak or overload picture. The AGC control also prevents any tearing and rolling of the picture and buzz in the sound output because of variations in the signal strength. It is located in the AGC section and may be varied if necessary.

Horizontal hold. It is located in the horizontal (line) sweep circuit to control its frequency. When set properly, it prevents any tearing or rolling of the picture. In some receiver designs, this control is made available on the front panel. In receivers employing digital means of generating line and field frequencies, both horizontal and hold controls are not necessary.

Horizontal linearity. It is also a part of the horizontal sweep circuit. It is a pre-set to obtain linear deflection of the beam across the width of the screen.

Horizontal centering. This pre-set in the line section of receiver is provided to center the beam on the raster area. This then ensures uniform deflection across width of the screen.

Width control. This is adjusted to control gain of the line (horizontal) amplifier for correct deflection to cover full width of the raster. Though a pre-set, it can be varied while servicing the receiver.

Vertical frequency control It is a pre-set which enables wide range variation of the vertical oscillator frequency. Any change of component in the oscillator section will need re-setting of this control. Essentially it is a part of the vertical deflection circuit. Certain receiver designs have large pull-in range and such a control is not provided.

Vertical centering. The enables shifting of the beams vertically to the central portion of the raster. It is varied while carrying out basic alignments during the manufacturing process.

Vertical linearity. It affects waveshape of the vertical deflection oscillator and is adjusted to obtain linear deflection vertically. This is set while checking formation of the raster.

Height. This control varies gain of the vertical output amplifier to obtain sufficient amplitude of the sweep waveform to fill the entire raster. It can be re-set if necessary while servicing the receiver.

7.26 COMMON FAULTS IN COLOUR RECEIVERS

Colour receiver servicing is given at the end of each chapter and also covered in Chapter 18. This section is devoted to introducing a broad approach to identifying the section or sections which can be the cause of some common faults. It must be kept in mind that a colour receiver also contains circuits of a monochrome receiver and many faults and their remedies are common to them. Thus it is advisable to first ascertain by turning down the colour saturation control if the monochrome picture is being correctly reproduced. If not, the trouble is in the common (monochrome) circuits. An exception to this rule is a defective colour picture tube or defective circuits supplying operating voltages to the tube. A defective gun (or beam) will make it impossible to see a picture in monochrome because one of the primary colours will be missing. Faults in colour picture tubes and their remedies are discussed in section 3.16.

(1) *Normal monochrome picture, no colour* (see Fig. 7.1). The following sections may be defective–

(a) Colour killer

(b) 4.43 MHz subcarrier oscillator

(c) Chroma band-pass amplifier

(d) Demodulators and associated circuits.

(2) *Abnormal Colour Intensity* (*Saturation*). Assuming adequate colour signal from the 2nd video amplifier the trouble can be in the

(a) ACC voltage (circuit)

(b) Band-pass amplifier.

(3) *Colour bars on the screen*. Assuming monochrome sync to be normal, such a fault is due to no colour sync. This is mostly causes by:

(a) Defect in the 4.43 MHz subcarrier oscillator

(b) APC circuit

(c) Colour burst amplifier.

(4) *All hues incorrect but normal monochrome picture*. Check controls in the chroma section. The faulty section can be one of the following:

(a) U and V demodulators

(b) APC Circuit

(c) Band-pass amplifier

(d) 4.43 MHz oscillator.

(5) *One hue missing, black and white picture normal*. By observation the missing hue can be easily identified. The trouble is in the channel of missing colour. The fault can be in the

(a) Picture tube

(b) Drive amplifier

(c) Matrixing circuits

(d) Pre-drive circuit (coupling components).

(6) *Changing hues, monochrome picture normal*. This is mainly due to weak colour sync. The trouble can be in the:

(a) Burst gate amplifier

(b) APC circuit

(c) Subcarrier oscillator.

For faults that are common with the monochrome picture refer chapter 30 "Receiver Servicing" of the book Monochrome and Colour Television by R.R. Gulati (*New Age Publication*) or any other suitable text on this topic.

REVIEW QUESTIONS

1. Describe how a varactor diode can replace a variable capacitor for tuning resonant circuits. Explain with a circuit diagram how four varactors in a VHF tuner enable selection of any channel in Band-I.
2. What do you understand by all electronic tuners? Explain with a suitable circuit diagram how switching diodes are employed to change over from band I to band III channels.
3. Explain briefly the frequency synthesis method of tuning control and how a microcomputer enables automatic selection of any channel.
4. Draw block diagram of a tuner unit and explain how AFT operates to keep the local oscillator frequency constant on selection of any channel.
5. What is a 'SAW' filter? Explain its basic principle of operation. Draw simple circuit of a 'SAW' filter especially designed for IF section of a TV receiver and explain how desired selectivity and waveshaping is achieved with it.
6. Why is sound IF signal detected ahead of video IF signal in a colour receiver? Describe various paths in which the composite video signal is directed after detection in the video detector. Why is it amplified before feeding to various sections of the receiver?
7. Draw block diagram of the luminance channel in a colour TV receiver. Explain briefly the operation of pedestal clamp circuit. Why black level clamp of the video signal is a must for proper reproduction of pictures?
8. Explain how contrast and brightness controls function in the luminance channel to control contrast between white and black parts of the picture and its brightness. Describe how retrace blanking pulses are added to the Y signal.
9. Describe basic principle of a comb filter. Draw block diagram of such a filter configuration to explain how luminance and chrominance signals are separated from the composite video signal without loss of any high frequency components.
10. Draw basic circuit of a colour bandpass amplifier and explain how the colour-killer circuit operates to keep this channel open during reception of colour programmes. What is the function of ACC circuit in the bandpass amplifier?
11. Describe in sufficient depth how a synchronous demodulator operates to detect V *i.e.*, (R–Y) colour-difference signals. What are the essential conditions that must be met for successful operation of such a demodulator?
12. Explain with necessary waveforms how different hues get detected in a synchronous demodulator. Illustrate your answer for blue and orange hues.
13. What do you understand by 'V' channel switching? Describe with a simple circuit diagram how this technique is employed to detect V *i.e.*, (R–Y) modulation product signal.

14. What is an 'Ident' signal and how is it developed? Explain how it is employed to direct the line-by-line inverted V modulation product signal to the demodulator for correct detection of (R–Y) colour-difference signals.
15. What are the two basic schemes of colour signal matrixing. Draw suitable block diagrams to explain their operation. What are the essential conditions that must be met in a matrixing circuit?
16. Explain why R, G, B matrixing is preferred in modern colour receivers. How is deweighting carried out in such a matrixing arrangement? Why is the gain of colour-drive amplifiers kept variable?
17. Describe with a block diagram how subcarrier is generated in the colour receiver and the manner in which APC circuits operates to keep it constant. How is its phase synchronized with the frequency generated at the transmitting end?
18. Describe how flip-flops are used for generating horizontal and vertical scanning frequencies. Why is the basic flip-flop operated at twice the line frequency?
19. Describe with a suitable block diagram the basics of a switched mode power supply (SMPS). Why has it superseded the most used series regulator type of power supply in TV receivers?
20. Enumerate common faults in the chroma section of a colour receiver and indicate defective sections or circuits which can cause them. Assume that a black and white picture is correctly produced.

8

COLOUR TV STANDARDS & MODERN RECEIVER PERSPECTIVE

INTRODUCTION

The picture standards of various motion picture systems all over the world are nearly the same and thus international exchange of film programming is comparatively straight forward. However, it is not so in the case of television programmes because it has not been possible to evolve universally accepted TV standards. The lack of compatibility between the three main systems in use (NTSC, PAL and SECAM) has its origin in many factors such as constraints in communication channel allocations and techniques, difference in local power source characteristics, pick-up and display technology and political considerations relating to international telecommunication agreements. The recommendations of XIVth Plenary Assembly of the International Radio Consultative Committee (CCIR) held in Tokyo, Japan in 1978 are thus in the form of a report where it is left to the controlling organizations of individual countries to make their own choice as to which standards to adopt.

This is understandable because one of the primary requirements for any colour television system is compatibility with the existing monochrome system, which in turn is dictated by such factors as local power line frequency relevant to field and frame rates, radio frequency channel allocations and pertinent telecommunication agreements if any. Thus, such technical factors as line number, field rate, video bandwidth, modulation technique and sound carrier frequency were predetermined and varied in many regions of the world. Therefore, the ease with which international exchange of program may be accomplished is hampered and carried out at present by means of standard conversion techniques or 'transcoders' with varying degree of loss of quality.

Ever since the inception of colour TV, receiver technology and manufacture has made rapid strides despite diversity in television standards. Today it is common to see receivers that perform all the functions with just a few integrated circuits compared to mainly discrete designs about a decade ago. A large number of ICs are now available that meet the requirements of different colour systems. It is thus desirable to be fully acquainted with different TV standards and learn about developments in receiver design that have taken place in the recent past before embarking upon circuit details. Accordingly, first part of this chapter is devoted to Worldwide Colour Television Standards and the later half describes state of the art of present day receiver technology.

8.1 MONOCHROME COMPATIBLE COLOUR TV SYSTEMS

As detailed in chapter 2, it is essential that the adopted colour system be fully compatible with the existing black and white system, that is, monochrome receivers must be able to produce high quality black-and-white images from a colour broadcast and colour receivers must produce high quality black-and-white images from monochrome broadcasts. The first such compatible colour system was developed in the United States of America. In December, 1953, the Federal Communication Commission (FCC) of U.S.A approved a set of transmission standards and authorised regular colour broadcast from 23rd January, 1954. This decision was culmination of the work done by the National Television Systems Committee (NTSC) upon whose recommendations the FCC action was based. Subsequently this 525 line-60 field colour system came to be known as the NTSC system. It was later adopted by Canada, Japan and many other countries.

The countries of Europe and U.S.S.R. delayed adoption of any television system till as late as 1967 because of their search for improved and compatible versions with their existing monochrome standards. An approach that later came to be known as SECAM was proposed by Herri de France of Campagnic de Television of Paris. Another colour system that received wide acceptance and was perfected almost at the same time as the French SECAM is PAL. It was developed by Walter Bruch of the Telefuncken Company of Germany.

Most of the basic techniques of SECAM and PAL are the same as in NTSC. For example, the use of wideband luminance and relatively narrowband chrominance is involved in all the systems. Similarly the concept of providing interleaving and off-setting for reducing dot visibility due to colour subcarrier is followed in each of the three systems. In addition, both the pick-up and development of colour-difference signals at the transmitting end and colour reproduction on the picture tube screen are nearly the same in NTSC, PAL and SECAM.

8.2 NTSC, PAL AND SECAM—OVERVIEW

As explained in chapter 6, the NTSC system is compatible with the American 525 line, 60 field monochrome system. In this, besides frequency interleaving to accommodate chrominance information, the colour-difference signals are modified to become 'Q' and 'I' for improved transmission characteristics.

In the PAL (Phase Alternation by Line) system the emphasis is on overcoming differential phase error that occurs in the NTSC system unless a high order of phase and amplitude integrity (skew-symmetry) of the transmission path characteristics around the subcarrier is maintained. The cancellation of such a phase error is accomplished in the PAL system by shifting phase of one of the colour signal components by 180° on alternate lines and later averaging it at the receiving end. If averaging is left to the observer's eye, as originally conceived in the system called 'SIMPLE PAL', a line flicker effect is noticed. However, this problem was soon circumvented by using a 1(one)-H delay line in the receiver where averaging is done before feeding colour signals to the picture tube. This technique gave PAL the designation of (PAL-DELAY) or PAL-D which is now the standard practice.

Public broadcast of PAL-D, as it is commonly called, began in Germany and United Kingdom in 1967 using two slightly different variants or standards. Later it was adopted by many European, Middle-East and Far-East countries. India too chose the PAL system recognising its superiority and compatibility with the existing 625 line, 50 field CCIR monochrome system.

The authors of SECAM system argued that if colour information could be relatively band-limited in the horizontal direction (as in the NTSC and PAL) it could also be band-limited in the vertical direction. Thus in SECAM, the two colour-difference signals are transmitted sequentially on alternate lines after frequency modulation with corresponding colour subcarriers. This technique avoids any possibility of unwanted crosstalk between colour signal components. A one line memory, commonly known as a 1-H delay element, is employed in the receiver to store one line signal to then become concurrent with the following line as received because of sequential transmission. This requires a special identification signal and is provided at the beginning of each line.

It is the sequential approach of colour signal transmission which gave the system the designation SECAM (Sequential Colour Avec Memoire; for sequential colour with memory). It was officially adopted by France and USSR where its broadcast began in 1967. Later many East-European and some other countries adopted it.

Basic Communication Channel

It is obvious from the above description of the development of three colour systems that the basic communication channel principle is the same in NTSC, PAL and SECAM. This is illustrated in Fig. 8.1 which is suitably labelled. In all the three systems, the same R, G and B pick-up devices are used where the basic camera function is to analyse spectral distribution of light from the scene in terms of red, green and blue components on a point-by-point basis as determined by the scanning rates. The three resulting signals are then transmitted over a band-limited communication channel to control the three-colour display device to make the perceived colour at the receiver appear exactly the same as that perceived by the eye on the scene.

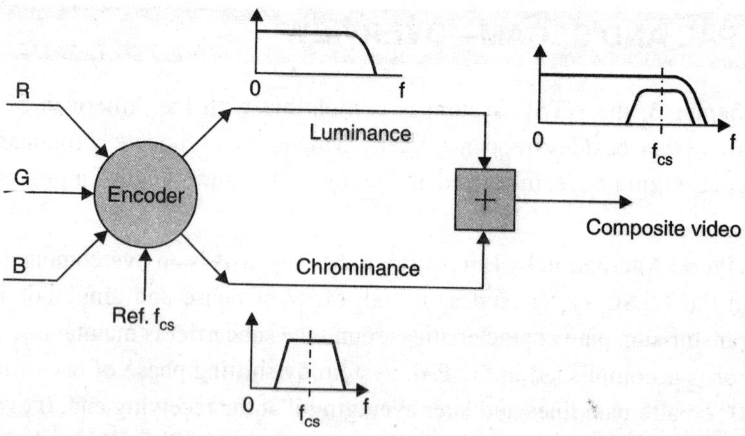

Fig. 8.1 Basic communication channel principle applied to colour television.

8.3 SIGNIFICANT FEATURES OF THE THREE COLOUR SYSTEMS

In order to fully appreciate different colour TV standards, significant features that distinguish NTSC, PAL and SECAM from each other are reviewed.

NTSC colour system. In this system the chrominance information is carried out as simultaneous amplitude and phase modulation of the colour subcarrier that is chosen to be in the high frequency region of the luminance 0 – 4.2 MHz video band. The subcarrier is specifically related to the scanning rates as an odd multiple of one-half of the horizontal rate. The colour-difference signals are rotated by 33° to become Q and I signals. This choice of Q and I colour modulation components relates to variation of colour registration characteristics of human colour vision as a function of the field view and spatial dimensions of objects in the scene.

At the encoder the Q signal component is band-limited to about 0.6 MHz while the I signal is allowed a bandwidth of about 1.3 MHz. The chroma signal expression within the frequency band common to both I and Q is given by

$$\text{NTSC} = \frac{(B-Y)}{2.03} \sin \omega_{sc} t + \frac{(R-Y)}{1.14} \cos \omega_{sc} t$$

At the receiver, quadrature synchronous detection is used to identify colour signal components. The identity of the chrominance information is maintained between the encoding and decoding process by transmitting a reference 'burst' signal consisting of nearly ten cycles of the subcarrier frequency at a specific phase (– (B–Y)) following each horizontal synchronising pulse. The choice of subcarrier at 3.579545 MHz is to reduce its visibility by its precise interlacing both with the luminance sidebands and beat frequency (about 920 KHz) occurring between it and average value of sound carrier. This necessitates shifting of line frequency to 15.734 KHz from 15.750 KHz and field rate to 59.94 Hz from 60 Hz during colour transmission and receptions.

PAL colour system. In the encoding of PAL the phase of 'V', *i.e.,* (R–Y) signal is reversed by 180° from line-to line to cancel certain colour errors that result from amplitude and phase distortion of the colour modulation sidebands caused by transmission path problems. The 'V' components is chosen for the reversal process because it has a lower gain factor than 'U' and therefore, is less susceptible to switching rate ($\frac{1}{2}f_H$) imbalances. In different PAL standards the bandwidth allotted to U and V modulation components varies because of the use of different luminance bandwidths and sound carrier frequencies. However, the maximum allowed bandwidth is about ± 1.3 MHz at 3dB. On encoding the PAL colour signal expression is :—

$$C_{PAL} = \frac{U}{2.03} \sin \omega_{sc} t \pm \frac{V}{1.14} \cos \omega_{sc} t \quad \text{where U and} \pm \text{V}$$

have been substituted for (B–Y) and (R–Y) respectively.

A 1-H delay line technique of averaging at the decoder produces cancellation of phase (hue) errors thus producing correct hue at all instants of scanning. As obvious, such a cancellation needs corresponding switching of the V signal at the decoder. In order to identify V signal switching sequence in the receiver, a swinging burst signal is transmitted during the back porch intervals. The swinging burst alternates the phase of subcarrier burst by ± 45° at the line rate. Since the burst is constituted from a fixed value of U phase and a switched value of V phase, the necessary switching 'sense' or identification information is available in it.

As in NTSC, the luminance components are also spaced at f_H intervals in the PAL system due to horizontal sampling process. Since the V components are switched symmetrically at half-the-line rate, only odd harmonics of V exist with the result that these are spaced at intervals of f_H. Thus, the V components are spaced at half-line intervals from the U components which in turn have f_H spacing intervals due to blanking. If half-line off-set were used as in NTSC (see Fig. 8.2 (*a*)) the U components would be perfectly interleaved but the V components would coincide with the Y (luminance) sidebands and thus not be interleaved, creating vertical stationary dot patterns. For this reason, in PAL, a ¼ line off-set for the subcarrier frequency is used as shown in Fig. 8.2 (*b*) and then f_{sc} become equal to 1135/4 f_H + ½ f_V. Now eight complete field scans must occur before a specific picture element 'dot' position is repeated.

As in all colour systems, the burst signal is eliminated during the vertical synchronising pulse period. Because in the case of PAL, the swinging burst phase is alternating line-by-line, some means must be provided for ensuring that the phase is the same for the first burst following vertical sync on a field-by-field basis. Therefore, the burst reinsertion time is shifted by on line at the vertical field rate by a pulse referred to as the 'meander' gate. This aspect is discussed fully in section 5.9 under the heading "colour burst gating" where Fig. 5.10 (*b*) shows corresponding 'meander' burst blanking gate timing diagram.

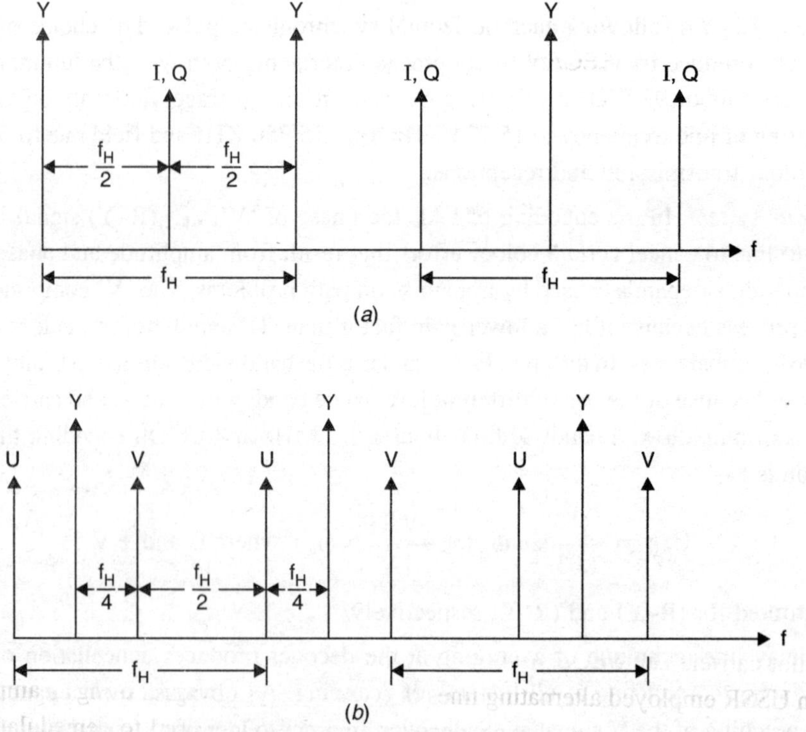

Fig. 8.2 NTSC and PAL frequency interlace relationship : (*a*) NTSC : ½ H off-set (4 field picture completion) (*b*) PAL : ¼ and ¾ H off-set (8 field picture completion).

SECAM colour system. As stated earlier the 'optimized' SECAM system called SECAM III is the system adopted by France and USSR. The manner in which colour information is modulated onto the subcarriers in this system is quite different from NTSC and PAL. Here (R–Y) and (B–Y) signals are transmitted alternately in time sequence from one successive line to the next — the luminance signal being common to every line. Since the total number of lines is odd, any given line carries (R–Y) information on one field and (B–Y) information on the next field. Also, the (R–Y) and (B–Y) colour information is conveyed by frequency-modulation of different subcarriers. Thus, at the decoder, a 1-H delay element, switched in time synchronization with the line switching process at the encoder is required in order to have simultaneous existance of the (B–Y) and (R–Y) signals in a linear matrix to form the (G–Y) component.

The (R–Y) signal is designated as D_R and (B–Y) as D_B. The centre frequency for the two subcarriers, respectively is determined by—

$$f_{sc}\ D_B = 272\ f_H = 4.250000\ \text{MHz}$$

$$f_{sc}\ D_R\ 282\ f_H = 4.406250\ \text{MHz}$$

These frequencies represent zero colour-difference information (zero output from the FM discriminator) for a neutral grey object in the television scene. The frequency deviation for D_R *i.e.*, ΔDR is ± 280 KHz and for D_B *i.e.*, ΔDB it is ± 230 KHz. The maximum allowable deviation, including pre-emphasis for DR = – 506 KHz and + 350 KHz while for D_B it is – 350 KHz and + 506 KHz.

As in PAL, the SECAM system must provide some means for identifying line switching sequence between the encoding and decoding processes. This is accomplished by introducing alternate D_R and D_B colour identifying signals for nine lines during the vertical blanking interval following equalizing pulses after vertical sync. The SECAM as practiced employs amplitude modulation of the sound carrier as opposed to FM modulation in other system.

8.4 STANDARDS AND SPECIFICATIONS OF COLOUR TELEVISION SYSTEMS

As stated earlier, it has not been possible to obtain total international agreement on 'universal' television broadcast standards. Even with the first scheduled broadcasting of monochrome television in 1936 in England, the actual telecasting started using two different systems on alternate days from the same transmitter. The Baird system was 250 lines (non-interlaced) with a 50 Hz frame rate, while the EMI (Electric and Musical Industries) system was 405 lines (interlaced) with a 25 Hz frame rate. These efforts were followed in 1939 in the United States by broadcasting a 441 line interlaced system at 60 fields per second (The Radio Manufacturers Association—RMA System).

Another two systems that were developed but never adopted in a big way are the ART (Additional Reference Transmission) and SECAM IV (NIR). The ART involved transmission of a continuous reference, pilot carrier. The SECAM IV or NIR (Russian National Institute for Research) which was developed in USSR employed alternating lines of (i) an NTSC like signal using an amplitude and phase modulated subcarrier and (ii) a reference signal having 'U' phase used to demodulate the NTSC like signal.

CCIR designations. The CCIR documents define recommended standards for worldwide colour television systems in terms of three basic colour approaches—NTSC, PAL and SECAM. The variations (atleast 13 of them)— are given alphabetical letter designations where some represent major differences while others signify only very minor frequency allocation differences in channel spacings or differences between VHF and UHF bands. The thirteen variations or subsystems are:– A, M, N, C, B, G, H, I, D, K, Kl, L and E. The systems A (405 lines), C (625 lines) and E (819 lines) are not recommended for new services. In early days the differences in power line frequency were considered as important factors and were largely responsible for the proliferation of different line rates versus field rates as well as a wide variety of video bandwidths. However, now the top priority is compatibility of any developing colour systems.

NTSC system characteristics. The most commonly used CCIR NTSC system is M (NTSC). It has been adopted by U.S.A., Japan, Canada, Mexico, the Philippines and several other Central American and Caribbean area countries. Another version of this is N (NTSC), which may be implemented either in the NTSC or PAL format. Many Latin American countries are adopting this system. The main characteristics of the two are given below in Table - I.

Table - I

CCIR Designation for NTSC System

S. No.	ITEMS	NTSC	
		M	N
1.	Channel bandwidth (MHz)	6	6
2.	Video (luminance) bandwidth (MHz)	4.2	4.2
3.	Frequency difference between video and audio (MHz)	+ 4.5	+ 4.5
4.	Vestigial sideband (MHz)	0.75	0.75
5.	Scanning Lines	525	625
6.	Line Frequency (Hz)	15734.26	15625
7.	Field Frequency (Hz)	59.94	50
8.	Video Modulation Method	AM (Negative)	AM (Negative)
9.	Audio Modulation Method	FM	FM
10.	Chrominance Subcarrier (f_{sc}) (MHz)	3.579545	—
11.	Chrominance Bandwidth—		
	(*i*) Q colour-difference signal (MHz)	± 0.6	—
	(*ii*) I colour-difference signal (MHz)	1.3 (Vestigial)	—
12.	Intermediate frequencies :		
	(*i*) Vision (MHz)	45.75 (58.75)	—
	(*ii*) Sound (MHz)	41.25 (54.25)	—

PAL system characteristics. The CCIR PAL colour designations that need mention are B, G, D, H, I and M. The B and G standards are most common in continental Europe, Middle-East and Asian countries. India also adopted 'B' (PAL) for transmission on VHF channels. The I (PAL) that is compatible to 8 MHz channel bandwidth monochrome standards is used in U.K., Hong Kong, Ireland and South Africa. China adopted D (PAL) because it is compatible to the country's black and white system.

Another version H (PAL) which is very similar to B and G standards is used in Belgium and Yugoslavia. A 525 line version of PAL is M (PAL) and adopted by Brazil. The main characteristics of PAL, B, D, G, and I, (that are widely used) are given below in Table - II.

Table - II
CCIR PAL Colour Designation

S. No.	Characteristics	PAL			
		B	D	G*	I
1.	Channel bandwidth (MHz)	7	8	8	8
2.	Luminance (video) bandwidth (MHz)	5	6	5	5.5
3.	Frequency difference between video and audio (MHz)	+ 5.5	+ 6.5	+ 5.5	+6.0
4.	Vestigial Sideband (MHz)	0.75	0.75	0.75	1.25
5.	Scanning Lines	625	625	625	625
6.	Line Frequency (Hz)	15625	15625	15625	15625
7.	Field Frequency (Hz)	50	50	50	50
8.	Video Modulation Method (Neg.)	AM (Neg.)	AM (Neg.)	AM (Neg.)	AM
9.	Audio Modulation Method	FM	FM	FM	FM
10.	Chrominance Subcarrier (f_{sc}) (MHz)	4.433618	4.433618	4.433618	4.433618
11.	Intermediate frequencies (MHz)				
	(*i*) Vision	38.9	37.0	38.9	38.9
	(*ii*) Sound	33.4	30.5	33.4	32.9
	(*iii*) Chroma	34.47	34.47	34.47	34.47

SECAM system characteristics. The SECAM system was initially developed to be compatible with the French 819 line, 50 field, 10 MHz channel bandwidth monochrome system. Its use was gradually withdrawn in favour of SECAM III and is not in use any more. Another SECAM system called SECAM IV or NIR that was developed in Russia did not go beyond the testing stage. Thus L (SECAM III) is now the most commonly used colour TV system that employs two subcarrier frequencies for colour signal transmission. Its main characteristics are given in Table - III.

*Note that the only difference between B(PAL) and G(PAL) is channel bandwidth. In G standards a gap of 1 MHz is provided between adjacent channels to prevent any mutual interference more so in the UHF range. Many countries like west Germany Netherlands, Norway, Sweeden, Switzerland, Union of Arab and Singapore use B(PAL) for channel allocations in the VHF band and G(PAL) for the UHF channels. India too has decided to opt for 'G' standards while allocating the UHF channels.

CHAPTER 8

Table - III

CCIR Designation for L (SECAM) *i.e.,* SECAM III

Base band–Luminance, Frequency Modulated Subcarriers, Line sequential D_R and D_B signals D_R-odd lines, D_B-even lines.			
(1)	Line/field	=	625/50
(2)	fh	=	15625 Hz
(3)	fv	=	50 Hz
(4)	Luminance bandwidth	=	6.0 MHz
(5)	Sound	=	6.5 MHz (AM)
(6)	f_{DR}	=	4.40625 MHz = 282 fH
(7)	f_{DB}	=	4.25000 MHz = 272 fH
(8)	Channel bandwidth	=	8 MHz

8.5 CCIR B (PAL) CHARACTERISTICS

The use of CCIR B (PAL) colour system is most common. As stated earlier G (PAL) is almost the same as B (PAL). A technical summary of B (PAL) is, therefore, given below in Table - IV.

Table - IV

Main Characteristics of B (PAL) Colour System

Frequency Range	—	VHF
RF Channel bandwidth	—	7 MHz
Video (luminance) bandwidth	—	5 MHz
Vision–sound carrier spacing	—	+ 5.5 MHz
Vestigial sideband	—	0.75 MHz
Spacing of vision carrier from nearest edge of channel	—	+ 1.25 MHz
Number of lines per picture	—	625
Line frequency (f_H)	—	15625 ± 0.016 Hz
Field Frequency (f_N)	—	50 Hz
Interlace ratio	—	2:1
Duration of line sync pulses	—	4.7 μs
Duration of line blanking pulses	—	12 μs
Back porch	—	5.8 μs
Field blanking period	—	20 lines
RF signal level	—	100%
RF blanking level	—	72%

(Contd.)

RF white level (residual carrier)	—	10%
Type of sound modulation	—	A5C (AM) negative
Type of vision modulation	—	F3 (FM)
Sound carrier relative to nearest edge of channel	—	0.25 MHz
Frequency deviation (FM) of sound-carrier)	—	± 50 KHz
Pre-emphasis (audio modulation)	—	50 μs
Vision/Sound power ratio	—	10:1 to 20:1
Luminance Signal	—	Y = 0.3R + 0.59G + 0.11B
Type of colour signal modulation	—	Suppressed carrier amplitude modulation of two subcarriers in quadrature.
Colour-difference (chrominance) signal	—	U = 0.493 (B–Y) V = 0.877 (R–Y)
Composite colour signal 'M'		
Amplitude of modulation	—	$Y + U \sin \omega_{sc}t \pm V \cos \omega_{sc}t$
Chrominance-sideband signals		$\sqrt{U^2 + V^2}$
Colour subcarrier frequency (f_{sc})	—	4433618.75 Hz ± 5 Hz
f_{sc} multiple of f_H	—	$1136 f_H/4 + f_V/2$
Chroma encoding	—	Phase and amplitude (QUAD) modulation (line-alternate)
Bandwidth of colour-difference (U and V) signals	—	f_{sc} + 0.6 MHz f_{sc} – 1.3 MHz
Colour switching IDENT	—	Swinging burst ± 45°
Duration of burst	—	10 ± 1 cycles
Phase of burst (U and ± V)	—	+ 135° for odd lines in 1st and second fields – 135° for even lines in first & second fields + 135° for even lines in 3rd and 4th fields, – 135° for odd lines in 3rd and 4th fields } Relative to + U axis
Additional Signals	—	Meander gate pulses ($f_{H/2}$)
Gamma	—	2.8
White	—	D 6500
Aspect Ratio	—	4:3

8.6 ADVANCES IN COLOUR RECEIVER TECHNOLOGY

Colour receiver technology had a breakthrough around 1975 when improved solid state devices and modern colour picture tubes became available. Almost at the same time special ICs were developed and this very much simplified receiver circuitry. The only vacuum tubes needed were those employed in vertical and horizontal circuits besides the high voltage rectifiers. In coming years special ICs also became available for the deflection circuits. In India colour receiver manufacture started quite late because colour transmission commenced only in 1982. However, taking advantage of the technological advances, receiver manufacturers in India opted for the latest receiver designs. The block diagram of a modern PAL-D colour receiver is shown in Fig. 8.3.

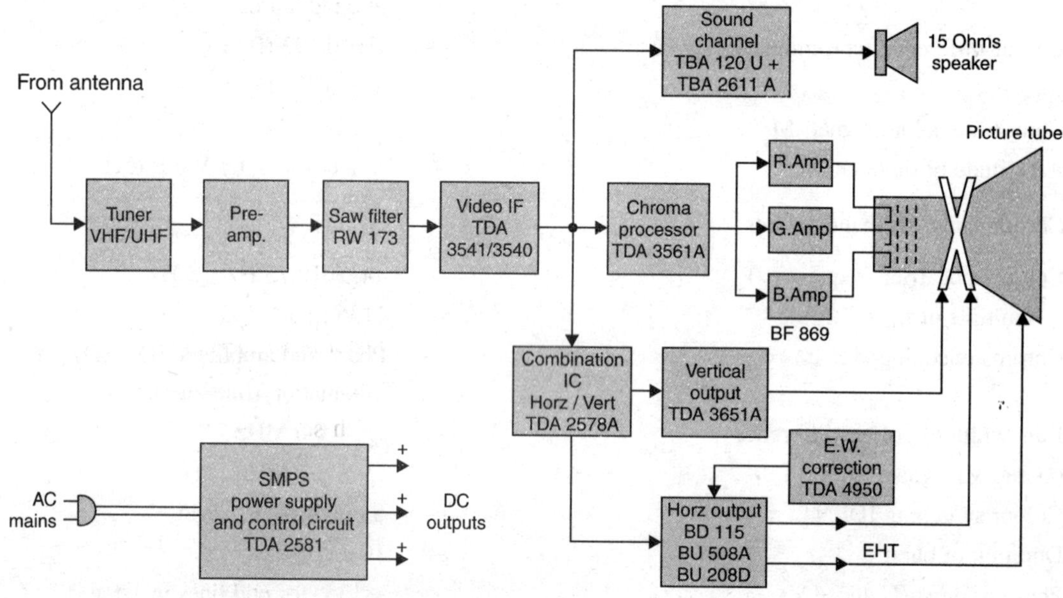

Fig. 8.3 Block diagram of a modern PAL-D colour receiver.

During the recent past, there has been considerable development in the area of channel selection and tuning, chroma signal decoding and matrixing, picture tube design and its peripheral circuitry. Frequency synthesized touch control tuning, is now common in colour receiver. Similarly, microprocessor based remote control and interface units are now easily available in module form for incorporating in TV receivers. In the chroma section, where earlier two or three ICs were necessary, a single chip integrated circuit is now available which performs the functions of colour signal decoding, matrixing and preamplification of R, G and B signals. In addition to all this, the design and control of Switch Mode Power Supplies (SMPS) has received due attention. Specially designed ICs are now available for accurate control and stabilization of dc sources for various sections of the receiver.

In latest colour picture tubes, dynamic convergence adiustments are no longer necessary. The P-I-L tubes do not need any N.S. correction and special ICs are available for E.W. correction. Small picture tubes designs up to a deflection angle of 90° have been so perfected that even E.W. pincushion correction circuitry is no longer necessary.

The latest arrival is the development of an Ic for the video output circuit that delivers enough R, G, B outputs to drive the picture tube. Sugh a receiver design employing SGS Ics is shown in Fig. 8.4.

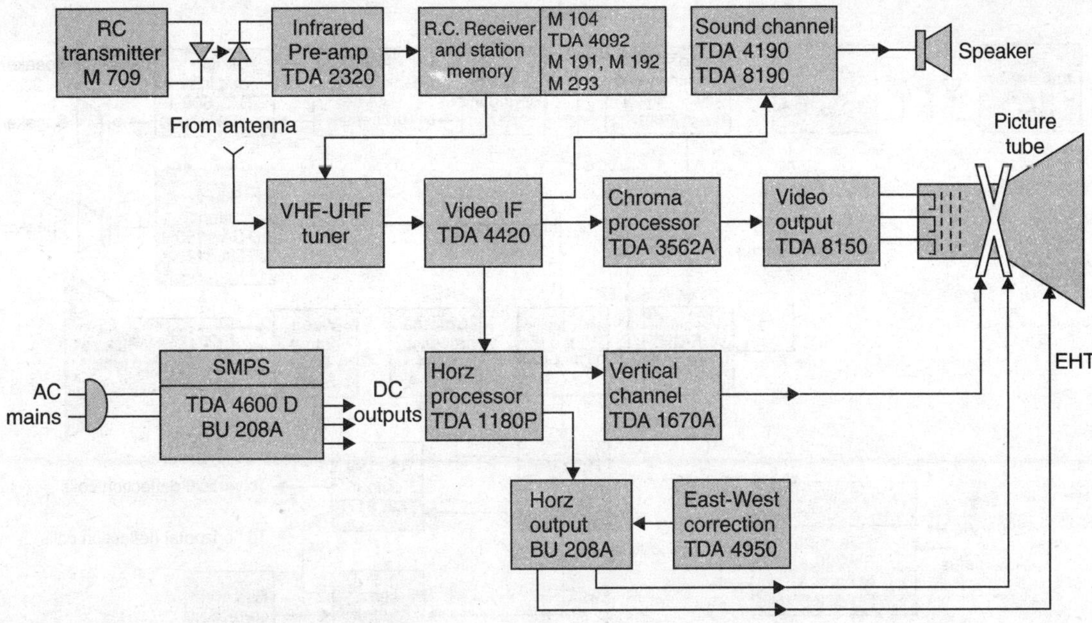

Fig. 8.4 Block diagram of a large screen colour receiver.

Stereo sound is yet another innovation that is finding its way in television transmission and reception. The block diagram of a high quality colour receiver with stereo sound option is shown in Fig. 8.5.

Most such and many other receiver designs have interfacing circuitry for videotext serivce. An external decoder enables viewing of paged data of various types of informations.

With a view to developing indigenous know-how, easy availability of components and a low-budget colour receiver, Bharat Electronics Limited (BEL), Bangalore, introduced in 1985 a simple colour receiver with full documentation for the benefit of small manufacturers. Its block diagram is shown in Fig. 8.6. As necessary, it conforms to PAL-B CCIR standards.

Sequencing of Circuit Description

As is obvious from the above developmental notes on receiver technology, an imperative needs exists to cover the subject in such a way that the reader acquires full knowledge of what is latest in receiver design and manufacture. It is equally important that the needs of those who are new to the subject are not ignored. Keeping these factors in view, first six chapters are devoted to basics of colour television and colour picture tubes. The following chapters 7 and 8 are designed to give details of the functioning of various sections of the receiver and colour TV standards. From chapter 9 onwards discrete and integrated circuits as employed in various sections of modern colour receivers are described. In chapter 18, a complete colour receiver circuit is examined and methods of alignment and various measurements explained.

With the facilities now available on INTERNET and MOBILE phones, the topics covered in chapter 19 do not appear to be very useful. As such this chapter has been replaced with a new chapter titled "Advances in Colour Receivers and Television Systems". Similarly, chapter 20 has been rewritten to explain the superior digital means of transmission and reception via satellites. Distribution of TV signals to subscribers by cables and other means is also explained in this chapter.

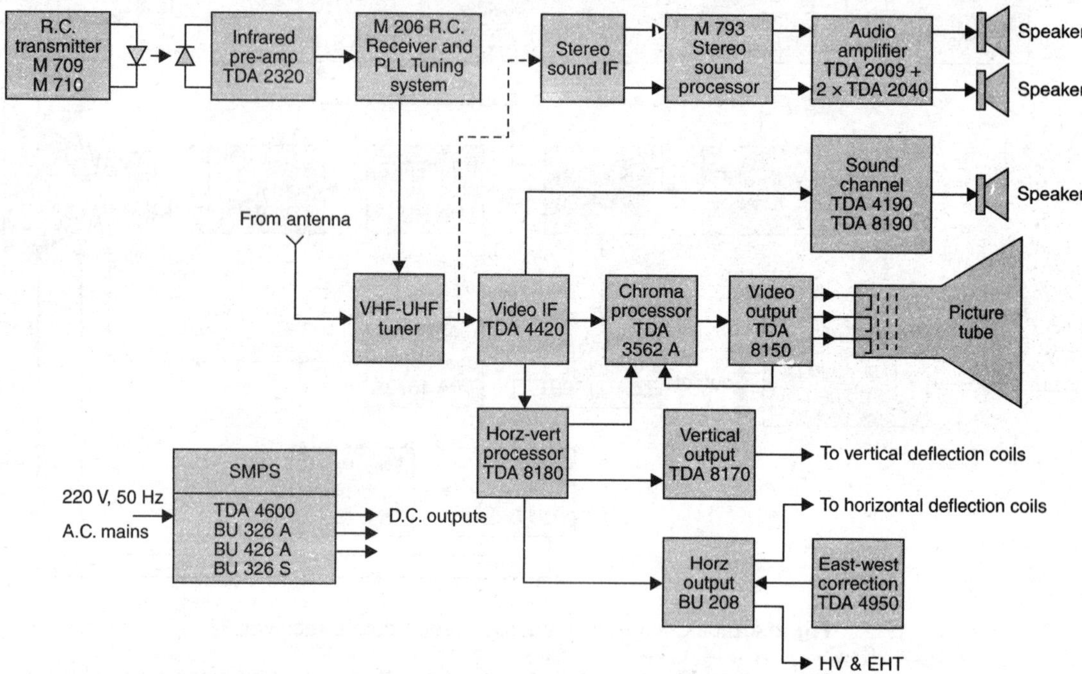

Fig. 8.5 Block diagram of a high quality colour receiver with stereo sound reception facility.

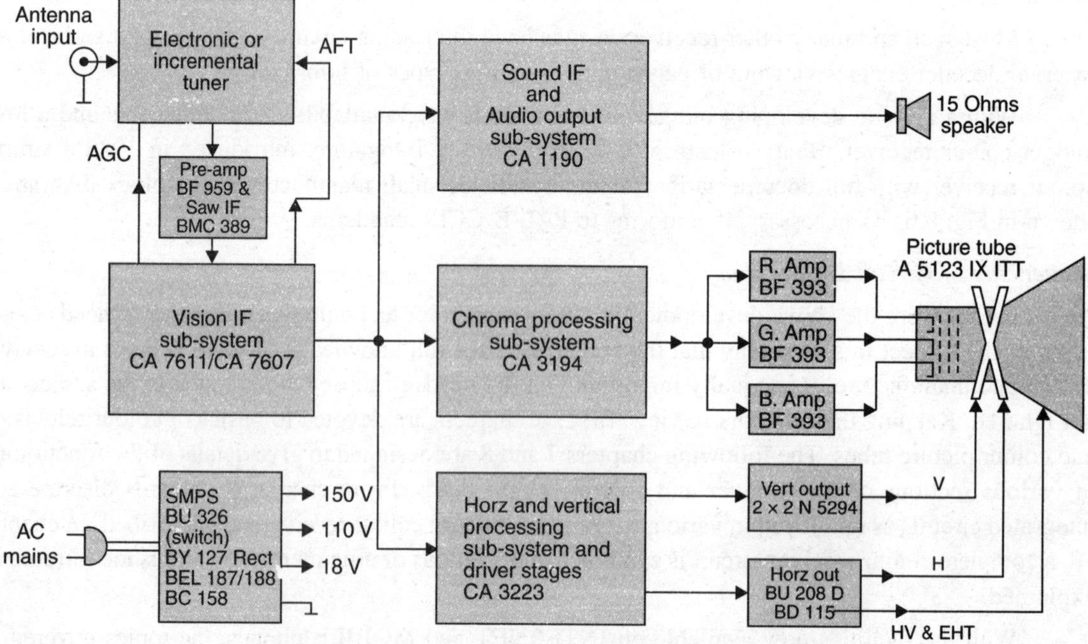

Fig. 8.6 Block diagram of a simple colour receiver.

REVIEW QUESTIONS

1. Explain briefly why it has not been possible to evolve universally accepted TV standards.
2. Describe the similarities and difference between NTS, PAL and SECAM.
3. List the main characteristics of M(NTSC) colour system. How is N(NTSC) different from this?
4. Compare the characteristics that distinguish PAL, B, G and I systems from each other. Explain why India chose PAL-B, CCIR standards for transmission in the VHF range.
5. Which is the most used SECAM colour system? Give its main characteristics and features. Why the 819 line French system has of late been discontinued?
6. Draw block diagram of a simple colour receiver and label all the sections giving IC numbers where necessary.
7. Draw block diagram of a modern high quality colour receiver having provision for remote control and stereo sound reception.

9

TV RECEIVER TUNERS

INTRODUCTION

The tuner or 'Front End' is a combination of VHF and UHF tuners with a common antenna input point. It selects desired station and rejects others, provides impedance match between antenna and receiver input, amplifies the selected signal to improve signal to noise ratio, converts received RF signal to IF signal as required in superheterodyne receivers and has an output circuit for proper impedance match to the IF sub-system.

Modern tuners are of electronic type and have no moving parts. Tuning is accomplished by varying dc voltage across varactor diodes and band selection by forward and reverse biasing switching diodes. In sophisticated designs digital techniques under control of a microcomputer are combined with varactor tuning for automatic selection and remote control operation. Such tuners do not need any fine tuning. The digital address system is mounted close to the tuner assembly on the same module for convenience of interconnections. The VHF and UHF tuners are electrically interconnected and operated as a composite unit. It is pretested at the manufacturing point and it is only necessary to make connections to its terminal strip after mounting it in position.

While more details of the tuner unit are shown in Fig. 7.1, its simplified block schematic is drawn in Fig. 9.1 for understanding the functions as performed by a television tuner.

The main functions assigned to a TV receiver tuner are:

(*i*) it selects desired stations and rejects others,

(*ii*) it converts all incoming channel frequencies to a common IF band,

(*iii*) it improves image rejection ratio and thus reduces interference from sources operating at frequencies close to the image frequencies of desired channels,

(*iv*) it prevents spurious pick-up from sources which operate in the IF band of the receiver and,

(*v*) it rejects any pick-up from stations operating in the FM band.

In the VHF tuner section (see Fig. 9.1), the function of channel selection is accomplished by simultaneously adjusting tuned circuits of all the three stages that form part of the tuner. The IF output

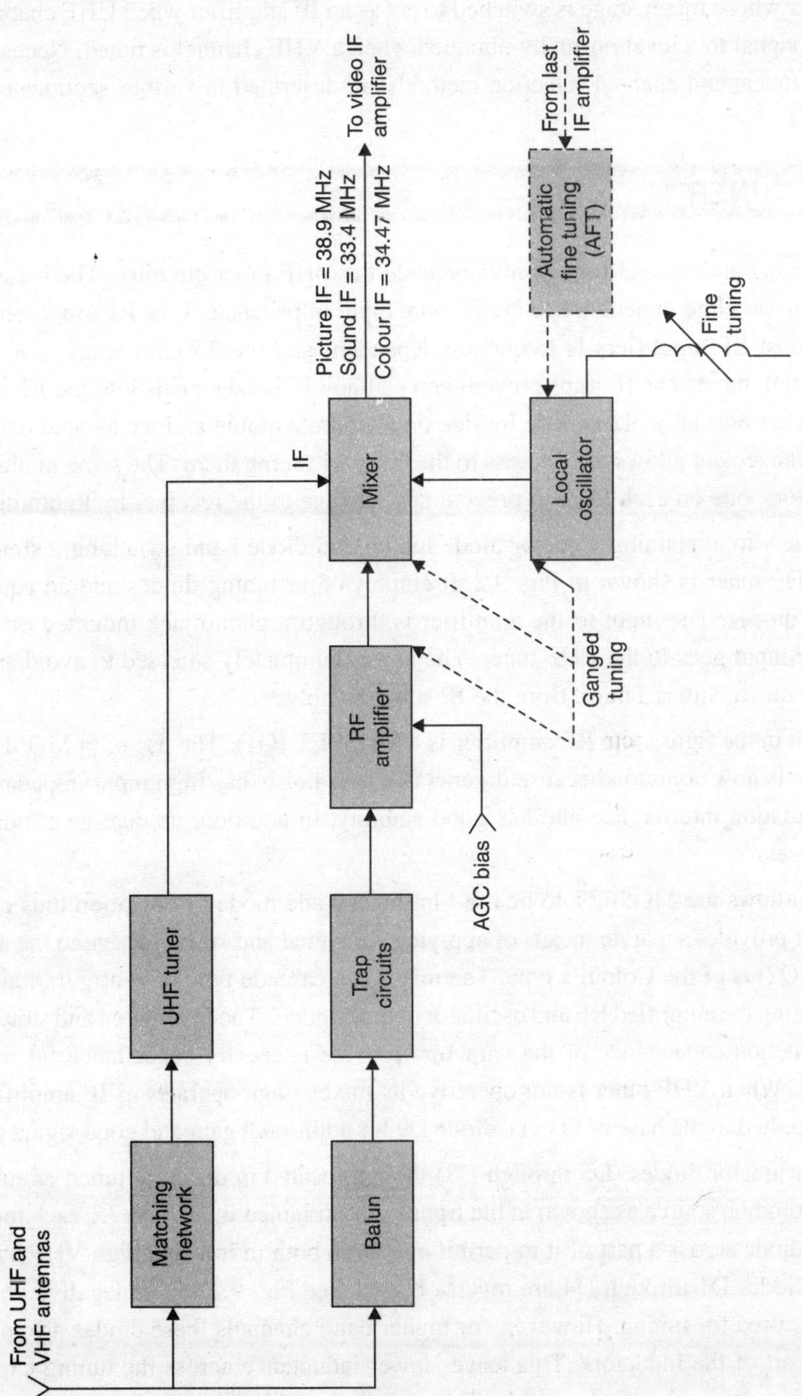

Fig. 9.1 Simplified block diagram of a TV tuner.

from the accompanying UHF tuner is low because of weak input from its antenna. It is therefore coupled to the VHF tuner whose mixer stage is switched to act as an IF amplifier when UHF channels are tuned. This boosts the signal to a level normally obtained when a VHF channel is tuned. Necessary details of VHF and UHF tuners and channel selection methods are described in various sections of this chapter.

9.1 VHF TUNER

A balun transformer and several trap circuits precede the VHF tuner circuitry. The balun matches the 300-ohm transmission line impedance to the 75-ohm input impedance of the RF amplifier. In fact input impedance of most RF amplifiers is frequency dependent and the 75-ohm value is a representative figure for mid-VHF band. The IF traps prevent entry of any IF band signals into the RF amplifier. The baluns and traps are normally placed side by side on a separate mount and are located outside the tuner unit. Such an arrangement allows easy access to the traps for tuning them. The same mount also carriers a pair of capacitors, one on each lead, to prevent any damage to the receiver by lightning.

With a view to explaining varactor diode tuning and diode band switching a simplified circuit diagram of a VHF tuner is shown in Fig. 9.2. It employs four tuning diodes and an equal number of band switching diodes. The input to the amplifier is through a phono-jack mounted on the shielding cover. The same input goes to the UHF tuner. The unit is completely shielded to avoid any direct pick up, in particular due to any radiation from the IF sub-assembly.

As shown in the figure, the RF amplifier is a MOSFET (Q1). The use of a MOSFET in the RF amplifier section is now common because it generates less noise, has high input impedance, produces little cross-modulation interference and has good stability. In addition, its dual-gate input serves two important purposes.

Firstly it allows the MOSFET to be used in the cascade mode of operation thus enabling more gain. Secondly it provides separate means of applying RF signal and AGC voltage to the amplifier. The local oscillator (Q2) is of the Colpitt's type. The mixer is a cascade type of configuration (Q3, Q4) for ease of heterodyning the amplified RF and oscillator output signals. The distributed and stray capacitances together with junction capacitance of the varactors provide necessary capacitance for tuning various resonant circuits. When VHF tuner is not operative its mixer stage operates as IF amplifier. The UHF tuner output is applied to the base of Q3 via diode D9 for additional gain and good signal to noise ratio.

The four varactor diodes (D5 through D8) though located in different tuned circuits are fed dc voltage from a common source as shown in the figure. As explained in Section 7.1 each tuning inductor has a switching diode across a part of it to permit operation both in low and high VHF bands. For low band operation diodes D1 through D4 are reverse biased (see Fig. 9.2) by a negative voltage and the entire inductor is used for tuning. However, for higher band channels these diodes are forward biased thus shorting a part of the inductors. This leaves lower inductance across the tuning capacitance and higher resonant frequencies become possible. Thus, in all-electronic tuners there is no need for variable capacitors for tuning and mechanical switches to change coils for different channels.

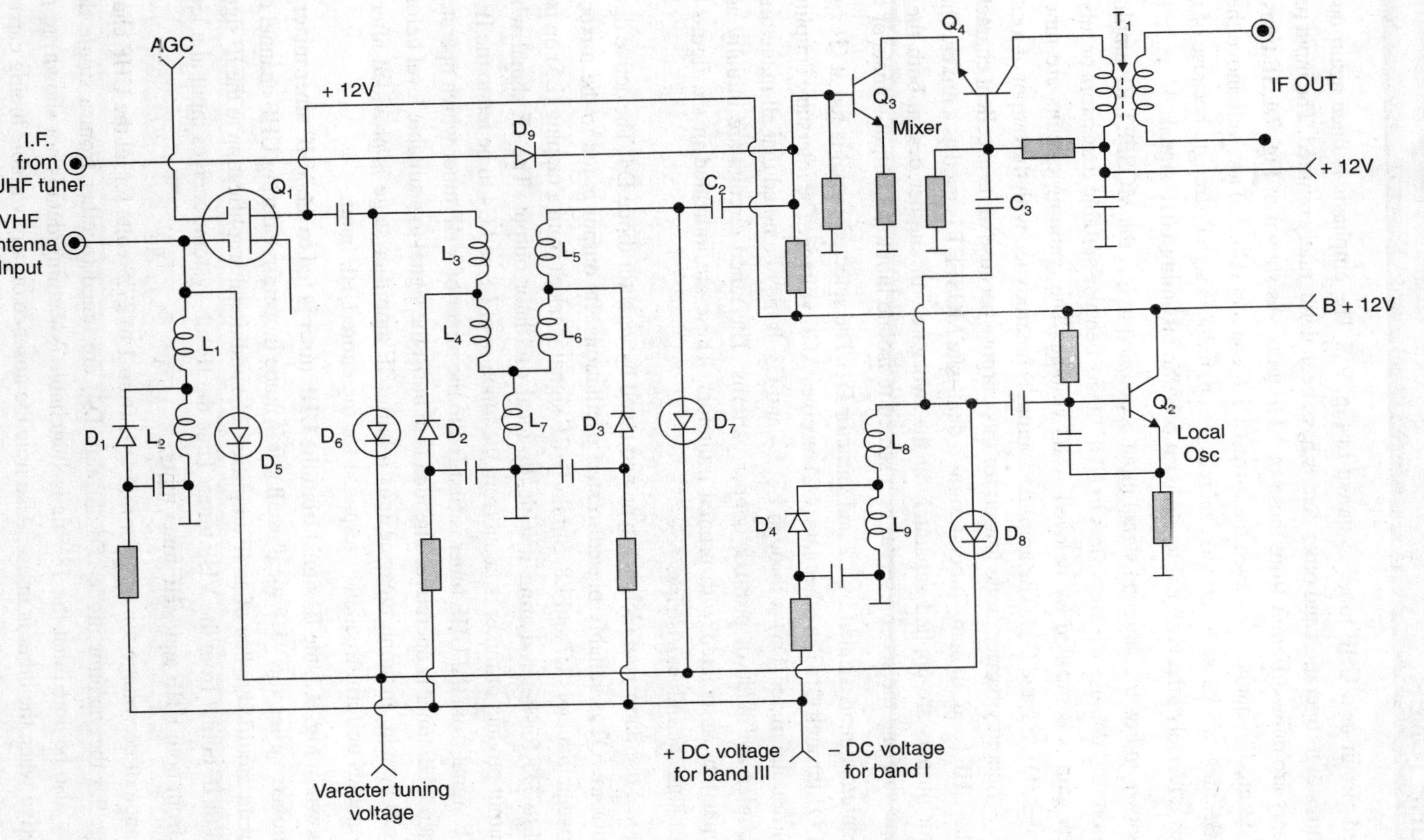

Fig. 9.2 Simplified circuit of a VHF tuner.

CHAPTER 9

9.2 UHF TUNER

A simplified circuit of a UHF tuner is shown in Fig. 9.3. The emphasis is once again on explaining special features of its operation and other details have been deliberately omitted. The input requirements of UHF tuners are quite different from those of VHF units. As shown in Fig. 7.1, IF traps are almost never used because frequencies of even the lowest UHF channel (Ch. 13) are well above the highest IF frequency. Besides, any passive network, such as a trap, reduces signal strength because of its insertion loss and thus adversely affects signal to noise ratio of the incoming UHF signal.

As shown in Fig. 9.3, antenna signal input is coupled to Q1, the MOSFET RF amplifier via L1. Earlier versions did not use RF amplifiers in UHF tuners because suitable transistors or tubes were not available. Its gain is controlled by delayed AGC voltage. The coupling circuits are tuned by three varactor diodes (D1, D2 and D3). The amplifier output is heterodyned with the output of local oscillator in diode D5. A Schottky barrier diode is used for this purpose because of its excellent characteristics as a mixer in the UHF band. In many tuner designs a dual-gate MOSFET is used as a mixer. The amplifier and oscillator outputs are applied separately at the two gates. In another design both the inputs are applied at one gate and the second serves as a screen to isolate input and output signals of the mixer.

The antenna input a tuned by L2 and varactor D1. The selected signal is fed at G1 (gate No. 1) of MOSFET Q1 through C1. The other gate (G2) receives AGC voltage. The interstage coupling between the RF amplifier and mixer (D5) is through L3, L4 and C3. It may be noted that all inductors shown as striped rectangles are made of printed copper patterns. The tuned circuits are initially adjusted by moving copper tabs associated with printed inductors. These are indicated in the figure by curved arrow heads along the inductor strips.

The local oscillator is of Colpitt's type and tuned by varactor diode D4. The active device of the local oscillator *i.e.* Q2 is suitably biased to start oscillations. Its output is fed to the mixer diode via coupling between inductors L7 and L8. Similarly RF signal is taken from a tapping (L5) on inductor L4 and coupled to D5 for heterodyning it with the local oscillator output. The IF signal which is the difference output product of mixer is taken from the junction of L8 and C4 to be fed to the IF amplifier. In fact, the IF signal from the UHF tuner is first fed to the mixer of VHF tuner which operates as an IF amplifier when UHF tuner is operative. The idea is to boost the signal to a suitable level before feeding it to the IF sub-system. In some tuner designs a buffer IF amplifier stage is provided after the UHF mixer. This enables uniform frequency response for the entire UHF band.

As shown in Fig. 9.2, the IF signal from the UHF tuner is fed to the VHF tuner mixer via diode D9. As mentioned earlier, the VHF local oscillator is made inoperative during UHF channel receptions and mixer circuit modified to that of an IF amplifier. The additional amplification to the IF signal brings it to the level of IF output from the VHF tuner. Thus the IF sub-system receives input at a level that is nearly same from both UHF and VHF tuner units.

The range of dc voltage for varactor tuning is from 1 to 27.5 volts for all the UHF channels. As shown in Fig. 9.3 the varactor diode (D1 through D4) are tuned together from a single dc voltage source. It may also be noted that the 18V dc is switchable (swicthing details not shown) to make one tuner inoperative when the other is in use. Also note the absence of any variable tuning capacitors. As stated earlier initial tuning is done by preset trimmers (copper tabs) and varactors supply additional capacitance necessary for correct tuning. The varactor diodes are supplied in matched set of four proper tracking correlation. When any one needs replacement a new set is obtained and all the four replaced.

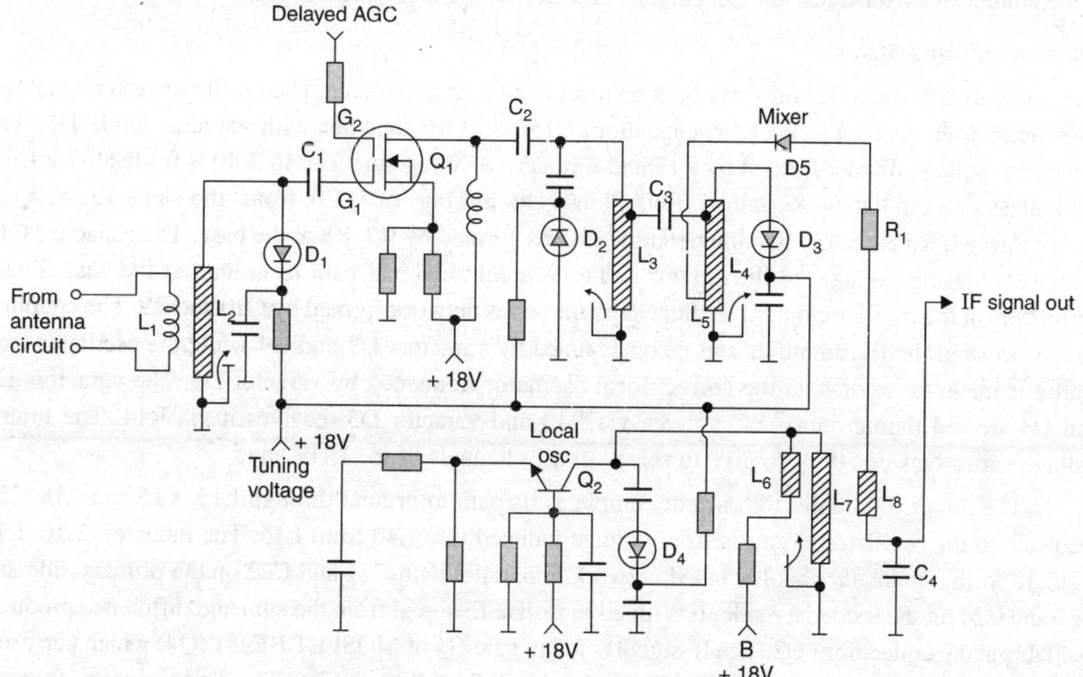

Fig. 9.3 Simplified circuit of a typical UHF tuner.

9.3 UHF-VHF TUNER CIRCUIT

After gaining familiarity with tuner operation a detailed study of a UHF-VHF tuner circuit is desirable. Such a circuit is shown in Fig. 9.4 (fold-out) and each tuner described separately.

UHF tuner. It employs PNP transistors BF679 and BF681 in the RF amplifier and mixer cum local oscillator sections respectively. The transistor BF679 (Q1) that operates as RF amplifier is connected in the grounded base configuration to provide uniform gain over the entire UHF band. The emitter to base junction of Q1 is biased from the + 12.35V dc source via, D1, R2, R1 and R3. The delayed AGC control voltage is fed at its base through R4. The two voltages combine to provide necessary forward bias and vary it depending on the magnitude of AGC voltage. The capacitor C5 is for effective ac ground at the base of Q1 and C6 serves as ac bypass across R1.

The input signal to the tuner passes through a dual high-pass filter consisting of C1, L3, C2 and C3, L4 and C4 before feeding at the emitter of Q1. The filter network prevent passage of any VHF channel signals and provides impedance match between antenna and emitter-base input of BF679. The collector circuit of Q1 is a wideband coupling circuit consisting of L7, L8, C7 and varactor D3 on the primary side and L9, L10, C9 and varactor D4 on the secondary side. The capacitor C14 is for mutual coupling. The collector current flows through L5 and L6 to complete its path to ground. The ac output voltage that develops across L6 is fed to the coupling network via C8. The capacitors C10 and C11 provide ac path to it to ground at the junction of L8 and L10. The amplified RF output signal is injected at the emitter of mixer transistor Q2 via C12 and L11 which is a winding on the coupling transformer.

Self Heterodyning Mixer

The transistor BF 681 (Q2) operates both as local oscillator and mixer. The oscillator resonant circuit consists of inductors L13 and L14, capacitors C15 and C16 together with varactor diode D5. The capacitive voltage divider formed by C17 and C18 across voltage induced in L40 is for feedback from the collector to emitter for sustained oscillations. The biasing of Q2 is from the same 12.35. V dc source through R6 at the emitter and potential divider formed by R7, R8 at the base. The capacitor C13 decouples biasing voltage to the emitter and C19 is for ac signal path to ground at the base. Phase correction of the oscillator output is affected by the series network formed by C20 and R9. The coupling circuit between the RF amplifier and mixer is tuned by varactors D3 and D4 while the oscillator tank circuit is made to resonate at the correct local oscillator frequency by varactor D5. The varactors D3 and D4 are fed tuning cum AFC voltage via R13 and varactor D5 receives it via R14. The tuning voltage varies between 0.5V to 30V to select all the channels of the UHF band.

The mixer (Q2) collector current completes its path to ground through L15, L16 and L18. The feedback to the oscillator circuit is from voltage induced into L40 from L15. The inductors L16, L17 and L18 form a mutually coupled tuned network in parallel with C21 and C22 on the primary side and C23 and C24 on the secondary side. It is tuned to isolate IF signal from the sum and difference products available at the collector of Q2. The IF signal is fed to gate G1 of MOSFET BF981 (Q4) which performs as mixer in the VHF tuner. The IF signal path is through R10, R39 and D6. The diode remains forward biased from the UHF B + source via L32 and R11. However, when the UHF tuner is inoperate there is no B + supply and D6 provides isolation between the two tuners. As explained earlier the VHF mixer operates as an IF amplifier during reception of UHF channels.

VHF tuner. The VHF tuner employs a MOSFET both in the RF amplifier and mixer stages. The use of MOSFET in the input stage (RF amplifier) improves cross-modulation rejection between stations transmitting on different bands or between stations transmitting on different channels within the same band. By using a FET in the mixer stage also, the desired characteristics are further improved. In addition, the signal acceptance and cross-modulation rejection of adjacent stations is greatly improved. The input stage works on a linear part whereas the mixer stage is operated on the non-linear (square law) part of the characteristic curve. This yields good results from both the stages.

RF amplifier. The antenna input circuit has two anti-parallel connected lightning protection diodes D20 and D2 to prevent any static discharge in the antenna circuit from damaging the MOSFET in the RF amplifier. Another very effective protection against destruction of the FET are the two zener diodes (formed within the integrated circuit of the device) connected from the two gates to source. To obtain best possible impedance match the antenna feeds at a trapping in the input circuit coil consisting of L19, L20 and L21. The high point of the inductor is connected to gate 1 (G1) of BF961 via capacitor C28 and resistor R19. Another inductor consisting of L22 and L23 is coupled to the input coils. This combined inductive-capacitive coupling of antenna to Q3 results in an almost constant bandwidth for the RF amplifier throughout the tuning range. The other gate (G2) receives AGC voltage via R31 from the same source as for pre-amplifier of the UHF tuner.

The drain of BF961 receives dc supply via coils L24, L25 and L26 from a dc source of 12.35V. The same source after the decoupling network L27, C29 and potential divider R20, R21 feeds dc biasing voltage at G1 of this FET. This dc source is the same as for the UHF tuner and is switched to VHF tuner when any lower channel is tuned. The diodes D7 and D8 are for changing over from lower VHF to higher VHF band channels. The tuning varactor D9 is for tuning any channel at the input circuit of the amplifier. Similarly output circuit of the amplifier has diodes D10 and D11 for switching bands and varactor D12 for effective tuning.

The band switching and tuning is done in a somewhat unconventional way. Both in the input and output circuits the switching diodes are fed with 8.5V for lower VHF channels and 12.35V for higher VHF channels through resistors R22 through R25. The ensuing damping effect across the inductors due to different forward biasing in different in each case and accounts for the desired change in inductance value. The tuning varactor D9 receives the same dc voltage through R22, D7 and L20 for providing broad tuning. The AFC voltage is connected to the source of Q3 through R26, R27 and associated network. The varactor D12 in the output circuit is fed AFC voltage via resistor R28 for tuning. The combination of various tuned circuit makes it possible to attain an image frequency rejection of 80 db in band I and 60 db in band III.

Local Oscillator

As shown in Fig. 9.4 the transistor BF939 (Q5) functions as an oscillator in the grounded base arrangement. Feedback from output to input side for obtaining oscillations is through internal transistor capacitances. However, at relatively lower RF frequencies the feedback is through an external network. To ensure approximately equal oscillator output amplitude in both the VHF bands, a higher current is permitted in the transistor circuit for band III channels. The base of Q5 receives different dc voltages at its base (12.35 V) for band I and 8.5 V for band III) through diode D18 and resistive network R56 and R57. The emitter is fed from the 12.35V source via R58 and it is disconnected to put the oscillator off when any UHF channel is tuned. The AFC voltage is applied to varactor D16 via R52 for controlling the oscillator frequency. The output of the local oscillator feeds at gate 2 (G2) of the mixer FET. The dc voltages to this gate and source of Q4 are fed from 12.5V dc source via resistors R42 and R41 respectively.

Mixer

As explained above amplified RF signal feeds at G1 and oscillator output at G2 of the mixer MOSFET BF981. Since it is operated in the non-linear region, heterodyning takes place between the two inputs and sum and difference products become available in the drain circuit. The required band *i.e.*, IF signal is separated by a tuned circuit (L38, L39 and C64) and fed at the input of IF sub-system. In fact, it feeds into a pre-IF amplifier (SL 1430 in Fig. 9.4) the output of which goes to the surface acoustic wave filter and then to the IF amplifiers in the corresponding integrated circuit. It may be noted as explained earlier that the mixer functions as IF amplifier for the UHF tuner. This is made possible by switching necessary dc voltage to Q4 and other changes in its input tuning network.

CHAPTER 9

9.4 PROGRAM SELECTOR

In any location a limited number of VHF and UHF channels are available. This number seldom exceeds twelve. Thus it is only necessary to be able to preselect limited number of channels so that later on any one of the chosen stations can be instantly tuned-in just by depressing a selector button. In receivers that employ non-digital but varactor diode tuning, each preselect position consists of a band selector switch, a channel selector switch and a potentiometer. However, receivers that employ computer controlled frequency synthesis technique of tuning, do not need such preselector units. Here the channel tuning is automatic either through an auto-select switch or under commands from a manual selector.

In most countries only a few channels are available in any one location and hence, an 8-fold program selector is more than adequate. The schematic of such a selector is shown in Fig 9.5. As illustrated, the band selector switch routes dc supply to the tuner and channel selector to the varactors as set on the associated potentiometer. Each preselect channel selector switch or button has a LED display (not shown) to indicate its serial number. There are thus eight LEDs and one of these (at a time) lights up on depression of channel selector switch and illuminates serial number of the preselector.

Preselection of Programs

It is desirable to make a list of the channels and other programs to be preselected to indicate the sequence in which these are to be memorized by the preselector unit. On switching on the receiver, the first band selector switch is thrown on the position (Band I, Band III or Band UHF) to which the first listed channel number belongs. For example if Ch. 4 is first in the list, band selector switch No. 1 will be moved to the location marked "Band I". This is illustrated in Fig. 9.5. Then the button of No. 1 channel selector switch is depressed. On doing so the associated LED will light up to indicate '1' as the serial number of the preselector. As is obvious from the circuit the first pole of the channel selector switch connects necessary dc voltages to the VHF band I tuner circuitry. The second pole applies dc voltage from another highly stabilized dc source (≈12V) to the associated varactor diodes. As shown, this supply is routed via a potentiometer. For tuning-in the desired channel it is varied till the required program appears on the screen. Thus for Ch. 4 that is under discussion, the pot will be slowly varied from minimum towards maximum voltage (clockwise) till this station program is seen on the screen.

The above procedure is repeated for all subsequent channels indicated on the list. For example, if the next channel to be selected is number 7, the 2nd band selector switch is thrown to Band III position as illustrated in the figure. The switch of the 2nd channel selector switch is then depressed and associated pot varied to tune in the program of Ch. 7. Similarly if the 3rd choice in the sequence happens to be Ch. 21, the 3rd bandswitch (see Fig. 9.5) is moved to Band UHF position and channel selected on closing the 3rd channel selector switch (button) and manipulation of associated potentiometer. It may be noted that the LED display illuminates only one location at a time depending on which button number is depressed.

It may be emphasized again that dc supply to the tuner unit is fed separately for different bands. This is necessary because for all channels in Band I, switching diodes are not to be forward biased while it is necessary to do so for channels of Band III. Similarly for the UHF channels only the UHF tuner is to be made operative with a limited supply to the mixer to perform as IF amplifier for the IF signal obtained at the output of UHF tuner.

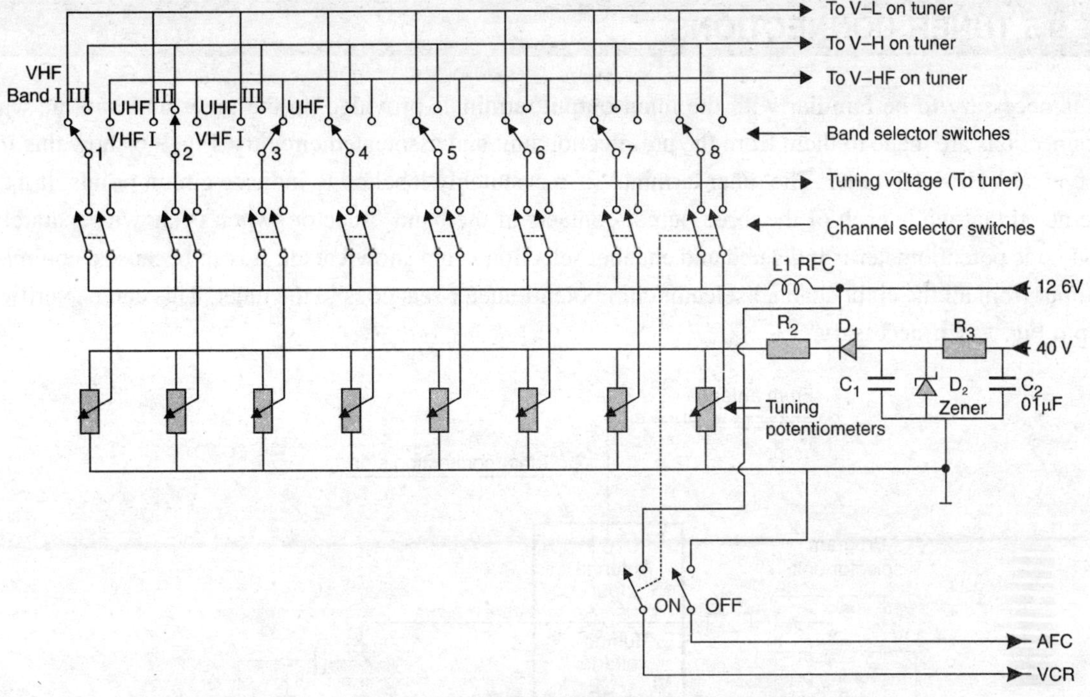

Fig. 9.5 Schematic diagram of an 8-fold program selector on different channels.

In Fig. 9.5 preselector positions 4 to 8 have been shown as idle with their hand selector switches left at Band I location. It is understood that these will be moved to necessary locations depending on the sequence of programs indicated in the list. The last selector position *i.e.*, 8th in Fig. 9.5 is usually recommended for connecting VCR output and as shown the VCR switch is connected to this location. There is also a provision for adding or removing AFC voltage to the dc source that feeds the varactor diodes.

It is necessary that a highly stabilized dc voltages is fed to the varactor diodes because it is only then that the selected channel can be correctly remembered and switched in when desired. Therefore, a preregulated supply is again regulated and decoupled before applying to the tuning potentiometers. It is equally necessary to feed a constant voltage to the band selector switch to ensure stable operation of the selected station. A diode is usually inserted in series with these supplies to make sure that any negative voltage does not get applied.

As mentioned earlier, the advantage of preselection is that any of the selected channels and other programs can be tuned in just on the depression of a selector button. However, it is necessary to remember the channels selected on different selector positions. It is not necessary to do so in digital selection methods to be described in the next chapter. There, the channel number is remembered by the memory of the computer and indicated by a nixey display either in a window on the front panel or just above the selector button.

9.5 TUNER CONNECTIONS

It is necessary to be familiar with the input-output terminals provided on the tuner unit and the way connections are made to them from the preselection unit and associated circuitry. Fig. 9.6 show this for a typical VHF-UHF tuner. The tuner terminal strip is suitably labelled to indicate circuit points. It may be noted that while each of the three output contacts of the band selector switch is shown separately, only one potentiometer is indicated and channel selection is not shown at all. It is so because a common output from all the eight channel selector cum potentiometer sets goes to the tuner. This can be verified from Fig. 9.5 if necessary.

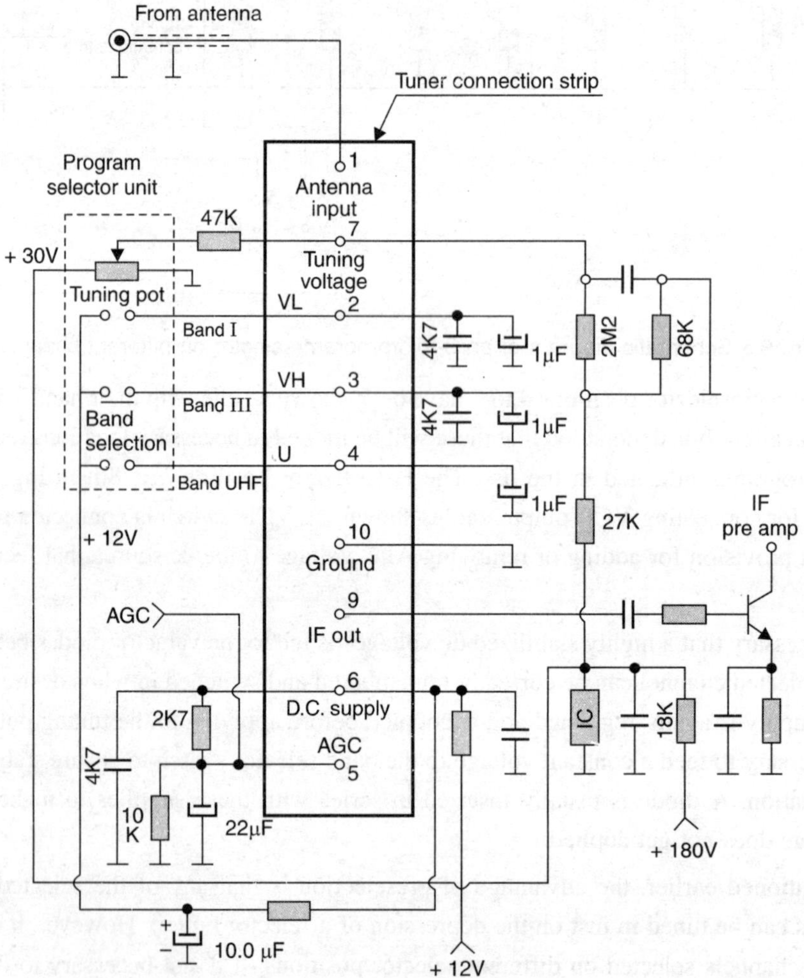

Fig. 9.6 External connections to a typical VHF-UHF tuner unit.

It is interesting to note the application of a small dc voltage through a 2.2. MΩ resistor at terminal 7 which receives variable point of the potentiometer. The aim is to maintain a small reverse bias (+ voltage) on the varactor diodes even when the movable point on any potentiometer is varied to touch its ground point. A low dc voltage (≈12V) is fed at terminal 6 as B + for the tuner RF amplifier and mixer cum oscillator circuits. The AGC voltage is applied at pin 5 which also receives a small positive voltage from the same regulated low voltage dc supply. Another source at 180V dc from the power supply is dropped and zener regulated to obtain 30V for feeding the channel selector switches via potentiometers. The signal from the antenna is connected at pin 1 for feeding to the input circuits of RF amplifiers of both VHF and UHF tuners.

9.6 COMMOM FAULTS IN VARACTOR TUNERS

Tuner manufacture is a specialized job and receiver manufacturers prefer to buy such units. At the manufacturing point the unit is tested in a jig which simulates all tuner functions. This process weeds out defective components before the assembly is made for use. Since there are no moving parts in varactor tuners, such units seldom give any trouble.

Simple test methods. A defect in the tuner section of a TV receiver will affect both video and sound signals. If signal does not pass through the tuner it will be indicated by a blank raster and no sound output. A simple rough test which will reveal whether or not a signal can get through the tuner is to turn the contrast control fully 'on' and momentarily short the receiver input terminals. Bursts of noise will be heard in the speaker and flashes of light will appear on the scanning raster if the tuner is operating normal. If these indications do not appear, it is obvious that signal is not passing through the tuner.

Another reason for no picture and sound output could be a break in the transmission line leading from antenna to the receiver. An ohm-meter check or visual inspection will reveal this fault. Signal injection is another way of finding out if the tuner is inoperative. Another method that is also used is to substitute the tuner with a known good one. Such units are specially designed for quick substitution to ascertain the condition of tuner in the receiver.

Common Symptoms

A defective tuner can have any of the following symptoms:

(1) erratic or unstable tuning

(2) no tuning capability

(3) AFT voltage drift.

For erratic or unstable tuning the cause may be (*i*) incorrect dc supply voltage, (*ii*) defective switch contacts (*iii*) defective tuning potentiometers and (*iv*) incorrect VHF tuner mixer and high low band switching voltages. For no tuning capability the possible sources of trouble are basically the same as for erratic or unstable tuning. A voltmeter check will show any AFT voltage drift. If present, the AFT system needs to be isolated and reviewed separately. If the fault persists the tuner should be sent to the manufacturer for servicing because the alignment procedure is quite critical and even a small twist on the end capacitors or leads while working on it may completely offset proper working of the entire unit.

REVIEW QUESTIONS

1. Draw a simplified circuit diagram of a VHF varactor tuner and explain how channel tuning is obtained. In addition explain the use of switching diodes for changing from Band I to Band III channels.
2. Explain special features of a UHF tuner. Draw its simplified circuit and explain how it operates to select different UHF channels. Why is the IF output fed to the mixer of the VHF tuner?
3. The circuit of a typical UHF-VHF tuner is given in Fig. 9.4. With reference to this, explain the function of (*i*) diodes D1 and D2, (*ii*) operation of transistor BF-581 as mixer and local oscillator (*iii*) diode switching for higher UHF channels and (*iv*) application of AGC and AFC voltages to the two tuners.
4. Draw schematic diagram of atypical program selector unit and explain fully how different programs can be preselected on different selector positions.
5. Draw a single circuit to explain the connections that are made on the tuner terminal strip. Why a highly regulated dc voltage is necessary for varactor tuners?
6. How will you proceed to check if the tuner is functioning or not? How does the operation of a preselector affect the functioning of the tuner? What is the procedure for finding out if the preselector is the cause of wrong selection of channels?
7. List common faults in tuner units and explain possible causes for such faults. Give methods of removing these faults.
8. Why is it not advisable to go in for a complete overhaul and alignment of a tuner unit? What precautions are necessary while replacing a bad tuner with a new unit.

10

FREQUENCY SYNTHESIZED TUNING, REMOTE CONTROL AND AFT

INTRODUCTION

The transition from discrete to integrated circuits and availability of dedicated microcomputers have totally transformed the method of channel selection and remote control in television receivers. In the new method of channel selection a frequency synthesizer is employed. It generates a number of crystal stable frequencies using only one crystal. These stable reference sources are used to lock the tuner oscillator frequency for each channel at its correct value by a phase-locked loop (PLL) control system. For ease of comparison and control, both the crystal reference and local oscillator frequencies are divided by suitable factors. The divide ratio depends on the actual local oscillator frequency which is different for each channel. Such a division is carried out by computer controlled digital pre-scalers and dividers. The computer memory stores necessary information (divide ratio) for all the wanted channels. This is put in the memory while deciding desired channels. Later, it is only necessary to press corresponding selector button on the program selector and the chosen channel is instantly tuned in. Thus, any VHF or UHF channel can be accurately tuned without the need of a fine tuning control and band selector switch.

The method of remote control of various receiver functions has also completely changed. The newer versions employ a microcomputer which carries out necessary functions on depression of various buttons provided on the remote control unit.

With the frequency synthesizer technique of channel selection, AFT is not necessary. However, it is provided but made optional by a switch on the front panel. It is useful to arrest any channel carrier drifs in MATV and other cable TV systems.

The PLL control, frequency synthesizer and microcomputers are newer concepts to many engaged in the television field. Therefore, the basics of these are explained before taking up digitally addressed channel selection, remote control and automatic frequency control methods.

10.1 PHASE-LOCKED LOOP (PLL) CONTROL

The PLL control compares two signals and develops a control voltage proportional to the frequency or phase difference between them. For channel selection the reference signal is the frequency obtained on dividing the stable crystal oscillator frequency. The other signal to be controlled is the local oscillator frequency. This is also divided by a suitable factor before feeding it to the comparator in the PLL circuit.

PLL Operation. The basic operation of a PLL control as used for channel selection can be explained with the help of a block diagram shown in Fig. 10.1. Assume that crystal accuracy of a 20 KHz tuned oscillator circuit is required. Precision crystals at 20 KHz are not easily available. However, high frequency crystals are readily available at low cost. Thus it is desirable to use a 20 MHz crystal to stabilize a 20 KHz oscillator. Such oscillators are always voltage controlled (VCO) configurations. As shown in Fig. 10.1 the output of 20 MHz crystal stabilized oscillator feeds into a divide by 1000 digital frequency divider. The output is a 20 KHz crystal stable signal and becomes reference input to the phase comparator. The output of the VCO to be controlled is also fed to the comparator. If the two signals are synchronized no correction is required. However, if the VCO drifts off frequency, the phase comparator develops a correction voltage proportional to the phase difference of its inputs signal. The output from the comparator is filtered and dc control voltage thus obtained applied to the 20 KHz VCO. The control voltage is such that it forces the oscillator to lock with the crystal stable frequency of 20 KHz. The PLL control thus enables crystal stable accuracy of a voltage controlled oscillator.

A similar circuit is used in electronic tuners employing frequency synthesized control to accurately select any VHF-UHF channel. This is explained in a later section of this chapter.

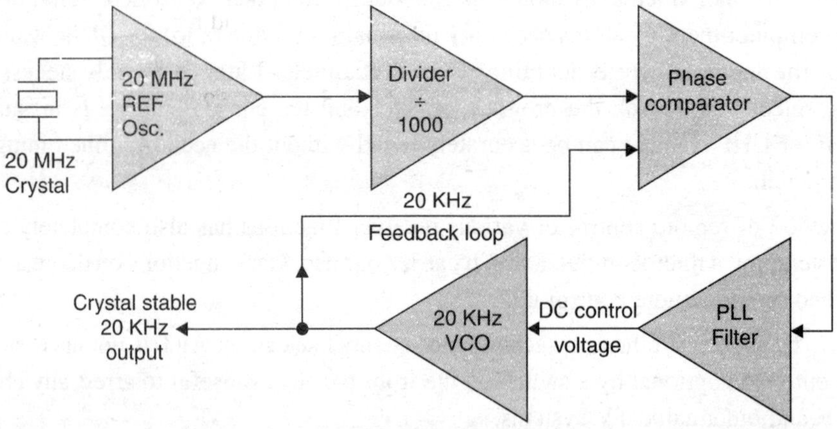

Fig. 10.1 Phase-locked loop (PLL) control of a 20 KHz voltage controlled oscillator (VCO).

10.2 DIGITAL SIGNALS & BINARY NUMBER SYSTEM

Electronic signals may be analog or digital. An analog signal like the output of a microphone can have any value within a wide range of frequencies. In contrast, a digital signal can have only two discrete values at any given instant. These are called logic or state values and termed 1 and 0. A digital number is a chain of 1's and 0's in any order. The signal processed by a hand calculator is of this type.

In electronic circuitry diodes and transistors can be used to represent 1 and 0, the two possible logic states. For example when a transistor is OFF, its collector voltage (V_{CE}) is equal to the supply voltage (V_{CC}) and when saturated, V_{CE} drops to near zero volt. In digital electronics these OFF and ON states are allowed two discrete voltage levels. These are generally 5V and 0 Volt. The 5V level represents 1 and 0V level a zero (0). An ON signal is referred to as 'high' (1) and OFF signal as 'LOW' (0). The 1 and 0 are called binary numbers because only these two states are possible.

Bits, Bytes and Words

A single binary digit is called a 'bit'. An information in a digital system is represented by a sequence of bits. An 8-bit sequence is called a 'byte'. The number of bits in the data sequences processed by a given computer is called its 'word size'.

In computers the processing of information in digital form requires special circuits, special numbering system and even a special algebra. Such circuits must provide for storing instructions and data to be processed, receiving new data, performing calculations, making decisions and communicating results. While details are beyond the scope of this book, elements of operations that have relevance to channel tuning and remote control are briefly introduced.

Binary numbers. The decimal system expresses numbers in power of 10 *i.e.*, to the base 10. For example, $134 = 1 \times 10^2 + 3 \times 10^1 + 4 \times 10^0 = 100 + 30 + 4$.

The same approach is used in the binary system. However, here the base is 2 and only two digits *i.e.*, 1 and 0 are employed to represent any number. Thus individual digits represent the coefficients of powers of 2 rather than 10 as in the decimal system. As an illustration decimal number 19 is written in the binary form as 10011 and read as 'one, zero, zero, one, one and not ten thousand and eleven. This representation is correct because:

$$10011 = 1 \times 2^4 + 0 \times 2^3 + 0 \times 2^2 + 0 \times 2^1 + 0 \times 2^0$$

$$= 16 + 0 + 0 + 2 + 1 = 19$$

Similarly it can be shown that decimal number 25 is equal to 11001 in binary form. For convenience of conversion Table 10.1 shows binary number equivalents for decimal numbers form 1 through 21.

As is obvious from Table 10.1, the binary equivalent becomes bigger and bigger as the number grows. For example the decimal number 10,000 in the binary form is 10011100010000. While it is too big still processing of such numbers does not present any serious problem because of the ease with which a very large number of logic elements (diodes, transistors and MOS units) can be contained in a tiny integrated circuit and the fast as light speed at which electronic circuits operate to carry out desired operations.

Table 10.1 Decimal to Binary Number Conversion Table

Decimal	Binary	Decimal	Binary
0	0000	11	1011
1	0001	12	1100
2	0010	13	1101
3	0011	14	1110
4	0100	15	1111
5	0101	16	10000
6	0110	17	10001
7	0111	18	10010
8	1000	19	10011
9	1001	20	10100
10	1010	21	10101

Binary to decimal conversion. As explained earlier each binary point corresponds to a power of 2 and so to convert a binary number to its decimal equivalent the corresponding decimal equivalents of I's should be added. Thus binary 110001 = $2^5 + 2^4 + 2^0 = 32 + 16 + 1 = 49$ in decimal. Similarly 110.11 = $2^2 + 2^1 + 2^{-1} + 2^{-2} + 4 + 2 + 0.5 + 0.25 = 6.75$.

Decimal to binary conversion. This can be easily done by reversing the above process. For example 9 can be converted to binary form as follows. Decimal (9) = 8 + 1 = 8 + 0 + 0 + 1 = 1001 (binary)

Alternatively, the conversion can be done by successive division as shown below :

2	9	Remainders
2	4	1....least significant bit
2	2	0
2	1	0
	0	1...most significant bit
		≅ 1001

10.3 BINARY ARITHMETIC

Arithmetic operations on binary numbers are done in exactly the same way as on decimal numbers. For example, in decimal addition on adding 6 to 5 we get a sum of 1 with a carry of 1 to the next column. Similarly, in binary adding 1 + 1 gives a sum of 0 and a carry of 1 since here 2 is the base instead of 10. Accordingly, the following rules apply:

Addition:

$$0 + 0 = 0 \text{ (sum = 0, carry = 0)}$$
$$0 + 1 = 1 \text{ (sum = 1, carry = 0)}$$
$$1 + 0 = 1 \text{ (sum = 1, carry = 0)}$$
$$1 + 1 = 10 \text{ (sum = 0, carry = 1)}$$

Subtraction:

$$0 - 0 = 0 \text{ (difference = 0, borrow = 0)}$$
$$1 - 0 = 1 \text{ (difference = 1, borrow = 0)}$$
$$1 - 1 = 0 \text{ (difference = 0, borrow = 0)}$$
$$0 - 1 = 1 \text{ (difference = 1, borrow = 1)}$$

Multiplication: Binary multiplication is done in exactly the same way as decimal multiplication. The following rules apply:

$$0 \times 0 = 0$$
$$0 \times 1 = 0$$
$$1 \times 0 = 0$$
$$1 \times 0 = 1$$

Example — Convert 22 and 6 to binary form and multiply.

22 (decimal)	=	10110 (binary)
× 6 (decimal)	=	× 110 (binary)
= 132 (decimal)		00000
		10110
		10110
		10000100 (binary)

As a check it is $= 2^7 + 2^2 = 128 + 4 = 132$ in decimal.

Division. Binary division is also carried out in the same way as for decimal numbers.

Example—Divide 12 by 2, both in decimal and binary forms. $(12 \xrightarrow{D\ \ B} 1100)$ & $(2 \xrightarrow{D\ \ B} 10)$

Decimal

$$2 \underline{| 12}$$
$$6$$

Binary

```
10)1100(110
   10
   --
    10
    10
    --
    00
```

Ans 110 in Binary and 6 in decimal

It may be noted that digital multiplication circuits are basically adding circuits. To multiply 22 by 6, 22 is added six times. Similarly division is carried out by repeated subtractions. Since the time taken by such operations is extremely small, repeated additions or subtractions hardly take any time.

CHAPTER 10

10.4 LOGIC ELEMENTS AND OPERATIONS

Logic is the science of reasoning. Here all statements are either true or false. In digital terms these two are represented by the binary states 1 and 0 which as explained earlier are represented by discrete voltage levels. Various logic operations are carried out by gates. A gate is a device that controls the flow of information, usually in the form of pulses. The three basic gates are 'AND', 'OR' and 'NOT'. These are briefly explained.

'AND' Gate

A simple two input diode AND gate is shown in Fig. 10.2 (*a*). An output appears only when there are inputs at A and B. The input pulses are in the form of positive voltage pulses with respect to ground. The inputs at A and B reverse bias both the diodes and no current flows through the resistance. Thus the output is equal to 1 (high) with magnitude of 5 volts. In general there may be several input terminals. If any on of the inputs is zero (0) current flows through the forward biased diode and the output is nearly zero (0). For two inputs varying with time a typical response is shown along the circuit of Fig. 10.2 (*a*). It may be noted that the output is also zero (0) when both the inputs are at (0) level.

'OR' Gate

A diode OR gate configuration along with typical response is shown in Fig. 10.2 (*b*). As is obvious the output is 1 (5V) if input A or input B is 1. For no input no current flows and output is zero. An input of 5V (1) at either terminal A or B or both (or at any terminal in a general case) forward biases corresponding diode, current flows through the resistance and output voltage rises to 5V (1).

'NOT' Gate

The NOT gate does inversion *i.e.*, (1) to (0) and (0) to (1). A simple transistor circuit which can do so is shown in Fig. 10.2 (*c*). With no input (0) the transistor switch is held open by the negative bias voltage and the output is + 5V *i.e.*, (1). A positive input voltage (1) forward biases the emitter junction, collector current flows, and output voltage falls to a few tenths of a volt ≈ (0).

'NOR' and 'NAND' Gates

A NOR gate is a combination of OR and NOT gates. Similarly a NAND gate is a combination of AND and NOT gates. These form very useful logic operation configurations and are commonly used for processing digital information. For example, the input conditions can be so set that the output obeys rules of addition, subtraction, multiplication or division as necessary.

Electronic logic circuits are classified in terms of the components employed. The diode logic (DL) is shown in Fig. 10.2. Similarly there are many other types like resistor-transistor logic (RTL) and transistor-transistor logic (TTL). The 'p' channel enhancement mode MOS transistors are commonly used in digital circuits because these need small area, are normally off and have high input resistance.

Memory Elements

The outputs of logic gates are determined by the present inputs or it may be said that in response to various inputs. These combinational circuits make 'decisions'. Along with these decision components we need memory components to store instructions and results. The outputs of such sequential circuits are affected by past as well as present inputs. Accordingly, a binary storage device must have two distinct states and it must remain in one state until instructed to change. It must change rapidly from one state to the other and the state value (0 or 1) must be clearly evident. The bistable multivibrator or FLIP-FLOP meets these requirements and is used in all types of digital data processing systems. A chain of such flip-flops can handle multi-input signals. The start and stop of such circuits is affected by PRESET and CLEAR input signals.

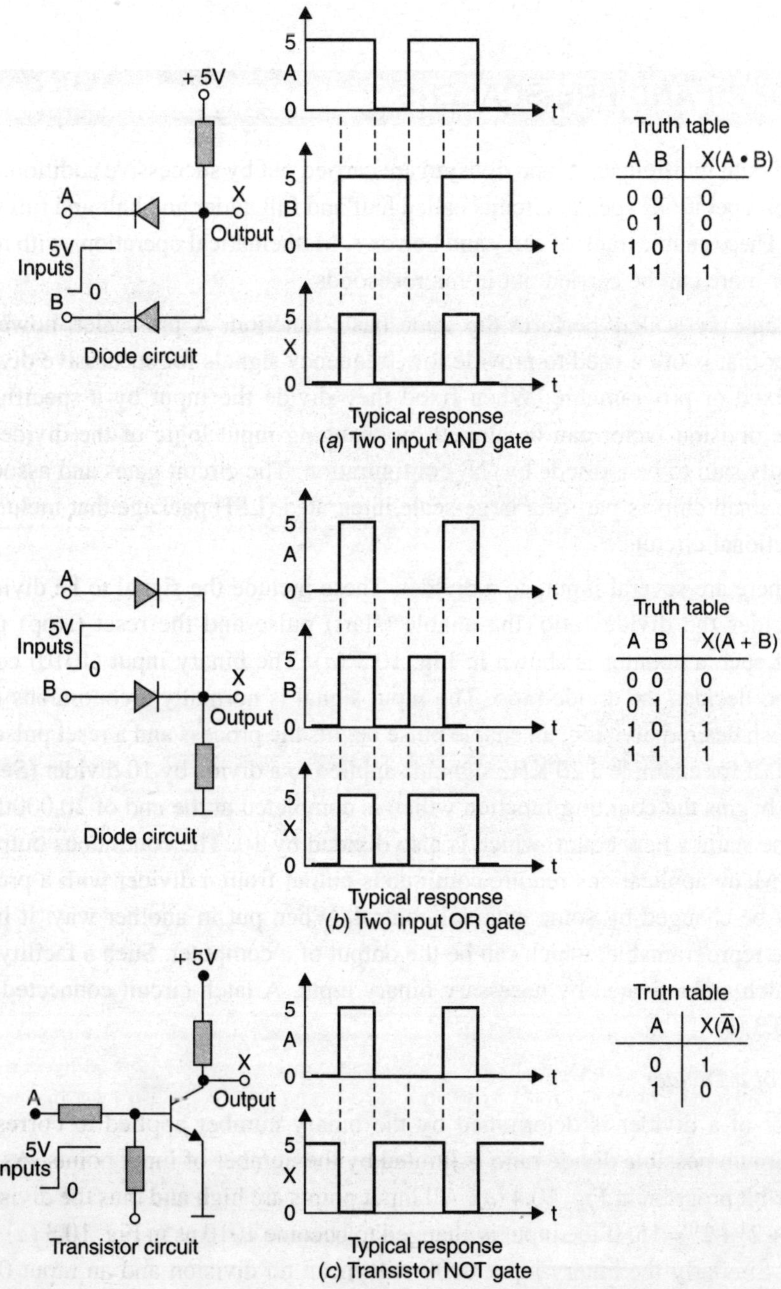

Fig. 10.2 Gate circuits (*a*) AND gate, (*b*) OR gate, (*c*) NOT gate.

Data Latch

In order to avoid any ambiguity an invertor is added to the basic flip-flop. Once the output goes low no change in the output is possible unless a new input pulse appears. The output is then said to be latched at the previous data value. This data latch is widely used as an element in digital systems, for example, a set of eight such latches could remember, the eight digits representing a number or an instruction.

10.5 DIVIDERS AND PRE-SCALERS

As already stated both multiplication and division are carried out by successive addition and subtraction methods. For such operations special circuits called half and full adder and half and full subtractor have been developed. These enable track of carry and borrows. Mathematical operations with numbers having digits up to 10 or more can be carried out in microseconds.

Dividers and pre-scalers perform the same basic function. A pre-scaler, however, is a high-frequency divider that is often used to provide low frequency signals for successive dividers. Dividers may be either fixed or programable. When fixed they divide the input by a specific factor. When programable, the division factor can be altered by changing input logic of the divider. This type of divider is generally said to be a divide by 'N' configuration. The circuit gates and associated circuitry is contained in a small chip as part of a large-scale integrated (LSI) package that includes many other devices and functional circuits.

Usually there are several inputs to a divider. These include the signal to be divided, the binary number that decides the divide ratio, the enable (start) pulse and the reset (stop) pulse. A block representation of such a counter is shown in Fig. 10.3 (*a*). The binary input (1010) consists of logic highs or lows and decides the divide ratio. The input signal is normally a continuous train of pulses. Thus to accomplish desired division, an enable pulse begins the process and a reset pulse re-establishes the starting point. If for example a 20 KHz signal is applied to a divide by 10 divider (See Fig. 10.3 (*a*)) the enable pulse begins the counting function which is completed at the end of 20,000th pulse. At this time, a reset pulse starts a new count, which is also divided by 10. The continuous output is therefore, a 2 KHz signal. Many applications require continuous output from a divider with a provision that the divide ratio may be changed by some external control. When put in another way, it implies that the divider should be reprogramable, which can be the output of a computer. Such a facility is obtained by latch circuits which are activated by necessary binary input. A latch circuit connected to a divider is shown in Fig. 10.3 (*b*).

Division Factor of a Divider

The divide factor of a divider is determined by the binary number applied to corresponding input points. The maximum possible divide ratio is limited by the number of input points. As an illustration, consider the four-bit program in Fig. 10.4 (*a*). All input points are high and thus the division factor is 15 (1111) $= 2^3 + 2^2 + 2^1 + 2^0 = 15$. If the input is changed to become 1010 as in Fig. 10.3 (*a*) the divide ratio will become 10. Similarly the binary input 0000 will mean no division and an input 0010 will cause division by the factor 2. Thus in a four-bit input any division factor form 1 through 15 can be obtained by properly programming the inputs point of this divider.

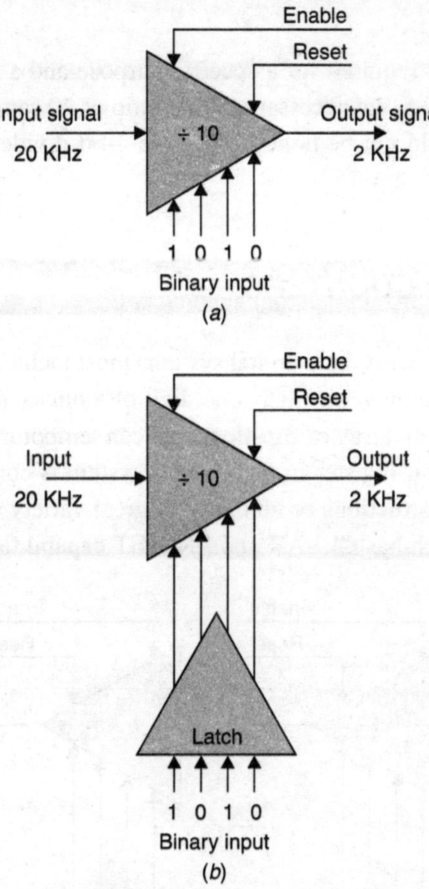

Fig. 10.3 Divider circuits (*a*) A divider circuit set to divide by a factor of 10(1010), (*b*) Divider with a latch circuit. It continuously supplies binary input until the latch is reset.

As another example consider an 8-bit divider as shown in Fig. 10.4 (*b*). If all input ports are at logic high (1), the division factor is 255 (11111111) = $2^7 + 2^6 + 2^5 + 2^4 + 2^3 + 2^1 + 2^0 = 255$. Thus proper programming can provide any desired division factor from 1 through 255. Actually divide factors of several thousand are not uncommon. For example a 12-bit divider can provide any division factor from 1 through 4095. Hence, if we have a N-bit divider, we can divide by any number between 1 to $2^N - 1$.

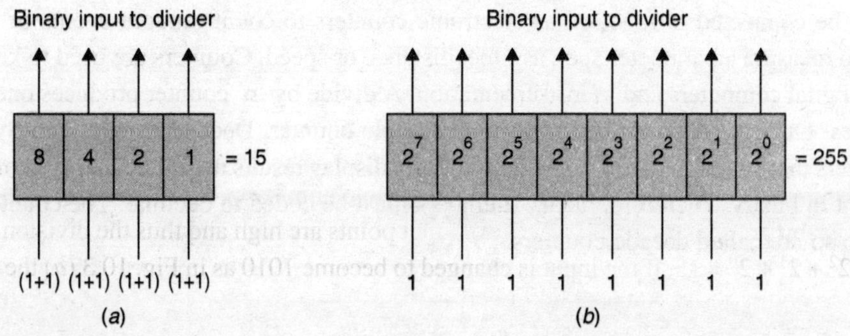

Fig. 10.4 Examples illustrating division by binary inputs (*a*) divide by 15, (*b*) divide by 255.

A Practical Divider Application

Assume that a 450 KHz signal is required for a specific purpose and a stable 9 MHz signal is readily available. As illustrated in Fig. 10.5, the necessary divide ratio of 20 can be obtained by employing two 4-bit dividers. Note that this could not be done in a single 4-bit divider since the maximum division available with this is $2^N - 1 = 2^4 - 1 = 15$.

10.6 REGISTERS AND COUNTERS

In addition to logic circuits that process data, digital systems must include memory devices to store data and results. A 'flip-flop' can store or 'remember' one digit of a binary number, *i.e.*, one bit (1 or 0) in one of its two states. A register is an array of flip-flops that can temporarily store data or information in digital form. For example, an 8-bit register in a computer system is composed of eight 'flip-flops' in parallel and can handle 8-digit instructions or numbers. A great variety of registers are available in IC form. A shift register is one which has CLEAR and PRESET capabilities.

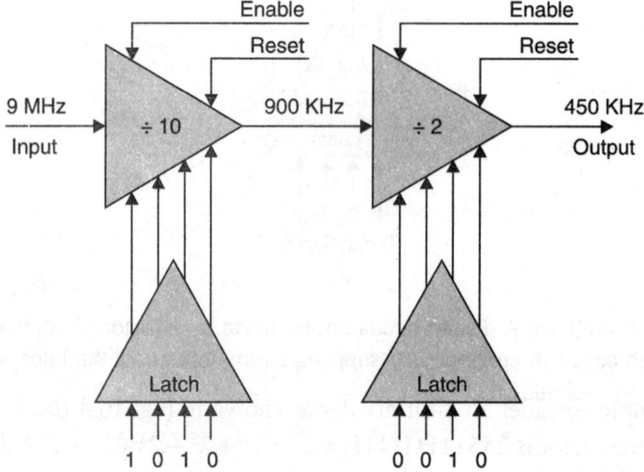

Fig. 10.5 Method of obtaining a 450 KHz stable output with two 4-bit dividers.

Counters

Flip-flops can be connected to function as electronic counters to count random events or to divide a frequency or to measure a parameter such as time, distance or speed. Counters are used to keep track of operations in digital computers and in instrumentation. A divide by 'n' counter produces one output for 'n' input pulses. Other type of counters are Binary ripple counter, Decade counter and Synchronous counter. Counters that communicate with humans usually display results in the decimal system. However, flip-flops count in binary. Therefore, binary numbers must be coded to decimal. The counters that are designed to do so are called decade counters.

10.7 COMPUTER ORGANISATION

An electronic digital computer is a vast assemblage of simple logic gates and memory devices organised to perform complex calculations at high speed by automatically processing information in the form of electronic pulses. The information consists of data and instructions. A complete set of instructions is called a Program. The functional organisation of a computer is illustrated in Fig. 10.6. Every computer contains the following five units:

Input Unit

Instructions and data from the outside world to the computer are entered into the machine through input devices that read data on punched cards, paper tapes, magnetic tapes or sensors. Thus all communication from the user (operator) to the computer is through the input section or unit.

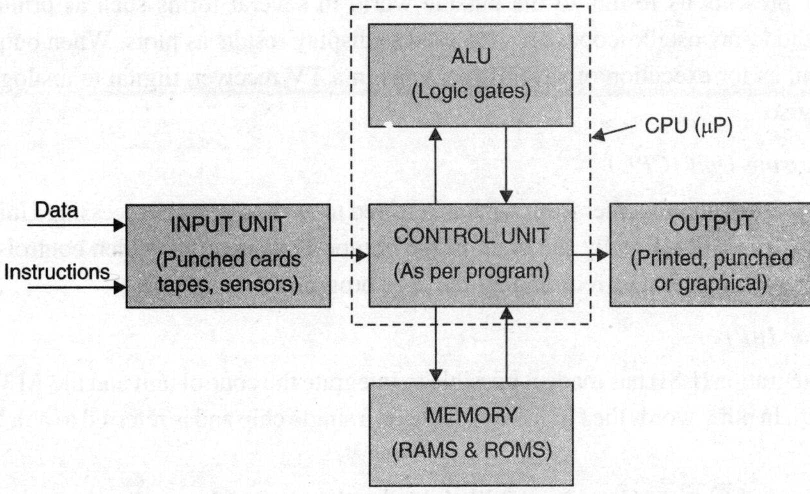

Fig. 10.6 Functional components of a general purpose digital computer.

Memory Unit

Each instruction is stored in the computer memory until it is retrieved for use. Flip-flops are used for temporary storage of data and instructions in active registers. There are two types of memories. A 'Read Only Memory' (ROM) stores special predetermined data which remains unchanged like instructions for logic operations etc. A 'Read and write Memory' (RAM) is capable of receiving and storing data which can be changed as required. The working computer memory usually consists of an array of integrated circuits capable of storing million of bits of information.

Control Unit

The flow of instructions, data, and results is governed by the control unit which sets various logic gates, feeds numerical data and provides clock pulses that regulate the speed of all operations. Instructions are received from the input device and stored in one part of the memory. Similarly data are stored in another part of the memory. In accordance with the program the control unit provides for a desired sequence of arithmetic operations on retrieving data from the memory.

As an elementary example of the operation of control unit, consider that stored numbers 'A' and 'B' of several bits are to be operated on in accordance with instructions in the form of four bit word. The instruction which can be 1011 is drawn from the memory and placed in a register associated with a logic circuit of several gates. The instructions set the logic gates and operation takes place. It is regulated by clock pulses.

Arithmetic Logic Unit (ALU)

The arithmetic logic or processing unit consists of logic gates arranged to perform such operations. The data to be operated upon can come to the ALU from the memory unit or input devices and the type of operation to be performed in indicated to the ALU by signals from the control unit. The result of operations performed in the ALU can be transferred to the memory unit or output devices.

Output Unit

The computer presents its results to the outside world in several forms such as printed, punched or graphical. Cathode-ray oscilloscopes are also used to display results as plots. When outputs are needed in analog form, as for execution of several functions in a TV receiver, digital to analog convertors are used for doing so.

Central Processing Unit (CPU)

The ALU and control unit together form what is referred to as the Central Processing Unit of a computer (See Fig. 10.6). The CPU is really the brain of the computer as it is this which controls all other units and does all the computation as required by the user program.

Microprocessor (μP)

Large scale integration (LSI) has made it possible to integrate the control unit and the ALU of a computer on a single chip. In other words the CPU is available on a single chip and is referred to as a Microprocessor (μP).

In addition, LSI technology has enabled the development of large capacity memories and many types of input/output (I/O) interface chips. Thus a relatively small number of such chips can be hooked together to perform specified functional operations. Such units as used in receiver channel selection and remote control circuits are called Microcomputers (μCs).

Program

In any μC, the controller must be told in great detail just what to do. In other words it must be programed. The programmer decides which tasks are to be performed and in what order. It defines the sequence in a written program. Thus a program is a detailed list of specific instructions in digital form each corresponding to an elementary step in the overall operation. Such instructions can be stored in the memory in a specific form (language). The details of programing are too involved and beyond the scope of this book. In fact the coverage of digital computer operations in this chapter is too elementary. However, an effort has been made so that the black box representation of the microcomputer as used in TV control circuits does not appear to be too black (empty) to any reader not familiar with digital techniques.

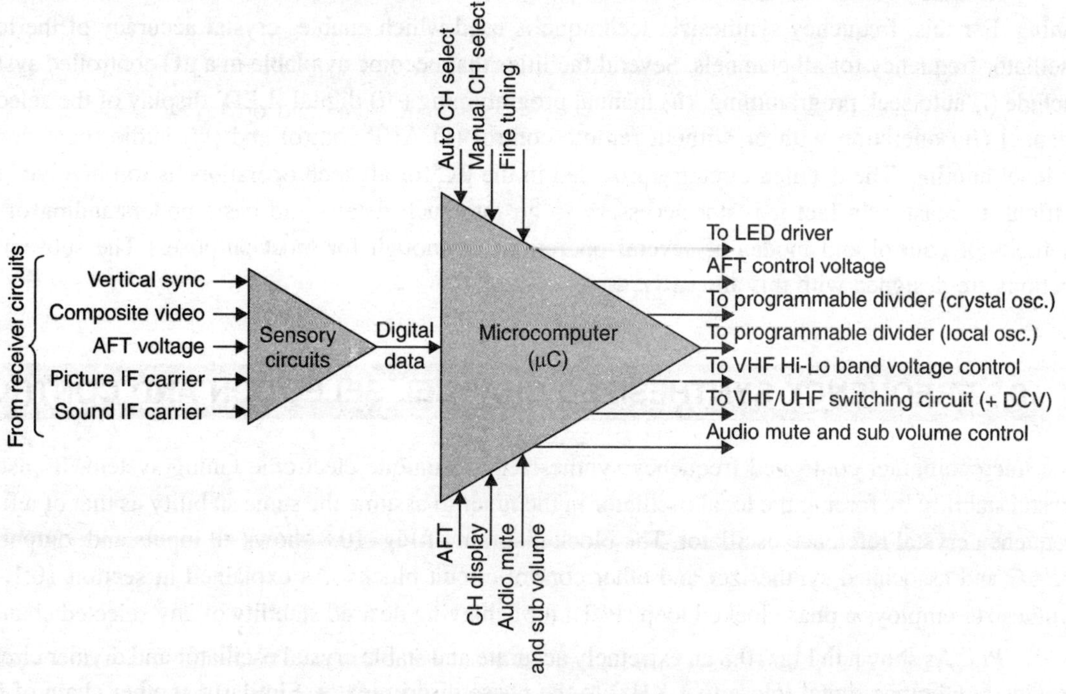

Fig. 10.7 Block diagram showing various inputs and outputs to a microcomputer specially designed for control of receiver functions.

10.8 MICROCOMPUTER CONTROL OF RECEIVER FUNCTIONS

The microcomputer based control of various receiver functions has become quite common. Out of the many systems that have been developed, some perform a limited number of control functions while others are designed to control all functions a viewer has to manipulate for optimum performance of the receiver. However, automatic channel selection and tuning is a common feature of all µC controlled systems. The inputs and outputs of a µC specially designed for a typical control system are shown in Fig. 10.7. One set of inputs is from the receiver control panel. As labelled, each input is for a particular operation. The user on depression of corresponding buttons sends out necessary signals to the µC in the form of high and low pulses. For example, depression of auto channel search button will send a train of pulses enabling the µC to initiate, search and tuning of all available channels in the area.

The second set of inputs is to give necessary feedback to the µC so that it can initiate control action for sustained correct operation of the receiver. Various signals like picture carrier, sound carrier and composite video signal are picked up from different sections of the receiver and fed to sensing or sensory circuits. The sensors employ discriminaters where picked up signals are compared with preset reference levels. The outputs are in the form of high or low voltage levels. The µC on receipt of such digital data takes necessary decisions and sends control signals to ensure proper operation of the receiver as desired (set) by the user.

As stated earlier the main thrust of all μC based control systems is channel selection and fine tuning. For this, frequency synthesizer technique is used which enables crystal accuracy of the local oscillator frequency for all channels. Several facilities that become available in a μC controlled system include (*i*) auto seek programming, (*ii*) manual programming (*iii*) digital 'LED' display of the selected channel (*iv*) operation with or without remote control (*v*) AFT control and (*vi*) audio mute during channel hunting. The detailed circuitry provided in the μC for all such operations is too involved and difficult to master. In fact it is not necessary to go into such details and basic understanding of the methods of control and modes of several operations is enough for most purposes. The subsequent sections are designed with this aim in view.

10.9 FREQUENCY SYNTHESIZED CHANNEL SELECTION AND CONTROL

The microcomputer controlled frequency synthesizer is a unique electronic tuning system. It ensures crystal stability by forcing the local oscillator in the tuner to assume the same stability as that of a fixed frequency crystal reference oscillator. The block diagram of Fig. 10.8 shows all inputs and outputs to the μC and associated synthesizer and other control circuit blocks. As explained in section 10.1, the synthesizer employs a phase locked loop (PLL) for achieving desired stability of any selected channel.

PLL As shown in Fig. 10.8 an extremely accurate and stable crystal oscillator and divider circuits provide a reference signal (close to 1 KHz) to the phase discriminator. Similarly another chain of pre-scalers and dividers reduce the local oscillator frequency to the same frequency as that obtained from the crystal reference oscillator. This becomes the second input to the discriminator. The μC changes the divide ratio in such a way that for any channel the local oscillator is reduced to the same value as the reference output to the discriminator. The discriminator compares the phase (or frequency) of the two inputs and develops output as necessary. If the two inputs are exactly same the control output is zero. However, if there is any discrepancy, a corrective signal appears as the discriminator output. A filter circuit provides a proportionate dc voltage which when fed to the varactor tuning diodes, along with the normal tuning voltage, shifts the local oscillator frequency to such a value that the two reference inputs to the discriminator become the same. As a result the tuner assumes accuracy and stability of the reference crystal oscillator.

Channel selection. The microcomputer controls tuning and other receiver functions by continuously scanning various inputs to it and making many computations based on details stored in the 'Read Only' Memory (ROM). This memory is contained in the metallic part of the IC and thus does not require standby batteries when power is removed. The μC also scans the keyboard some 200 times per second and examines the auto-seek, fine tuning and other control button switch outputs. It thus continuously looks for new command signals. It also receives input from the sensor circuits that determine whether or not a station signal is present and if so, how accurately the set is tuned.

When a new channel is selected by appropriate entry on the keyboard, the μC loads correct binary oscillator dividers. It also determines whether the new channel is in the low VHF, high VHF or UHF band and applies correct dc voltages to the band selector switch. In addition, another output applies necessary dc voltage to VHF high/low band switch. A step signal also goes to the sound IC to mute the sound output during channel selection. The LED drivers are also fed with necessary binary input. All these functions occur simultaneously.

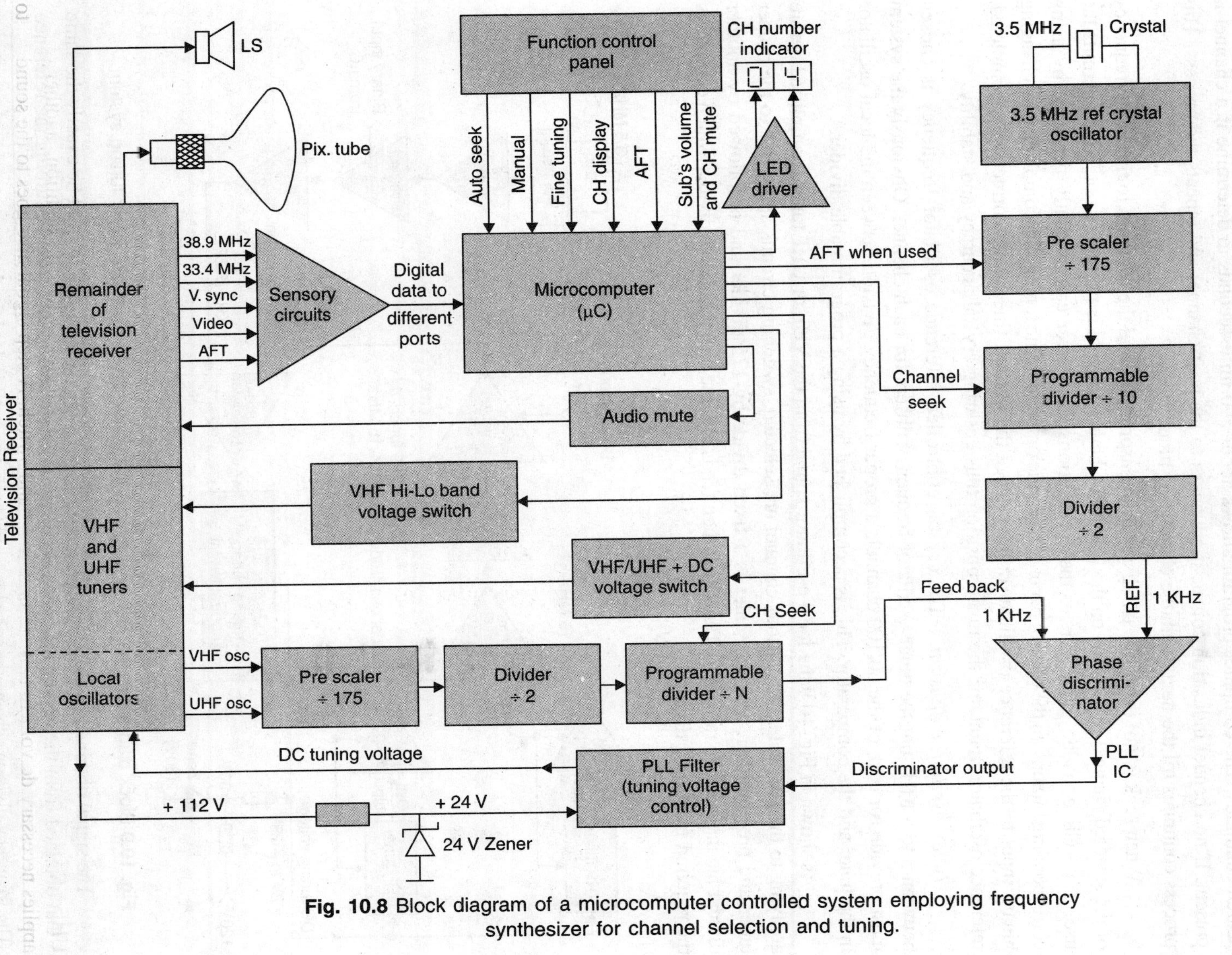

Fig. 10.8 Block diagram of a microcomputer controlled system employing frequency synthesizer for channel selection and tuning.

When auto-seek is in progress and sensor inputs to the µC indicate a no signal condition, the search continues. The computer program allows about 300 milli-seconds to determine if a channel is present. If no channel indication comes, it advances to the next channel and repeats the process. This process continues till the next available channel is tuned in.

When seek (auto) is off, the µC disregards sensor inputs and the set tunes to the exact frequency of the selected channel. It remains on this channel whether a signal is present or not, which permits the use of a VCR or video game. Under these conditions the viewer can manually fine tune by using corresponding buttons on the control panel. Automatic or manual fine tuning is accomplished by changing binary input to the reference oscillator divider. Once the desired fine tuning correction is completed, reference oscillator maintains the tuned position thus ensuring crystal accuracy and stability.

Tuner oscillator division. The crystal controlled reference oscillator frequency is chosen between 3 to 5 MHz but the tuner oscillator frequency differs for each channel. Obviously the system must provide for each of the 60/100 channels a proper frequency division between the local oscillator in the tuner and the comparator. This is accomplished by using a programmable divider.*

As shown in Fig. 10.9 the local oscillator signal from the VHF or UHF tuner is amplified before applying to the pre-scaler. As various channel are selected, switching circuits apply B + to the proper tuner and pre-amplifier. The pre-scaler is a fixed divide-by-175 divider and is followed by another divider having a divide ratio of 2. The signal then feeds into the programmable divider which completes the required frequency division.

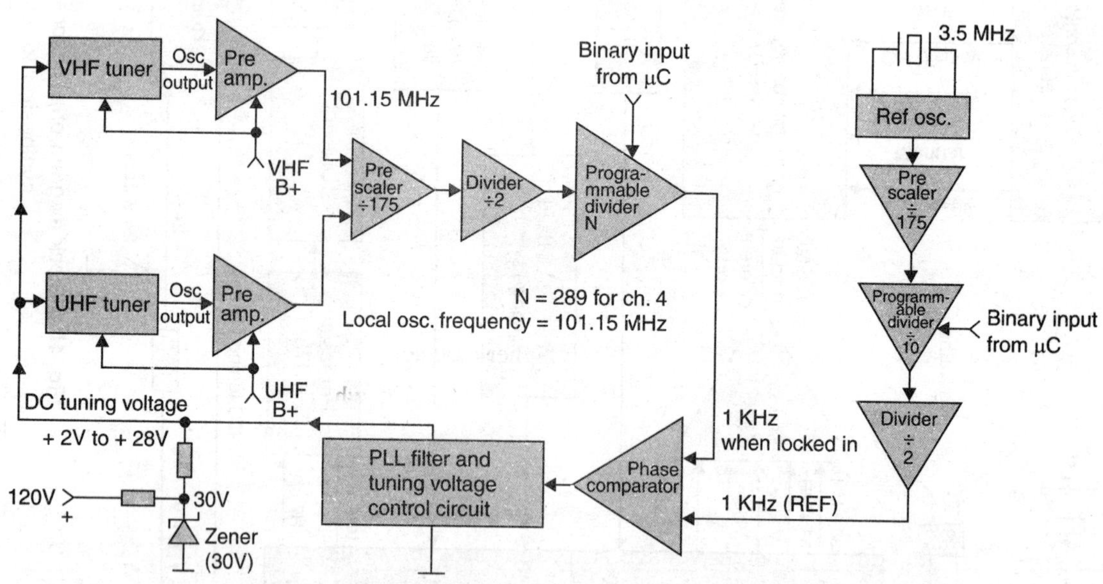

Fig. 10.9 Block diagram of the PLL of a typical frequency synthesizer electronic tuning system.

The signal obtained by frequency division of the local oscillator output must be exactly the same as that obtained from the crystal reference oscillator after successive division. Assuming the reference

* The frequency division scheme described is to illustrate the methods and does not pertain to any commercial control system. However, the frequency and divide ratios used in practice are close to those mentioned in the narration.

oscillator frequency to be 3.5 MHz, a divider chain consisting of three binary dividers ($175 \times 10 \times 2 = 3500$) scales it down to 1 KHz. Note that one of the dividers is programmable. As labelled the µC feeds necessary input to obtain a divide ratio of 10 that is suitable for lower VHF channels. It can change to another suitable number for higher VHF and UHF channels if the reference frequency is changed to become higher. In any case reference frequency is maintained close to 1 KHz. For example in the NTSC system it is chosen to be 976.5625 Hz. This is obtained from the reference crystal oscillator having a fixed frequency of 3.581055 MHz. Such odd numbers are chosen because this enables the value of N to become just equal to the channel local oscillator frequency expressed in MHz.

As explained earlier for proper operation of the PLL, the second input to the discriminator must also have the same value as that fed from the reference source. Thus the local oscillator frequency which is different for each channel must be divided by a suitable factor to obtain 1 KHz in the system under discussion. As shown in Fig. 10.9, two of the three dividers in the chain are fixed while the third is programable from the microcomputer. It feeds necessary binary input for each channel. The decimal equivalent of this input *i.e.*, N must be such that satisfies the expression:

$$\frac{\text{Local Oscillator Frequency}}{175 \times 2 \times N} = 1 \text{ KHz (1000 Hz)}$$

$$\text{Thus the value of } N = \frac{\text{Local Oscillator Frequency}}{175 \times 2 \times 1000}$$

Consider the lower VHF channels. For channel No. 2 the local oscillator frequency is 87.15 MHz. Similarly for channel 3 and 4 it is 94.15 MHz and 101.15 MHz respectively. Therefore, the value of N for these channel can be found as under:

$$N_{(2)} = \frac{87.15 \times 10^6}{175 \times 2 \times 1000} = 249$$

$$N_{(3)} = \frac{94.15 \times 10^6}{175 \times 2 \times 1000} = 269$$

and $$N_{(4)} = \frac{101.15 \times 10^6}{175 \times 2 \times 1000} = 289$$

Note that the value of N increases by 20 for each higher channel.

Similarly for higher VHF and UHF channels a suitable whole number factor (N) for the programable counter can be obtained. For this, it may be necessary to change the programable divide ratio in the reference oscillator chain of dividers.

PLL filter and tuning voltage control. As explained earlier any difference in the two inputs to the discriminator results in a dc output voltage. This voltage is fed to the VHF or UHF tuner oscillator (VCO) to correct its local oscillatory frequency. Actually the output of phase comparator controls another circuit which is turn supplies dc tuning voltage. This circuit is shown in Fig. 10.10. The B + voltage at about 120 volts is clamped down to 30V by zener diode D_1. This establishes the maximum varactor tuning voltage. If there is a frequency difference between the two phase comparator input signals, the comparator produces a series of dc pulses at its output. These are filtered (integrated by R_2, C_3 and R_3, C_4) and become base bias voltage for transistor Q_1. The voltage that develops across R_5, the emitter resistance of Q_1, feeds the base of Q_2 which functions as a variable resistance in conjunction with resistor R_7 between the varactor tuning voltage and ground. Increasing the drive decreases the tuning voltage. This voltage normally varies between 2 to 24 volts.

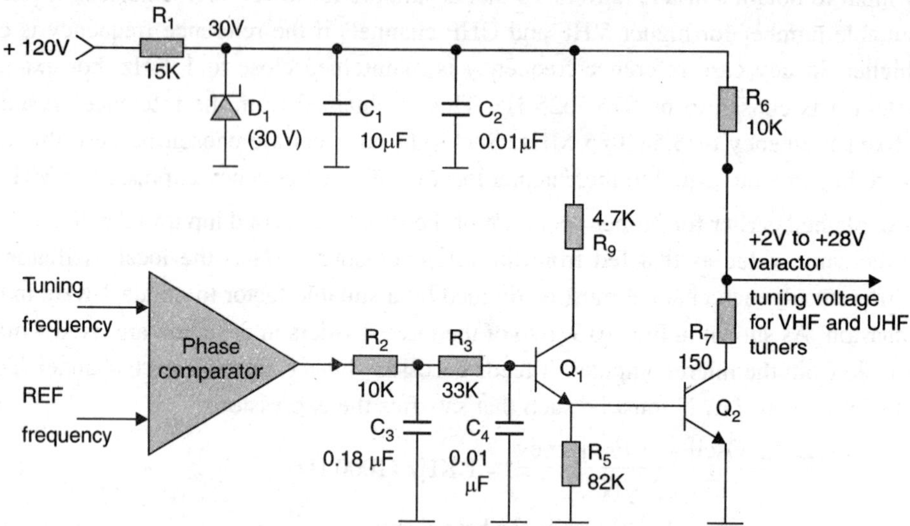

Fig. 10.10 PLL filter and tuning voltage control circuit.

10.10 SENSORY AND SWITCHING CIRCUITS

As already explained, specially designed sensors are used to process receiver data and feed it to the microcomputer. Similarly, outputs from the µC have to be processed before feeding them to the HIGH/LOW band switching, UHF band switching and audio mute circuits. Such circuits form part of ICs and hence are quite complicated. However, a good grasp of their functioning can be obtaind by considering simplified block diagram and discrete circuits.

Sensing circuits. Two comparator ICs are used to determine whether a station signal is present and if so how accurately it is tuned. Five different signals from the receiver module are picked up for feeding various sensor circuits. The chosen signals are picture IF, sound IF, composite video, AFT and vertical sync. These are converted to proportionate dc voltages before applying to corresponding sensor input terminals. The comparators (1 to 8) shown in Fig. 10.11 are actually differential amplifiers and their outputs are either low or high depending on the level of input signals. When a properly tuned channel is operating the outputs of corresponding sensors are high (5V). For seek mode there is a continuous input to the µC from the sensing circuits. It may be noted that this circuitry continues to give outputs even when manual operating mode (seek off) is in use but the computer disregards such inputs and functions in accordance with commands received from the control unit.

As shown in Fig. 10.11 the outputs of comparators 1, 2 and 3 are common. Thus the absence of sound, compositive video or picture signal results in a low logic (0V). The logic becomes low because when any one of these inputs becomes less than its reference value, the comparator changes state. Thus a low inputs to the µC from this set of sensors indicates absence of any one of these inputs. The computer on receipt of such an information sends out signals to correct the situation.

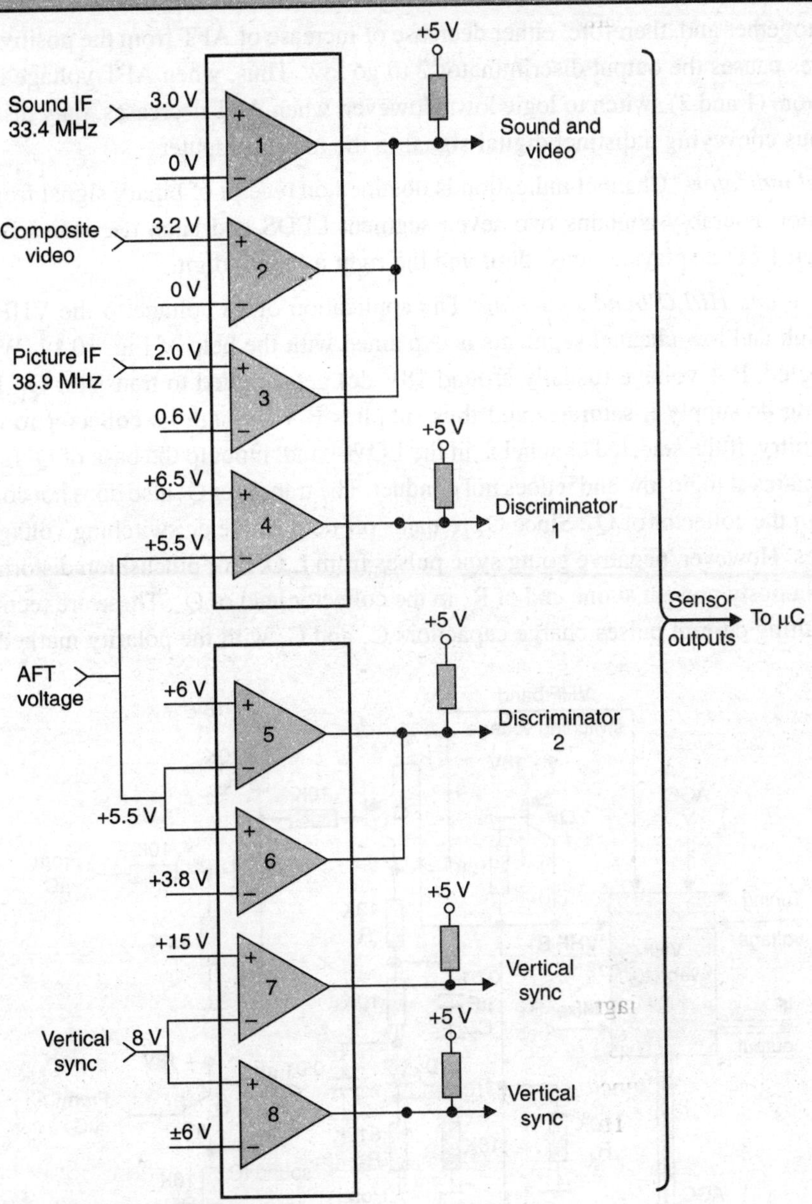

Fig. 10.11 Simplified functional diagrams of various sensing circuits.

The dc derivative of vertical sync applied at the negative input terminal of comparator 7 and positive input terminal of comparator 8 produces two different type of outputs. For example, when the station is properly tuned, nearly 8V dc gets applied to the two comparators. This causes both the comparator outputs to go high. If the receiver is badly mistuned, the input voltage becomes vary low causing comparator 8 to go low. In the absence of vertical synchronization, input voltage to both the sensor units increases and output of comparator 7 switches to a logic low.

The AFT voltage is applied to comparators 4, 5 and 6. Their outputs determine whether frequency deviation of the local oscillator is too high or too low. On receipt of such information the computer reacts to these sensor signals and provides fine tuning correction. Note that outputs of comparators 5 and 6 are tied together and, therefore, either decrease or increase of AFT from the positive and negative reference values causes the output discriminator 2 to go low. Thus, when AFT voltage increases, both the discriminators (1 and 2) switch to logic low. However, when AFT decreases, only discriminator 2 is at low logic thus conveying a distinct digital signal to the microcomputer.

Channel indicators. Channel indication is obtained on receipt of binary signal from the μC. The channel indicator assembly contains two seven segment LEDS and latch decoder driver ICs. When activated the left LED displays a 'tens' digit and the right a 'units' digit.

VHF B + and HI/LO band switching. The application of dc voltage to the VHF tuner and its switching to high and low channel segments is explained with the help of Fig. 10.12. When any VHF channel is selected, B + voltage (usually around 18V dc) gets applied to transistor Q_1. It is so biased that on receiving dc supply it saturates and thus supplies B + through its collector to the associated VHF tuner circuitry. If the selected channel is in the LOW band, input to the base of Q_2 (see Fig. 10.12) from the μC remains at logic low and it does not conduct. The transistor Q_3 also does not conduct because its base is tied to the collector of Q_2. Since Q_3 remains off no positive dc switching voltage is fed to the tuner coil diodes. However, negative going sync pulses from L.O.T. or pincushion distortion correction board are continuously present at one end of R_3 in the collector lead of Q_3. These are rectified by diode D_1 and the resulting current pulses charge capacitors C_1 and C_2 with the polarity marked across them.

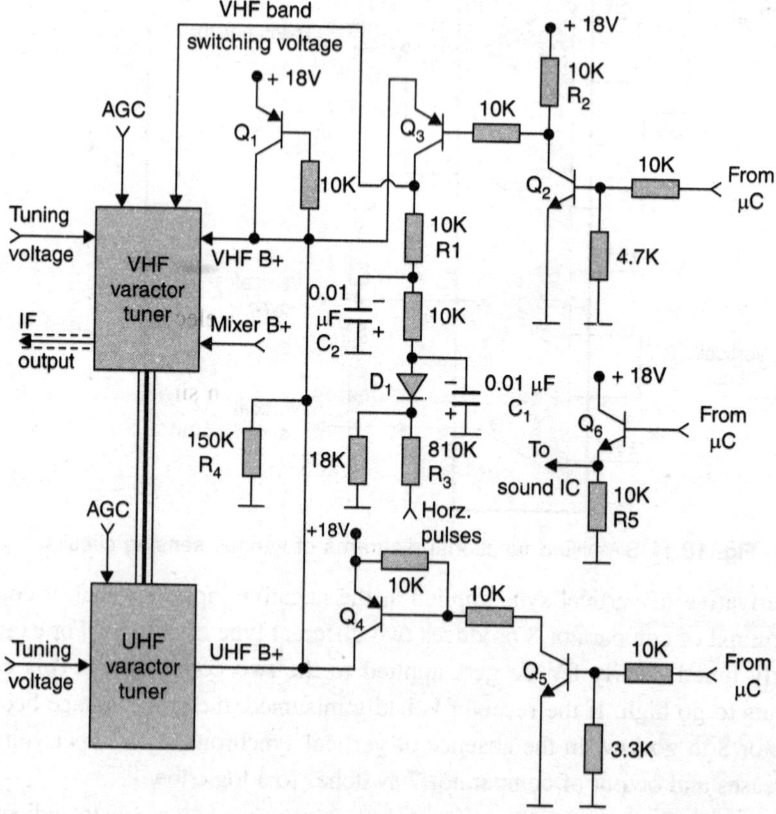

Fig. 10.12 Simplified circuit diagram of VHF B+, HIGH/LOW VHF band and UHF band switching circuitry.

Thus negative dc voltage is applied through R_1 to reverse bias band switching diodes provided across a section of the coils in the VHF tuner.

When a HIGH Band VHF channel is selected, input to Q_2 from μC goes high and this saturates it. The excessive collector current causes enough voltage drop across R_2 and transistor Q_3 also saturates. The full conduction of Q_3 enable application of B + from collector of Q_1 to the band switching diodes. The conduction of Q_3 also discharges C_1 and C_2 and hence no negative voltage can reach the band switching diodes. The band switching diodes when forward biased, short part of the coil windings and hence higher VHF channels can be selected.

UHF band switching. The necessary circuitry for UHF band switching is included in Fig. 10.12. When any UHF channel is selected, dc supply from the emitter of Q_1 switches over to the emitter of Q_4 thus removing dc (B +) supply from the VHF tuner and connecting it to the UHF tuner. The μC removes positive step voltage from the base of Q_2 and instead applies it to the base of Q_5. This saturates Q_5 which in turn saturates Q_4. Thus + 18 V is applied to the UHF tuner. The collector of Q_4 is also connected to the base of Q_1 and hence a high positive voltage at this electrode reverse biases Q_1. Thus no B + voltage can get applied to the VHF tuner. A potentiometer is often used instead of resistor R_4 and AGC reference is tapped from here. Such details have deliberately been omitted from the circuit.

Audio-mute switching. An output from the μC mutes audio in the absence of a station signal. The necessary circuit is included in the control IC (see Fig. 10.12). When there is no station signal a positive step voltage from the computer provides base drive to transistor Q_6. It conducts and supplies a dc voltage (that develops across R_5) to a pin on the sound IC. This voltage drives the IC into cut-off and thus mutes the audio. When a signal is present *i.e.*, a station is tuned in, the input Q_6 from the μC goes low and it goes off. This removes application of any voltage to the sound IC and hence, it functions normally.

10.11 PROGRAM PRESETTING

After having learnt the essentials of computer controlled channel selection and tuning it is instructive to review the procedure to be followed for automatic and manual programming. In recent receivers while the tuning process remains the same the sequencing has been simplified. Some such details are given in chapter 17. A typical control panel having provision for remote control has been chosen for this purpose. The location of various controls is shown in Fig. 10.13. With this panel, 12 channels including other program sources such as a video tape recorder can be selected. The programs can be selected by pressing program select buttons located on the receiver control panel or by using the hand-held Remote Commander. The available VHF and UHF channels can be selected in numerical sequence automatically using the automatic program search system or by manual programming when it is desired to select channels in a order other than the numerical. With manual mode a channel with a signal too weak to be picked up by the automatic search system can also be selected. Manual presetting can also be used to cancel a channel preset at a particular position and substitute another channel.

Panel Layout

As shown in Fig. 10.13, There are 12 program select buttons. Each button has its indicator which lights up to show program number being selected. There are 10 channel indicator LEDS located below the volume and power buttons. These LEDS glow to indicate approximate number of the channel being hunted for during auto selection. These LEDS also act as guides when manual up-down (+, –) program buttons are operated. The AFT switch is of sliding type and can be set at either 'ON' or 'OFF' position. The auto-program, manual program and reset buttons are covered by a search switch gate (not shown in Fig. 10.13). The inner side of the search switch gate has a pin which slides into corresponding hole (shown) located below the auto program button. With gate closed the pin presses and keeps the initiate switch off. As soon as the gate (search switch) is opened the contacts of 'initiate switch' close and set the system ready for channel selection. This is indicated by the flashing of 1st program indicator LED. The entire panel has a cover lid with a window. It is held by hinges and only opened for channel selection and manipulation of other controls.

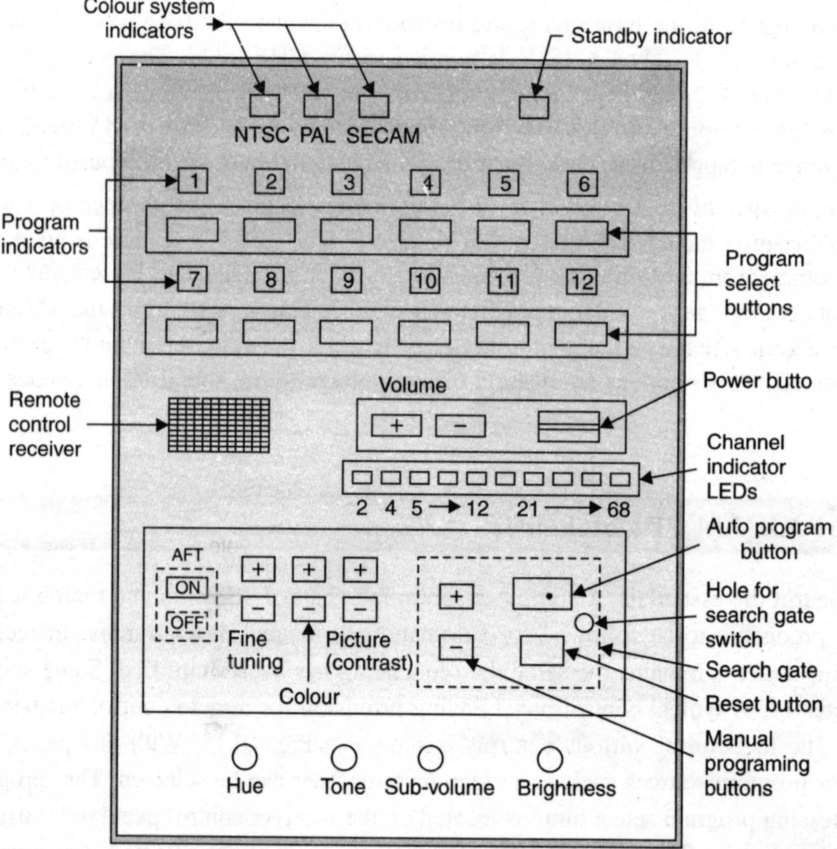

Fig. 10.13 Television receiver control panel showing location of various switches, and allied indicators.

Automatic Programming

As stated earlier the system under discussion is designed to select automatically all receivable channels in the VHF and UHF bands. The procedure for auto programming is as follows:

(1) The first step is to open the panel door and push power button to switch 'on' the receiver. The first program button is then pressed and search switch gate opened to expose program pre-setting controls. The program indicator, LED No. 1 starts blinking.

(2) On pressing the 'auto-programming' button, channel selection process commences. The first channel found is preset in the first program select position. The process repeats automatically in a sequence till no channel can be found. The auto-programming stops at the highest UHF channel. This is indicated by the glowing of last two or three channel indicator LEDS.

The µC has in its memory necessary divide ratios to be fed to the programmable dividers of the frequency synthesizer for all VHF and UHF channels. It does so sequentially. Thus the local oscillator frequency in the tuner changes in turn for each successive channel. As shown in Fig. 10.8 and 10.9 the divide ratio of the programmable divider in the reference oscillator divide chain also alters under command from µC, for example when channel selection moves from VHF to UHF range. The comparator in the PLL loop ensures that the local oscillator attains correct frequency for each channel. If a channel is available the sensors inform the µC accordingly and it is preset on the first available program selector. The next selector indicator starts blinking and this process continues till the last available channel is preset. Note that µC tries to tune-in all channels. However, for channels which are not available in the area, the sensors send opposite data to it and the same are ignored. The entire process takes a few seconds and stops at the last (69th) channel (100th in most present receivers).

(3) As soon as channel selection is over, it is indicated by the channel indicator LEDs. The last program indicator stops blinking. The search gate switch is then closed.

(4) When presetting is over, channels are selected by the viewer in numerical sequence by pushing program select buttons to check that channels have been properly selected.

Manual Programing

With manual programing any desired channel and other sources like VCR and TV games can be preset in any sequence. A list is usually made of the programs to be preset and the sequence in which these are desired. The procedure to the followed is as under:

(1) On switching on the receiver and opening the search switch gate the first program indicator LED starts blinking. The desired select button is pressed.

(2) Referring to picture on the screen (+) and (–) manual program buttons are pressed to locate desired channel number. The (+) manual program button leads to higher channels and (–) button to lower channels. These buttons are not to be held in but just pressed and released.

A good guide to channel number being hunted for is indicated by the channel indicator LEDs, for example when the first two LEDs light up to shows that lower VHF channels are being searched. Similarly, glowing of 3 to 5 indicator LEDs shows that upper VHF channels are being tried. The lighting of last five (6 to 10) LEDs indicates that selection in the UHF range was going on. Thus a decision to press (+) or (–) manual program button can be taken after looking at the channel indicator LEDs.

(3) The next desired preselect button is pressed and step number 2 repeated. This is continued till all listed programs are selected. In strong signal areas, searching may stop between channels. When this happens the manual program button is pressed again. The search gate switch is closed on completion of all desired selections.

Reset Button

The reset button is provided for erasing any memory on a particular program position and to move that channel to a frequency below channel 2's frequency. With the search switch gate closed the program position on which any memory was earlier erased will be skipped. The desired channel can thus be reached quickly.

AFT switch. The AFT switch is kept 'ON' for operation in the automatic mode. The AFT locks-in the correct signal and secures a sharp picture.

Fine tuning. In case of external interference or for an extremely weak station, fine tuning control can be used. For this, the search switch gate is closed (program indicator does not blink) and AFT switch set to 'OFF'. The manual fine tuning buttons (+, –) are then operated until a clear picture is obtained. The AFT switch must stay on at OFF while fine tuning for the channel.

Other controls. The picture (+, –) buttons are actually for contrast control. With colour control set for near no colour, the two buttons can be used to get a picture with correct contrast of black and white areas. Similarly, the colour or picture buttons are provided to obtain a well defined picture with proper colours. If the colours are light, the (+) colour button is pushed. In the other case *i.e.*, when colours are too deep the (–) colour button is operated.

On switching on the receiver to any program, the sub-volume control can be varied to get desired listening level. The sound volume can be further changed with the (+), (–) volume buttons. However, once the power is switched off the sound volume gets automatically reset to the level of 'sub-volume' control and the receiver always turns on to that level regardless of the level previously set by the volume buttons. The brightness control functions in the usual way.

The tone control is varied to get desired level of treble in the sound output. The hue control is preset at the factory for PAL reception. It can be varied to get good colours when the receiver is set to receive NTSC transmission or when playing a tape on the VCR recorded for NTSC 4.43 colour system.

10.12 REMOTE CONTROL OF RECEIVER FUNCTIONS

Two schemes of remote signalling to the television receiver are:

*(1) by mechanically or electrically generated ultrasonic waves in the range of 34 to 54 KHz.

(2) by infrared wave transmission and reception using an infrared transmitting LED and a receiving sensor.

All modern receivers employ infrared transmission technique for channel selection and remote control of other functions. The complexity of any such system depends on the number of things it does. Its most important function is changing of stations. Next it is useful to be able to alter the sound level preferably in a series of small steps. Turning the receiver ON and OFF is also a desired control function. The remote control of contrast and brightness are also useful. Finally it is also desirable to be able to alter colour intensity from a distance. As such many remote control systems have been developed to suit different requirements. The remote circuitry is quite complex and integrated circuits are used to

* For details refer chapter 27 of the book "MONOCHROME AND COLOUR TELEVISION" by R.R. Gulati (New Age Publication).

process various inputs before infrared wave transmission. A sketch of the remote control transmitter unit which goes with the channel selection system described in section 10.11 is shown in Fig. 10.14. The unit has 12 buttons for channel selection including one (12) for VCR connections. There are also volume, picture and colour buttons. In addition it has a power button, a sound mute button and a normal level button. The unit operates with four 1.5V small cells.

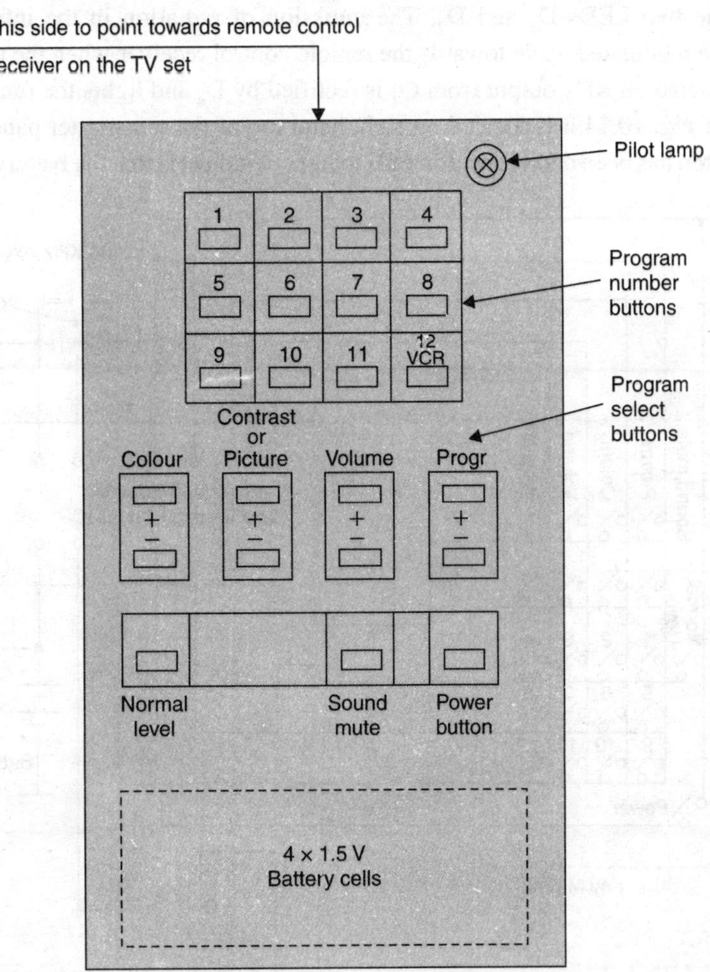

Fig. 10.14 Front view of a typical infrared wave remote control transmitter.

Since transmission to the TV receiver is by infrared waves, it is somewhat directional. The surface labelled as such should point towards the receiver. The range of useful operation is around 7 meters within a radius of ± 30° from the remote control receiver on the control panel of the TV set.

Remote Transmitter

A simplified schematic diagram of the remote transmitter of Fig. 10.14 is shown in Fig. 10.15. It needs 24 different commands to carry out all the listed function. Thus 24 different pulse patterns are generated and when a button is pressed, the infrared waves are modulated via a 48 KHz carrier by one of the 24 pulse patterns.

The pulse output (high or low) on depression of any one key button activates special individual circuitry in the encoder. It produces a distinctive pulse train which modulates the crystal controlled 48 KHz carrier. After due amplification it is fed to the LED driver which passes proportionate current pulses through the two LEDs-D_1 and D_2. The emission of radiation in the infrared region causes transmission of the modulated wave towards the remote control receiver when the unit is aimed at it. In addition the modulated 48 KHz output from Q_1 is rectified by D_4 and lights the function indicator LED (D_3). As shown in Fig. 10.14 it is located on right hand top of the transmitter panel. This LED shows that a function button has been depressed. If it fails to light, it indicates that the battery needs replacement.

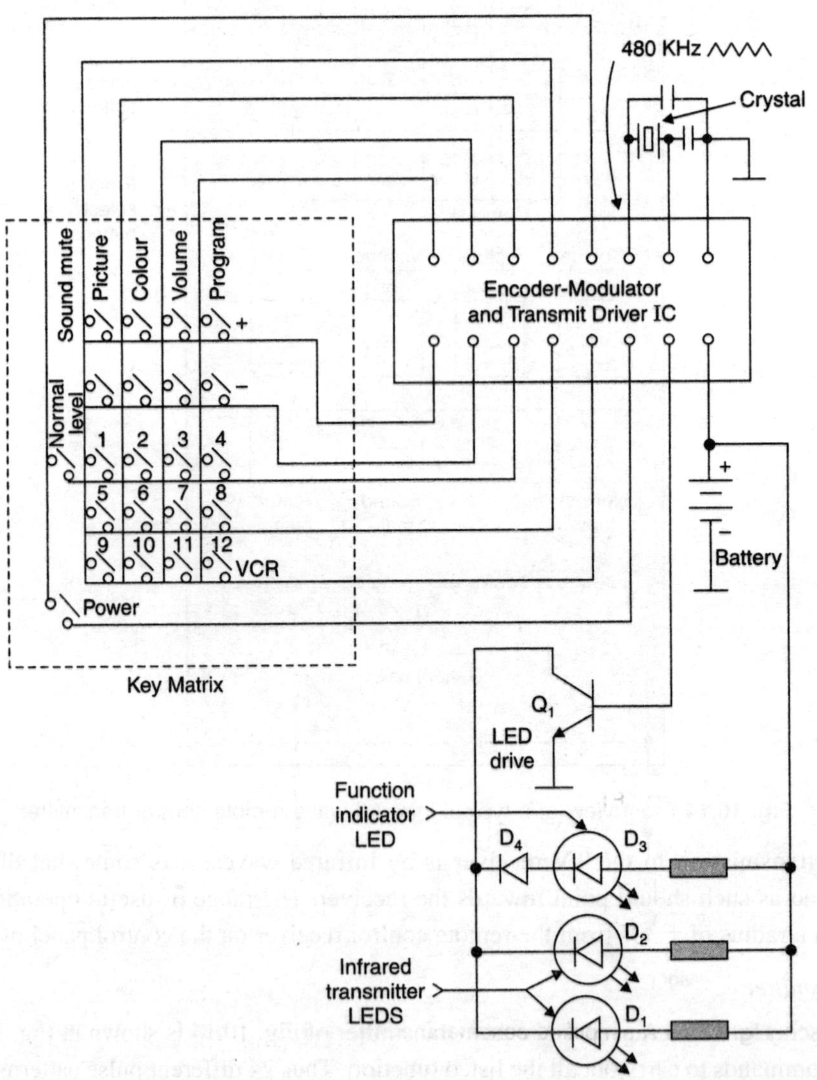

Fig. 10.15 An oversimplified schematic diagram of a remote infrared transmitter.

Remote Control Receiver

A schematic diagram of the remote control receiver is shown in Fig. 10.16. The pulse-modulated infrared signals from the remote infrared transmitter are picked up by the remote infrared sensor D_1. The picked up ac signal feeds into the FET (Q_1) amplifier, for some amplification. It is then coupled to the remote processor and pulse shaper via emitter follower Q_2. The modulated signals are further amplified and then demodulated. The recovered pulse train then feeds into the pulse shaper which is designed to generate pulse patterns (ramps, high, low etc.). Each pattern is different and corresponds to the function button pressed on the remote control unit. The shaped individual pulse pattern is then dc coupled to the microcomputer (see Fig. 10.16). The μC on receipt of different inputs produces appropriate outputs to perform the ordered function.

To turn power 'ON', the power button is pressed (see Fig. 10.14). For this the μC output is a step (+) voltage at one of the pins of its IC. This positive voltage is fed through a resistance (33K in Fig. 10.16) to the base of transistor Q_3 which on conduction lights the lamp of the optical coupler. When the lamp lights its radiation reduces resistance of the associated LDR (Light Dependent Resistor). The low resistance of the LDR enables application of 220V ac to the gate of *TRIAC. Since the triac conducts on both positive and negative half cycles of the ac line voltage, it acts as an ac power switch.

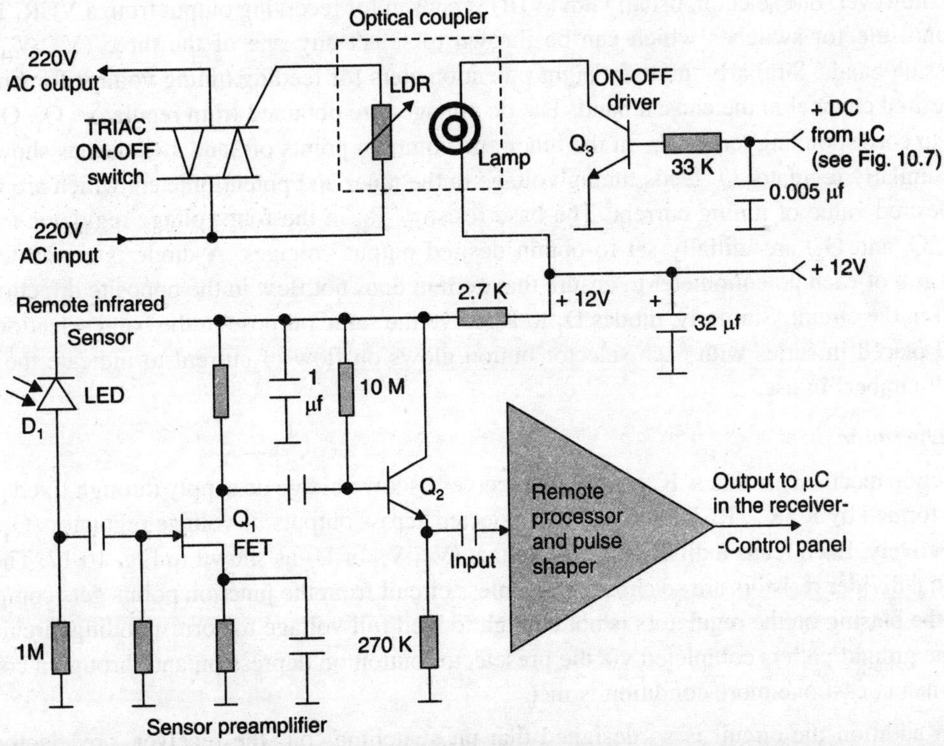

Fig. 10.16 A simplified schematic diagram of the remote control receiver.

*The triac is a solid state bi-directional gate controlled switch that provides full-wave control of the ac power. It may be noted that a thyristor provides one way control only.

CHAPTER 10

To turn the receiver 'OFF' power switch on the remote controller is pressed again. This results in a pulse output that finally removes positive voltage at the base of Q_3. The lamp in the optic coupler goes off and the LDR assumes its high resistance. The triac then ceases to conduct and ac supply to the receiver is cut-off. Another depression of the power button will again switch on supply to the receiver. The sequence is repeated on successive depression.

10.13 ELECTRONIC TOUCH TUNING

Electronic touch tuning is quite common in present day receivers. The switches are of feather touch type and interlocked in such a way that only one button (switch) can remain 'on' (depressed) at a time. The moment another switch is depressed by a light touch on its button, the one already 'on' pops-up to assume its 'off' position. The channel selection and tuning is affected through a specially designed switching IC. The integrated circuit is essentially a gating circuit which provides a closed circuit when several inputs are simultaneously present.

An oversimplified circuit of a typical electronic touch tuning system is shown in Fig. 10.17 (pull-out). A set of eight (I to VIII) soft touch buttons enable selection of eight different channels. In practice, however, one selector, usually 8th (VIII) is chosen for receiving output from a VCR. There are eight band selector switches which can be thrown to select any one of the three (V_L, V_H, U) TV transmission bands. Similarly there are eight potentiometers for feeding tuning voltage for fine tuning of the desired channel in the chosen band. The dc voltages are obtained from regulators Q_1, Q_2 and Q_3 and fed to corresponding terminals on the tuner and common points of band switches as shown in the figure. Similarly regulator Q_4 feeds tuning voltage to the tuner and potentiometers which are varied to obtain desired value of tuning current. The base resistors R_B in the four voltage regulator transistors (Q_1, Q_2, Q_3 and Q_4) are initially set to obtain desired output voltages. A diode is connected in the variable arm of each potentiometer to ensure that current does not flow in the opposite direction due to any fault in the circuit. Similarly, diodes D_1 to D_8 serve the same purpose in the band selection circuit. An LED placed in series with each selector button glows on flow of current to indicate the selector position (number) in use.

Circuit Operation

The selection mechanism that is IC controlled receives its low voltage dc supply through fixed potential dividers formed by R_1-R_2 , R_3-R_4 and R_5-R_6 connected across outputs of voltage regulators Q_1, Q_2 and Q_3 respectively. Each feeds a different band contact (V_L, V_H or U) as shown in Fig. 10.17. The values of potential divider resistors are so chosen that unless circuit from the junction points gets completed to ground, the biasing on the regulators is not enough to feed full voltage to corresponding circuits in the tuner. The ground path is completed via the preselector button on depression and through a contact on the IC when atleast one more condition is met.

In addition the circuit is so designed that on switching 'on' the receiver, preselector button (switch) No. 1 provides alternative path for flow of current even when it is not depressed. This is possible because as soon as dc voltages becomes available on switching 'on' the receiver, current flows from the junction of R_5-R_6 via channel selector switch No. 1 (shown at contact for band V_L), diode D_1, LED_1 emiter of Q_5 to earth terminal (G) on the IC thus completing the circuit. The same potential (V_{CE} of Q_5) also appears at pin 1′ of the switching IC. Another circuit gets completed from Q_4 via tuning potentiometer No. 1, Q_5 and earth terminal G, of the IC. The enhanced potential now available at pin 1′

of the IC is enough to close the gate configuration at that pin which then provides paths for channel selection and tuning currents from corresponding dc sources.

Initially band switches are set to select desired bands and potentiometers varied in turn after depressing corresponding feather touch buttons to obtain desired channels. Later, a light touch on any one of the selector buttons tunes in instantly the preselected channel.

On switching on the receiver, if the channel available on No. 1 is not desired, the selector switch of the wanted channel, say No. II is depressed (touched). This disconnects button I contact that pops-up., the moment No. II is closed. Now current flows from dc source (Q_4) via 2nd band selector switch shown on V_H), D_2, LED_2, closed contact of selector switch No. II, R_4 and pin G. (ground) of the IC. The same voltage also appears at pin 2′ of the IC and the action explained earlier repeats to select the preset channel. This sequence also repeats when any other selector button is depressed, thus enabling instant selection of any one of the pre-set channels.

10.14 AUTOMATIC FREQUENCY TUNING (AFT)

The need for AFT control and its principle of operation were explained in section 7.3. It may be stressed again that proper tuning of a colour receiver is critical because the colour subcarrier must be properly positioned on the IF response curve. This is illustrated in Fig 10.18 (*a*). Consider that due to faulty tuning the local oscillator frequency changes to become high by 1 MHz. This as shown in Fig. 10.18 (*b*) will shift the picture, subcarrier and sound-IF frequencies by the same amount causing the low frequency video side-bands to become weaker and chrominance side-bands stronger. The highly selective traps in the IF amplifier will reject the vision-IF and place it at zero level on the IF response curve. Such a mistuning can cause a weak picture with interference on colours due to heterodyning between colour subcarrier and sound frequencies.

If, however, the local oscillator frequency changes to become less by 1 MHz, the opposite will happen as shown in Fig. 10.18 (*c*). As labelled on the IF response curve, picture IF will shift to the maximum level and subcarrier reduced in amplitude. As a result of decrease in colour level and in particular loss of colour burst, only a black and white picture will be seen on the receiver screen. From the above discussion it is clear that proper tuning is absolutely necessary for high quality colour TV reception. Improper fine tuning can result in loss of colour, weak colour, wrong colours, and colour interference. Thus it is essential that an automatic fine tuning circuit must be incorporated in colour receivers. Receivers employing synthesized frequency tuning may not need such a control but it is always provided to take care of any channel frequency offsets and inaccurate cable frequencies.

AFT circuits. The circuit diagram of an AFT circuit employing discrete components is shown in Fig. 10.19. The IF frequency signal is picked up from the 3rd IF amplifier and fed to the Foster-Seely discriminator through tuned amplifier Q_1. The transformer T_1 is tuned to the correct picture IF frequency *i.e.*, 38.9 MHz. The diodes D_1 and D_2 are then fed equal signals and develop equal and opposite voltages across R_3 and R_4. Hence, no correction voltage is developed. With any deviation in local oscillator frequency the IF frequencies also change accordingly. This makes the voltages fed to D_1 and D_2 different and hence a net positive or negative dc control voltage is developed at the output of discriminator. It is amplified by Q_2 and fed to the varactor diode that is located in the oscillator circuit of the tuner. A typical response curve of the discriminator is drawn in Fig. 10.19. As obvious, any deviation in the IF frequency results in a corresponding dc control voltage to pull the oscillator back to the correct frequency.

CHAPTER 10

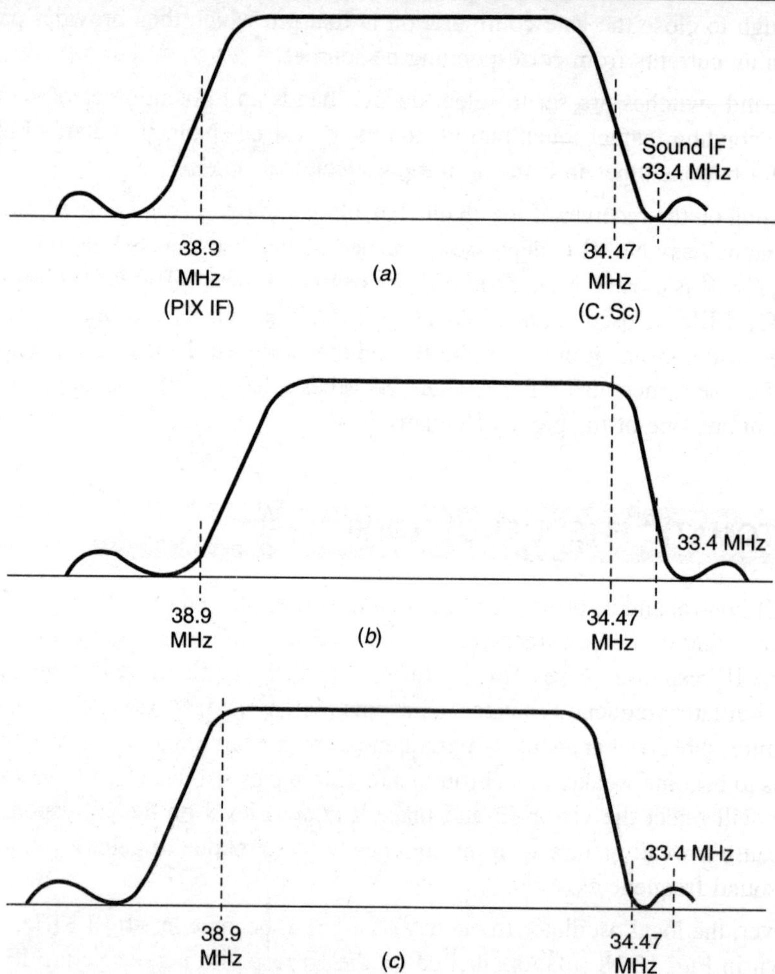

Fig. 10.18 Effect of changes in fine tuning on the location of picture, colour sub-carrier and sound IF frequencies on the IF response curve (*a*) Correct local oscillator frequency (*b*) high, and (*c*) low.

The AFT control is actually an automatic frequency control (AFC). All AFC circuits have a 'pull in' range and a 'hold in' range. The 'pull in' range is the frequency farthest from the normal frequency that the AFT circuit can lock into when operative. An example of this is when channels are switched. The 'hold in' range is that frequency range, starting from locked AFT condition, in which the AFC circuit can stay locked even if there is oscillator drift or fine tuning control is changed. The pull-in range of the AFT circuits is about ± 50 KHz and the hold-in range around ± 1 MHz.

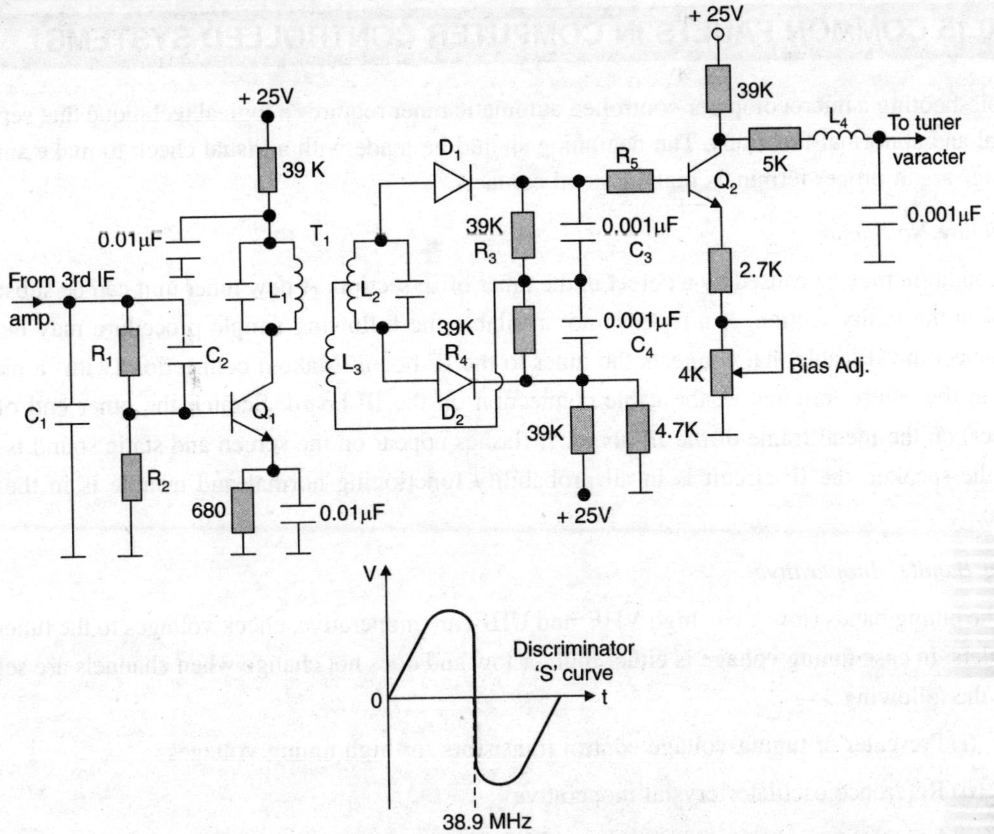

Fig. 10.19 An AFT frequency sensitive detector and amplifier circuit.

Integrated AFT circuit. Modern receivers employ an integrated circuit for AFC. It is either a separate IC or part of the IF sub-system. Such a circuit is shown in Fig. 10.20. Note that the 38.9 MHz input tuned circuit, the discriminator transformer and decoupling R-C networks are the only components that are external to the IC. The input as earlier is from the 3rd IF amplifier and dc control output feeds across the tuner varactor.

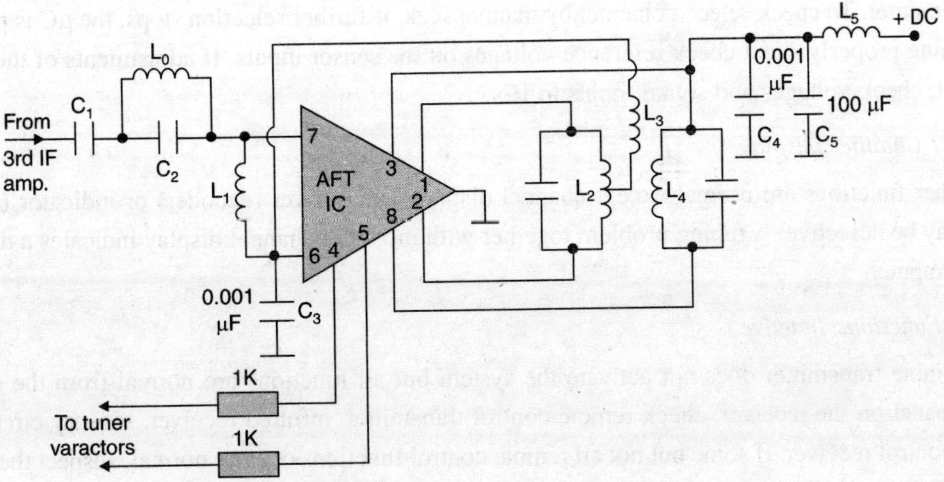

Fig. 10.20 An AFT circuit employing integrated circuitry.

10.15 COMMON FAULTS IN COMPUTER CONTROLLED SYSTEMS

Troubleshooting a microcomputer controlled automatic tuner requires a logical technique that separates normal and abnormal functions. The beginning should be made with a visual check to make sure that all plugs are at proper terminals making good contacts.

No Picture No Sound

This condition may be caused by a defect in the tuner of IF section. A new tuner unit can be substituted to isolate the faulty section. If a tuner is not available the following simple procedure may be used. Disconnect the IF cable that connects the tuner to the IF board. Make a connection (with a piece of wire) to the centre terminal of the cable connection on the IF board. Scratch the other end of wire (jumper) on the metal frame of the IF board. If flashes appear on the screen and static sound is heard from the speaker, the IF circuit is in all probability functioning normal and trouble is in the tuner system.

Tuning Band(s) Inoperative

If all the tuning bands (low VHF, high VHF and UHF) are inoperative, check voltages to the tuners and prescalers. In case tuning voltage is either high or low and does not change when channels are selected check the following :—

(*i*) Prescaler or tuning-voltage control transistors for high tuning voltages

(*ii*) Reference oscillator crystal inoperative

(*iii*) Low tuning voltage at the output of PLL filter.

If only one tuning band is inoperative check discrete switching circuits and μC outputs to these circuits (refer figures 10.7 through 10.12).

Automatic programming incorrect. Automatic programming or seek function is active only when 'auto programming' button is depressed. Continuous channel selection may be caused by a defective sensor IC, misadjusted threshold control levels (reference dc voltages in Fig. 10.11), or a bad microcomputer. To check select a channel by manual seek, if further selection stops, the μC is probably functioning properly. Next check reference voltages on the sensor inputs. If adjustments of these have no effect, check voltages and signal inputs to ICs.

Incorrect Channel Display

If all other functions are normal except channel display, LED driver (decoder) or indicator (see Fig. 10.8) may be defective. A tuning problem together with incorrect channel display indicates a defective microcomputer.

Remote Functions Inactive

If the remote transmitter does not activate the system but all functions are normal from the receiver control panel on the receiver, check remote control transmitter, infrared receiver, sensory circuits and remote control receiver. If some but not all remote control functions operate normal, suspect the remote control transmitter. In any case, measure transmitter battery voltage and if necessary replace weak cells. An oscilloscope if available can be used to check the infrared system. If operative a modulated 46 KHz wave (≈ 1V PP) will appear on the scope on depression of any button of the remote commander. The vertical input of the C.R.O. can be connected through a 10:1 probe at any suitable point in the infrared modulator circuit.

10.16 FAULTS IN AFT CIRCUITS

The AFT system is a closed loop system. Locating problems in such a system is difficult because output signal depends on the input signal and the input signal depends on the output signal. It is necessary to open the loop to locate the fault. That is why an AFT defeat switch is often provided at its output.

The AFT control though normal otherwise may not function if 'pull in' and 'hold in' ranges are exceeded. Manufacturers usually give typical operating voltages and test points for measuring them. In general a weak or no AFT performance can be caused by a defective transistor or transformer in its discrete circuitry. In the integrated version the IC may be defective. A misalignment of the associated transformers can also be the cause of a weak AFT control.

When there is no picture, no sound and raster is otherwise normal, it can be the result of an inoperative local oscillator due to some defect in the AFT circuit. This can be a shorted varactor or a varactor forced into forward bias by some defect in the dc amplifier (see Fig. 10.12). In general a large tuning error on all channels is caused by a shift in varactor bias. Such a shift can be caused by a defective dc amplifier or a defective varactor diode. Thus a very careful analysis and voltage checks at several points with AFT switch 'ON' and 'OFF' are necessary to localize a fault in the AFT and associated circuits.

REVIEW QUESTIONS

1. Explain with a suitable block diagram how the output frequency of a voltage controlled oscillator (VCO) can be maintained to the accuracy of a crystal controlled oscillator.
2. What is the significance of base 2 in the digital system of writing numbers ? Write digital equivalents of numerical numbers :

   (*i*) 21, (*ii*) 134, (*iii*) 10^3 and (*iv*) 10^5.
3. Enumerate basic rules of binary addition, subtraction, multiplication and division. Explain the functions of AND, OR, NOT and NAND gates that are employed to preform the above mathematical operations.
4. Explain the need for binary inputs to a digital frequency divider. Illustrate with a block diagram the operation of a binary divider using a latch control circuit.
5. Draw functional schematic diagram of a simple digital computer and explain briefly the functions of (*i*) ALU (*ii*) Control Unit, (*iii*) CPU (*iv*) ROM and (*v*) RAM. How is the ROM memory different from a RAM memory.

6. Enumerate various methods of feeding input to a computer ? Explain how the output unit is interfaced to provide output results in different forms.
7. What do you understand by frequency synthesized method of channel selection ? What is the role of a microcomputer (μC) in selection of various channels? Illustrate your answer with a suitable block diagram.
8. Explain with simple circuit diagrams how the picture IF, sound IF, composite video and AFT signals from the receiver are compared with reference voltages in the comparators (sensors) to obtain necessary digital data to be fed to the microcomputer.
9. Explain how a μC operates to control VHF band switching, channel display and audio mute circuits.
10. Explain with simplified discrete circuitry how the command signals in binary form are converted to suitable dc voltages for controlling various operations like application of dc voltages to varactor diodes in the tuner and audio mute signal to the sound IC.
11. Explain a basic key board organisation of an 12 channel electronic 'TOUCH TUNING' channel selection and identification system.
12. A simplified circuit of a typical electronic touch tuning is shown in Fig. 10.17. Explain how channels can be preselected on different locations of the eight position preselector.
13. A control unit (window) is provided on the front panel of receivers equipped with a microcomputer for preselection of channels. Sketch a typical panel and explain the procedure for auto and manual selection of channels. Why is the AFT switch provided in such units?
14. Draw block schematic diagram of a typical remote control unit provided with TV receivers. Describe simply its basic mode of operation and the manner in which infrared waves are used to transmit various command signals. How are these signals decoded by the infrared receiver located on the front panel of the receiver?
15. What is the function of AFT control in a TV receiver ? Why is it necessary to incorporate such a control in colour receivers ? Draw circuit diagram of a typical AFT module and explain how it operates to keep the local oscillator frequency locked at the correct value. Is it necessary to provide AFT in a frequency synthesized system of channel tuning? If so, why?

11 VISION IF SUBSYSTEM

INTRODUCTION

The tuner output feeds into IF amplifier where the small signal is processed to provide large gain, high selectivity desired wave shaping, vestigial sideband correction and rejection of adjacent channel signals. In modern receiver a SAW filter is used to obtain desired selectivity characteristics. Its use in place of conventional L-C filters offers many advantages like low component count, lower assembly costs and consistent performance over large periods of time. However, the use of SAW filter causes an insertion loss of about 20 dB. This is made up by providing a wideband amplifier that preceds the filter configuration.

The IF amplifier forms part of the vision IF IC that also includes AGC, AFT (AFC), video detector and preamplifier circuitry. A large number of ICs like TDA 2540, TDA3540 and CA7611 for n-p-n (incremental) tuners and TDA2541 TD3541 and CA7607 for the *p-n-p* (electronic) tuners are available to perform these functions. These ICs are monolithic integrated circuits in 16-lead dual-in-line plastic packages and have identical pin connections. The ICs TDA3540 and TDA3541 are, however, improved versions of TDA2540 and TDA2541 respectively and hence preferred in improved receiver designs. The integrated circuits CA7611 and CA7607 introduced by BEL have nearly the same performance characteristics as those of TDA3540 and TDA3541. Another IC that is commonly used in many receivers is TDA4420. It is also a monolithic IC suitable for electronic turners but in a 18-lead package where both positive and negative video outputs are simultaneoulsy available on two separate pins. The VIF ICs used in some earlier receiver designs include TA7607AP, TA7611AP and HA1144OA. These are also in 16-lead packages like CA7607/CA7611 and have similar pin designations except that there is no provision for independent AFT ON/OFF and IF AGC, VCR switches. An IC CX 2005 which has 28 pins and combines VIF and SIF stages is also used in some receivers.

The circuit description of various section of the vision IF IC that follows is based on CA 7611/ CA7607 which has operating characteristics and package similar to TDA 3540/TDA 3541. Later sections of this chapter are devoted to VIF subsystems employing some other ICs.

11.1 VISION IF INTEGRATED CIRCUIT

The block schematic of vision IF subsystem employing IC CA 7611/CA 7607 is shown in Fig. 11.1. It is housed in a modular box and soldered to the main PCB to avoid any RF interference. The IF signal from the tuner is coupled to the preamplifier through a 75 ohm RF cable. A matching network at the amplifier input ensures required bandwidth for vision and sound carriers without sacrificing tuner gain. The preamplifier, as stated earlier is to compensate for the loss in SAW filter.

As shown in the figure, IF signal from the SAW filter is applied to the input pins 1 and 16 of the IC. It delivers video and intercarrier sound IF (SIF) outputs at pin (12), AFC output at pins 5 and AGC voltage for the tuner at pin 4. The tuner AGC delay is set by a potentiometer connected at pin 3. The VCR switch is connected at pin 14. It may be noted again that these pin numbers pertain to ICs TDA3540/ TDA3541 and CA7611/CA7607.

Preamplifier

A large number of IF SAW filters are available to match different VIF ICs. Similarly, a few integrated preamplifier units have also been developed. A typical combination of a discrete preamplifier and SAW filter feeding on to the IC is shown in Fig. 11.1. The amplifier is built around the high frequency transistor BF959. The matching SAW filter is BMC389. The network comprising of L_{201},C_{201} and C_{202} and output impedance of mixer circuit forms a double-tuned circuit which is designed to provide impedance match and have bandwidth enough to accommodate both video and SIF signals. The transistor BF959 has features like large signal handling capability, higher f_T (f_T is the frequency at which the short circuit current gain attains a value equal to unity) and lower noise figure. The resistors R_{201}, R_{202} and R_{205} determine quiescent operating point of the transistor. The capacitor C_{204} across R_{205} bypasses signal frequencies that lie at the upper edge of IF signal bandwidth thus enabling gradual decrease in negative feedback at higher frequencies necessary for optimum bandwidth characteristics. The parallel combination of L_{202} and R_{206} constitutes the collector load while the series combination formed by C_{203} and R_{203} is the feedback network from collector to base. The components R_{204}, C_{206} and C_{207} provide decoupling across the V_{CC} supply. The closed loop gain of the preamplifier is around 26 dB. The amplified IF signal from the collector of BF959 is coupled to the SAW filter through C_{205} thus isolating dc voltage at input terminals of the filter.

SAW Filter

The basic structure and operating principle of a SAW filter are explained in section 7.6. Its main features as VIF filter are as under:

(*i*) enables significantly improved picture quality due to its excellent phase characteristics.

(*ii*) has high performance equivalent to several conventional coils.

(*iii*) saves circuit assembly hours when compared to discrete circuitry.

(*iv*) is small in weight.

(*v*) has very little deviation in characteristics with changes in temperature.

(*vi*) is adjustment free and hence cannot be tempered with during servicing of the receiver.

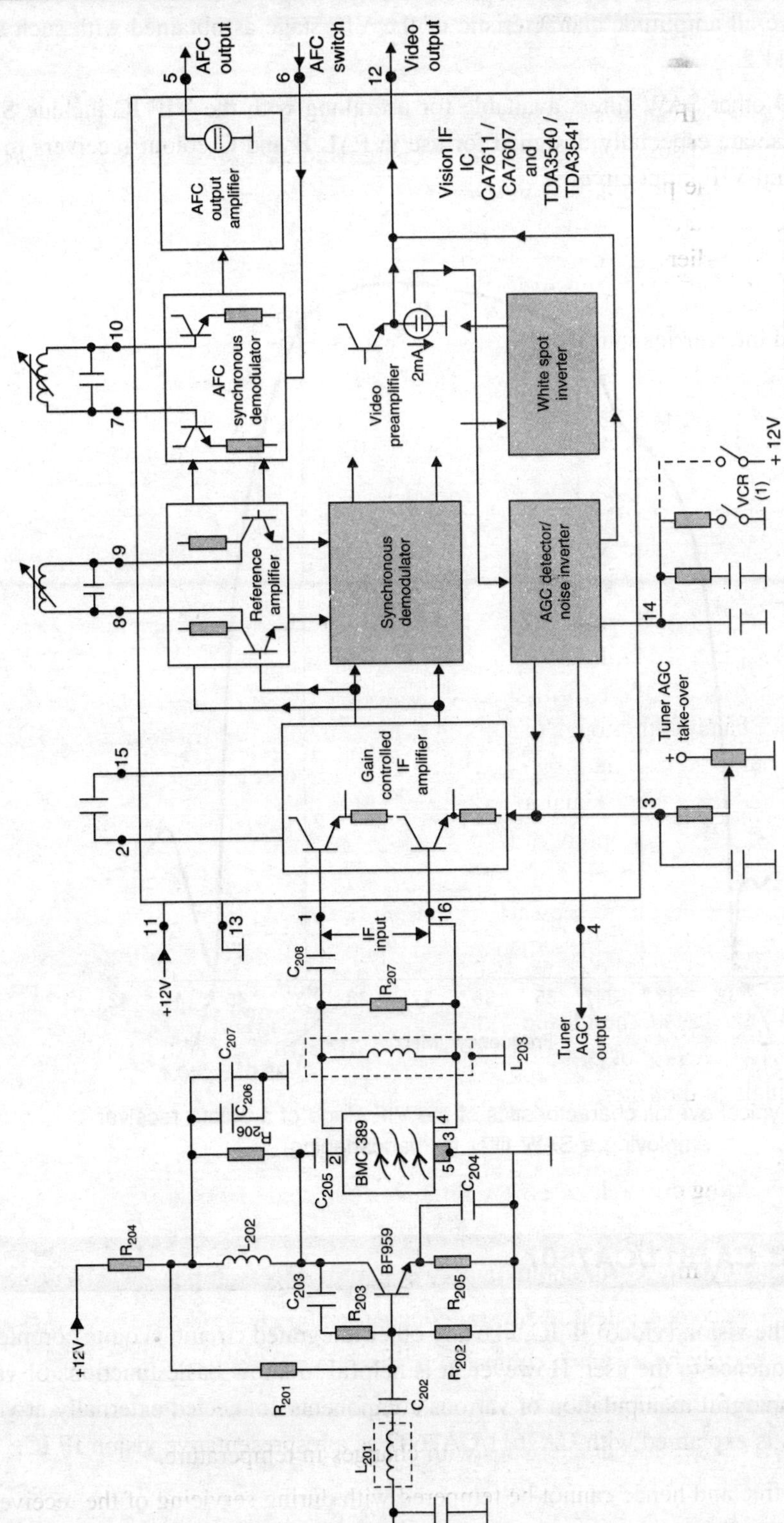

Fig.11.1 Block diagram of a vision IF subsystem employing IC CA7611/CA7607 or TDA 3540/TDA3541.

CHAPTER 11

The improved overall amplitude characteristic of the VIF stage as obtained with such a SAW filter are shown in Fig .11.2.

Besides BMC389 other SAW filters available for use along with the VIF IC include SW173, F1034 and F10488. These are especially designed for use in PAL B and G colour receivers to match different tuner outputs and VIF input circuits.

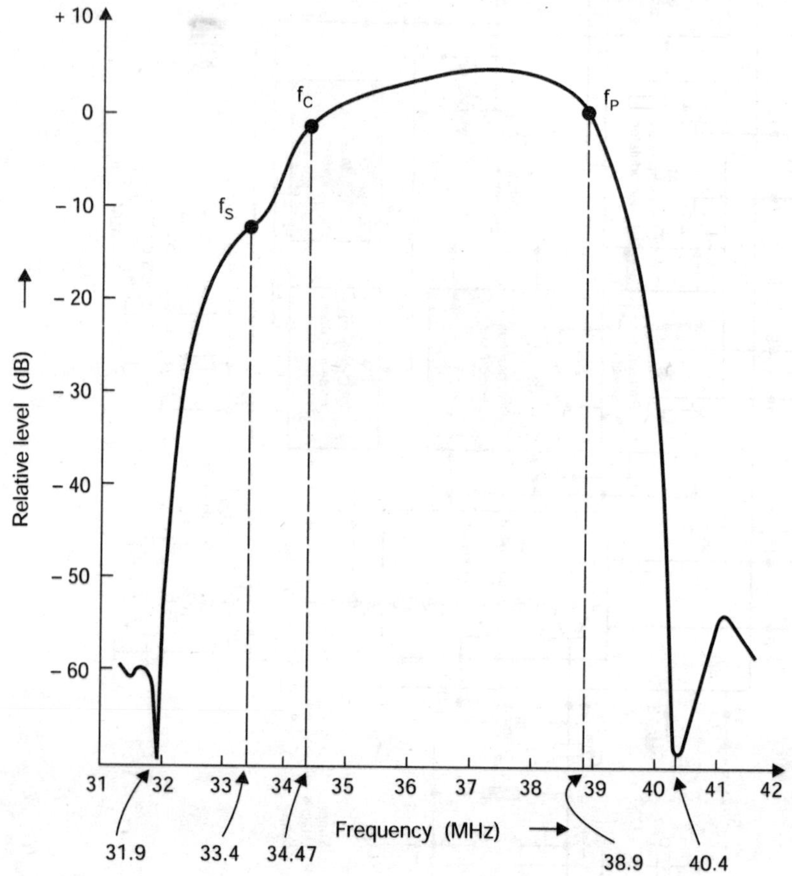

Fig. 11.2 Typical overall characteristics of the VIF stage of a colour receiver employing a SAW filter for bandshaping.

11.2 VISION IF IC CA7611/CA7607

The detailed circuitry of the vision (video) IF IC, like any other integrated circuit, is quite complex and infact of not much consequence to the user. However, it is helpful to know basic functions of various sections of the IC or meaningful manipulation of various components connected externally at various pins of the package. This is explained with CA7611/CA7607 as a respresentative vision IF IC.

IF Amplifier

It is a cascaded three stage wideband amplifier as shown in Fig. 11.3. It amplifies IF output received from the SAW filter to a level suitable for detection. Pins 1 and 16 are the input pins for the first differential amplifier formed by transistors Q_1 and Q_2 with Q_5 and Q_6 as their dynamic loads. The two emitter followers Q_3 and Q_4 are to increase input impedance at the input points of the differential amplifier. The AGC function is obtained by varying gain of these amplifiers by changing impedance of the dynamic loads. The operating points of Q_5 and Q_6 (dynamic loads) are changed to vary their impedance by applying AGC voltage at their emitter leads. Overall gain of the amplifier is stabilized by negative feedback from buffer amplifiers Q_{11} and Q_{12} provided at the output of three stage amplifier. The feedback is through an R-C network and an external capacitor connected across points 2 and 15 (see Fig. 11.3). A value of 1 KPF for C_{209} provides high frequency roll-off necessary for desired bandwidth of the IF stage. The inductor C_{203} (see also Fig. 11.1) is varied to be tuned at 35.15 MHz (central IF frequency) with output capacitance of the SAW filter, input capacitance across pins 1 and 16 and other stray capacitances. Better video frequency response is obtained by having a high loaded Q (quality factor) for L_{203}.

The maximum overall gain of the IF amplifier is kept at about 60 dB to obtain a sensitivity of nearly 100 µV at 38.9 MHz at the onset of AGC. A wide AGC range equal to the gain of amplifier is obtained by applying control voltage (AGC) sequentially starting from the third amplifier. This takes care of the wide range of input signal levels and enables ideal signal to noise performance. The input stage of the 1st IF amplifier is so designed that input impedance of the IC stays constant at about 3K-ohm despite AGC voltage variation thus preventing any deterioriation of circuit performance.

Video detector. The object of this stage is to recover composite video signal from the output of IF amplifier. The usual diode detector through simple, is not used because of its following limitations:

(*i*) it does not provide any output unless the modulated input signal amplitude exceeds barrier threshold voltage of the diode. This means that the IF amplifier gain should be of the order of 80 dB with a minimum usable antenna signal of 100 µV.

(*ii*) with low input signal level, the diode operates on the non-linear portion of its characteristics to generate sum and difference frequencies of picture IF, sound IF and colour IF carriers. For example the 33.4 MHz SIF can beat with 34.47 MHz colour IF subcarrier to generate the unwanted beat interference frequency of 1.07 MHz. Similarly other combinations result in more beat frequencies needing many trap circuits to filter them out.

(*iii*) this detector does not provide any gain and actually causes a signal loss to be made up by setting higher gain of the video amplifier.

Synchronous Detector

The above stated disadvantages are overcome by providing a synchronous detector in the VIF IC. Its major advantages are:-

(*i*) with no barrier potential to be overcome it provides appreciable gain thus eliminating the need of another (4th) video IF amplifier stage.

(*ii*) operation is in the linear region of the devices used and hence, no interfering sum and difference frequencies are generated thus making it unnecessary to provide IF traps.

(*iii*) In the absence of any sum products at the output of the detector, no radiation interference is caused at the tuner and hence elaborate shielding is not necessary.

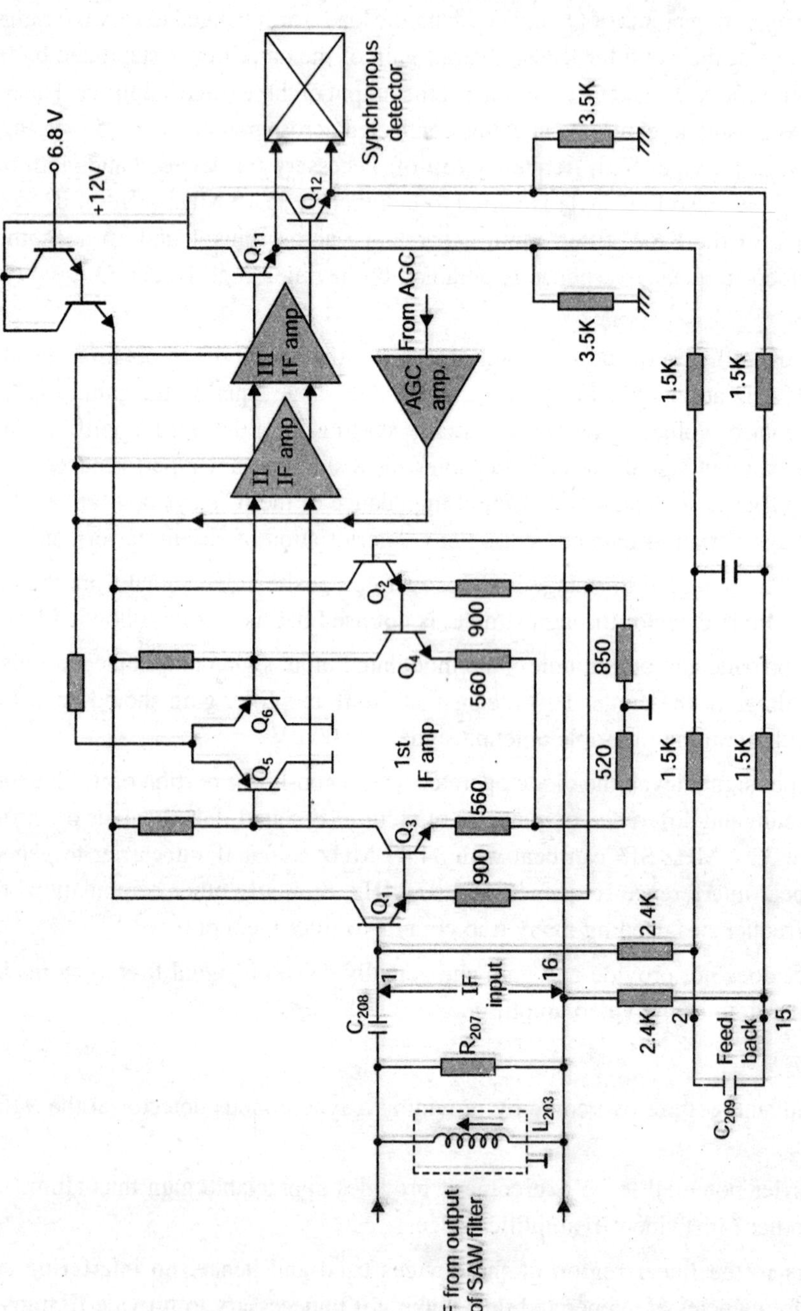

Fig.11.3 A three stage wideband AGC controlled IF amplifier that forms part of the vision IF IC CA 7611/CA 7607.

Synchronous Detector Principle

A synchronous detector (as explained in chapter 7) requires two separate inputs unlike the simple diode detector. For detecting composite video signal it needs (*i*) the usual modulated IF signal and (*ii*) a reference signal at the picture IF carrier frequency and phase. A simplified block diagram of such a detector is shown in Fig. 11.4. Basically it consists of two dual-differential amplifiers and two modulator transistors. The reference picture IF carrier is fed in a push-pull fashion to the differential amplifiers and their outputs are connected in parallel. This results in a push-pull input and a parallel output configuration. The two modulating transistors which are connected from the emitters of differential amplifiers to ground also have their inputs connected in a push-pull manner. As a result these can vary output currents of differential amplifiers according to modulation of IF signal. The resulting current variations represent video modulation as it occurs at the transmitting end. The detector circuit is called 'synchronous' because it detects only IF modulation information that is in synchronism with frequency and phase of the reference picture IF carrier. The picture carrier itself is cancelled in the differential amplifiers which act as a balanced demodulator. Thus when synchronization is correct, the only output from the detector is at the modulating frequencies.

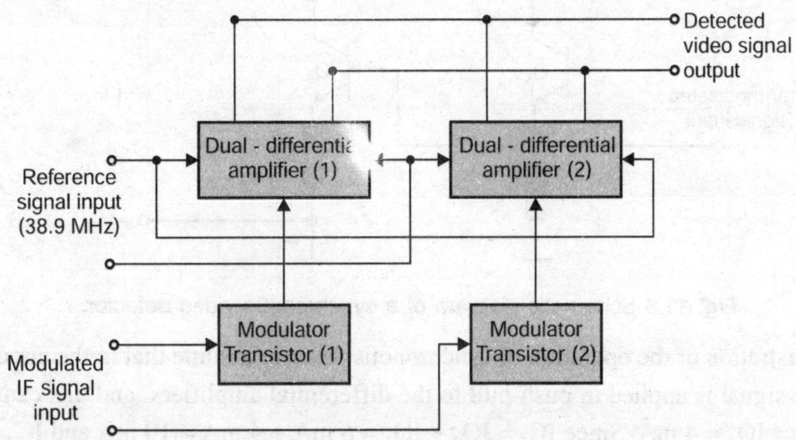

Fig. 11.4 Block diagram of synchronous video detector.

Schematic diagram. The circuit diagram of the synchronous detector of Fig. 11.4 is shown in Fig. 11.5. The transistors Q_1, Q_2, constitute one differential amplifier while Q_3, Q_4 form the other. Similarly, Q_5 and Q_6 are the two modulating transistors. The 38.9 MHz reference signal to be fed at the differential amplifier inputs is obtained by passing the complete IF signal through a narrow band amplifier called the reference amplifier (see Fig. 11.1). This is tuned to 38.9 MHz and thus most IF sideband frequencies are filtered out leaving only an amplified picture IF reference signal. Any remaining AM sidebands do not interfere with the operation of detector because reference signal amplitude into the detector is around 300 mV which over-rides the differential amplifier thereby cliping-off any amplitude variations.

Referring again to Fig. 11.5, the equal loads ($RC_1 = RC_2 = RC$) are so paralleled that the push-pull inputs to the two differential amplifiers do not produce any output voltage. Similarly any unmodulated signal impressed at the inputs of Q_5 and Q_6 does not develop any output voltage. However, with modulated signal as the inputs, voltage drops across the load resistors become unequal and a differential output

CHAPTER 11

voltage becomes available that is proportion to the AM envelope and thus to the modulating composite video signal. Any increase in amplitude of the carrier on modulation causes a positive going output while its decrease produces a negative going output. Thus the modulating video signal polarity is correctly reproduced. As explained earlier no other output except the differential video is produced at the output terminals of the synchronous detector.

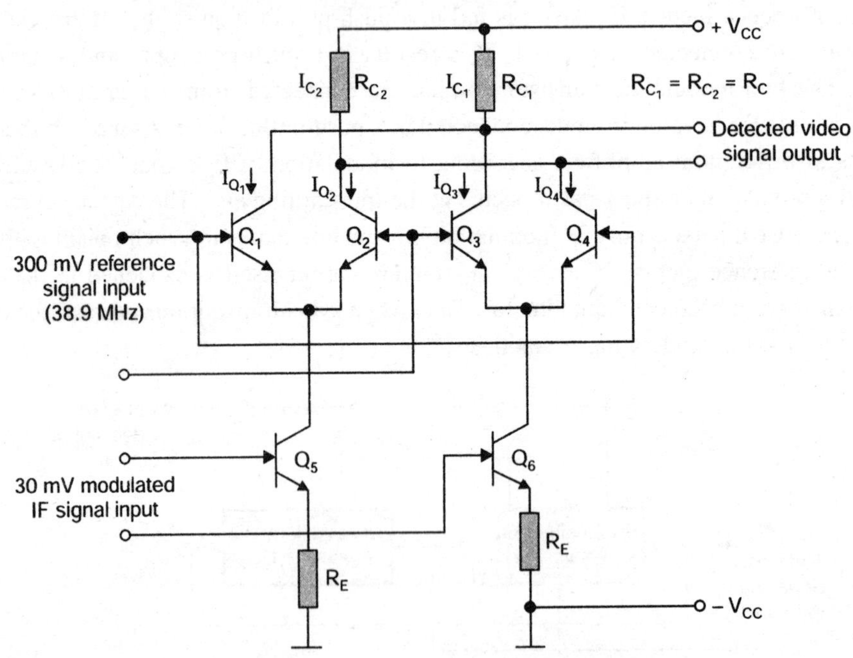

Fig. 11.5 Schematic diagram of a synchronous video detector.

As an illustration of the operation of synchronous detector assume that in the circuit of Fig. 11.5 only a reference signal is applied in push-pull to the differential amplifiers and this causes $IQ_1 = IQ_4 = 6$ mA and $IQ_2 = IQ_3 = 4$ mA. Since $IC_1 = IQ_1 + IQ_3 = 6$ mA + 4 mA = 10 mA and $IC_2 = IQ_2 + IQ_4 =$ 4 mA + 6 mA = 10 mA, the voltage drops across the two equal loads ($RC_1 = RC_2 = RC$) remain equal and no differential output is developed. Next assume that a modulated signal of small amplitude is impressed at the inputs of Q_5 and Q_6 in push-pull. A positive going signal at the base of Q_5 will increase IQ_1 and IQ_2 while corresponding negative swing at the base of Q_6 will decrease IQ_3 and IQ_4 by corresponding amounts. Let the new values be $IQ_1 = 6.6$ mA, $IQ_2 = 4.4$ mA, $IQ_3 = 3.6$ mA and $IQ_4 = 5.4$ mA. This will make $IC_1 = IQ_1 + IQ_3 = 6.6$ mA + 3.6 mA = 10.2 mA while $IC_2 = IQ_2 + IQ_4 = 4.4$ mA + 5.4 mA = 9.8 mA. The difference in load currents IC_1 and IC_2 equals (10.2 – 9.8 = 0.4) 0.4 mA and this causes a differential output voltage across the otherwise equal load resistors. During the next half-cycle when signal polarity at the input of Q_5 and Q_6 will reverse it will cause an output voltage with the opposite polarity. Since the above is true for any modulation amplitude, the video output is a faithful reproduction of the modulating video signal.

Integrated Demodulator Circuit

The integrated circuit arrangement of a synchronous demodulator and preamplifier circuit is shown in Fig. 11.6. The video modulated IF signal is multiplied by its own carrier to recover the video signal. For

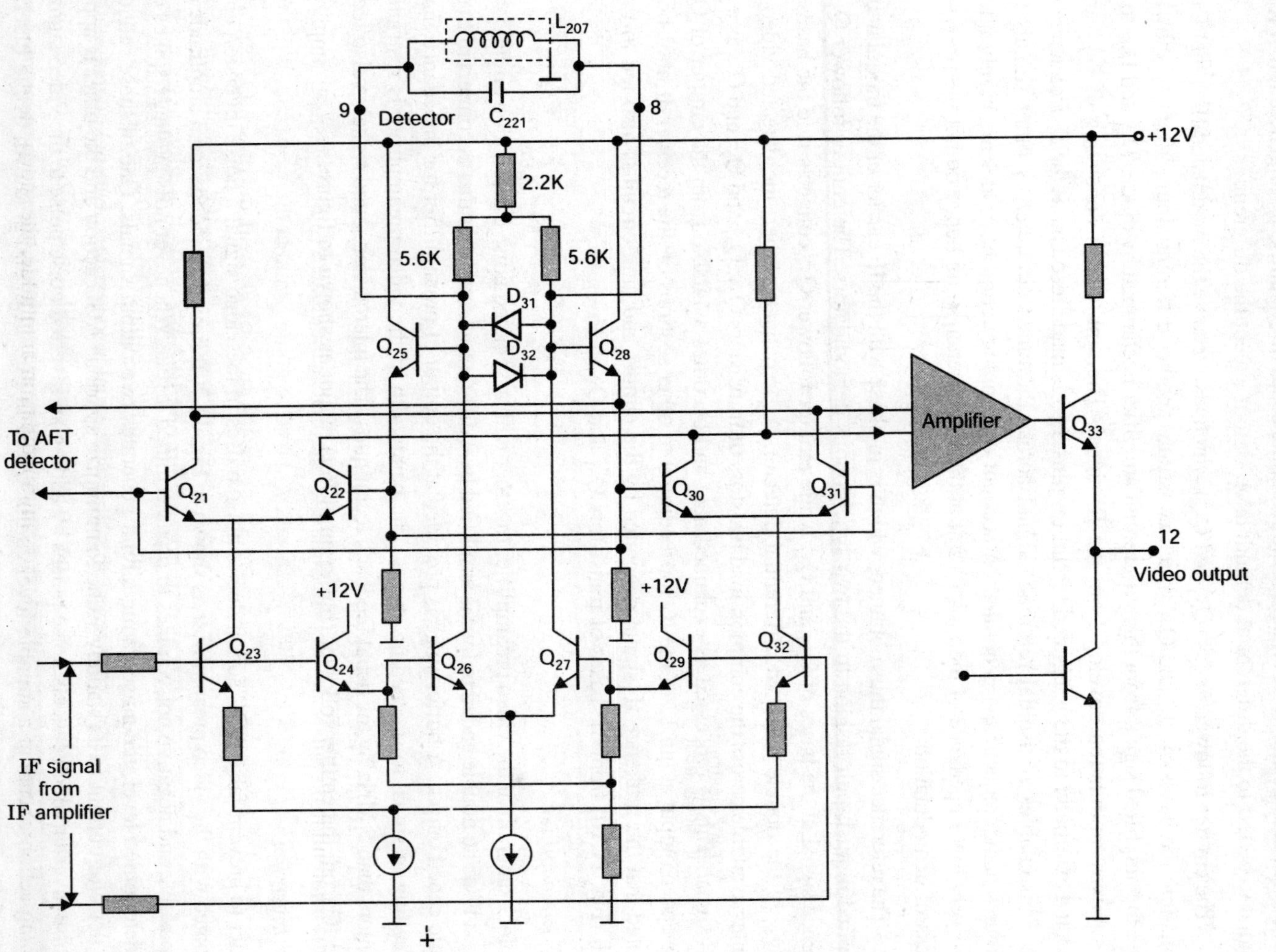

Fig.11.6 The integrated version of synchronous demodulator.

this a double balanced analog multiplier consisting of transistors Q_{21}, Q_{22}, Q_{23} and Q_{30} , Q_{31}, Q_{32} is provided. The modulated signal from the IF amplifier is fed at the bases of Q_{23} and Q_{32} in push-pull for demodulation action. The same signal is also fed at the bases of Q_{24} and Q_{29} for generation of reference IF signal to be fed to the differential demodulator.

The emitter follower stages (Q_{24} and Q_{29}) which also receive the modulated IF signal feed their outputs into the bases of Q_{26} and Q_{27} which constitute another differential amplifier. As shown in the figure, the amplified signal from this differential amplifier is clamped by diodes D_{31} and D_{32} to a level that removes modulation envelop from the IF signal. The tank circuit consisting of L_{207} and C_{221} connected externally to the integrated circuit (at pins 8-9) is made resonant at the IF frequency of 38.9 MHz. This enables sinusoidal reference signal across the tuned circuit. Any other signals (mostly harmonics of IF frequencies) generated on account of clipping action of diodes are highly attenuated due to very low impedance of the tank circuit at these frequencies and hence do not interfere with the process of demodulation.

The reference signal thus obtained, which is in phase with the IF carrier of the modulated signal is applied in push-pull fashion to the two synchronous demodulators. The emitter follower Q_{25} feeds reference signal at the bases of Q_{22} and Q_{31} while emitter follower Q_{28} connects it to the bases of Q_{21} and Q_{30} of the double balanced differential detector. As explained earlier, the detected video output becomes available across the common load resistors of transistors Q_{21}, Q_{31} and Q_{22} and Q_{30} as a double ended signal. This is processed as explained next and becomes available from the emitter of Q_{33} as a single ended output at pin 12 of the IC for feeding it to other sections of the receiver. It may, however, be noted that the reference IF signal generated by the clipper and tank circuit is also fed to the AFT circuit (Fig. 11.10) from the bases of transistors Q_{21} and Q_{30}.

Video Preamplifier

The block diagram of the video preamplifier together with associated noise cancellation circuit is shown in Fig. 11.7. The double ended output is amplified by a differential amplifier that is connected to deliver a single ended output. A buffer stage is provided at the output of preamplifier for impedance matching and feeding of signal to succeeding stages. The bandwidth of the video preamplifier is restricted to a little more than 5 MHz by the usual feedback techniques. The intercarrier sound signal is also fed to the sound IC from the emitter follower (buffer amplifier) output as shown in figure.

Noise Inversion

In order to prevent false AGC action due to noise pulses in the video signal a noise inversion circuit is associated with the video preamplifier as shown in Fig. 11.7. It limits the video signal amplitude during both positive and negative noise spikes to predetermined levels. Whenever noise voltage spikes exceed a predetermined level, the noise gating circuit generates two output signals. One of these reduces the AGC detector current to its standby value to minimize spurious AGC action and the other is applied to the noise inverter which cancels noise pulses by adding an inverted noise pulse to the video signal. The clamping action occurs at a suitable level as illustrated in Fig. 11.8, and is known as the black noise clamp level.

White spots on the screen are caused by excessive peak beam currents due to overmodulated noise signals which are faithfully detected by the synchronous demodulator. Such white spots are

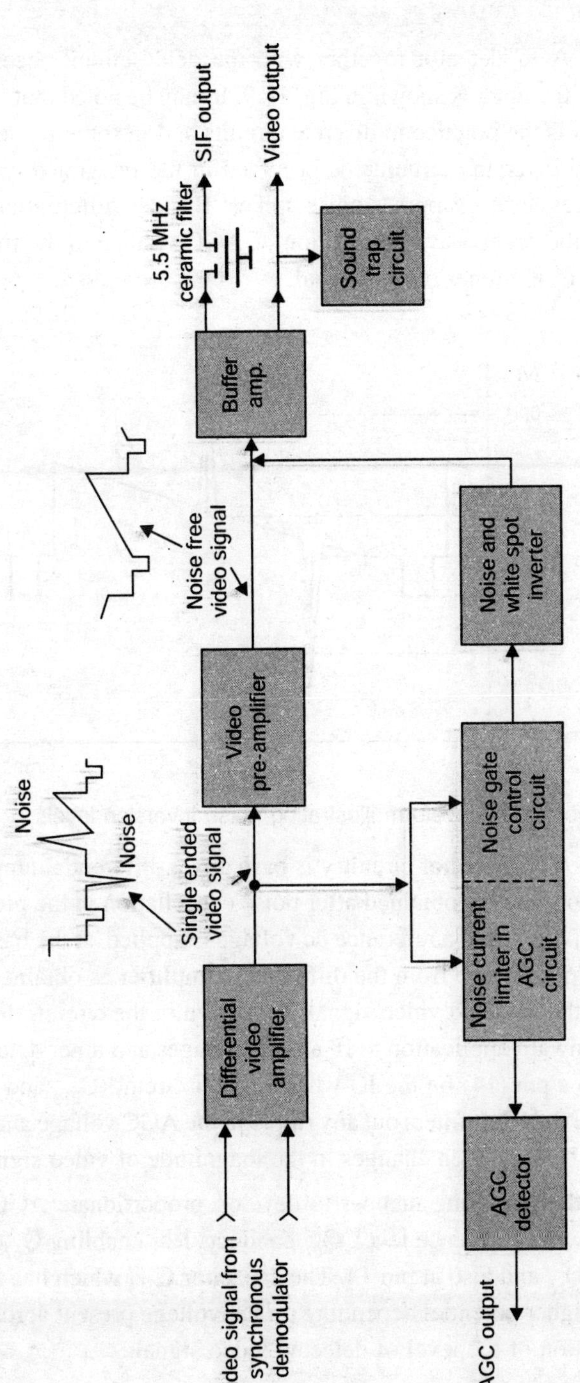

Fig. 11.7 Block diagram of the video preamplifier and noise cancellation circuit.

eliminated by a white spot inverter that forms part of the inverter circuit. It detects white noise spikes exceeding a fixed level and clamps them to a safe amplitude. This is also illustrated in Fig. 11.8.

AGC Detector and Delay Circuit

The circuit schematic of the IF AGC detector together with the delay circuit necessary for feeding automatic gain control voltage to the tuner is shown in Fig. 11.9. It may be noted that flyback pulses are not used in the detector circuit as is the practice in discrete circuits and in some IC versions. This is so because elaborate noise gate and inversion circuitry is provided in the integrated circuit and as such flyback pulses are no longer necessary to suppress noise spikes. Besides differential detector circuits eliminate unwanted signals and the progressive application of AGC voltage to the three IF amplifiers enables noise-free amplification of the tuner output signal.

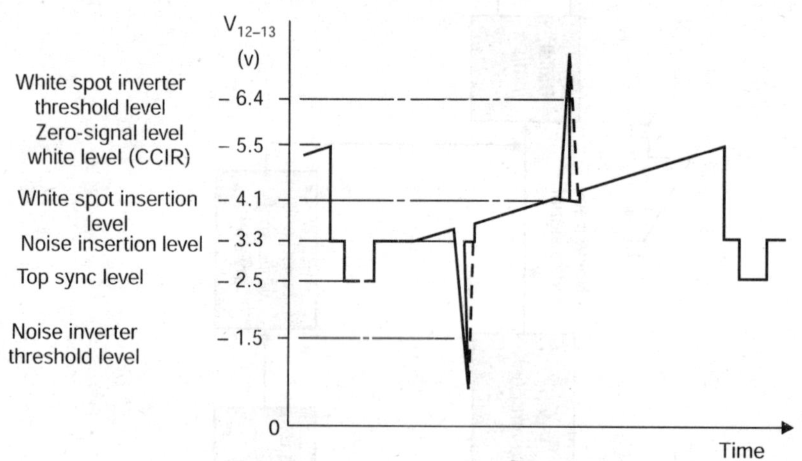

Fig. 11.8 Video signal waveform illustrating noise inversion levels.

As shown in Fig. 11.9 the AGC detector circuitry is basically a differential amplifier formed by transistors Q_{61} and Q_{62}. The video signal as obtained after noise cancellation in the preamplifier circuit (see Fig. 11.7) is fed at the base of Q_{61} while a reference dc voltage is applied at the base of Q_{62} through the emitter follower Q_{64}. The output voltage from the differential amplifier as obtained at the collector lead of Q_{62} is then a function of the detected video signal. As shown in the circuit, this output voltage is applied at the base of Q_{65} for onward application to IF ampifier stages and tuner delay control circuit. This voltage is also connected to a pin (14) on the IC where a filter circuit (C_{213} and R_{214}) of suitable time constant is connected outside the IC. It filters out any ripple in the AGC voltage and also determines how fast the AGC action should be to sudden changes in the magnitude of video signal.

The detector operates in the following manner to develop proportionate AGC voltage. When video signal amplitude falls below the reference level, Q_{61} conducts less enabling Q_{62} to conduct more reducing potential at the base of Q_{65} and also at pin 14. The capacitor C_{213} which has 12V dc supply at one end charges to a somewhat higher potential depending on the voltage present across R_{214}. Thus the potential across C_{213} is an indication of the level of detected video signal.

A decrease in positive potential at the base of Q_{65} (*p-n-p* transistor), causes more conduction in its emitter lead and hence more current through Q_{66}. This in turn means more base current to the *n-p-n* transistor Q_{67} and hence higher current through 1-k ohm resistor to the gain determining dynamic

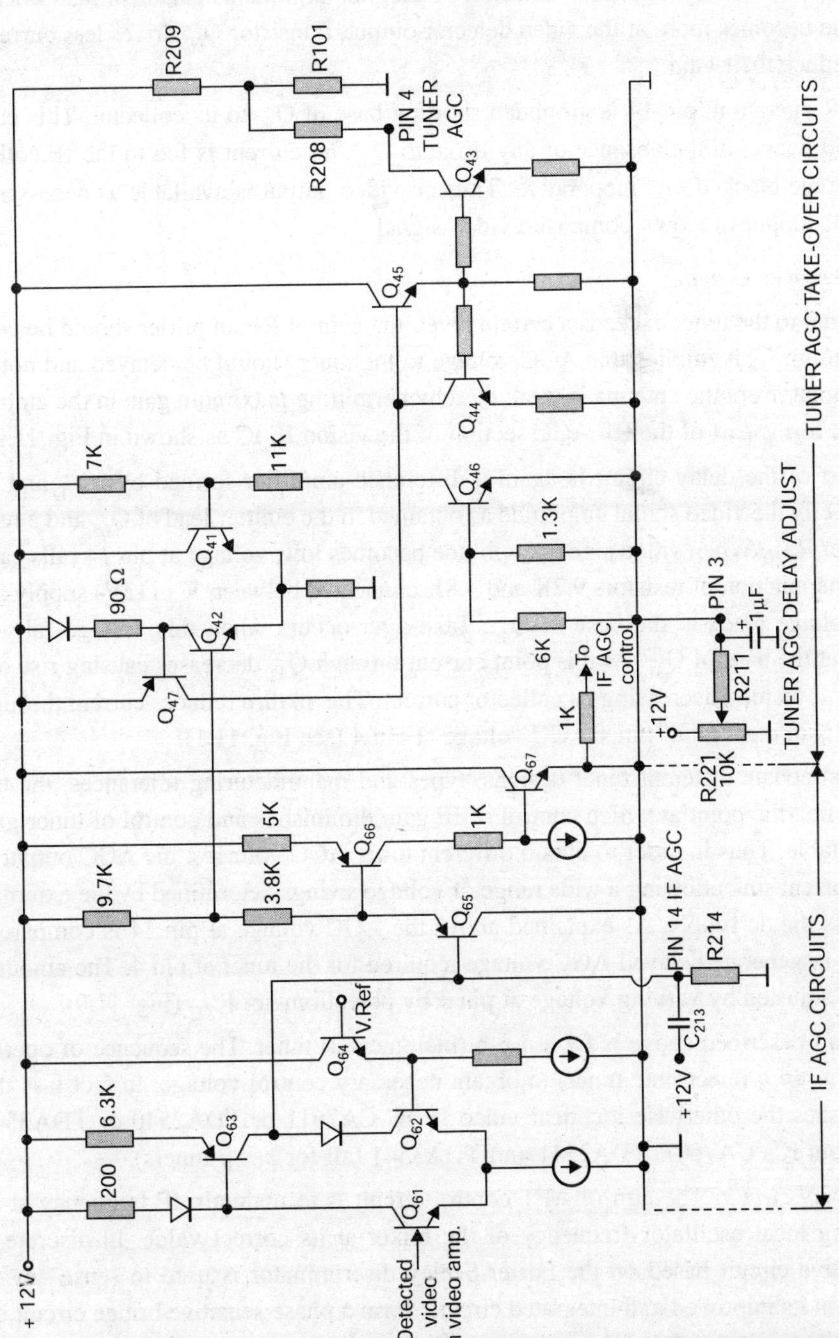

Fig. 11.9 IF AGC and tuner AGC take-over circuits.

CHAPTER 11

collector loads of the IF amplifiers (see Fig. 11.3). The AGC action thus raises their gains to deliver higher video signal to the synchronous detector. Analogous arguments establish that when the video signal amplitude becomes more at the video detector output, transistor Q_{67} drives less current to the IF amplifiers to reduce their gain.

For VCR operation, pin 14 is grounded shorting base of Q_{65} to its collector. This cuts-off both Q_{65} and Q_{66} and hence, in the absence of any drive to Q_{67}, no current is fed to the IF collector loads which then become blocked *i.e.,* inoperative. Thus no video output is available as necessary for VCR operation which supplies its own composite video signal.

Tuner AGC Take-over Circuit

When input signal to the tuner exceeds a certain level, the gain of RF amplifier should be controlled to prevent overloading. This implies that AGC voltage to the tuner should be delayed and not applied so long as input signal from the antenna is weak thereby permitting maximum gain in the amplifier. Such a control circuit forms part of the RF AGC section of the vision IF IC as shown in Fig. 11.9.

The heart of the delay circuit is again a difference amplifier formed by Q_{41} and Q_{42} which compares indirectly the video signal amplitude as obtained in the emitter lead of Q_{65} and a reference set by potentiometer R_{221}. When video signal amplitude becomes low, voltage at pin 14 falls causing drop in potential at the junction of resistors 9.7K and 3.8K connected between V_{cc} (12V) supply and emitter of Q_{65}. This voltage feeds at the base of Q_{42}. Take-over occurs when this voltage falls below the reference level at the base of Q_{41}. At this point current through Q_{42} decreases causing rise of potential at the base of Q_{47}, thereby decreasing its collector current. This in turn reduces current through Q_{45} and Q_{43} which results in increase of tuner AGC voltage at pin 4 (see Fig. 11.1).

To accommodate different tuner designs, types and manufacturing tolerances, the tuner AGC take-over point i.e., the point at which control of IF gain diminishes and control of tuner gain begins, should be adjustable. Thus in order to obtain different tuner AGC voltages, the AGC output at pin 4 is designed as a current sink allowing a wide range of voltage swings determined by the external network connected across the dc supply. As explained above the AGC voltage at pin 14 is compared with the voltage at pin 3 to generate delayed AGC voltage required for the tuner at pin 4. The amount of delay required can be adjusted by varying voltage at pin 3 by potentiometer R_{221} (Fig. 11.9).

The circuit described above is for a n-p-n (incremental) tuner. The sequence of operations will be opposite for a *p-n-p* (electronic tuner) to obtain necessary control voltage. In fact it is this feature which distinguishes the otherwise identical video IF IC CA7611 or TDA2540 or TDA3540 (all for *n-p-n* tuners) from ICs CA7607, TDA2541 and TDA3541 (all for *p-n-p* tuners).

AFT control circuit. The aim of AFT control circuit is to maintain IF frequency at its chosen value by keeping local oscillator frequency of the mixer at its correct value. In discrete circuits a frequency sensitive circuit based on the Foster-Seeley discriminator is used to sense any frequency error. However, in its improved and integrated circuits form a phase sensitive bridge circuit employing differential circuitry is used to develop error voltage. It is applied to the tuner for keeping local oscillator frequency steady at its correct value. Such a circuit which forms part of the vision IF IC is shown in Fig. 11.10.

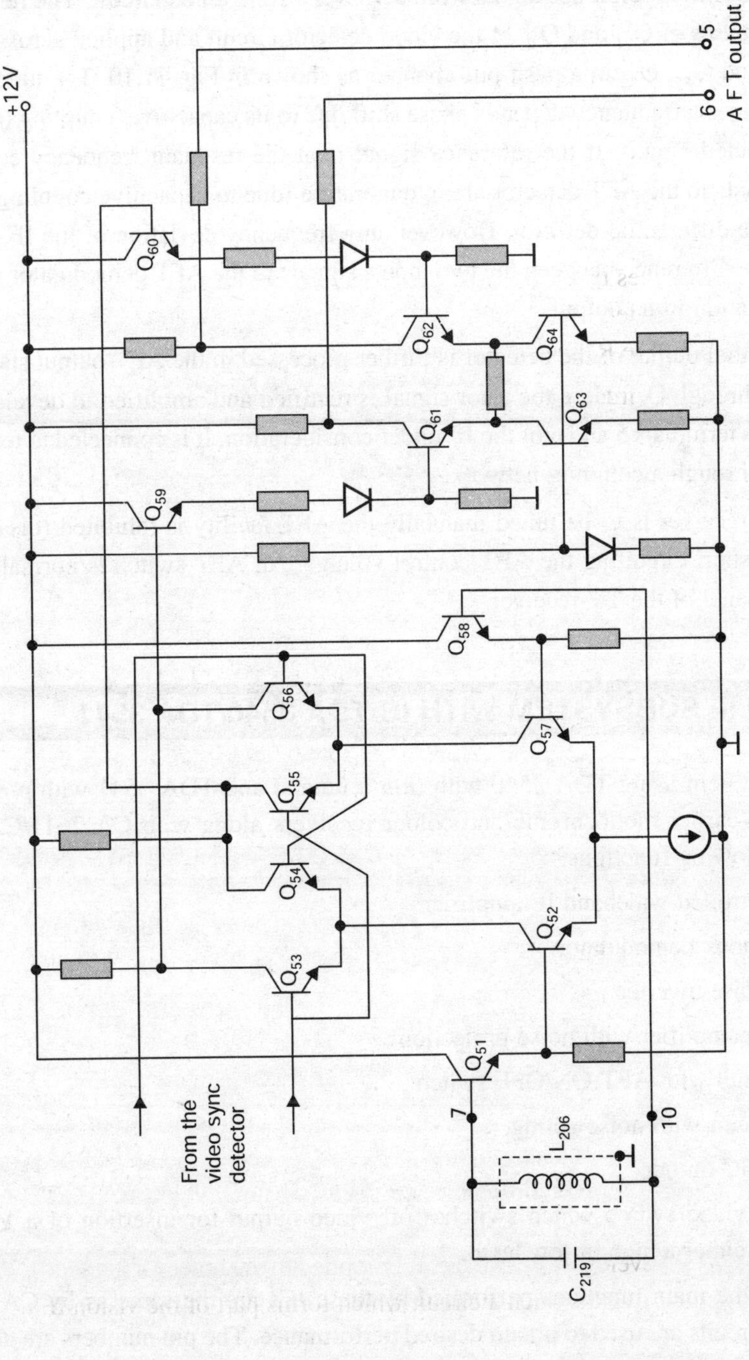

Fig. 11.10 AFT detector and error voltage amplifier circuit.

The AFT detector, the centre of which is a double balanced phase comparator, consists of transistors Q_{51} through Q_{58} . It senses reference IF signal and develops an output proportional to the frequency error. The inputs to the detector are the reference signal from the video detector (see Fig. 11.6) and the phase shifted reference signal from across the AFT tuned circuit. The reference signal is obtained from the bases of Q_{21} and Q_{30} of the video detector circuit and applied across the differential pairs of Q_{53}, Q_{54} and Q_{55}, Q_{56} in a push-pull manner as shown in Fig. 11.10. The tuned circuit across pins 7 and 10 provides a frequency depends phase shift due to its capacitive coupling (see Fig. 11.1) to the demodulator tuned circuit. If the reference signal is at the resonant frequency equal to IF (38.9 MHZ), the two inputs to the AFT detector are in quadrature (due to capacitive coupling) and no output is developed by the differential detector. However, any frequency deviation of the IF carrier causes a proportionate phase difference between the two inputs signals to the AFT demodulator which results in corresponding variation in its output.

The differential output of the detector is further processed in the AFT output stage that consists of transistors Q_{59} through Q_{64}. Here the error signal is rectified and amplified to develop a suitable dc error voltage across terminals 5 and 6 of the IC under consideration. It is connected to terminals marked AFT on the tuner through a coupling network.

In case the receiver is to be tuned manually the AFT facility is inhibited (disabled) either by open circuiting or short circuiting the AFT control voltage. An AFT switch is normally provided for this purpose on a panel of the TV receiver.

11.3 VISION IF SUBSYSTEM WITH IC TDA 3540/TDA 3541

The vision IF subsystem using TDA 3540 with (*n-p-n* tuners) and TDA 3541 with *p-n-p* tuners were commonly used in earlier monochrome and colour receivers along with CA 7611/CA 7607. These incorporate the following functions:

(*i*) gain controlled wideband IF amplifier

(*ii*) synchronous demodulator

(*iii*) White noise inverter

(*iv*) video preamplifier with noise protection

(*v*) AFT circuit with AFT ON/OFF switch

(*vi*) AGC circuit with noise gating.

(*vii*) tuner AGC output

(*viii*) external video switch which switches off video output for insertion of a VCR playback signal at either a high or !ow level.

As obvious the main functions performed by these ICs are the same as by CA7611/CA7607. Similar integraed circuits are used to obtain desired performance. The pin numbers are also same in the two sets of ICs.

11.4 VISION IF SUBSYSTEM EMPLOYING ICs SL1430 AND TDA 4420

Another vision IF subsystem employing IC SL 1430 as preamplifier, SAW 173 as IF bandshaping filter and TDA 4420 for processing the small signal is also used in many receiver circuits. The interfacing circuity can vary from chassis to chassis but essential components remain the same.

IF IC TDA 4420. The vision IF IC TDA 4420 is a monolithic integrated circuit in 18 lead D-I-L plastic package. Its various pin designations are shown in Fig. 11.11. The functions in-corporated in the device are:

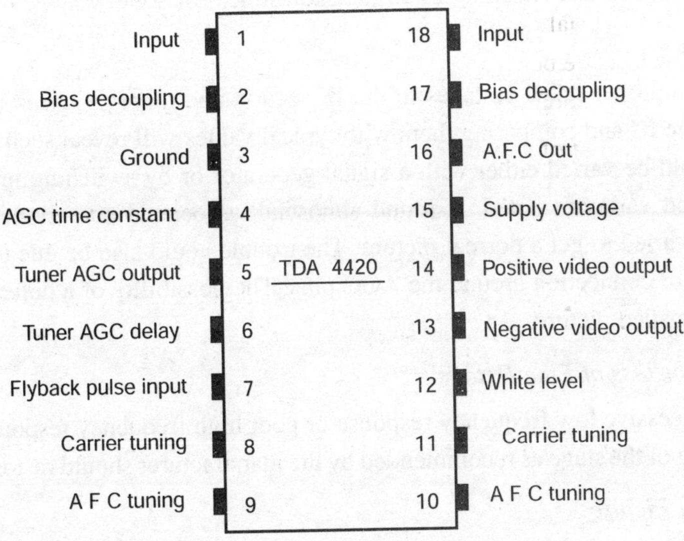

Fig. 11.11 Pin designations of IC TDA 4420.

(*i*) gain controlled vision IF amplifier

(*ii*) video demodulation controlled by picture carrier

(*iii*) AGC detected with gating facility

(*iv*) AGC amplifier for tuner drive with variable delay

(*v*) phase comparator for AFC (AFT) current generation

(*vi*) electronic AFC switch controlled by a dc threshold detector

(*vii*) thermally compensated push-pull AFC output voltage.

11.5. COMMON FAULTS IN VIF SUBSYSTEMS

The television receiver designs have progressed with time though the essential features remain the same. The subsystems described in this and following chapters continue to be in use in many receiver makes. However, with, the coming in of multinationals from Japan, South Korea and some other countries, the receiver industry in India has changed hands and the now popular receivers employ different ICs and somewhat modified corresponding subsystems. But the servicing and fault finding procedures remain nearly the same; as described at the end of this and other chapters. A list of common faults in the VIF subsystem, their probable causes and trouble shooting are as follows:

No Picture No Sound but Raster Normal

This could be due to no dc supply to any one block in the system (preamplifier, SAW filter and IC). A bad input cable from the tuner can also cause such a fault. Besides, an improper AGC voltage to the tuner or IF amplifier section can also interrupt the passage of signal through the VIF subsystem. Trouble shooting includes checking of dc supply voltage to the three sections of the subsystem, locating any faulty cable connection and checking of any misadjustment of AGC voltage to the tuner.

Over Loaded Picture

This can be due to improper AGC voltages to the IF section and tuner. A check of these voltages at appropriate pins of the IC and comparing them with typical values will reveal such a discrepancy. The input RF signal should be varied either with a signal generator or by switching in a distant and local station alternately and variation noted. If found abnormal, associated potentiometers like the AGC take-over should be varied to get a normal picture. The trouble could also be due to a shorted or open circuited component of connection around the AGC pins. The possibility of a defective IC can also be the cause of an overloaded picture.

Smearing of Picture or loss of Fine Details

This can be due to excessive low frequency response or poor high frequency response due to improper alignment. Realigning of the stage as recommended by the manufacturer should restore a normal picture.

Sound Interference in Picture

This as obvious is due to improper attenuation of the 5.5 MHz SIF (intercarrier) signal. The trouble shooting includes readjustment for minimum sound interference of the trap coil used to bypass the SIF signal from the composite video signal or replacing the 5.5 MHz bypass ceramic filter if used instead of a discrete trap circuit.

Ringing (Appearance of Ghosts that are Tunable with Fine Tuning Control)

This is caused by excessive high frequency response. The remedy is realignment of the subsystem as detailed in the receiver's service manual.

REVIEW QUESTIONS

1. Draw simplified block schematic diagram of a vision IF subsystem and explain briefly how each section operates to process the input signal.
2. Why is an IF amplifier necessary before the SAW filter that preceds the VIF IC? Draw its circuit diagram and explain the function of input and output circuits.
3. Draw simplified circuit diagram of the IF section of a vision IF IC and explain how AGC operates to keep the gain constant over a wide range of input signals.

4. Explain with the help of a circuit diagram the operation of a synchronous demodulator that enables simultaneous detection of video and intercarrier sound signals. Illustrate your answer with assumed values of currents in the differential detector.
5. How is the AGC voltage developed in the VIF IC and made proportional to the magnitude of input signal? How is the delay incorporated in the path of AGC voltage for the tuner?
6. Draw a simplified circuit of the AFT (AFC) detector and explain how error voltage is developed and applied to the tuner to keep necessary local oscillator frequency constant for any channel.
7. A faulty VIF subsystem can cause any of the following faults. Explain how would you proceed to trouble shoot and rectify them:

   (*a*) overloaded picture

   (*b*) smearing of picture

   (*c*) sound interference in the picture

   (*d*) no sound, no picture.

12

THE FM SOUND SYSTEM

INTRODUCTION

In television systems FM is preferred over AM for sound signal transmission because it enables almost noise-free reception even under adverse conditions. The noise that tends to cause a low signal to noise ratio at higher audio frequencies is overcome by providing pre-emphasis at the transmitting end and de-emphasis in the receiver circuitry.

The sound signal channel is the same in both monochrome and colour receivers. In the tuner, FM sound signal is first amplified along with the AM picture signal and then transferred to the sound IF frequency from the incoming channel carrier frequency. The SIF signal thus obtained is passed through IF amplifier without much amplification to avoid any interference with the video signal. In earlier colour receivers that employed discrete circuitry, separate diode detectors were employed for video and audio signal detection to avoid beat-interference effects. However, in modern receivers that employ integrated circuits, a synchronous demodulator is incorporated in the vision IF IC which detects both vision and intercarrier SIF signals simultaneously without any beat-note output.

The sound section of the TV receiver was the first to be produced in the integrated form. Earlier ICs contained IF amplifier-limiter, FM detector, dc volume control (attenuator) and audio preamplifier stages. The ICs TBA570, CA3065, TA7176AP, AN241, HA1124A and TDA2790 are common examples of this type of integrated circuits. A discrete audio power amplifier is necessary with these ICs to deliver audio power to the loudspeaker. Later, separate ICs became available for the audio power amplifier. Then combinations like CA3065 + CA810 and TBA120U + TDA2611 became quite popular and remained in use for quite sometime. Now, we have ICs which also contain AF amplifier stage in the same package and the speaker is connected directly across appropriate pins of the integrated circuit. Such ICs include TDA1701, TDA1035T, TDA1190Z, TDA3190, CA1190, TDA4190, and TDA8190. These ICs need no screening and employ a coincidence (differential) peak detector needing only one tuned circuit. There is also a provision for audio muting so that no noise is heard in the speaker while selecting channels or with no signal input to the receiver. The IC TDA4190 provides separate VCR input and output pins. High quality colour receivers with stereo sound output use ICs like M793 as sound processor and TDA2009 in the amplifier section. Separate ICs which function as phase-locked loop (PLL) and digital FM detectors are also now available. These eliminate the use of any transformer or tuning coil in the sound section of the TV receiver.

12.1 SOUND SECTION SUBSYSTEM*

The block diagram of a typical sound subsystem as found in modern receivers employing ICs is shown in Fig. 12.1. Its various sections are:

(1) Input circuit (external to the IC)

(2) Regulated power supply

(3) IF amplifier-limiter

(4) Active filter

(5) FM detector

(6) DC volume control

(7) AF power amplifier.

(1) Input Circuit

The synchronous video detector in the vision IF IC detects both video and intercarrier SIF signals. These outputs that becomes available at a pin of the IC are routed to many sections of the receiver. As shown in Fig. 12.1, it is fed to the sound system IC through an impedance matching and filtering network. The double tuned transformer (TR1) providies impedance match between the signal source and input circuit of the IC (pins 1 & 2). The transformer and associated capacitors combine to provide good filtering of unwanted video frequencies present in the input signal. The signal that feeds at pin 1 has a 3dB bandwidth of about 150 KHz centered around 5.5 MHz.

The use of a ceramic filter instead of coils and capacitors is preferred in many receiver designs. Such filter configurations (like BMC 389) are specially designed to select SIF and need no alignment. In some input circuits a discrete bandpass filter precedes the ceramic filter to narrow the signal bandwidth fed to it. A typical input circuit employing a ceramic filter is also shown in Fig. 12.1. The amplitude of SIF signal that feeds into the sound IC is around 400 mV (peak-to-peak).

(2) Regulated Power Supply

A unique feature of the sound section ICs is their centrally regulated dc supply. A zener diode establishes a control voltage which serves as reference to various voltage regulators that feed independent dc voltages to the IF amplifier-limiter, FM detector, preamplifier and dc volume control circuits. The audio power amplifier sections are fed directly form the 24V dc source that enters the IC at a particular pin (14). It is also suitably decoupled as shown in the figure. Such a provision of feeding independent dc supply to various sections of the IC reduces unwanted coupling between different circuits and thus ensures overall stability of the sound section circuitry.

* (*i*) For basic functions of different section of the TV sound system employing discrete circuitry refer chapter 21 "Sound System" of the book 'Monochrome & Colour Television' by R.R Gulati (New Age International).

(*ii*) The pin number labelled on the circuit diagrams of this section (12.1) pertain to IC CA1190.

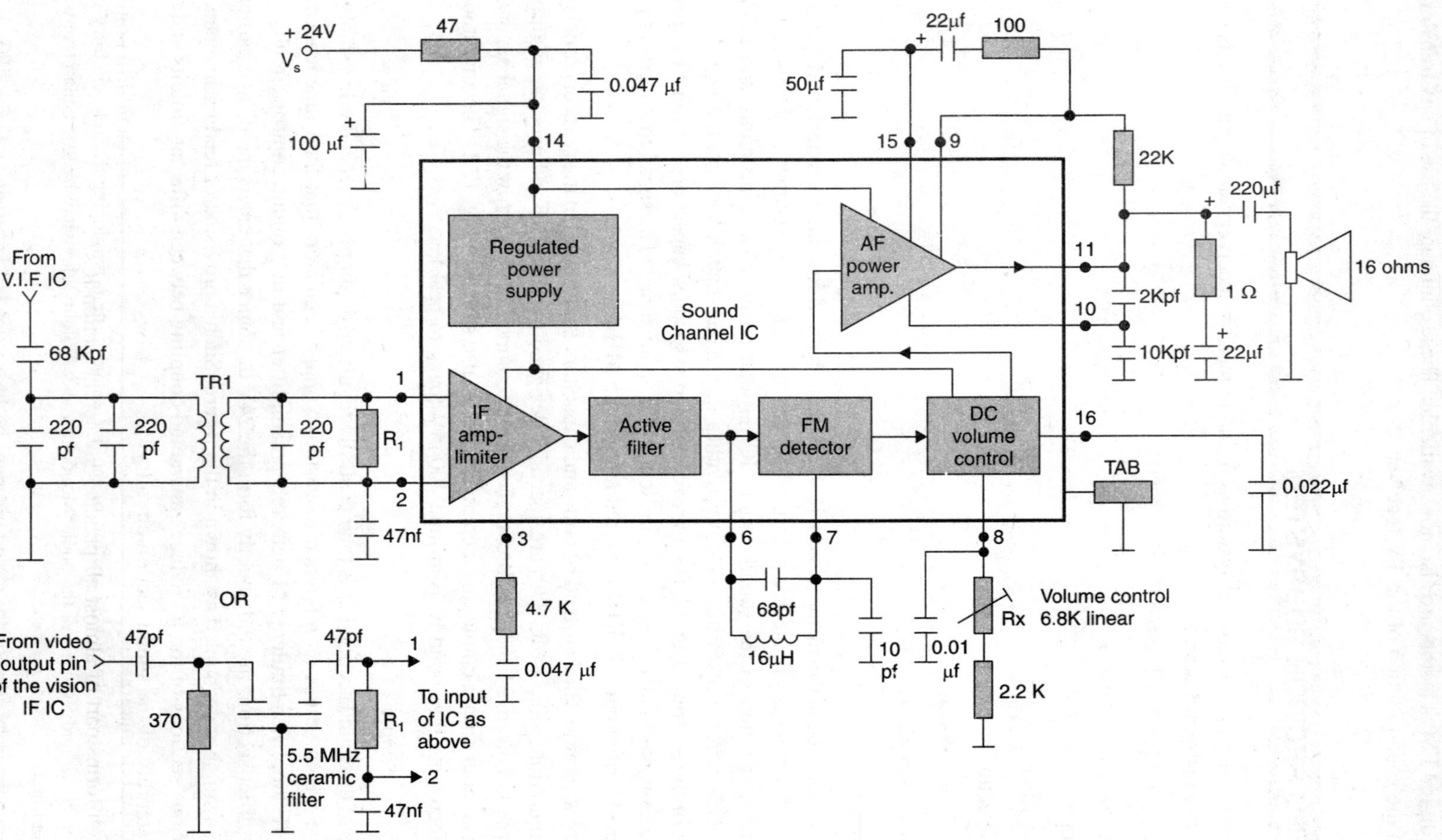

Fig. 12.1 Block diagram of the sound subsystem of a TV receiver together with basic interface circuitry.

(3) IF Amplifier Limiter

The FM SIF signal as it enters the sound IC is quite weak and has some amplitude variations due to noise pulses and somewhat unequal amplification at the RF and IF stages of the receiver. Therefore, this stage is designed to provide large gain and remove amplitude variations by limiting. A typical circuit of such an amplifier-limiter is shown in Fig. 12.2. The transistors Q_3 to Q_{14} together with associated load resistors constitute a six stage differerntial amplifier. The devices Q_{15} to Q_{20} act as individual constant current sources to these amplifiers. The base drives to these current sources is maintained at a fixed level by regulator transistors Q_1 and Q_2. Any change in the emitter current of Q_2, that supplies base currents, also affects base drive to Q_1. This has a feedback effect on the base current of Q_2 which then maintains a fixed base drive to transistors Q_{15} to Q_{20}.

The intercarrier SIF signal is fed through C_1 at the base of Q_3 of the first differential amplifier. The base of the other transistor (Q_4) in this stage is maintained at ground potential for the ac signal through capacitor C_3 connected externally (pin 3) to the IC. The base voltages of Q_3 and Q_4 are maintained at about 3V dc with suitable negative feedback from the emitter loads R_{21} and R_{22} of buffer amplifiers (Q_{21} and Q_{22}) which precede the active filter. As shown in the figure, voltage across R_{21} feeds through R_{24} at pin 2 and then on to the base of Q_3 via R1. Similarly, voltage across R_{22} feeds via R_{23} (pin 3) at the base of Q_4. Subsequent IF stages are biased by connecting their repective input base terminals to the collectors of preceding differential stage. With a dc supply of 4.5V to the amplifier and 3 volts at the base, the effective V_{CB} when collector current flows, reduces to near zero thus ensuring early saturation of the differential amplifiers necessary for good limiting. Once the signal amplitude (peak-to-peak) exceeds the dynamic range, limiting action commences to deliver a constant output voltage. The quiescent current of transistors employed in the differential amplifiers are so adjusted that as the signal amplitude increases on successive amplification, best possible limiting and AM rejection characteristics are obtained. The amplifier output feeds into corresponding buffer amplifiers Q_{21} and Q_{22} which as shown are conneted to be emitter followers.

(4) Active Filter

The active filter (see Fig. 12.2) is built around transistor Q_{23}. Its feedback network is so designed that it provides a near 40 dB/decade attenuation to frequencies beyond the intercarrier SIF. Thus the filter removes all harmonics caused by limiting action of the preceding If amplifier section and feeds a sine wave SIF signal to the demodulator with sidebands not exceeding beyond about ± 100 KHz of 5.5 MHz.

(5) FM Detector

There has been considerable development in the design of FM detectors with a view to reduce number of components and eliminate the need for tuned circuits as far as possible. The Foster-Seeley discriminator and many ratio detector circuits were commonly used in earlier versions. The quadrature detector was the first to be produced in integrated circuit form. This was followed by the more efficient differential peak-detector which forms part of almost all present-day sound section ICs.

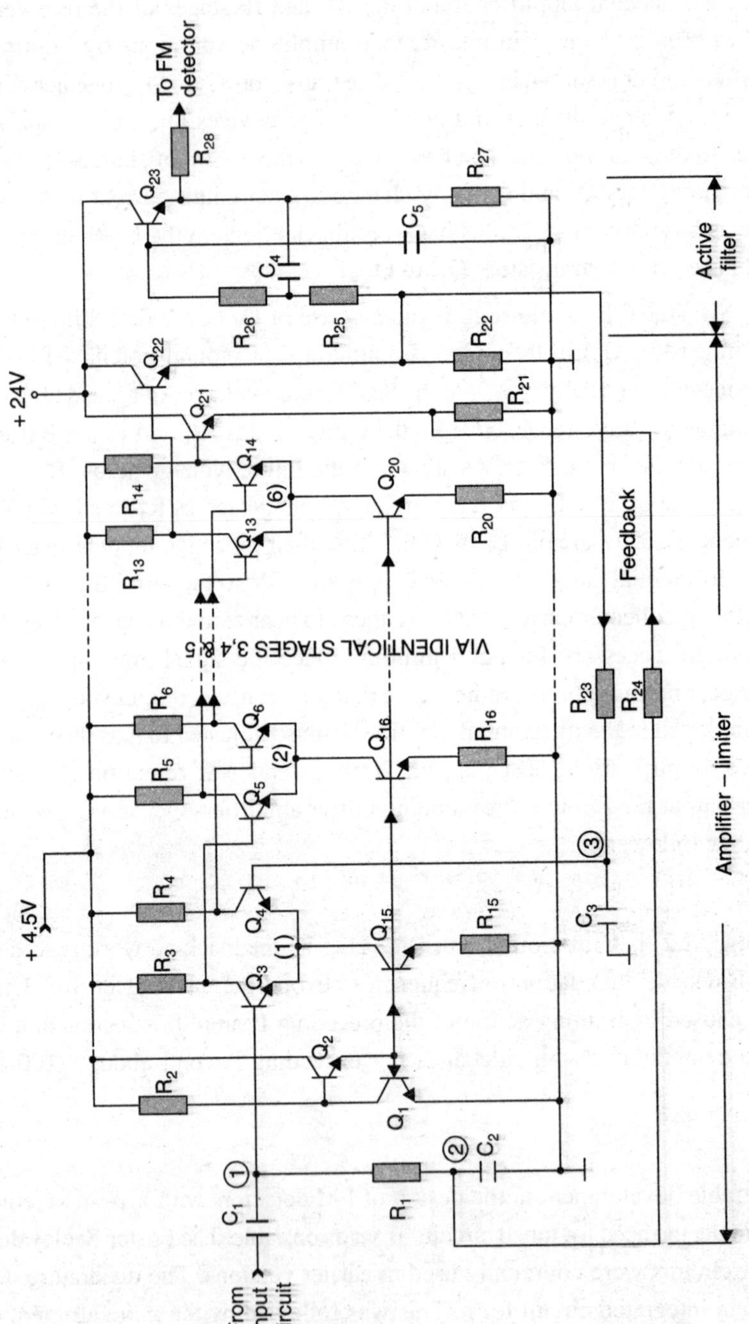

Fig. 12.2 A typical multistage high gain SIF amplifier-limiter and active filter circuit commonly used in sound subsystem integrated circuits.

Differential Peak Detector

A somewhat simplified circuit of the integrated differential peak detector together with associated preamplifier and dc volume control technique is shown in Fig. 12.3. The SIF signal from the output of active filter feeds at the base of Q_1 through R_1 (R_{28} in Fig. 12.2). The signal is also fed at the base of Q_2 through the tuned circuit network) L_1, C_4, C_5) connected externally at two pins (6 and 7) of the IC. The transistors Q_1 and Q_2 act as peak detectors and charge capacitors C_1 and C_2 respectively, (on alternate half-cycles) to the peak value of the voltage that appear at their input points. The differential amplifier consists of transistors Q_3 and Q_4 with Q_5 as the constant current source. The voltage drop cross D_1 provides a fixed base current to Q_5. The differential configuration formed by Q_6 and Q_7 is the active load for Q_3. Similarly, the differential pair Q_8 and Q_9 is the active load for Q_4. The single ended audio output that becomes available at the collector of Q_9 is fed to the audio voltage amplifier through R_{10} as shown in the figure. The de-emphasis capacitor C_6 is also connected at this point but external to the IC (pin 16).

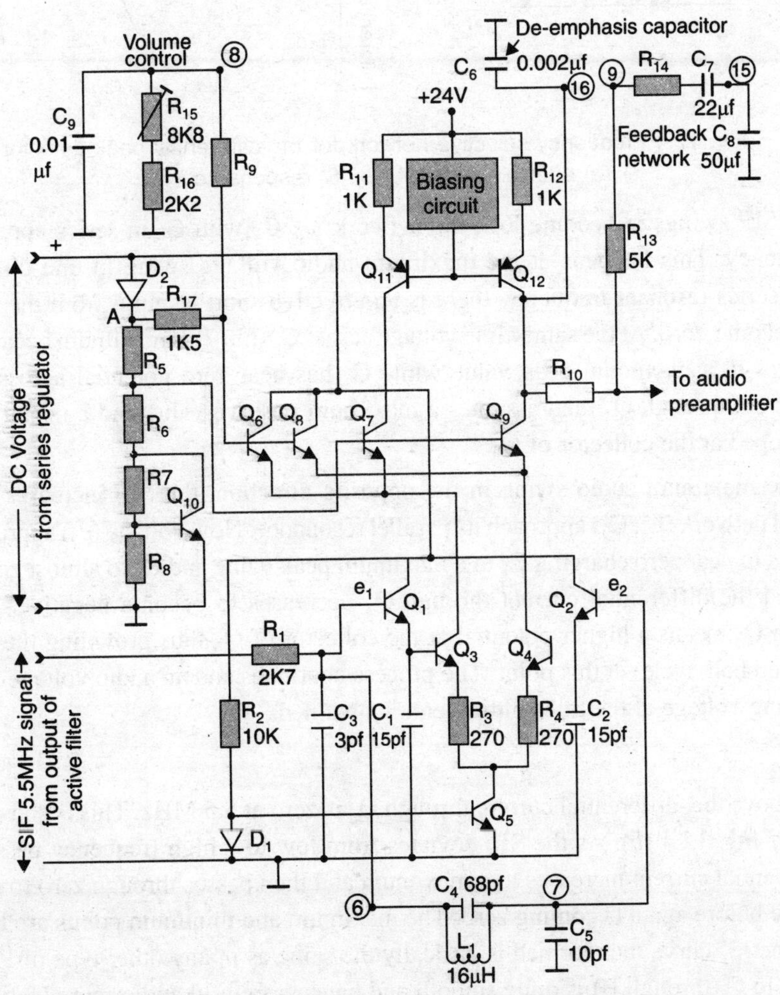

Fig. 12.3 Simplified circuit diagram of differential peak detector and dc volume control sections of the sound IC.

Detector Operation

The frequency selective network is drawn separately in Fig. 12.4 (*a*) for convenience of explaining the detection mechanism. The SIF signal feeds at pin 6 (base of Q_1) while the voltage that develops across C_5 (pin 7) gets applied to the base of Q_2. The component values of L_1, C_4 and C_5 are so chosen that at the centre frequency (5.5 MHz) the voltages (e_1) and (e_2) are equal. Thus with no audio modulation (f_c = 5.5 MHz) C_1 and C_2 charge to the same value and hence the differential output current is zero. This is indicated by point B on the response curve shown in Fig. 12.4(*b*).

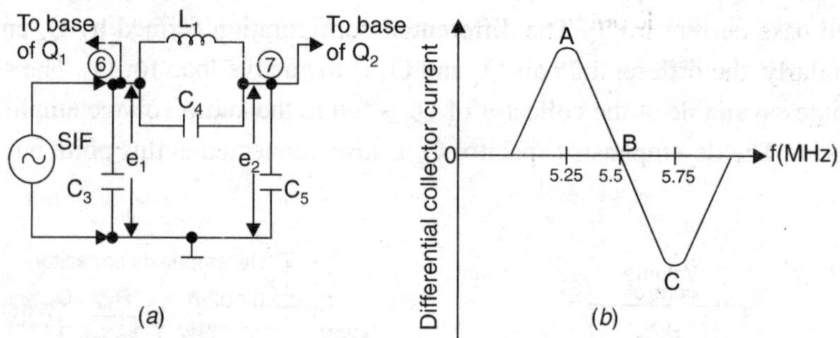

Fig. 12.4 Frequency selective network for the differential peak detector (*a*) equivalent circuit, (*b*) 'S' response curve.

As the SIF swings to become low, the network L_1, C_4 with C_5 in series approaches its series resonant frequency. This happens at the maximum audio voltage swing in one direction (first half cycle). At the series resonant frequency there is an effective short at pin 6 from the ac point of view causing e_1 to become zero. At the same time voltage across C_5 (pin 7) is maximum because of resonance. Thus C_2 charges to a maximum peak value while C_1 has near zero potential across it. The ensuing differential current through Q_4 then assumes a maximum positive value and a negative audio voltage swing is developed at the collector of Q_9.

With the maximum audio swing in the opposite direction, the SIF increases above 5.5 MHz causing parallel network (L_1, C_4) approach its parallel resonance. Now voltage (e_1) approaches maximum while (e_2) drops to near zero charging C_1 to a maximum peak value and C_2 to almost zero. The opposite now occurs and the differential current through Q_4 decreases to become negative. The decrease in current through Q_4 means a higher potential at the collector of Q_9 thus providing the desired positive audio swing (2nd half cycle) at this point. The process continues and an audio voltage that corresponds to the modulating voltage at the transmitting end is obtained.

The 'S' Curve

As explained above the differential current through Q_4 is zero at 5.5 MHz. This is shown by point B on the 'S' curve of Fig. 12.4 (*b*). As the SIF deviates from low to a high frequency passing through 5.5 MHz, the differential current increases to a maximum and then passes through zero to attain a negative maximum value before again becoming zero. The maximum and minimum points are labelled A and C respectively. The 'S' curve thus formed is basically the same as in any other type of FM detector. The change from A to C (through B) is quite smooth and hence a faithful audio reproduction occurs in the differential peak detector. The audio voltage swing at the collector of Q_9 is opposite to that of current shown in Fig. 12.4 (*b*). As mentioned earlier, the boost given to higher audio frequencies at the transmitting end is removed by the de-emphasis network connected across the audio output point.

(6) DC Volume Control

The dc volume control is affected by controlling current through the active loads of differential detector transistors Q_3 and Q_4. As shown in Fig. 12.3, the two differential amplifiers Q_6-Q_7 and Q_8-Q_9 are the active loads to Q_3 and Q_4 respectively. The audio signal whose amplitude is to be controlled is available at the collector of Q_9. The bases of Q_6 and Q_8 (one each of the differential pairs Q_6, Q_7 and Q_8, Q_9) are tied together and returned to a dc potential which can be varied by an externally connected potentiometer. Similarly, the bases of Q_7 and Q_9 (the other transistors of differential pairs) are also tied together but returned to a fixed dc potential at the junction of R_5 and R_6. This voltage is maintained constant by the regulatory action of transistor Q_{10}. The potentiometer R_{15} forms part of the potential divider connected between point A and ground. Thus by changing the value of R_{15} the bias voltage applied to the bases of Q_6 and Q_8 can be varied. For any setting of the external (pin 8) volume control, the currents it transistors Q_6 and Q_8 are equal. Similarly currents in Q_7 and Q_9 are also equal because their bases are returned to the same fixed potential.

Control Action

As shown in the circuit of Fig. 12.3, transistor Q_{11} is the active load to Q_7 and Q_{12} to Q_9. A biasing network (details not shown) supplies suitable base drive to Q_{11} and Q_{12}. When volume control resistance (R_{15}) is at its minimum value, maximum current flows from point A to ground. The resulting high voltage drop across R_{17} lowers bias drive to Q_6 and Q_8 to the extent that they are starved of collector currents. However, at the same time currents in Q_7 and Q_9. are maximum thus causing maximum differential current variations leading to highest audio voltage swings at the collector of Q_9. When the volume control potentiometer is moved to its maximum resistance, less current flows through R_{17} enabling maximum drive to Q_6 and Q_8. Since total currents for the differential pairs are constant, collector currents of Q_7 and Q_9 decrease to almost zero. Thus no output is developed at the collector of Q_9 and audio output drops to a very low level. Any position of the dc volume control other than the two extremes will enable different value of sound output. A dynamic range of nearly 90 dB is achieved in such a method of volume control. This technique is considered superior to earlier methods because no audio signal is present across the volume control and so there is no possibility of hum or noise pick-up as can happen in conventional methods of volume control.

(7) AF Power Amplifier

The audio amplifier consists of a preamplifier, a complementary symmetry type drive and a power amplifier. A somewhat simplified circuit of this integrated section is shown in Fig. 12.5. The transistors Q_1, Q_2 together with their active loads Q_3, Q_4 and constant current source I_1 constitute the differential preamplifier. The audio signal from the output circuit of FM detector is fed at the base of Q_1 via R_{10}, while the base of Q_2 (non-inventing input of the differential pair) is maintained at a fixed dc potential obtained across zenar Z_1 through R_1. The base of Q_2 is maintained at zero ac potential by grounding it through a large capacitor (C_7) connected externally to the IC at a particular pin (15). The single ended audio output available at the collector of active load Q_4 is fed directly at the base of driver transistor Q_5. The feedback capacitor C_2 is for maintaining desired bandwidth of the audio signal.

The complementary driver pair is formed by transistors Q_6 and Q_7. The constant current source I_2 and Q_8 provide base drive to this pair. The *p-n-p* transistor O_9 that is biased by diode D_4 amplifies output of O_6 and feeds it to the base of power transistor Q_{10}. Similarly, transistor Q_7 provides necessary drive to Q_{11}, the other power transistor. The biasing circuit formed by constant current source I_2, diodes D_2, D_3 and resistors R_4, R_5 is so designed that a standby current flows through Q_{10} and Q_{11} in the absence of any audio input. Thus the output stage is biased for class AB operation to prevent any cross-over distortion. As shown in the figure, the speaker is connected at the junction of emitter and collector of transistor Q_{10} and Q_{11} respectively through C_3 a large electrolytic capacitor. The two power transistors conduct alternately to deliver audio power to the loudspeaker. The 24V dc supply to the IC is the Vcc source for Q_{10} while the charge stored across C_3 when Q_{10} conducts, becomes the Vcc source for Q_{11}.

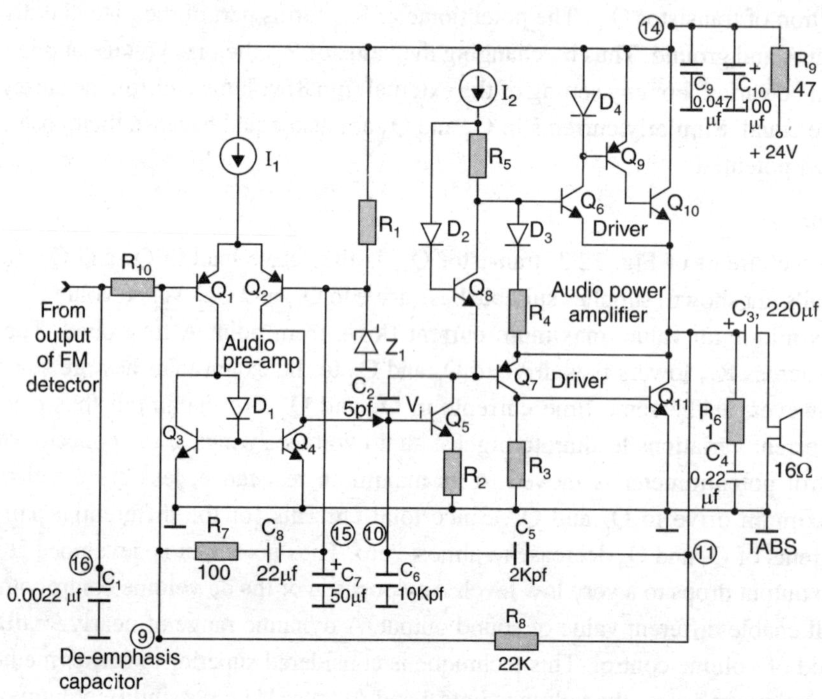

Fig. 12.5 A typical audio amplifier circuit that forms part of many sound section ICs.

Circuit Operation

As the audio signal V_i at the base of Q_5, swings to become negative its collector current that flows through Q_8 and comes from constant current source I_2 decreases. Since I_2 is constant, any decrease of current through Q_8 means more base drived to Q_6. At the same time, the fall of current through Q_8 raises potential at the base of Q_7 to reverse bias it. This nearly cuts-off Q_7 and hence, there is no base drive to Q_{11} which in turn does not conduct to allow emitter current of Q_{10} to flow through it.

The forward biasing of Q_6 caused by the negative voltage swing at the base of Q_5 represents audio signal. On amplification through Q_6, it causes more drive to transistor Q_9, the collector current of which is the base drive to output transistor Q_{10}. Thus, a current, that is proportional to the audio signal amplitude flows through the base of Q_{10} which turns on to conduct heavily from the 24V supply connected at a pin (14) of the IC from the receiver's regulated power supply. Since Q_{11} is in the non-conducting state, the emitter current of Q_{10} flows through C_3 and voice coil of the loudspeaker to complete its path. The capacitor C_3 thus charges to nearly 24V with the polarity marked across it and as mentioned earlier becomes the dc source for Q_{11} which conducts next.

When V_i changes to become positive during the next audio half-cycle, both Q_5 and Q_8 conduct more thus depriving base drive to Q_6 and hence forward biasing Q_7. The reduced drive to Q_6 cuts-off Q_9 thus turning-off Q_{10}. However, forward biasing of Q_7 and the consequent increase in its collector current causes enough voltage drop across R_3 to turn Q_{11} 'on'. With Q_{10} turned-off the charge earlier stored in C_3 passes large current through Q_{11} which completes its path through the speaker coil in the opposite direction. This current, as obvious, is proportional to positive half-cycle of the audio signal.

The capacitor C_3 is again charged when Q_{10} conducts during the negative swing of the following negative half-cycle of V_i. This sequence continues to provide sound output from the loudspeaker in accordance with the impressed audio signal at the input of driver (Q_7).

Gain Stability and Frequency Compensation

The overall negative feedback via R_8 connected externally to the IC (between pins 11 and 9) and that around the preamplifier through R_7 (pin 9) provides stability and determines gain of the amplifier. The capacitor C_8 in series with R_7 can be chosen to obtain desired tone characteristics. The R_6-C_4 nertwork (across pin 11 and ground) is meant for loudspeaker impedance compensation at higher audio frequencies. The capacitors C_5 and C_6 (at pins 11 and 10) provide overall high frequency compensation to the audio signal. The capacitor C_1 (pin 16) is the de-emphasis capacitor. The decoupling network consisting of R_9, C_9 and C_{10} (pin 14) is to prevent any undue coupling between the audio section and other circuits in the receiver. The cooling is provided by heat sink tabs as in other similar integrated circuits.

12.2 PHASED-LOCKED LOOP AND DIGITAL FM DETECTORS

The phase-locked loop and digital FM detectors are also available in IC form for SIF signal detection. These are superior in the sense that no tuning coils are needed for demodulation action. The two are described separately.

PLL FM Detector

The basic PLL consists of four sections as shown in Fig. 12.6 (*a*). These are (*i*) Phase detector, (*ii*) low-pass filter, (*iii*) DC amplifier and (*iv*) voltage controlled oscillator (VCO). The phase detector receives two inputs (*a*) SIF signal and (*b*) signal from the VCO. The *PLL operation seeks to keep the VCO frequency the same as that of the intercarrier frequency modulated signal. Since frequency of the FM sound IF signal continuously changes depending on audio modulation, the VCO frequency also changes in an effort to 'catch up' with the SIF signal frequency. In order to do so, the phase detector puts out a dc error voltage and feeds it to the VCO through the dc amplifier. The polarity of the error (audio) voltage is determined by the instantaneous frequency difference between the SIF and VCO as the FM sound signal deviates to become, more or less from 5.5 MHz. Thus the error voltage itself is the audio signal. When SIF is at the carrier (centre) frequency, the VCO is at the same frequency. At this time, the phase detected output error is zero and hence no audio output is developed. The amplitude of the error

*For PLL operation refer chapter 10 "Frequency synthesized tuning, remote control and AFT".

voltage is a function of the instantaneous difference between SIF and VCO frequencies. The rate of change of IF deviation determines the error i.e., audio frequency.

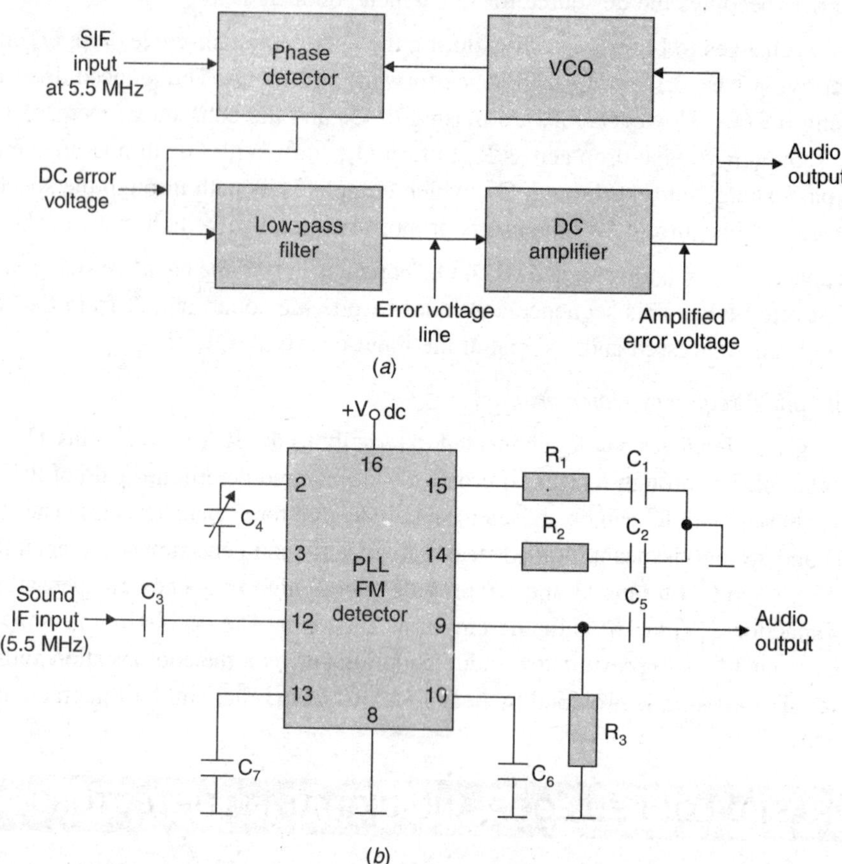

Fig. 12.6 Phase-locked loop (PLL) FM detector (*a*) block diagram (*b*) schematic diagram.

Circuit Details

The external circuit components to the IC are shown in Fig. 12.6 (*b*). The sound IF signal at 5.5 MHz is coupled to the IC through capacitor C_3 at pin 12. The capacitor C_4 connected across terminals 2 and 3 is used to set the frequency of VCO close to 5.5 MHz. The networks R_1-C_1 and R_2-C_2 that are connected at terminals 15 and 14 form the low-pass filter. The audio output is obtained from terminal 9 through R_3-C_5. The capacitor C_6 at pin 10 is the de-emphasis capacitor which along with a resistor inside the IC forms the 50 μs time-constant network of the de-emphasis circuit.

Digital FM Detector

The digital FM detector uses no coils, is relatively independent of the IF and does not require any alignment. As shown in Fig. 12.7 the signal first passes through a bandpass filter which restricts it to the FM sidebands around 5.5 MHz. The filter is usually ceramic to avoid any inductors. The IC (1) contains the amplifier-limiter stage. The IC (2) is a digital logic, monostable (one shot) multivibrator. This circuit produces one pulses each time it is triggered by the AM limited SIF signal pulses. The IC

(2) output pulses have a constant amplitude and duration. However, at any given instant their amplitude and repetition rate is the same as that of the deviated frequency of the SIF which in turn represents audio modulation. There are two complementary outputs from IC (2). Each output is fed to its own integrator formed by R_1-C_1 and R_2-C_2. The integrators average the outputs to obtain a push-pull signal. This audio signal is fed to the differential inputs of an operational amplifier contained in IC (3). The output of IC (3) is the audio signal which drives a conventional discrete or integrated audio output amplifier to deliver output to the loudspeaker.

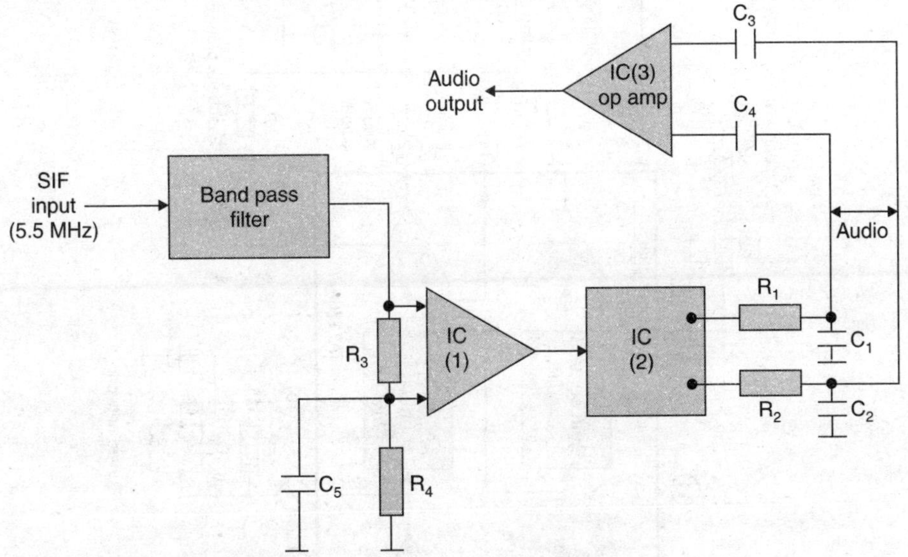

Fig. 12.7 Block diagram of a digital FM detector.

12.3 SOUND SUBSYSTEM WITH IC 7176AP

The integrated circuit 7176AP performs as amplifier-limiter, FM detector and audio preamplifier. It is pin-to-pin same as the BEL CA 3065 which has similar performance characteristics. Both need a separate audio power amplifier. The IC 7176AP together with a shunt regulated discrete power amplifier as used in some colour receiver designs in shown in Fig. 12.8. The power amplifier is a push-pull configuration employing two identical n-p-n power transistors. The use of output transformer T_1 reduces power consumption, improves sound quality and eliminates heat sinks for the output transistors Q_1 and Q_2.

The various sections of the IC function in the same manner as explained in section 12.1. The necessary interface circuitry to the IC is suitably labelled in the figure. The audio signal available at pin 12 of the IC is applied to the push-pull amplifier via network C_3-R_7. In the absence of any audio input, current flows from the 112V regulated dc source through R_1, R_2, R_3, and R_4 to feed base current to Q_1 and collector current to Q_2. The base drive to Q_2 is from the junction of R_5 and R_6 that are connected from the emitter of Q_1 to ground. The value of R_6 is so chosen that Q_2 is suitably forward biased. Any current flow in the collector lead of Q_2 causes voltage drop across R_4 and thus reduces base drive to Q_1.

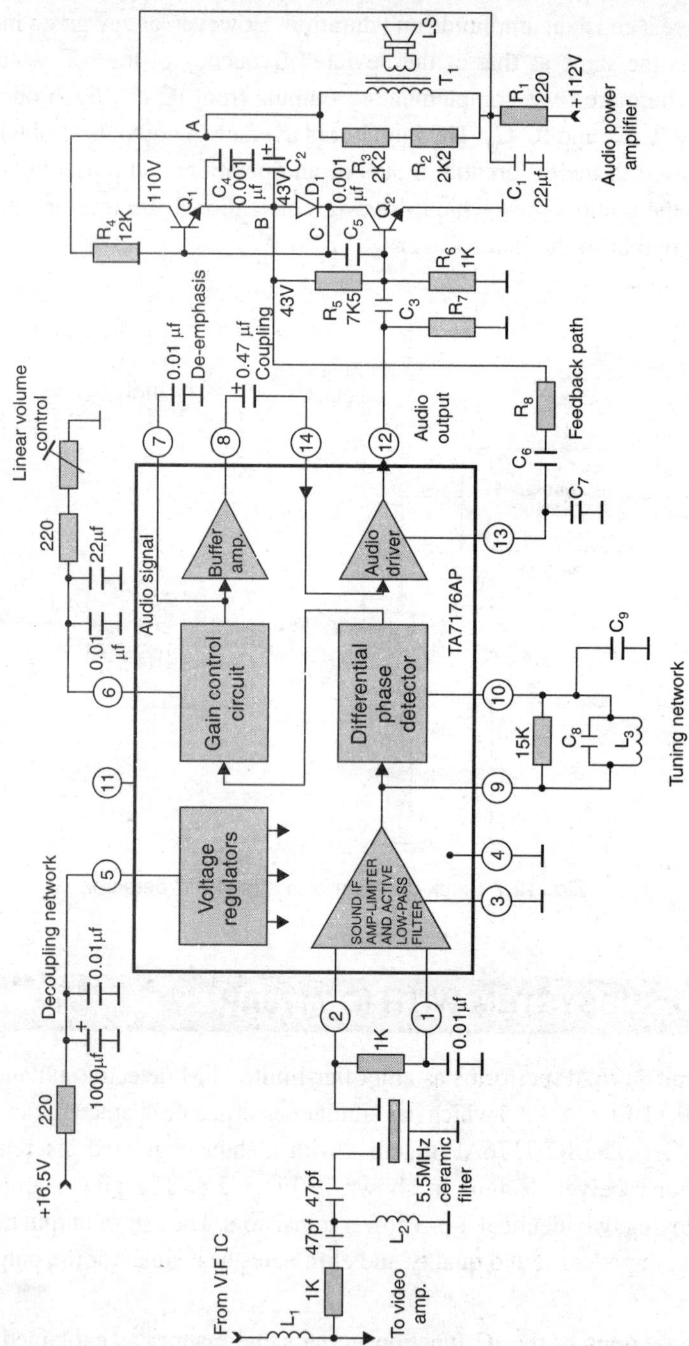

Fig. 12.8 Sound channel employing IC 7176AP and a discrete audio power amplifier.

The resulting reduction in the emitter current of Q_1 affects base drive to Q_2 which tends to reduce its collector current. Such a negative feedback action holds the emitter of Q_1 at a voltage that keeps the diode D_1 near cut-off allowing only sustaining currents through the two output transistors. This takes care of the crossover distortion and also prevents unnecessary heat dissipation in the transistors. A part of the standing current that flows through the primary winding of output transformer (T_1) provides necessary saturation of its core to keep operation in the linear region of (B-H) curve when audio current flows through it. In steady state, the capacitor C_2 is charged to a potential that exists across points A and B and with the polarity marked across it.

During positive swing of the audio signal, Q_2 is fully turned 'on' causing a large voltage drop across R_4. This not only reduces voltage at point 'C' (collector of Q_2) but also lowers voltage at the base of Q_1 to a level that is enough to cut it off. The diode D_1 which is now forward biased due to reduced potential at point 'C' permits C_2 to charge to a higher value. The charging current that flows from the dc source through the primary winding of T_1 is proportional to the potential change at point 'C' which in turn is a function of the positive swing of input audio signal. Thus the voltage induced in secondary winding of the output transformer, that feeds the loudspeaker, represents positive half-cycle of the input audio signal.

During negative half cycle of the audio input signal the reverse happens and Q_2 turns-off allowing Q_1 to conduct via R_5 and R_6. This enables C_2 to discharge to its oirginal value with current flowing from 'A' through the primary winding to T_1 and transistor Q_1 to point B. This reverse surge of current through T_1 induces voltage across its secondary winding and represents negative half cycle of the input signal. Such a sequence repeats to deliver amplified audio signal to the loudspeaker.

12.4 SOUND CHANNEL EMPLOYING TWO ICs

With the advent of integrated circuit for audio power amplifier the combination CA3065 + CA810 came into use. Soon after this the use of TBA120U as SIF IC and TDA2611A for power amplifier became popular with receiver designers. The interface circuitry around the combination of these two ICs is shown in Fig. 12.9. The IC TBA120U has IF amplifier-limiter with a symmetrical FM demodulator and an AF amplifier with adjustable output voltage. The AF amplifier is also provided with an output for volume control and an input for VCR operation. The TDA2611A is a monolithic IC in a 9-lead single-in-line (SIL) plasitc package with a high d.c. supply voltage audio amplifier. It is easier to mount because of SIL package and needs very few external components.

Circuit Operation

The SIF signal from the VIF IC enters TBA120 U through a ceramic filter at pin 14, and audio output becomes available at pin 8. The magnitude of this output can be varied by connecting a suitable resistance between pins 4 and 5 (see Fig. 12.9). The supply voltage to the IC is from a 12V regulated source. The admissible range is 10-18V with an output AF voltage between 1.0 to 1.2 V (r.m.s.). The audio muting is affected through diode D_1 at pin 2. So long as there is signal at the input of TV receiver a dc voltage from the combination IC (sync separator) remains applied at the collector (cathode) of D_1 thus keeping it reverse biased. However, in the absence of any input signal and during channel selection this voltage drops to a low value thus forward biasing the diode. This causes current flow from within

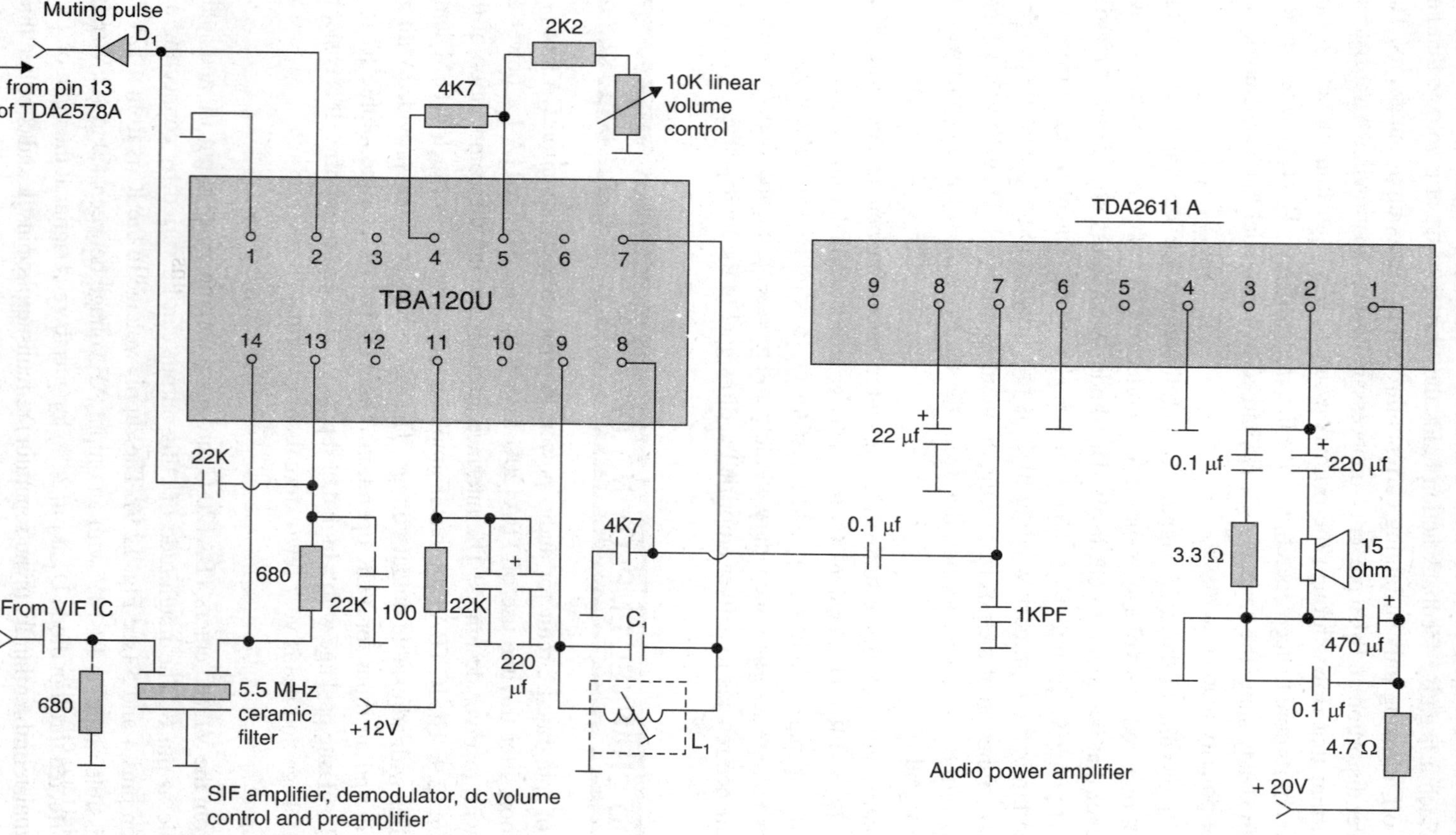

Fig. 12.9 Sound channel employing IC TBA120U and IC TDA2611A.

the IC dropping dc potential to a very low value at the non-inverting input (pin 2) of the first SIF differential amplifier. This reverse biases the stage which then cuts-off to prevent any noise output from the loudspeaker.

The AF signal from TBA120U feeds at pin 7 of IC TDA2611A through a coupling capacitor. The low audio from the VCR can also be injected at this pin. The amplified output becomes available at pin 2 where the speaker is connected via a 220μF capacitor. The 20V dc is applied at pin 1 through a decoupling network. The 22μF capacitor at pin 8 is another decoupling capacitor. The speaker is shunted by an R-C network to improve frequency response as in other similar circuits. Input impedance of the amplifier can be increased by connecting a suitable R and C series combination across pins 5 and 9 of the IC. Typical dc voltage data for the two ICs is as under:

IC TBA120U		*IC TDA2611A*	
Pin No.	*DC voltage (V)*	*Pin No.*	*DC voltage (V)*
1	0 (ground)	1	20
2	0.5 to 1.8	2	10
3	2.8	3	N.C.
4	4.7	4	ground
5	3.2	5	N.C.
6	2	6	ground
7	2.7	7	1.5
8	3.9	8	10.5
9	2.7	9	N.C.
10	1.7		
11	10.8		
12	6.0		
13	1.6		
14	1.6		

12.5 SOUND SUBSYSTEM EMPLOYING TDA1035T/TDA1701

The ICS TDA1701 and TDA1035T which are identical pin-to-pin (though in a different package) were the first to be produced containing complete sound subsystem. The IC TDA1035T which found wide acceptance is encapsulated in a plastic package with 13 connection pins and contains all the stages for the sound channel of a television receiver. The various sections as shown in Fig. 12.10 are nearly the same as described under section 12.1 and hence such details are not repeated. However, various pin connections and interface circuitry is described in this section.

The SIF signal is fed to the IF amplifier-limiter circuit across pins 1 and 2 either through a coupling transformer as shown in Fig. 12.10 or through a ceramic filter. The 10 nF capacitor at pin 2 is for decoupling in the feedback loop. The switch shown at pin 2 after a 2.7K resistor is to turn-off the IF amplifier electronically by grounding it when the IC is to be used as an audio amplifier for inputs other than detected by the demodulator. Pin 3 is for connections to a sound recorder. It provides direct low distortion demodulator output which is not effected by the electronic volome control. Pins 4 and 5 are for connecting L-C tuning circuit to the demodulator. The dc volume control potentiometer is connected at pin 6 as shown in the figure. Any audio input from a tape recorder, record player or VCR (audio) can be fed at pin 7 for amplification and reproduction by this IC. While doing so switch S_1 is kept closed. For gain stabilization and frequecny compensation, output is returned at pin 8 through a 10 K resistor. The 100 ohm resistor with 10μF capacitor in series and grounded from this pin provide a fixed tone adjustment. A different feedback network can be used instead for providing manual tone control. The output amplifier is a series push-pull configuration and feeds the speaker from pin 9 through a large electrolytic capacitor. The network across the speaker is for compensation to the voice-coil impedance. The 24V supply voltage enters the IC at pin 10. It has a built-in stabilizing circuit. Overload protection is provided to the power stage within the IC. At a chip temperature of 150°C the AF voltage at the drive transistor is short-circuited. The smoothing capacitor at pin 11 is for decoupling in the output amplifier circuit. The de-emphasis capacitor is connected at pin 12. The last pin *i.e.*, 13 returns to ground.

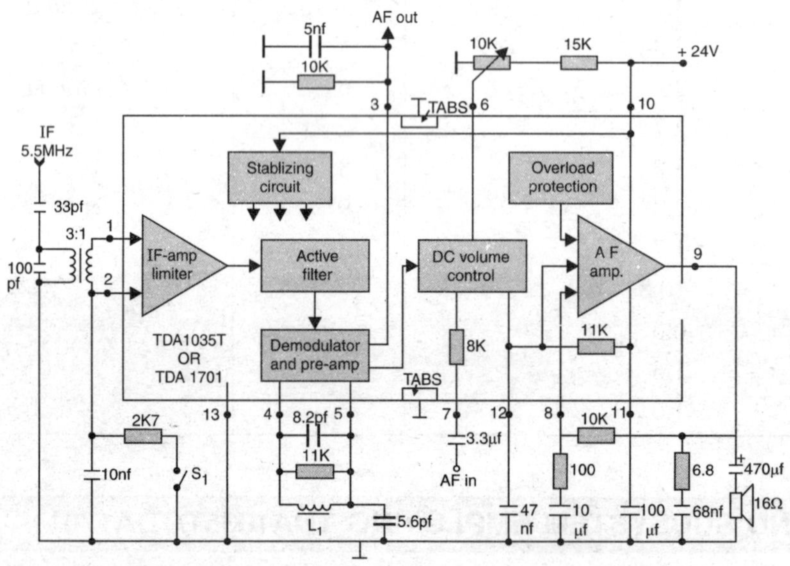

Fig. 12.10 Operating circuit of the sound channel IC TDA 1035T. This IC is identical to TDA1701 pin-to-pin.

The cooling tab projecting from the case on either side are sufficiently large for a 2W output power without additional heat sink. If these tabs are cooled further by soldering them to a large copper clad area of the printed circuit board, an output power upto 4W can be obtained.

12.6 SOUND SUBSYSTEMS WITH CA1190Z AND TDA 4190

The sound subsystem ICs TDA 1190Z and TDA 3190 which are pin-to-pin replacement of each other become available after TDA 1035T. These can deliver an output power of 4.2 W (distortion within 10 %) into a 16 ohm load at Vs = 24 V or 1.5 W into an 8 ohm load at Vs = 12V. Soon after CA 1190 was introduced by Bharat Electronics Limited (BEL) as replacement to the sound ICs described in the previous three sections. While its pin numbers arc thc samc as that of TDA 3190 it is claimed to have somewhat superior performance. The internal circuitry of CA-1190 is fully described in section 12.1. As explained there, BEL CA 1190 combines sound IF and audio subsystem on a single monolithic integrated circuit to provide the entire television sound system. This (CA 1190Z) third generation IC includes a multistage IF amplifier-limiter, FM detector and an audio power amplifier that is designed to drive the loudspeaker directly. The salient features of this IC are (*a*) nominal power output of 4 watts at 24 volt V_s into a 16 ohm load (*b*) wide power supply range of 9 to 28 V. (*c*) low quiescent current (*d*) good (3dB) limiting sensitivity *i.e.*, excellent AM rejection and (*e*) electronic volume control with large dynamic ragne. In addition, its integral heat sink makes the design of printed circuit board and heat sink arrangement quite simple and economical. It is available in 16-lead dual-in-line plastic package with integral bent down wing tab heat sink. A maximum power dissipation of about 2.5W can be handled by the IC even at an ambient temperature of 60°C. This, however, needs exposure to free air by vertically mounting the PCB in the TV receiver.

Sound System TDA4190

The TDA4190 is yet another good IC for the TV sound channel with dc tone and volume control plus an internally switched VCR input/output. Mounted in a powerdip 16 + 2 + 2 package, the device delivers an output power of 4W into 16 ohm with distortion less than 10%, at V_s = 24V. Included in IC TDA4190 are: IF amplifier-limiter, active low-pass filter, AF preamplifier and power amplifier, turn off muting, VCR switch, mute circuit and thermal protection. Its high output, high sensitivity, excellent AM rejection and low distortion make the device suitable for use in TVs of almost every type. Further, no screening is necessary because it is free of radiation problems.

ICTDA8190 is an improved version of TDA4190 with more flexibility of operation like separate VCR intput and output pins. Otherwise it is nearly the same as TDA4190.

Television receivers which are now being marketed by Japanese and South Korean companies use their own sound system ICs. However, their performance characteristics are almost same as described in this chapter for various sound subsystems.

12.7 COMMON FAULTS IN THE SOUND CHANNEL

All modern receivers have ICs in the sound channel. Some common faults in such systems are as under:

(1) No Sound, Picture Normal

In receivers that employ discrete AF amplifier begin with touching centre terminal of the volume control with a screw driver and if a click/hum is heard in the loudspeaker, audio output stage is operating. Next proceed to check sound IC as detailed in the next paragraph. However, if no click is heard check for (*i*) no or low dc supply to amplifier, (*ii*) open coil of the loudspeaker (*iii*) open volume control potentiometer or associated connections, (*iv*) defective power transistors (*v*) improper base and collector voltages of the devices and (*vi*) defective coupling or by pass capacitors. If an audio output transformer is in use, check its windings for open or short circuit.

In receivers employing an IC for the complete sound channel check for dc supply to the IC, voltages at various pins of the IC-in particular at AF amplifier output pin, volume control connection pin, and SIF input pins. Also check loudspeaker coil, its associated circuitry, tuning coil to the demodulator and windings of the input circuit coils for open/short. Lastly replace IC which may be defective. Also check VIF IC to find out if it is delivering SIF signal to the input circuit of the sound IC.

(2) Low Sound Output

If dc supply to the subsystem is normal, low sound output could be due to improper alignment of detector coil and coils of the coupling transformer. Realign these coils to restore sound output. A defective speaker can also cause low sound output.

(3) Buzzing and Random Noise in Sound

Assuming proper alignment of RF and IF stages, buzzing may be due to improper alignment of detector and input coils. Realign these coils as per procedure given by the manufacturer.

(4) Distortion in Sound

This can be due to improper alignment of tuning coils around the sound subsystem, defective IC or loudspeaker. Realign coils, check IC and loudspeaker by replacing them. Make sure that the fine tuning control is set properly for the selected channel.

(5) Sound Bars on the Picture

This is not a fault of the sound system as such. Due to faulty alignment of trap circuit that prevent passage of audio signal in the video channel, the sound signal also reaches the picture tube input to cause such bars. Realign such trap circuits at the output of VIF IC and input to the chrome section. Also ensure that wave traps are aligned properly. The fine tuning control should also be checked for correct setting of the incoming channel signal.

REVIEW QUESTIONS

1. Draw block diagram of a typical sound subsystem IC with basic interface circuitry and label all the blocks and pin outputs.
2. Draw basic circuit of the IF amplifier-limiter cum active filter section of a sound IC and briefly explain how it functions to provide a constant amplitude sinewave FM signal to the demodulator.

3. Describe with a suitable circuit diagram how the FM differential peak detector operates to produce audio signal. Why is the de-emphasis capacitor connected at the point of audio output?
4. What do you understand by dc volume control? Explain how such a control operates to regulate the input of audio signal to AF power amplifier.
5. Draw circuit diagram of a commonly used AF amplifier in sound ICs and explain how it operates to produce sound output in the speaker. How is the gain stability and desired bandwidth obtained?
6. Describe with suitable block diagrams the operation of (*i*) PLL FM detector and (*ii*) digital FM detector. Why are these considered superior to earlier methods of detection?
7. Draw circuit diagram of a shunt fed AF power amplifier employing output transformer to feed the loudspeaker. Explain how it operates with least heat dissipation in the two series output transistors.
8. Explain how would you proceed to trouble shoot the following in the sound channel of a TV receiver.

   (*a*) no sound but picture normal

   (*b*) distortion in sound output

   (c) buzz in sound output.

13

FRAME AND LINE DRIVE INTEGRATED CIRCUITS

INTRODUCTION

In order to produce a picture on the receiver screen as generated at the transmitting end it is essential to first create a synchronised raster. For this, it is necessary to first establish proper scanning rates (15625 Hz for line and 50 Hz for field) by oscillators synchronized with sync pulses received as part of the composite video signal. Earlier* receivers employed discrete circuitry, both for drive and output of frame and line deflection stages. However, with improvements in the operating efficiency of modern colour picture tubes and technological advances in integrated circuit techniques it became possible to contain both vertical and horizontal deflection drive circuits in an integrated package. Later, ICs containing the entire frame deflection stage became available. In fact, now it is only the line output stage which is not manufactured in an integrated circuit form.

The ICs that provide drive pulses to the frame and line output stages are called "Combination ICs". These also make available "Sandcastle Pulse" for the burst gate and other circuits in the chroma section of the receiver. A separate output at a pin of such ICs is also available for muting sound output with no RF input to the tuner.

In latest designs of combination ICs, digital means are used to count-down line and field scan rates from a fixed high frequency reference source thus eliminating the need for external hold control potentiometers. In addition it is possible to use such ICs both for 625 line-50 field and 525 line-60 field TV standards. The circuitry automatically switches over to signals of either system. In order to gain familiarity with application circuits of various combination ICs, several representative integrated circuits are described in this chapter.

*For basic principles and discrete circuitry of field and line deflection stages refer chapters 15, 17 and 18 of the book "MONOCHROME AND COLOUR TELEVISION" by R.R. Gulati (New Age Publication).

13.1 FRAME AND LINE DRIVE IC-TA7609P

This is one of the earlier combination ICs. Its block schematic and external circuitry is shown in Fig. 13.1. Different sections of TA7609P function in the following manner:

SYNC Separator

A transistor with double time-constant leak-bias input circuit is used as sync separator to prevent blocking during strong noise pulses. As shown in the figure, negative going composite colour video signal obtained either from a pin of the vision IF IC or as output of first video amplifier is applied through isolating resistor $R_1$to the input circuit. The output available at the junction of R_3 and C_3 is fed via pin 16 to the base of sync separator transistor, the operating point of which is set primarily by the time constant $C_1 \times R_3$ ($1\mu F \times 56K = 56$ ms). During field sync periods when signal amplitude rises to become positive, base current flows through R_1,C_1 and D_1 to turn 'on' the transistor.

In this operation, C_1 charges with polarity shown and later discharges slowly when the pulse train passes away. Because of large time-constant of $R_2 \times C_1$, charge across C_1 leaks slowly thus maintaining negative bias to the transistor till the next pulse recharges it to cause base current flow.

The line sync pulses being of much shorter duration (high frequency) pass through the small value capacitor C_2 to cause base current and hence produce sync output. The charge that builds across C_2 quickly discharges through D_1 when each sync pulse passes away. The double time-constant circuit also prevents any noise pulses from holding the device at cut-off for longer periods. The transistor operates with a small Vcc supply and large load resistance so that it bottoms quickly to provide clean sync pulses. The sync output (see Fig. 13.1) is fed directly to the AFC circuit and after amplification to the vertical and chroma sections of the receiver from pin 14.

AFC Circuit

A single ended discriminator similar to the 'Gruen' AFC detector is built in the IC to prevent any phase or frequency deviation of the line oscillator. The two inputs to this circuit are the sync pulses and line flyback pulses. The flyback pulses are shaped into sawtooth shape by the integrating capacitor C_5 before feeding from pin 1 to the AFC circuit. The small resistor R_5 in series with C_5 is to limit peak charging current to a safe value and obtain linear output. A small dc potential is also applied at pin 1 through R_8, R_9 and R_6 to obtain no error voltage output condition when the line oscillator is on correct phase and frequency. On any frequency deviation, control voltage from the AFC circuit is in the form of pulses at the oscillator rate. It is smoothened by the external filter networks R_6-C_6 (pin 1) and R_7-C_7 before applying through R_9 to the line oscillator at pin 2.

Horizontal Oscillator

The line oscillator circuit is around an operational amplifier where the feedback network is for sustained oscillations and input circuitry to determine frequency. The capacitor C_8 at pin 2 forms part of this circuit. The frequency is set at the correct value by varying dc voltage to the oscillator circuit (pin 2) from the 12V dc rail through R_8, R_{10} and R_{11}. Thus potentiometer R_{11} acts as 'hold' control for the line deflection frequency.

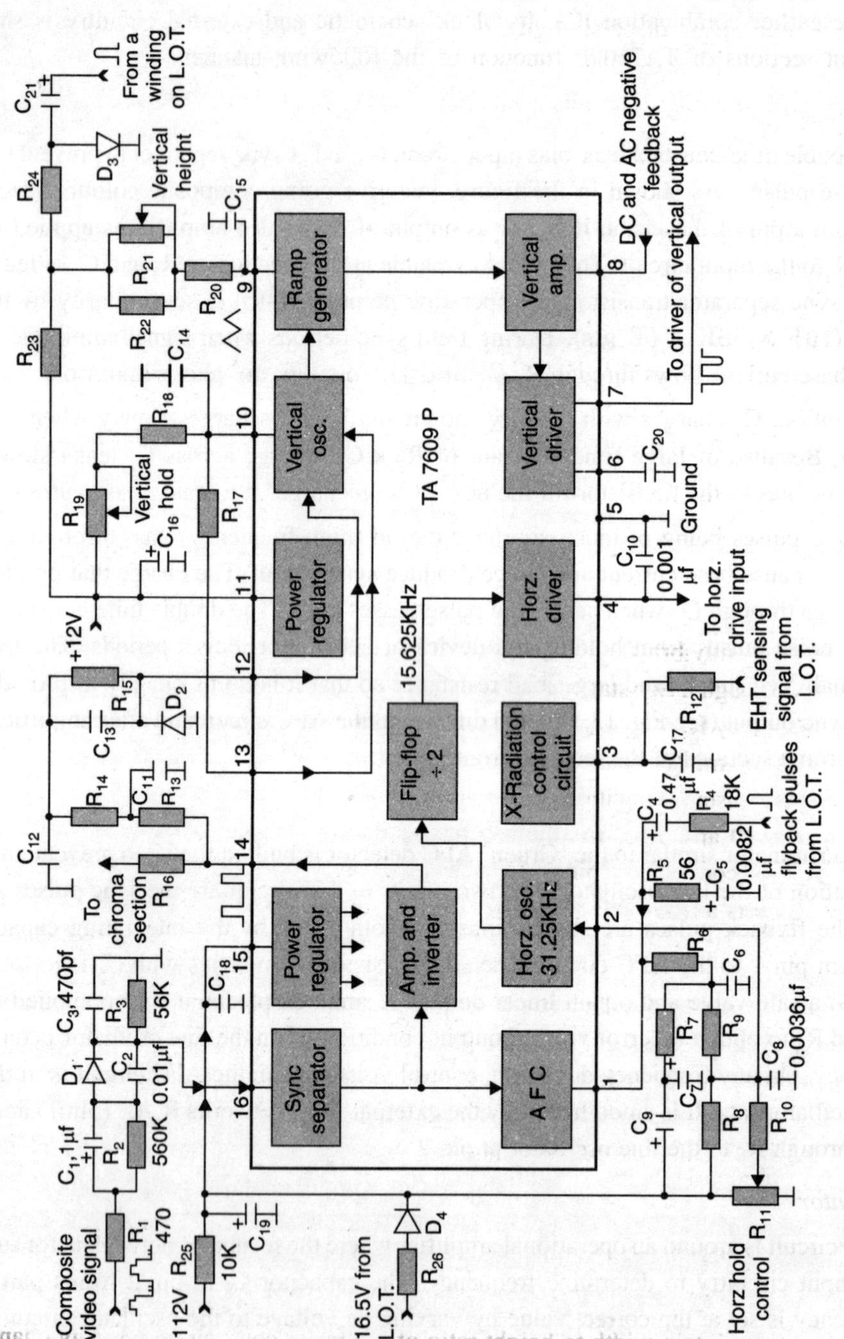

Fig. 13.1 Schematic diagram and external circuitry of the frame and line drive ICTA7609P.

The oscillator is operated at 31.25 KHz (twice of 15.625 KHz) and then a flip-flop circuit divides it by 2 to obtain the desired 16525 Hz rate. The reason for fixing the oscillator frequency at 31250 Hz instead of 15625 Hz is two fold. Firstly, the divide circuit indirectly prevents any variations in the width of output pulses thus minimising dispersion effects in the line output circuit. Secondly, the magnetic induction effect from line to frame oscillator in the ICs is identical during both fields of a frame thereby ensuring good interlaced scanning than when alternate fields are affected differently with oscillator operating at 15625 Hz.

X-Ray Radiation Control

The X-ray radiation prevention circuit is designed to disable the line oscillator should the EHT voltage exceed a limit that can cause radiation from the screen face. The signal to sense this is obtained from a section of the line output transformer and fed to the control circuit at pin 3 through R_{12} and C_{17}. The control operation is fully explained in section 3.13 with the help of its circuit diagram.

Horz Driver

The driver that is fed from the flip-flop output is a single-ended differential amplifier which besides waveshaping amplifies line drive pulses. The capacitor C_{10} (0.001μF) connected at pin 4 is for waveshaping as necessary. The output from the driver amplifier is fed to the line output stage from pin 4 as shown in the figure.

Vertical Oscillator

The vertical oscillator is a multivibrator circuit triggered by vertical sync pulses. These pulses on obtaining from pin 14 are passed through a two stage integrating network comprising of R_{13}-C_{11} and R_{14}-C_{12}. The output is then waveshaped by R_{15}-C_{13} and diode D_2 before feeding to the oscillator control circuit. The differential amplifier of the multivibrator circuit performs current switching function and generates a positive going output of constant amplitude at a frequency of 50 Hz. A negative feedback through R_{17} connected between pins 10 and 13 is to stabilise the oscillator operation. The time-constant circuit formed by C_{14}, R_{18}, R_{19} and connected at pin 10 from the 12V source determines frequency of oscillations. Thus potentiometer R_{19} acts as the vertical 'hold' control.

Ramp Generator

A ramp shaped drive is necessary for the vertical deflection amplifier. It is generated by charging C_{15} through a transistor driven by oscillator's output pulses. The charge and discharge i.e., retrace and trace periods are controlled by the external R-C network connected at pin 9. With a positive pulse of constant width at the base of ramp generating transistor, capacitor C_{15} charges quickly to provide the retrace period. As the input pulse drops, the transistor turns-off enabling the capacitor to discharge through resistive network formed by R_{20}, R_{21}, R_{22}, R_{23} and input resistance of dc supply. The time-constant of this discharge path is large enough to provide a near linear trace period. The amplitude of ramp thus generated can be controlled by changing magnitude of dc voltage applied at pin 9. This as shown in the figure, is done by varying R_{21}.

It is necessary to maintain width to height ratio of the raster constant at 4:3. Any change in dc voltage to the line output stage will affect width of the reproduced picture. It then become necessary to cause corresponding change in height of the picture to maintain the aspect ratio at 4:3. This is obtained by altering height of the ramp in an interesting way. Any deviation in the amplitude of line output pulses is sampled from a section of the winding on the line output transformer. These positive going pulses are rectified by diode D_3 to produce a proportionate dc voltage across C_{21}. The negative end of this voltage (see Fig. 13.1) is connected to R_{21} via R_{24} at the point where dc supply is connected to it. Any change in the magnitude of negative voltage caused by line pulse amplitude variations will affect net dc voltage applied to the ramp capacitor. For example, any decrease in voltage across C_{15} which reflects reduction in width of the raster will cause more current flow through R_{23} from the 12V dc source causing more voltage drop across it. The consequent reduction in voltage at pin 9 will slow down charging and discharging of C_{15} thereby reducing height of the ramp in the fixed pulse width time. This in turn reduces drive to the frame output amplifier and hence, height of the raster. The reverse will occur if the raster width increases for any reason thus maintaining correct aspect ratio despite mains supply fluctuations or other factors that directly affect the line output circuit.

Vertical Amplifier

This is a single ended differential amplifier designed to amplify ramp output to a level necessary to drive the vertical output stage. Negative feedback, both dc and ac is applied to it from different points in the output stage amplifier to obtain correct centering and linear deflection in the vertical direction.

Vertical Driver

It is an emitter follower for current gain and impedance matching between the drive amplifier and vertical output stage. The capacitor C_{20} at pin 6 is for decoupling purposes in the vertical drive section of the IC.

DC Supply

As shown in the figure, dc supply to the IC is from two sides. A dc supply at 12V from the L.O.T. side feeds at pin 11 to a regulator that feeds vertical section blocks in the integrated circuit. Another dc voltage from L.O.T. at 16.5V is connected to a different dc rail via voltage dropping resistor R_{26} and diode D_4. This is paralleled with a source obtained from ac mains side at 112V and dropped through R_{25} to obtain nearly 10V dc on the dc rail. This common dc rail is connected through pin 15 to another voltage regulator for supplying dc voltages to various line oscillator blocks in the IC. Initially, dc supply to the IC is from the 112V dc source but once line output stage becomes operative, auxiliary dc supplies become available. The diode D_4 which is initially reverse biased due to 10V from the 112V supply line conducts to connect the 16.5V auxiliary supply to the dc rail. The resistance values are so chosen that when 16.5V dc becomes available, the dc line has a little higher voltage sufficient to block current drain from the mains side (112V dc). This improves efficiency of the line output stage and decreases power drain from ac mains.

13.2 HORIZONTAL COMBINATION IC-TDA 2593

The ICs TDA2593, TDA2594 and TDA2595 are horizontal (line) oscillator combination ICs with nearly similar operating characteristics. These employ improved techniques for phase and frequency control of line oscillator and output pulse shaper circuits. The IC TDA2593 has been chosen to illustrate such a control operation with its schematic shown in Fig. 13.2. It combines line oscillator and phase

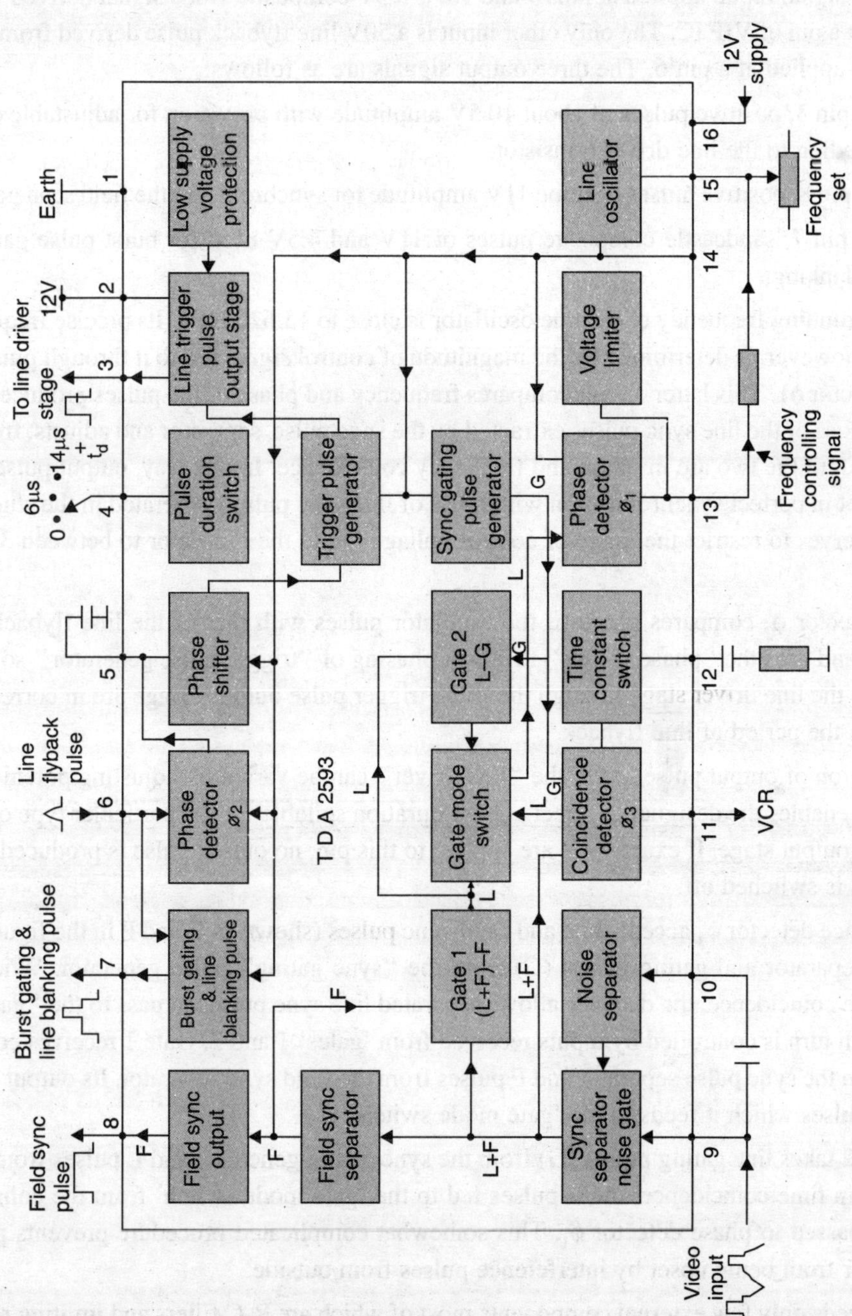

Fig. 13.2 Block schematic of the horizontal (line) combination IC TDA 2593.

comparator stages of the line scan generator and also includes stages such as line and field sync pulse separators.

The main signal input, applied at pins 9 and 10, is a 3V composite video signal derived from the vision detector at a pin of VIF IC. The only other input is a 50V line flyback pulse derived from the line output stage and applied at a pin 6. The three output signals are as follows:

(*i*) From pin 3, positive pulses of about 10.5V amplitude with provision for adjustable duration for feeding to the line driver transistor.

(*ii*) From pin 8, positive pulses of some 11V amplitude for synchronising the field scan generator.

(*iii*) From pin 7, sandcastle composite pulses of 11V and 4.5V used for burst pulse gating and line blanking.

The free-running frequency of the line oscillator is close to 15.625 KHz. Its precise frequency at any given time, however, is determined by the magnitude of control signal fed to it through pins 13 and 15 by phase detector $\phi 1$. This latter circuit compares frequency and phase of the pulses produced by the oscillator with those of the line sync pulses extracted by the snyc pulse separator and adjusts, frequency of the oscillator until the two are in phase and frequency coincidence. In this way, output pulses of the oscillator are kept in perfect synchronization with those of line sync pulses generated in the studio. The voltage limiter serves to restrict the range of control voltage fed to the oscillator to between 3.8V and 8.2V.

Phase detector ϕ_2 compares phase of the oscillator pulses with that of the line flyback pulses applied at pin 6 and uses the "phase shifter" to adjust phasing of "trigger pulse generator" so that the pulses it feeds to the line driver stage through the "line trigger pulse output" stage are in correct phase relationship with the period of line flyback.

The duration of output pulse fed to the "line driver" can be varied by adjusting potenial at pin 4 of the IC. This enables the designer to select a pulse duration suitable for the particular type of circuit used for the line output stage. If exactly 4V are applied to this pin, no output pulse is produced and the line output stage is switched off.

Coincidence detector ϕ_3 accepts line and field sync pulses (shown as L and F in the figure) from the sync pulse separator and gating pulses (G) from the "sync gating" pulse generator. When these pulses are in time coincidence, the detector allows separated line sync pulses to pass to the "gate mode switch", which in turn is controlled by inputs received from 'gates' 1 and 2. Gate 1 receives combined L + F pulses from the sync pulse separator and F pulses from the field sync separator. Its output consists of separated L pulses which it feeds to the 'gate mode switch'.

The gate 2 takes line gating pulses (G) from the sync gating generator and L pulses from gate 1. When these are in time coincidence, the L pulses fed to the 'gate mode switch' from the coincidence detector ϕ_3 are passed to phase detector ϕ_1. This somewhat complicated procedure prevents phase of the line oscillator from being upset by interference pulses from outside.

The IC needs only few external compo.ients most of which are R-C filters and limiting resistors. The variable preset at pin 15 is for adjusting frequency of the line oscillator. An external switch connected at pin 11, when closed, lowers the time-constant thus enabling the coincidence detector to operate in conjuction with a video cassette recorder.

13.3 HORIZONTAL AND VERTICAL OSCILLATOR IC-TDA2578A

The IC TDA2578A separates horizontal and vertical sync pulses from the composite TV signal and uses them to synchronize vertical and horizontal oscillators. It also contains driver amplifiers for line and frame deflection stages. Its operation is similar to TDA-2593 and finds use in many makes of present day colour receivers. A typical application circuit is shown in Fig. 13.3 and a brief description of its operation follows.

Sync Separators

Two separate sync separators are used to separate line and frame sync pulses. A sufficient band-gap is kept between the two to aviod any mutual interference. The positive going composite colour video signal obtained from the video detector output is applied through R_{10} and C_9 at pin 5 for feeding the two sync separators. The resistor R_8 between pins 6 and 7 is for fixing the slicing level of line sync separator at the middle of sync pulse amplitude. The associated noise inverter is designed to be amplitude selective and becomes operative at an amplitude equal to 0.7V. The capacitor C_8 at pin 6 is for decoupling purposes while the time-constant components R_7 and C_6 maintain the slicing level independent of input signal amplitude. Similarly dc potential at 4 obtained via R_2 from the 12V dc supply fixes slicing level for the vertical sync separator. The capacitor C_{11} decouples the dc voltage at pin 4.

Horizontal Oscillator and Output Circuit

The line oscillator produces a sawtooth output at 15625 Hz. Its frequency is determined by the R-C network connected at pin 15 and adjusted to the correct value by varying R_{13} connected at the same pin in series with R_{14}. Synchronization is achieved through a phase lock loop (ϕ_1) that compares oscillator output with line sync pulses to do so. The feedback is from the horizontal drive pin 11 through R_{15} and C_1 to C_4 which integrates the output pulses before feeding to the comparator circuit at pin 8 through R_3. The detector (ϕ_1) is designed to have good noise immunity by choosing a particular time-constant at its input with the capacitor C_5 (100 KPF) connected at the same pin.

The oscillator output is also compared with flyback pulses obtained from the L.O.T. and fed at pin 12 through peak current limiting resistor R_6. The central loop circuit (ϕ_2) and associated pulsewidth modulator compare the oscillator output with period of flyback pulses and apply corrective feedback to keep their phase relationship at the correct value. The capacitor C_{16} and series combination R_{12} and R_{11} connected in parallel at pin 14 are for adjusting the phase relationship. R_{11} is initially varied to obtain optimum condition. The time-constant obtained through these components is chosen for fast compensation of switch-off delays in the horizontal output stage.

The phase and frequency synchronized line signal obtained at the output of pulsewidth modulator is fed to the horz output stage where it is amplified before feeding from pin 11 to a pin of the horizontal control IC or to the driver transistor of the line output stage.

Vertical Oscillator

The vertical oscillator operates in synchronism with frame sync pulses to produce a 50 Hz sawtooth output. The capacitor C_7 at pin 3 charges and discharges periodically to convert the oscillator pulses to a sawtooth form. The potentiometer R_{18} is varied to set the frequency at its correct value. Stability of operation is obtained through a comparator which receives both the sawtooth output and frame sync

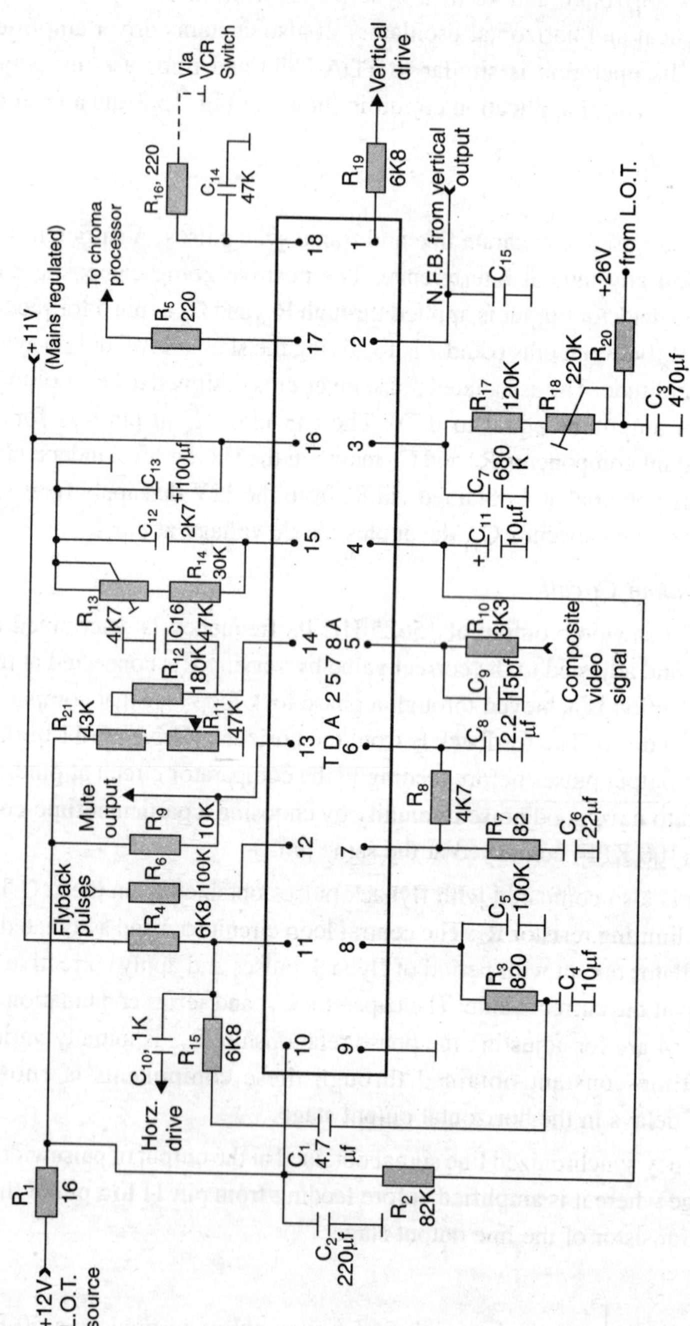

Fig. 13.3 Application circuit of the combination IC TDA 2578 A.

pulses. The 50 Hz frame drive signal from the pulsewidth modulator and comparator is fed to the vertical driver which is an emitter follower. It provides current gain and low output impedance necessary for feeding to the next stage. The output from pin 1 is fed via R_{19} to either a pin on the frame (vertical) output IC or to input circuit of the discrete vertical output amplifier. For overall stability and linear deflection, negative feedback from the output stage is connected via pin 2 to the comparator and vertical guard circuits. The capacitor C_{15} at pin 2 is for waveshaping the feedback signal.

Sandcastle Pulse Generation

The sandcastle pulse generator receives inputs from vertical blanking generator, vertical guard circuit, horizontal oscillator (through burst key) and horizontal flyback input. These are timed and level adjusted as illustrated in Fig. 13.4. The highest level (11V) can be used for burst gating and black level clamping.

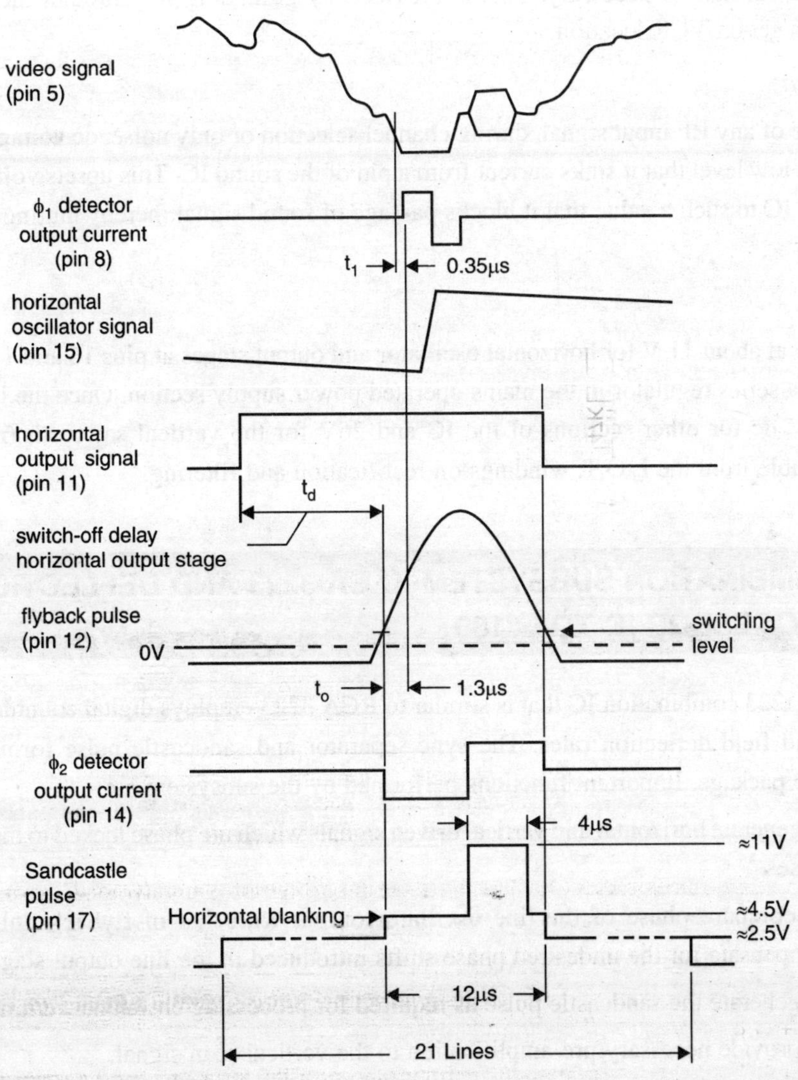

Fig. 13.4 Timing diagram of the IC TDA 2578 A to illustrate formation of Sandcastle Pulse.

The second level (4.6V) is obtained from the horizontal flyback pulse applied at pin 12 and used for horz blanking. The third level (2.5V) is used for vertical blanking. It is derived by counting horizontal frequency pulses. For 50 Hz the blanking pulse duration is equal to 21 line period. The IC also incorporates a vertical guard circuit which monitors the vertical feedback signals at pin 2. If the level is below 3.35V or higher than 5.15V the guard circuit will insert a continuous level of 2.5V into the sandcastle output signal that becomes available at pin 17. This will result in complete blanking of the screen if the sandcastle pulse is used for blanking the TV receiver.

VCR Operation

The in-sync, out-of-sync or no video-condition is determined by the coincidence detector (ϕ_3). Its gating or time-constant is partly determined by the external components (R_{16}, C_{14}) at pin 18. For VCR operation a shorter time-constant is necessary. This is provided by grounding R_{16} through the selector switch contacts when set on VCR location.

Muting Circuit

In the absence of any RF input signal, during channel selection or only noise, dc voltage level at pin 13 falls to such a low level that it sinks current from a pin of the sound IC. This upsets voltage level in that section of the IC to such a value that it blocks passage of sound signal thereby muting output from the loudspeaker.

DC Sources

The dc supply at about 11 V for horizontal oscillator and output stages at pins 16 and 11 respectively is derived from a series regulator in the mains operated power supply section. Once the line output stage picks up, 12V dc for other sections of the IC and 26V for the vertical sawtooth (ramp) generator become available from the L.O.T. windings on rectification and filtering.

13.4 COMBINATION SUBSYSTEM IC CA 3223 AND DEFLECTION PROCESSOR IC TDA 8180

The BEL CA 3223 combination IC that is similar to RCA 3223 employs digital countdown methods to obtain line and field deflection rates. The sync separator and sandcastle pulse forming circuits are external to the package. Important functions performed by the subsystem are:-

(1) To generate horizontal and vertical driven signals which are phase locked to the synchronizing pulses.

(2) To compare phase of the line oscillator output with that of flyback pulses in order to compensate for the undesired phase shifts introduced in the line output stage.

(3) To generate the sandcastle pulse as required for processing chrominance/luminance signals.

(4) To provide necessary pre-amplification to the vertical scan signal.

(5) To facilitate acceptance of non-standard signals generated by equipments like video cassette recorder.

Deflection Processor TDA8180

The TDA 8180 deflection processor integrates the singal processing functions for horizontal and vertical deflection in TVs and monitors. It generates drive wave-forms for external deflection power stages plus super-sandcastle and vertical blanking signals for the chroma processor. Additionally it includes CRT protection circuits and an internal voltage regulator. A 5V supply is used and only a series resistor is required for higher supply voltages. A sophisticated sync separator, phase-locked loops and countdown circuitry guarantee high precision and eliminate all frequency and phase adjustments. A low cost 503 KHz ceramic resonator defines all circuit timings. The TDA 8180 works with both 625 line/50 Hz and 525 line/60 Hz standards and switches automatically, correcting the vertical sawtooth amplitude. A 50/60 Hz identification output is provided.

13.5 COMMON FAULTS IN COMBINATION ICs

Some common faults that occur due to faulty operation of the combination IC and associated circuitry are as under:

(1) No Raster

All receivers obtain EHT and dc voltages from line output and if it becomes inoperative no raster is produced on the receiver screen. One of the causes for this can be absence of line drive signal from the combination IC. For this measure dc supply to the IC and voltages (dc and ac) at the horizontal oscillator and line output pins. If necessary, check all components connected to these pins on the IC. In case dc supply is normal and no break is noticed in the external components, the absence of line drive can be due to a faulty IC.

(2) Picture Tearing

This happens if the horizontal sync is not available. Measure ac voltage on the sync output pins, and if missing check input video signal to the IC. A defective IC can also be the cause of such a fault.

(3) Stable But Out of Phase Picture

This can be due to absence of line flyback pulses. Check if these pulses from the L.O.T. are available on the IC.

(4) Vertical Rolling of Picture

This happens when the vertical sync is absent. Check ac voltage on corresponding pin of the IC. If necessary, ensure that the discrete integrating circuit that forms vertical sync pulses is not defective. Also measure dc voltages and try replacing the IC.

(5) Only Horizontal Line on Screen

The probable causes are (i) dc voltage to the vertical sections in the IC is not available (ii) vertical drive output is missing. Check dc voltage on the vertical section pins and if necessary replace the IC.

REVIEW QUESTIONS

1. See Fig. 13.1 of the IC TA 7609P and explain briefly how it operates to provide line and frame drive output. Why is the line oscillator operated at $2f_H$? How is the aspect ratio maintained as 4 : 3?
2. Draw block diagram of the IC TDA2593 and label it fully. Explain briefly how the two phase lock loop circuits function to synchronise the line oscillator with sync pulses and relative picture with flyback pulses.
3. With reference to Figs. 13.3 and 13.4 explain how the sandcastle pulse is generated. Describe what circuit change is necessary for VCR operation and how is it made possible in the external circuitry of IC TDA 2578-A.
4. Describe how a faulty combination IC can cause the following fault in a TV receiver.

   (*i*) stable but out of phase picture

   (*ii*) vertical rolling of the picture

   (*iii*) only horizontal line on the screen.

14 CHROMA PROCESSING SUBSYSTEM

INTRODUCTION

Colour signal generation, formation of composite colour signal and encoding it for transmission formed the subject matter of first 8 chapters of this text where different TV standards and various CTV systems are also described. Chapter 7 is exclusively devoted to functioning of a colour receiver where signal processing from antenna input to formation of coloured picture is fully explained. As necessary, more emphasis there is on describing the decoder because it has no counterpart in the monochrome receiver. Later chapters describe how various integrated circuits function to recover composite colour signal at the output of video detector from the RF signal.

Now we turn to investigate in depth, how Y, R, G and B signals are recovered from the composite colour signal. While a single IC does the entire job, it is felt that the complex process of decoding and matrixing be first explained with the help of discrete circuits for a fuller understanding of the techniques employed to do so. It will thus be a continuation of chapter 7 where a block diagram approach was adopted. The reader is, therefore, advised to revise relevant sections of that chapter before proceeding to learn more about the decoder and associated circuits.

With the above sequence in view, initial sections of this chapter are devoted to discrete circuitry of decoder circuit blocks and later section describe functioning of various 'Chroma Processor' integrated circuits. As in other chapters, the last section is devoted to trouble shooting of common faults in this part of the receiver.

14.1 LUMINANCE SIGNAL PROCESSING

In modern colour receivers a video preamplifier forms part of the vision IF IC and the composite colour signal that becomes available at a particular pin of this IC is usually fed via emitter follower to two different pins on the chrome IC through appropriate filter configurations. In one case the chroma signal is selected while in the other it is rejected to obtain the Y signal. In most receivers, L-C configurations are used as filter elements but in recent designs the use of SAW and Comb filters is becoming more common. These are explained in chapter 7 and may be referred to if necessary.

Video Preamplifier

A typical circuit suitable for luminance signal processing is shown in Fig. 14.1 where Q_1 is the preamplifier designed to provide an overall gain of about three. The amplified and inverted signal that becomes available at the collector of Q_1 is fed via a delay line at the base of Q_2. The in-between series trap formed by C_1, R_4 and L_1 is to bypass the chroma signal. This avoids the danger of dot patterning and other forms of distortion on the displayed picture. The delay time, typically 0.6μs, is chosen to ensure that both luminance and chrominance signals arrive at precisely the same instant at the modulating electrodes of the picture tube. This, as explained fully in an earlier chapter, is necessary to avoid blurring of colours especially at the edges of various objects in the reproduced picture.

Delay Line

The delay line has a characteristic impedance (Zo) of 1000 ohms and a delay time of 600 nano-seconds (0.6μs). It is constructed on the principle of distributed L-C elements and consists of a number of copper strips extending full length of the line (≈ 100 mm) and disposed around an insulated rod. These strips in conjunction with a very thin sheet of metallic foil give the distributed capacitance. Similarly, distributed inductance is obtained by winding nearly 1400 turns of very thin wire on the same rod. Each of the small L-C sections thus formed progressively delay the passage of signal due to inherent charge and discharge phenomenon to provide a total delay of 600 nanoseconds.

Luminance Amplifier

The output from Q_2 is fed to Q_4 via emitter follower Q_3 which acts as buffer to prevent any mutual interference between contrast control and black level clamp circuits. There is some attenuation of signal between the collector of Q_2 and emitter of Q_3 because of contrast control network and less than unity gain of the emitter follower.

Contrast Control

The contrast control elements R_{10}, R_{11} and R_{12} form part of a bridge circuit where potentials at points A and B are set at the same level. The dc voltage at point 'A' is nearly 3V and therefore values of R_{13} and R_{14} are so chosen that voltage at point 'B' is also equal to 3V. The result is that when R_{10} is varied to adjust amplitude of Y signal at the output of luminance amplifier, it does so without affecting either the dc content of signal or the amplitude of line and field blanking pulses which are applied at both ends of the control via resistors R_{11} and R_{12}.

The negative going blanking pulses that appear at the base of Q_3 via contrast control elements are of such an amplitude that it turns-off for these intervals. The pulses being coincident with flyback periods of the line and field scans ensure that the signal fed to the picture tube is held at black level during retrace intervals. Thus alternate application of Y signal and blanking pulses at the base Q_3 enables contrast control without affecting dc content of the luminance signal and making sure that retrace lines do not appear on the screen.

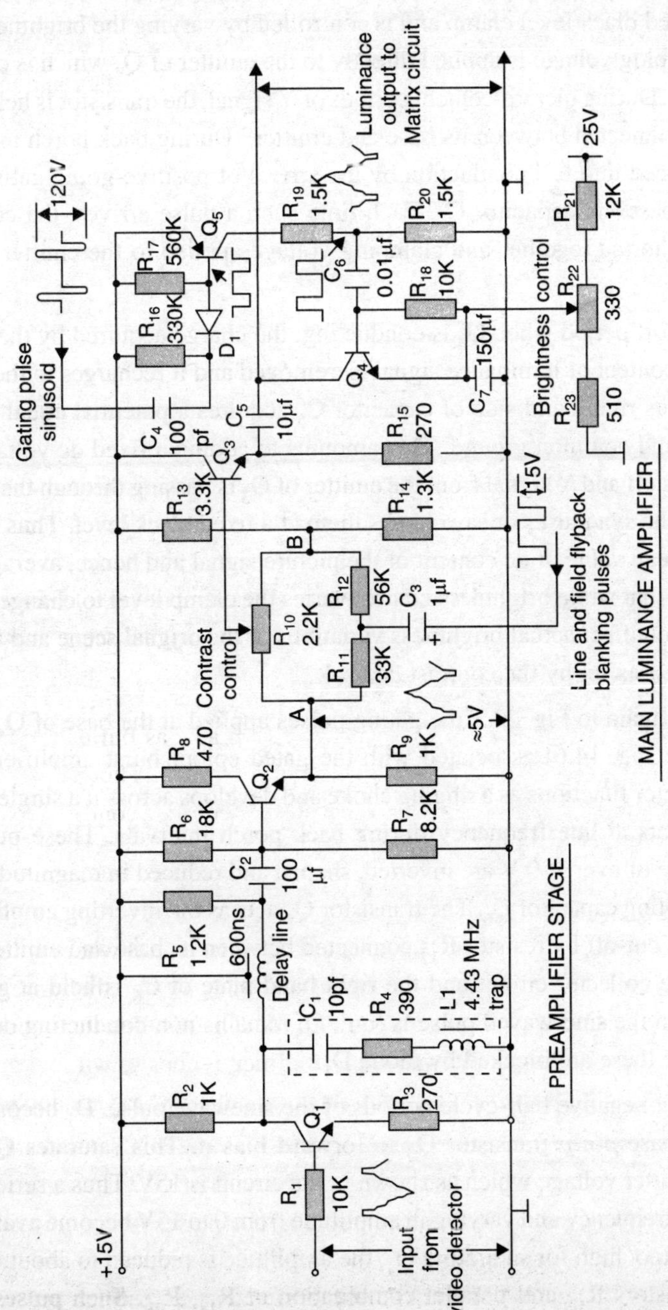

Fig.14.1 A typical discrete circuit designed for processing luminance (Y) signal and providing contrast and brightness controls.

CHAPTER 14

Brightness Control (Black-Level Clamp)

The average value of luminance signal fed to the matrix unit determines mean brightness of the picture appearing on the screen. It is thus convenient to vary black level of the signal to control average brightness of picture. This is called black level clamp and is controlled by varying the brightness control as shown in Fig. 14.1. The clamping voltage is applied directly to the emitter of Q_4 which is connected to operate as an electronic switch. During picture-content periods of Y signal, the transistor is held in non-conducting state by resistor R_{18} connected between its base and emitter. During back porch intevals of the signal, Q_4 is momentarily forced into full conduction by the arrival of positive-going gating pulses applied at line frequency to its base via capacitor C_6. Each time such a pulse arrives, the collector and emitter electrodes of Q_4 are shorted together and clamping voltage applied to the emitter also appears at the collector.

During each short period when Q_4 is conducting, the charge acquired by the coupling capacitor (C_5) from the picture content of luminance signal is removed and it recharges to the potential set at the brightness control. Thus right hand side of capacitor C_5 acquires a potential equal to that between the variable point of R_{22} and common ground. This amounts to adding a fixed dc voltage to ac part of the Y signal when Q_4 turns off and Y signal from the emitter of Q_3 is passing through the coupling capacitor. Such an action aligns the syn pulses or say clamps them at a fixed black level. Thus it is the clamp level that dertermines the mean value of dc content of the picture signal and hence, average brightness of the scene. It may be noted that while brightness control varies the clamp level to change average brightness of the picture it does not affect actual brightness variations of the orignal scene and also does not in any way change the contrast as set by the contrast control.

Referring once again to Fig. 14.1 the gating pulses applied at the base of Q_4 originate from the small inductor L_3 (see Fig. 14.6) associated with the gated colour burst amplifier described later in section 14.4. The inductor functions as a ringing choke and develops across it a single cycle of sinusoidal waveform which repeats at line frequency during back porch intervals. These pulses which have a peak-to-peak amplitude of over 100 V are inverted, shaped and reduced in magnitude beore applying at the base of Q_4 via coupling capacitor C_6. The transistor Q_5 acts as the inverting amplifier. In the absence of sinusoid it is held at cut-off by resistor R_{17} connected between its base and emitter. In this condition no current flows in the collector circuit and the right hand plate of C_6 is held at ground potential by resistor R_{20}. Even when the sine-waved pulse is 'on', Q_5 remains non-conducting during positive half-cycle intervals because these are blocked by diode D_1.

However, during negative half-cycle periods of the sinewave pulse, D_1 becomes forward biased and pulses reach the base *p-n-p* transistor Q_5 to forward bias it. This saturates Q_5 and its collector voltage rises to the emitter voltage which as shown in the circuit is 15V. Thus a series of positive going pulses occuring at line frequency and varying in amplitude from 0 to 15V become available for switching Q_4. Since 15V level is too high for saturating Q_4, the amplitude is reduced to about 4V by the potential divider formed by resistors R_{19} and parallel combination of R_{18}, R_{20}. Such pulses as shown in the figure attain a flat top duc to the limiting action of base-emitter junction of transistor Q_5.

The black level set by the brightness control usually varies from 0.9V to 14V and this is enough to provide a wide range of average brightness in the reproduced picture.

14.2 THE PAL-D DECODER

The purpose of a colour decoder is to recreate the original red, blue and green camera signals and feed them to the picture tube. The first job of the decoder is clearly to extract chroma and colour-burst signals from the composite signal and to reject any other signals that may be present at that point. Next it must extract from the colour signal the two sets of amplitude modulation which represent U and V signals and then restore the amplitude of both to that of the original (B – Y) and (R – Y) colour-difference signals. Lastly it must generate (G – Y) signal by a suitable matrix ciicuit involving (B – Y) and (R – Y) signals.

The block schematic of a PAL-D decoder is shown in Fig. 14.2. The letter 'D' is added because the decoder incorporates a chroma delay line and is different from PAL-S decoder which being inferior in performance is no longer used in modern colour receivers.

As shown in the block schematic the final outputs of decoder are applied to RGB amplifiers along with the luminance (Y) signal. This latter signal, as explained in the previous section, is obtained from the main luminance amplifier.

General Survey

The chroma signal and colour burst are separated from the incoming composite colour video signal by the chroma signal selection circuit. It essentially consists of a bandpass circuit whose centre frequency is chosen to be equal to that of the chroma subcarrier itself *i.e.*, 4.43 MHz. Its passband is approximately 2 MHz measured at the conventional 3 dB down points of the frequency response curve. It is followed by a bridged-T circuit which offers very high impedance at 5.5 MHz to prevent break through of the intercarrier sound signal into the colour decoder. In many cases the 4.43 MHz band-pass filter is preceded by a 2.2 MHz rejection filter configuration. Its purpose is to prevent passage of 2.2 MHz frequencies which might be present in the luminance signal. Such a rejection is considered necessary because the 2nd harmonic (2 × 2.2 MHz) of such frequencies would fall within the pass-band of the filter circuit which will passs them on as part of genuine chroma signal. The output from these filter networks which are connected in tandom is fed into an emitter follower which provides due isolation and feeds chroma and colour burst signals to the first chroma amplifier from its emitter.

As already mentioned and described in section 7.10, cumb filters are considered more suitable for saparation of luminance and chroma signals. The filter circuits as used in colour receivers are described in sections 14.11 and 14.12 which deal with signal processing by integrated circuits.

On separation, the chroma and burst signals are amplified by the first chroma amplifier which is gain controlled by the voltage developed by the Automatic Chroma Control (ACC) amplifier. The output from the amplifier goes to both the second chroma amplifier and burst pre-amplifier. The second chroma amplifier incorporates colour saturation control circuit. The output of colour killer also feeds into it. As shown in Fig. 14.2 the output of 2nd stage is fed to the delay line and also, simultaneously, to addition and subtraction circuits. On separation the U and V colour video signals are fed to the U and V demodulators respectively where the original colour-difference signals are recovered from the subcarrier. These two latter signals are then passed on to the RGB amplifiers along with the Y signal to recover original colour camera signals.

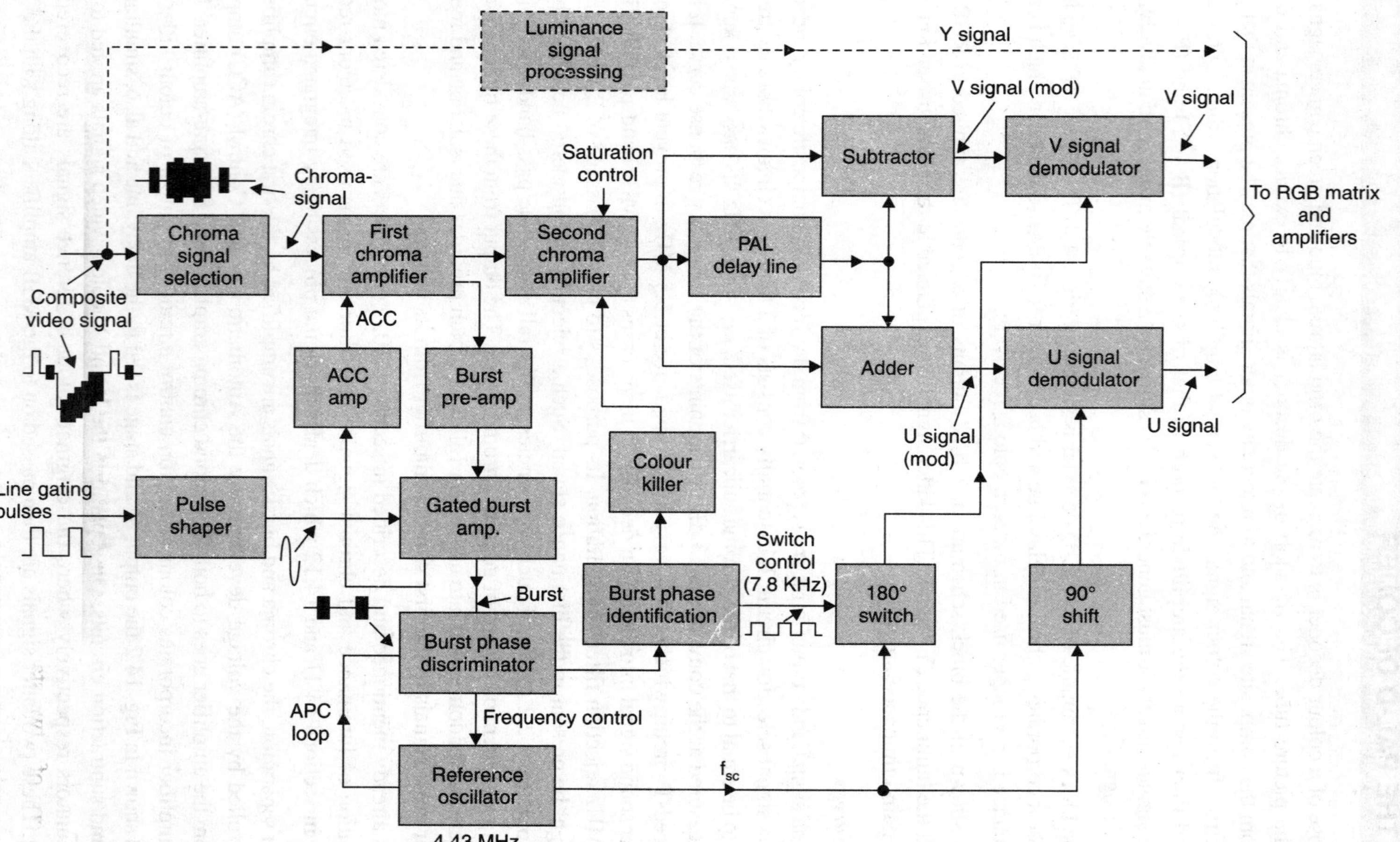

Fig. 14.2 Block schematic of the PAL-D decoder.

Going back to the first stage of chroma amplifier, its second output is fed to the burst pre-amplifier which forms part of a two-stage amplifier. The second stage is gated by pulses coincident in time with the line flyback pulses which are applied to this stage through a pulse shaping network. The purpose of these two stages is thus to separate burst pulses and amplify them to a level suitable for operating the 'burst phase discriminator' which is sensitive to burst pulse only. It is designed to detect any difference which might exist between the phase of burst pulse and that of the reference oscillator. It produces at its output a dc voltage whose magnitude and polarity are proportional to the magnitude and direction of the detected phase difference. It is used to control the frequency of reference oscillator to keep it stable at 4.43 MHz. The control circuit is represented by the APC loop in the block diagram.

A second output from the gated burst pre-amplifier is converted to a dc voltage by a rectifier circuit and then fed to the ACC amplifier. The magnitude of the voltage so fedback is proportional to the amplitude of burst and therefore to the amplitude of chroma signal itself. After amplification in the ACC amplifier, the voltage is used to control gain of the first stage of chroma amplifier is such a way as to ensure constant chroma signal amplitude at its output.

A second output from the burst phase discriminator is fed to a circuit which is able to identify phase relationship of the colour burst. It may be recalled that phase of the burst alternated by ± 45° relative to the phase of – U signal and that it is this phase difference which enables the decoder to differentiate between U and V signals. The circuit *i.e.*, the 'burst phase identification' has two outputs. One of these is used to control operation of the 180° electronic switch which periodically inverts the waveform fed from the reference oscillator (RO) to the V signal demodulator. It is important that this switching shall occur in the correct phase and in synchronism with a similar switching operation which took place in the PAL encoder at the transmitting end and hence the need for such a circuit. The waveform fed from the Reference Oscillator (R.O.) to the U-signal demodulator is phase shifted by a fixed 90° in order to make its phase coincide with that of the subcarrier which was similarly phase-shifted before being modulating by the U signal.

Another output from the burst phase identification circuit is fed to the 'colour killer'. This is no more than a half-wave rectifying circuit which produces a steady dc potential from the succession of burst pulses. This dc potential is fed to the second chroma amplifier to enable it (keep it operative) thereby allowing the chroma signal to reach the two demodulators. In the absence of burst pulse, which will be the case when a monochrome picture is being transmitted, this dc potential is missing and 2nd stage of the chroma amplifier is inhibited (blocked). The advantage of this is that colour noise will be prevented from appearing on the picture tube when a black and white picture is being received. This is especially desirable in conditions of poor signal strength.

This completes the intial survey and we next proceed to examine individual decoder blocks in greater detail.

14.3 CHROMA SIGNAL AMPLIFIERS

The curcuit of a two stage chroma amplifier together with chroma delay line connections is shown in Fig. 14.3.

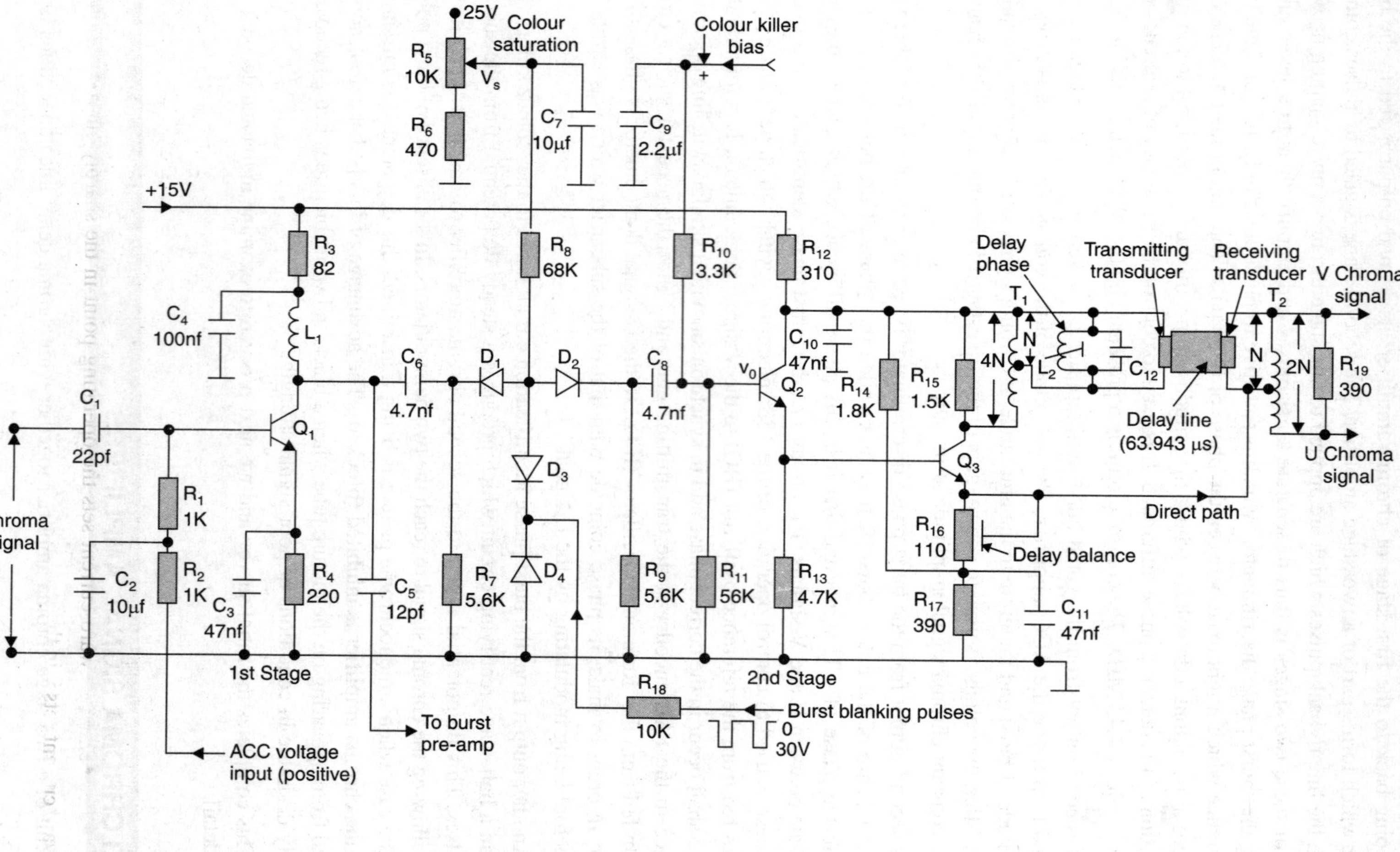

Fig. 14.3 A typical circuit of 1st and 2nd chroma signal amplifiers feeding into the chroma delay line.

1st stage. The first stage (Q_1) functions as a common emitter inverting amplifier and receives chroma signal input at its base via coupling capacitor C_1. With C_4 providing an ac short to the chroma signal frequencies the gain of this stage is determined by the reactance of inductor L_1 and magnitude of ACC controlling voltage fed at the base through resistors R_2 and R_1. The junction of these two resistors is decoupled by C_2 to prevent undesirable feedback between Q_1 and ACC amplifier. The forward bias and hence the collector current of Q_1 is mostly determined by the ACC voltage. Whenever a strong signal arrives the ACC voltage becomes more positive to cause more collector current. Similarly with weak input signal the collector current is reduced . Since beta (β) of a transistor decreases with increase in collector current, the stage gain becomes less on strong signals to keep the output voltage constant. Thus, the ACC control is just the same as forward-AGC control used in IF section of the receiver. The ouput signal voltage at the collector of Q_1 is typically 0.5V peak-to-peak and is fed via C_6 to the second stage and through C_5 to the burst pre-amplifier.

2nd stage. The output signal from the first chroma amplifier is fed to input of second stage through colour saturation control network and colour killer bias circuit. The amplifier comprises of transistors Q_2 and Q_3 where Q_2 is connected as an emitter follower and provides a low-impedance chroma signal drive for the input of Q_3. The transistor Q_3 functions as a phase splitting amplifier whose purpose is to provide anti-phase chroma signal for the delay line.

Colour Saturation Control

The purpose of colour saturation or colour control is to form a variable attenuator to vary the magnitude of chroma signal fed to U and V signal demodulators and thus the amount of colouring information fed to the picture tube. The attenuator is formed by diodes D_1, D_2 in conjunction with resistors R_7, R_9 and variable dc voltage obtained from a separate dc source through variable resistor R_5. As shown in Fig. 14.3, the chroma signal is applied to the saturation control via C_6 and taken out by way of C_8. These two capacitors are a short for the high frequency chroma signal but prevent any dc flow through them. Thus effective ac load on the output side of the circuit is a parallel combination of R_9, R_{10} and R_{11}. Since the value of R_{11} is much greater as compared to R_9 and R_{10}, the effective load is R_9 parallel $R_{10} \approx 2$ K ohms. Next consider dc operating conditions of the control circuit. The diodes D_1 and D_2 are connected back-to-back and get a forward bias from the variable point of R_5 through R_8. The component values are so chosen that forward bias current through the diodes can be varied from a few tens of μA to hundred of μA. The decoupling capacitor C_7 isolates Vs from the chroma signal and hence only dc current flows in the wires that connect the circuit on PCB to colour potentiometer located on the front panel. The idea is same as for dc volume control in the audio section of the receiver. The high frequencies are prevented from flowing through the long leads which would otherwise cause undesired feedback and loss of high frequency content of the signal.

The magnitude of forward current in a diode determines the operating point on its forward characteristics. A small forward current sets the operating point in the sharply curved region while a large forward current sets it in the steeper but less curved region of the characteristics. Thus ac resistance at low forward current values is much higher than when more current flows through the diode. The two diodes then act as variable ac resistances in the path of chroma signal. They form a potentiometer with

$R_9 \parallel R_{10} = R_L$ to attenuate the chroma signal depending on the forward bias *i.e.*, setting of colour control potentiometer R_5. Thus the chroma signal (v_o) developed across R_L can be expressed as:

$$V_0 = \frac{R_L \times V_i}{R_L + (rac_1 + rac_2)}$$

where v_i is the magnitude of input chroma signal and rac_1, rac_2 are ac resistances of the two diodes. Since the two diodes are identical.

$$V_0 = \frac{R_L \times V_i}{R_L + 2\,rac}$$

Substituting 2K for R_L

$$V_0 = \frac{2000 \times V_i}{2000 \times 2\,rac}$$

Typical values of rac range from 20 ohms for a high colour saturation setting to about 1000 ohms for a low one. Thus chrome signal transmission can be varied from some 50% to 90%. The use of two diodes connected back-to-back tends to cancel non-linearity of their characteristics and thus a near linear overall control of the chroma signal transfer is obtained.

Burst Pulse Blanking

The purpose of diodes D_3, D_4 and resistor R_{18} shown in the colour control circuit is to prevent colour burst pulses from getting through second stage of the chroma amplifier. If allowed to pass through, these would get demodulated (rectified) at the U and V demodulators alongwith the chroma signal, and result in a positive pulse being created by the V-demodulator, and a negative pulse by the U-demodulator, both coincident with the burst pulses and thus occurring during back porch regions of the line blanking periods.

This would be most undesirable because it is during these periods that black level clamping is applied to the U and V signals. The presence of such pulses would cause the clamping level to vary every time the colour saturation control was adjusted which in turn would cause inaccurate colour variations in the reproduced picture. Another disadvantage of letting the burst pulse to get through to the U and V demodulators is that more effective blanking would be needed to ensure that the pulses do not become visible on the screen during flyback periods.

Returning back to the operation of burst blanking circuit, a series of negative pulses of about 30V magnitude are obtained from a winding on the LOT. These are time coincident with flyback periods and as shown are applied at the junction of D_3 and D_4 via R_{18}. During no pulse intervals the right hand side of R_{18} is effectively at ground potential and this enables D_3 to conduct because of positive-voltage at its anode due to Vs on the colour control potentiometer. However, this has little effect on the chroma signal transmission because the shunting path has very high impedance due to the 10 k.ohm value of R_{18}.

When a blanking pulse arrives, D_4 becomes forward biased and provides a low resistance path to current through D_3. Diode D_3 now gets heavily forward biased by the current supplied through R_8. A near short circuit is thereby placed across the chroma signal path and it is prevented from reaching the second stage of chroma amplifier.

Colour Killer Control

Resistors R_{10} and R_{11} determine forward bias applied to the base of Q_2 and R_{13} controls its steady emitter current. The upper end of R_{10} is returned to the positive bias voltage developed by the colour killer circuit on rectifying burst pulses. When a colour signal is being received, the bias voltage is typically 12V, which is more than adequate for forwarding biasing Q_2 by the potentiometer formed by resistors R_{10} and R_{11}. This ensures free passage of the chroma signal to Q_3.

However, when a monochrome signal is being received, no colour burst pulses are present and the killer bias applied to R_{10} falls to less than 2V. This is not enough to forward bias Q_2 which, therefore, remains in the non-conducting state. Thus no signal of any sort is fed to Q_3 in the absence of colour signal.

Chroma Delay and Signal Separation

As shown in Fig. 14.3 the phase-splitter transistor Q_3, produces two anti-phase chroma signals, one at its collector and the other at its emitter. The collector signal (greater of the two) is developed across the primary of autotransformer T_1 which forms the collector load. The turns ratio of 4:1 between primary and its secondary is chosen to match input impedance of the chroma delay line input transducer. On transmission through delay line, the chroma signal is fed between one end and centre tap of autotransformer T_2 through the receiving transducer.

Similarly the chroma signal appearing at the emitter of Q_3 is applied direct (with no delay) to centre point connection of T_2 and reappears at either end of its 2N winding as two-in-phase signals of identical amplitude. The delayed chroma signal originating from the previous line of scan is developed across the primary (N) winding of T_2 and reappears as two antiphase signals at the two ends of the 2N winding. Transformer T_2 thus receives simultaneously two chroma input signals, one direct and the other delayed. Since the direct signal appears without phase difference at the two ends of winding (2N) whereas the delayed signal appears there in antiphase, it follows that at one of the end points of the winding the two output signals will add to give a U-only output while at the opposite (other) end they will subtract to give a V-only output. The transformer connections are so arranged in practice that the ± V signal appears at the upper end of the winding while + U appears at the lower end.

Separation of Chroma Signals

The above explained method of addition and subtraction to obtain ± V and U chorma signals is further illustrated in Fig. 14.4 which is essentially the same as Fig. 6.7(*a*). To be more precise the chroma signal may be expressed as U + jV, U – jV, U + jV etc. on successive line scans. The chroma delay is designed to introduce a delay of 63.943μs and not 64 μs for reasons explained in section 6.9.

The signal separation technique is based on the assumption that picture signal information carried on two successive lines changes so little that it can be regarded as identical for both. Consider a point in time at which information from line No. 50 is being applied to the circuit. It is fed both into the delay line and also directly to adder (+) and subtractor (–) circuits where it arrives at exactly the same moment as chroma information from the previous line (No. 49) reaches these points after being delayed by one line period. Assuming that 50th line signal is U + jV and 49th line signal is U – jV, the adder circuits will then produce an output = + 2U while the subtractor will give an output = – 2jV.

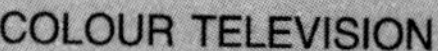

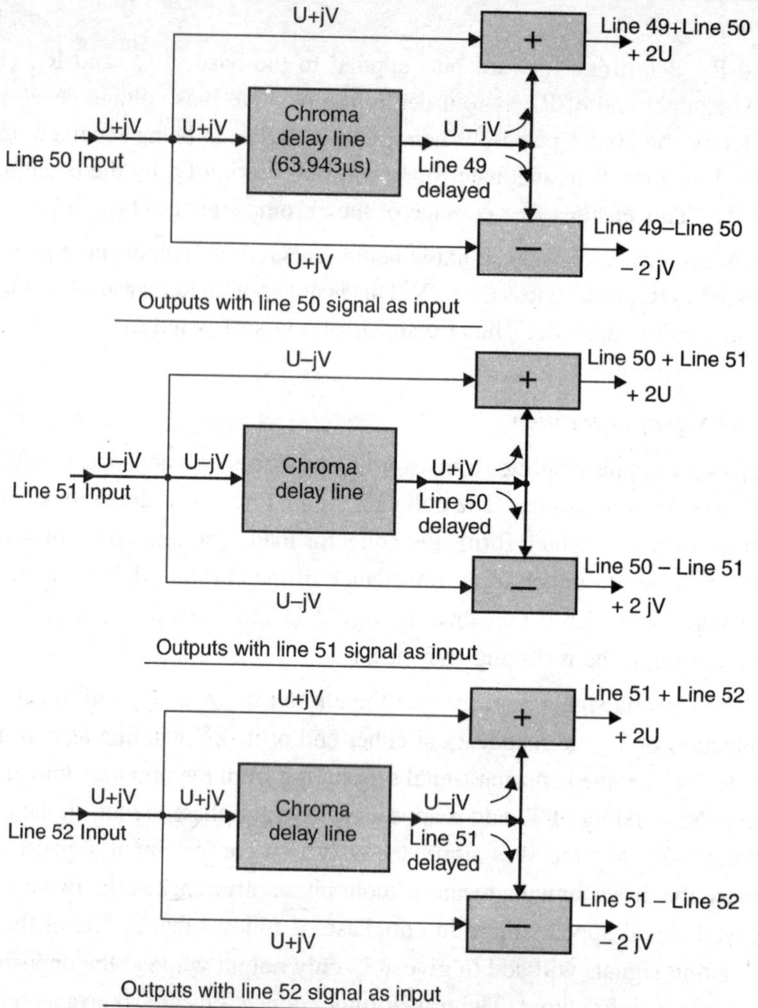

Fig. 14.4 Separation of U and V chroma signals.

Now consider the outputs obtained during the next (No. 51) line scan. Note that now signals of line Nos. 51 and 50 will add and subtract to produce + 2 U and + 2jV as the outputs. If the process is repeated, as it actually does, then ouputs with signal U + jV for the 51st line as input and U – jV at the output of delay line will yield + 2U and – 2jV as outputs from the adder and subtractor circuits. Note that the subtractor reverts back to – 2jV as its output. Thus in general, the circuit as a whole produces two quite different outputs that from the adder being the U chroma signal of fixed polarity and that from the subtractor being the V chroma signal whose polarity alternates from line to line in sympathy with the phase reversal of this signal in the encoder at the transmitter. Note that in the circuit of Fig. 14.3, addition and subtraction is carried out through transformer action and 'j' in the above expressions indicates a relative phase shift of 90° between the U and V chroma signals.

Ultrasonic Delay Line

A typical multipath ultrasonic delay line is shown in Fig. 14.5. It consists of a block of special glass fitted with two ultrasonic transducers. One transducer functions to convert electrical signal into an ultrasonic signal whose mean frequency and amplitude variations are identical to those of the electrical signal while the other converts back the ultrasonic signal into electrical form at the same frequency.

In the delay line application of Fig. 14.3 the electrical signal from the chroma amplifier with a mean frequency of 4.43 MHz is coupled into the transmitting transducer where it is converted into an equivalent ultrasonic signal having the same mean frequency and pattern of amplitude variations. The signal travels through the glass block in a multiple path due to reflections from the polished surfaces before it reaches the receiving transducer. As illustrated in Fig. 14.5 here it is converted back into its original electrical form but delayed in time with respect to the signal at the input of transmitting transducer.

The delay of signal is caused by low velocity of the ultrasonic signal as it passes through glass of the delay line. This is typically, 3 mm per μs as compared to the hundreds of thousands of millimetres per μs traversed by an electrical signal passing along a length of wire. The glass used for such delay lines has to be of a special type known as 'isopaustic' which means that the propagation characteristics are virtually unaffected by the physical expansion or contraction of glass caused by temperature variations. It is about 50 mm in length and the overall delay of 63.943 μs is the sum of six transmit times t_1 to t_6.

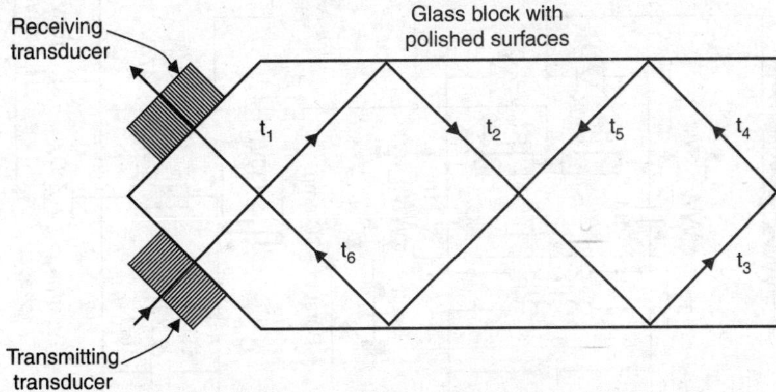

Fig.14.5 Ultrasonic multiple-fold delay line principle.

14.4 COLOUR BURST AND PHASE DISCRIMINATOR CIRCUITRY

Before proceeding to U and V demodulators it is necessary to examine the reference oscillator, 180° switch and 90° phase shifter because all these are connected to the demodulators. But, since the latter are effectively controlled by the colour burst signal it is logical to first examine burst separation and associated circuits. The circuitry in Fig. 14.6 comprises burst preamplifier, pulse shaper, gated burst amplifier and burst phase discriminator.

Burst Pre-amplifier

The chroma input signal derived from the first stage of chroma amplifier (see Fig. 14.3) is fed to the base of Q_4 which functions as a straight-forward amplifier with a gain of about 8 (eight). The amplified chroma signal appearing at the collector of Q_4 is fed via C_{13} to the base of Q_5, the gated burst amplifier.

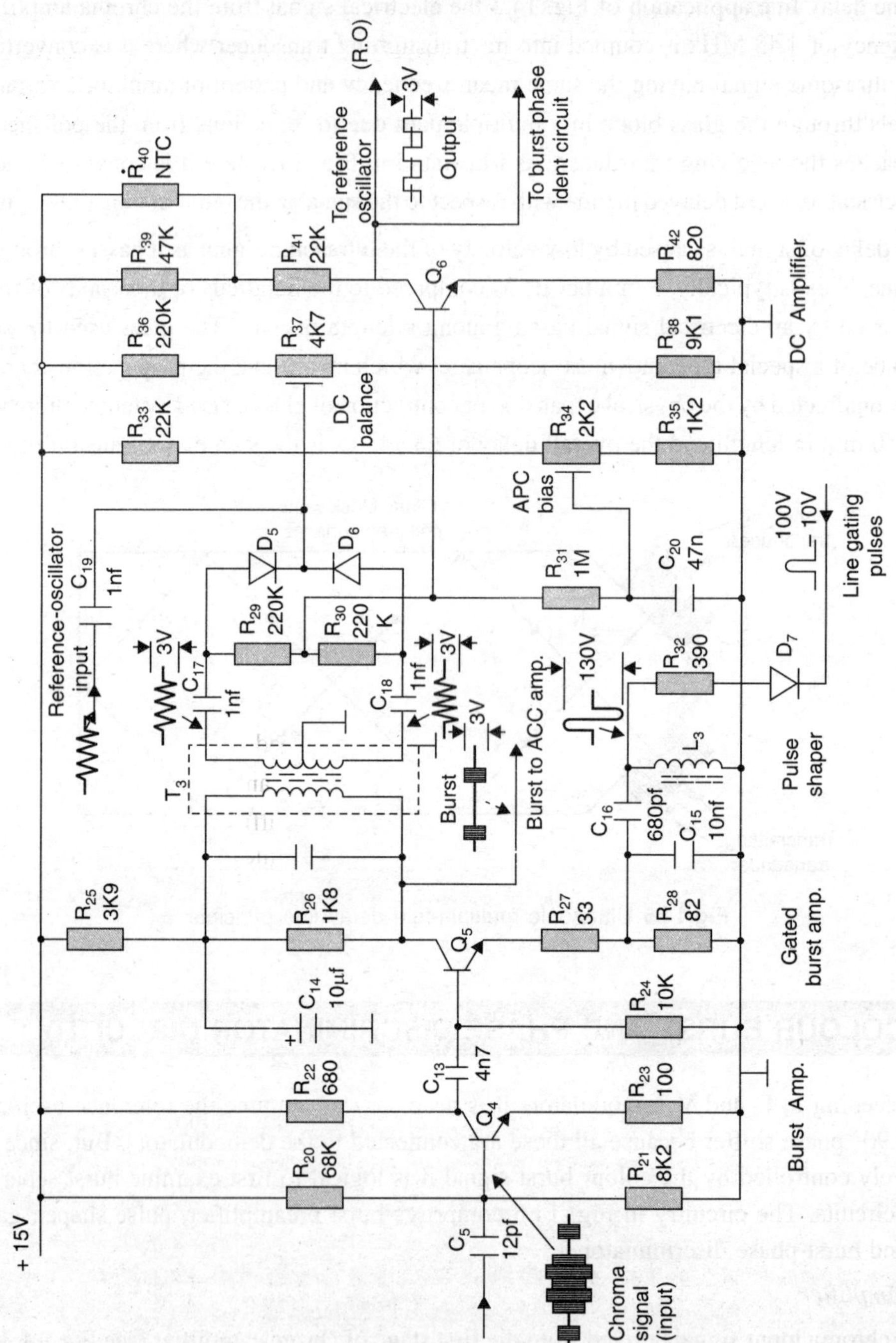

Fig. 14.6 Burst pre-amplifier, gated burst amplifier and burst phase discriminator circuitry.

Gated-Burst Amplifier

The gated burst amplifier transistor (Q_5) is normally non-conducting because it is not supplied with any standing forward bias as its base point is returned to ground through R_{24}. The chroma input signal on its own is not large enough to force Q_5 into conduction. However, it is switched 'on for short periods by gating pulses that are time coincident with the burst pulses. These pulses are negative going and are applied at the junction of R_{27}, R_{28} and derived from the pulse shaper.

Pulse Shaper

The gating pulses used to force Q_5 into conduction during the period of colour burst are the same as used by the black level clamp circuit of Fig. 14.1. Each pulse is generated by the inductor L_3 and its associated components C_{15}, C_{16} and R_{28} from a positive pulse obtained from the line timebase circuit via D_7 and R_{32}. The pulse has a peak-to-peak amplitude of nearly 100V and is coincident with the line flyback period. During line scan periods the base of this pulse remains at – 10V causing D_7 to be forward biased and to conduct current through L_3. This current causes a steady voltage drop of about - 1V across L_3 because of its dc resistance. When the 100V line gating pulse arrives, the cathode of D_7 is suddenly raised to + 90V causing it to cut-off. This sudden interruption of current flow through L_3, causes onset of self-oscillations due to energy stored in it. The natural frequency of self-oscillations is arranged to be at 98 KHz so that its one full cycle occurs during the presence of line gating pulse and the negative going half-cycle occurs coincidentally in time with the back proch of the line blanking period. This means that the negative going-half cycle which is ultimately used to turn on Q_5 embraces the colour burst centred within the back porch period. This is illustrated in Fig. 14.7.

The amplitude of self oscillatory sinusoid is too high to be applied directly to Q_5 and is, therefore, reduced to about 6V by the reactive attenuator formed by C_{15}, C_{16}, and R_{32}. These components not only help to produce the desired ringing frequency but to also cause sufficient damping of L_3 to ensure that it remains unaffected by spurious overshoots on the line gating pulse.

Gated Burst Amplifier

The direction of current flow in L_3 is so chosen that the first half-cycle of itself-oscillation is positive going. This half-cycle has no effect on Q_5 but the next *i.e.*, negative going half-cycle forces Q_5 into a short period of conduction by making it forward biased on account of voltage drop across R_{28} that is positive towards ground end. The conduction period is marked in (*b*) of Fig. 14.7. Since only colour burst occurs during this period in the chroma signal, it is only passed and amplified. The burst output that appears at the collector of Q_5 is fed to both the burst phase discriminator and ACC amplifier circuits.

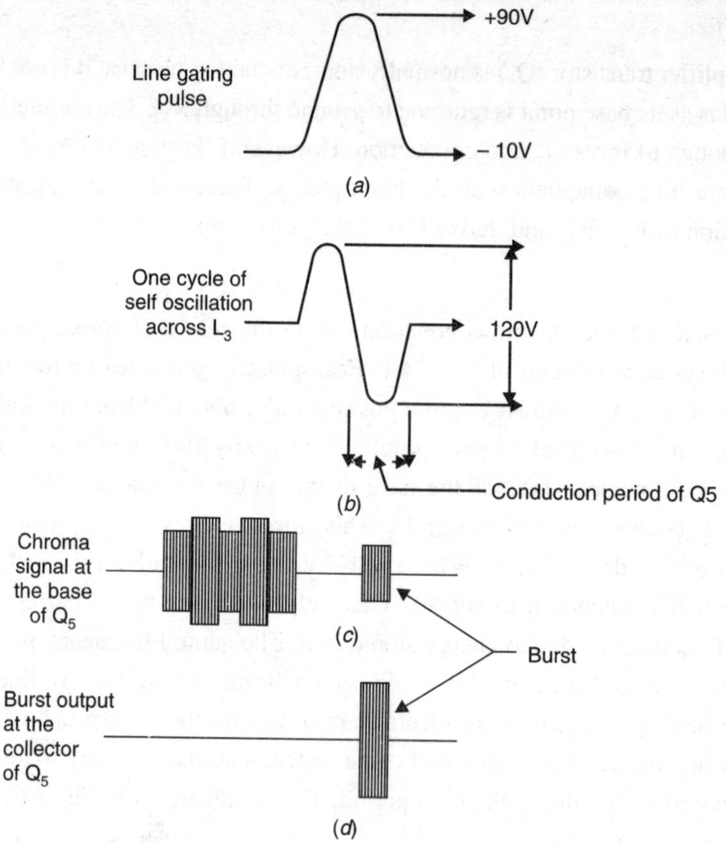

Fig. 14.7 Separation of colour burst signal by time coincident trigger pulses obtained from the pulse shaper.

Burst Phase Discriminator

The phase and frequency stability of the reference oscillator are of crucial importance to the proper functioning of U and V signal demodulation circuits. This job is assigned to the burst phase discriminator, the circuit of which is shown in Fig. 14.6. It works by comparing the phase of waveform produced by the reference oscillator with that of the burst amplifier. Any difference between the two is used to generate a controlling voltage which is fed to the reference oscillator. In this way, the frequency and phase of the oscillator are kept locked in synchronism with those of the colour burst. The controlling circuit is shown as 'Automatic Phase Control' (APC) loop in the block diagram of Fig. 14.2.

With reference to Fig. 14.6, every time the burst arrives, waveforms of equal amplitude but opposite phase appear at the two ends (with reference to ground) of the secondary of transformer T_3. The transformer primary is tuned to pass the burst signal. These then feed into the discriminator circuit formed by C_{17}, C_{18}, R_{29}, R_{30} and diodes D_5, D_6. As shown, reference oscillator output is made input via C_{19} at the junction of diodes D_5 and D_6. The discriminator functions in the same way as any other similar configuration to develop an output voltage that is proportional to deviations in phase between the two inputs. In the absence of any reference oscillator input, the diodes conduct to charge the capacitors to equal voltage values. These discharge during the non-conducting half-cycle to develop equal voltages across R_{29} and R_{30}. The polarity of these voltage drops are such that the two cancel and the potential at their junction is zero. Since R_{29}, R_{30} and D_5, D_6 form a bridge, the potential at the junction of D_5 and D_6 is also zero when only burst phase is being received. Also note that there is no voltage drop across R_{31} the upper end of which is connected to the base of Q_6.

Next consider that both burst and reference oscillator signals are being applied to the discriminator. Recall before proceeding further that phase of the transmitted burst alternates ± 45° relative to the phase of – U signal or when put in another way the phase of burst changes by 90° (from + 45° to – 45°) from line to line. Also note that the colour burst has about 10 cycles at a frequency of 4.43 MHz.

When the frequency of reference oscillator (RO) is exactly the same as that of antiphase burst signals being applied at the free ends of D_5 and D_6, and when its phase also coincides exactly with that of the unswitched + V signal (which is not present in this section of the decoder), the successive 90° phase shifiting of the burst which occurs from line to line will cause a symmetrical, nearly square, waveform to be produced at the junction of resistors R_{29} and R_{30}. The frequency of this waveform is equal to half the line frequency because positive voltage during one line and negative during the next cause one cycle of this waveform. Since the currents through R_{29} and R_{30} complete their path through R_{31} and via ground, the same square wave voltage also asppears across R_{31} and hence, gets applied to the base of Q_6.

Since the waveform is symmetrical, its positive and negative half cycles effectively cancel each other over the time and the mean potential at the junction of R_{29} – R_{30} remains exactly as it was before the RO signal was applied to the circuit. The same is true of the voltage drop across R_{31}. The waveform representing this condition is shown in Fig. 14.8(*a*).

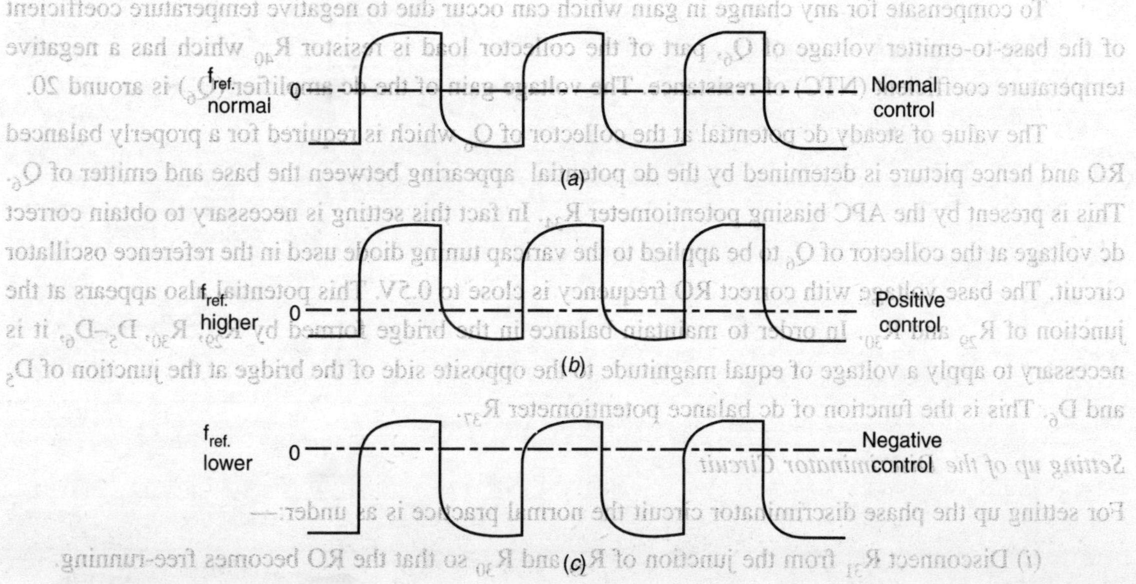

Fig. 14.8 Burst phase discriminator output waveshapes. (*a*) When RO frequency has the same phase and frequency as that of the colour burst subcarrier. (*b*) when RO frequency is higher, (*c*) when RO frequency is lower.

Next consider that the frequency of RO were for some reasons to increase relative to the frequency of colour burst subcarrier. Then the phase relationship with the burst signals would change and the 90° phase shifting of the burst signals would cause unequal voltage drops across R_{29} and R_{30} with the result that an unsymmetrical wave, though, again nearly square shaped will be generated. Its positive half-cycle will be predominant as illustrated in Fig. 14.8 (*b*).

This asymmetry would cause the potential at the upper end of R_{31} (*i.e.*, junction of R_{29} and R_{30}) to become more positive causing a frequency correcting signal of the appropriate value to be fedback to RO by way of dc amplifier Q_6. This continues till the frequency of RO decreases to become equal with the frequency and phase of the subcarrier and by then the correcting waveform also returns to its zero mean value.

The reverse would happen when the frequency of RO falls below that of the colour burst subcarrier. This time it is the negative half cycles of the asymmetrical wave which are predominant and the resulting negative going signal is used to increase the frequency of reference oscillator until it is restored to its correct value. The waveshape for such a condition is shown in Fig. 14.8(*c*).

DC Amplifier

The nearly square waveform developed across R_{31} is applied to the base of Q_6. The output at its collector is of similar shape having typically a peak-to-peak amplitude of about 3V superimposed on a mean dc level of near 6V. It pertains to condition (*a*) of Fig. 14.8 and would vary somewhat for conditions (*b*) and (*c*). This waveform, (see Fig. 14.6) feeds into the reference oscillator (RO) and burst phase ident circuits.

To compensate for any change in gain which can occur due to negative temperature coefficient of the base-to-emitter voltage of Q_6, part of the collector load is resistor R_{40} which has a negative temperature coefficient (NTC) of resistance. The voltage gain of the dc amplifier (Q_6) is around 20.

The value of steady dc potential at the collector of Q_6 which is required for a properly balanced RO and hence picture is detemined by the dc potential appearing between the base and emitter of Q_6. This is present by the APC biasing potentiometer R_{34}. In fact this setting is necessary to obtain correct dc voltage at the collector of Q_6 to be applied to the varicap tuning diode used in the reference oscillator circuit. The base voltage with correct RO frequency is close to 0.5V. This potential also appears at the junction of R_{29} and R_{30}. In order to maintain balance in the bridge formed by R_{29}, R_{30}, D_5–D_6, it is necessary to apply a voltage of equal magnitude to the opposite side of the bridge at the junction of D_5 and D_6. This is the function of dc balance potentiometer R_{37}.

Setting up of the Discriminator Circuit

For setting up the phase discriminator circuit the normal practice is as under:—

(*i*) Disconnect R_{31} from the junction of R_{29} and R_{30} so that the RO becomes free-running.

(*ii*) Adjust R_{34} until zero colour beat *i.e.*, minimum patterning on the screen is observed. When this condition is attained the RO will be operating at exactly the correct frequency.

(*iiii*) Reconnect R_{31} and accept a colour signal *i.e.*, tune in a colour channel. Now APC loop takes over until the discriminator circuit is pulled into a phase-locked state.

(*iv*) Finally adjust R_{37} to balance the output, a procedure which makes symmetrical the pull-in-range for the control loop as a whole.

14.5 ACC AMPLIFIER

The amplified burst signal that becomes available at the output of gated burst amplifier (Q_5 of Fig. 14.6) is also fed to diode D_8 of the automatic colour control amplifier circuit shown in Fig. 14.9. The diode together with filter circuit (C_{21}, R_{43}, C_{22}) acts as a half wave rectifier-cum-filter to develop at the base of Q_7 a negative going dc voltage that is proportional to the strength of received signal. As already mentioned C_{21}, R_{43} and C_{22} form a low-pass filter to smooth any 4.43 MHz variations present in the rectified signal. The output voltage at the collector of *p-n-p* transistor Q_7 is a positive voltage which increases or decreases with the strength of chroma signal. This positive voltage is typically 7V under normal signal strength conditions. As shown in Fig. 14.3 it is taken to the first stage of chroma amplifier to control its gain by means of forward bias. The resistor R_{44} provides an adjustable reverse bias for Q_7 to delay its conduction until the chroma signal exceeds a given threshold. The potentiometer formed by resistors R_{46} and R_{47} is to obtain correct steady bias for the first chroma amplifier. It is also necessary for making the collector of *p-n-p* transistor Q_7 negative with respect to its emitter.

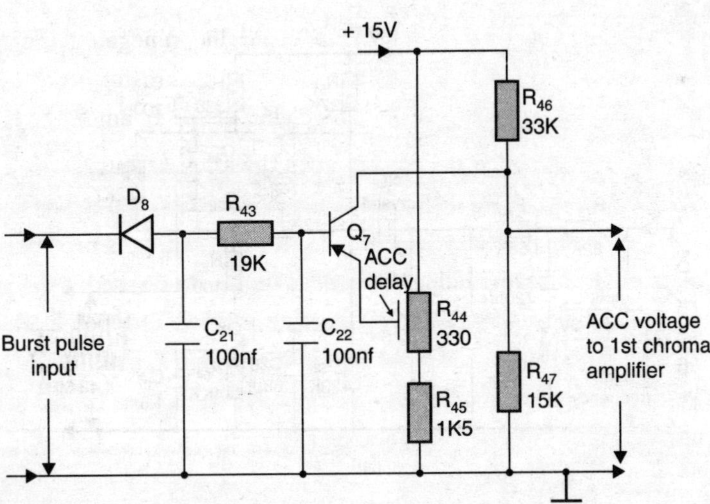

Fig. 14.9 ACC amplfier circuit.

It may be argued that with AGC control applied at the IF stage why was it necessary to incorporate ACC in the decoder. The reason is that in certain conditions the ratio of chroma subcarrier to the main vision (picture) carrier, which carries both the luminance signal and chroma subcarrier may not be constant. Although the strength of the overall picture carrier may be steady, it is sometimes possible for the chroma subcarrier associated with it to vary considerably. Thus, it is necessary to have this additional mean of maintaining the chroma signal strength constant during reception of any programme.

14.6 THE REFERENCE OSCILLATOR

The U and V chroma signals are separately produced at the transmitting end by what is known as double-balanced suppressed-carrier modulator. Thus it is necessary to regenerate the subcarrier in the receiver to affect demodulation of colour-difference signals. The reference oscillator circuit that generates a sinusoidal ouput at 4.43 MHz is shown in Fig. 14.10(*a*). The oscillator circuit is formed by transistor Q_8 and the positive feedback which occurs between its collector and base electrodes through the tuned transformer T_4. The frequency is determined by the centre-reasonant frequency of the crystal (XL) which in this circuit is approximately 4.43 MHz. Small frequency errors are corrected by the trimming capacitor C_{24} and by the varicap diode D_9. The value of 68 PF of capacitor C_{26} is sufficiently large to swamp any likely variations in the input capacitance of Q_8 ,which could otherwise affect tuning of the crystal. Resistor R_{51} and R_{52} fix the forward bias and R_{53} and R_{54} determine collector current. The components R_{55} and C_{25} form a decoupling network to ensure that oscillator ouput is not fedback into the power supply. The resistor R_{56} provides damping of the tuned circuit in the collector load and thus any variations in the output impedance of Q_8 have no significant effect on tuning of the crystal. The output waveform is taken from the capacitive potential divider formed by C_{28} and C_{29}. This arrangement limits output voltage to about 1.3V peak-to-peak.

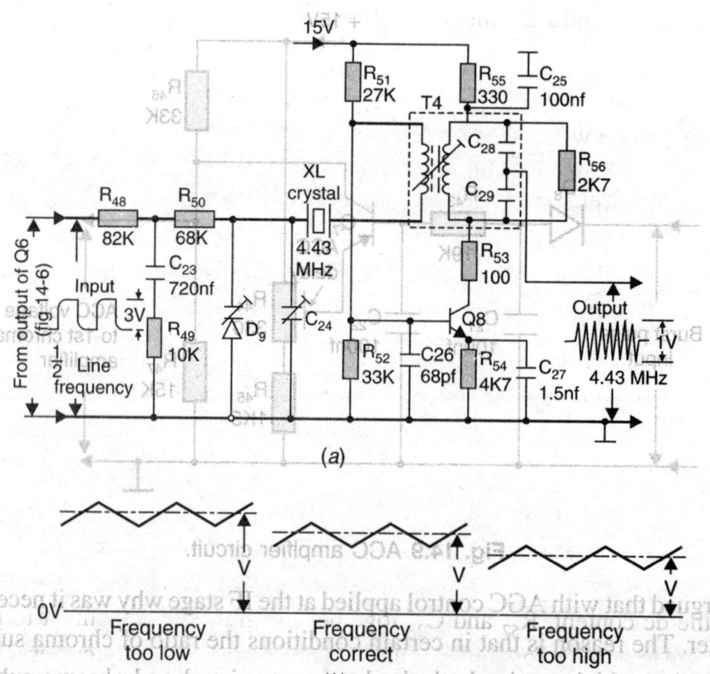

Fig. 14.10 Subcarrier reference oscillator (*a*) typical circuit (*b*) control voltage magnitudes.

Frequency Control

The nearly square waveform (Fig. 14.8) produced at the output of Q_6 of the burst phase discriminator (Fig. 14.6) is fed to the varicap diode D_9 of the reference oscillator through an integrating network formed by R_{48}, C_{23} and R_{49}. This smoothes out the 7.8 KHz (1/2 line frequency) and produces at the junction of R_{48}-R_{50}, a dc voltage whose value is proportional to mean level of the waveform.

When frequency (4.43361875 MHz) and phase of the reference oscillator are correct, the positive and negative half-cycles of the square wave become symmetrical and its mean value being zero does not add or subtract to the dc voltage (≈ 6V) set by the APC bias control. This then causes the right reverse bias voltage across D_9 and hence correct capacitance value for an output frequency of 4.43361875 MHz.

Should the frequency of RO for some reason become too high, the mean value of control waveform as obtained at the collector of Q_6 will drop to lower the reverse bias voltage across D_9. This will result in an increase in the varicap capacitance which in turn causes the RO frequency to fall by an amount which will eventually compensate for the unwanted increase.

In the same way a reduction in the frequency of RO causes an increase in the reverse bias across D_9. This in turn causes an increase in the reverse current flow in the varicap diode and a reduction in its effective capacitance. The end result is an increase in the RO frequency to make up for the unwanted reduction. The three cases are illustrated in Fig. 14.10 (*b*).

The resistor R_{49} in series with C_{23} provides some damping for the response time of APC loop. This is helpful when channels are being switched by the viewer because it prevents any undesired hunting by the RO while settling to its correct frequency and phase. The thus caused sluggish response time by the RO will cause the pull-in range of frequency control to be small, but it will at the same time, give good protection against noise (visual disturbance) which often accompanies the arrival of subcarrier burst.

A short response time will on the other hand give a wide pull-in range, but poor immunity from noise and a greater tendency for the circuit to 'hunt'. A compromise is the answer and a pull-in range of ± 400 Hz is representative of present day receiver designs.

14.7 IDENT AND COLOUR KILLER CIRCUITS

The burst phase 'Ident' circuit accepts the near square-wave waveform produced by the burst phase discriminator and amplifier circuit (see Fig. 14.6) and amplifies its ac content with a high gain, narrow-band amplifier and feeds it as a phase identifying signal to the 180° PAL bistable switch. The amplifier circuit is shown in Fig. 14.11, which also includes the rectifier-filter circuit to produce colour-killer output voltage.

The 7.8 KHz signal is fed via R_{57}-C_{28} to the base of Q_9 of the high gain narrow-band amplifier. While C_{28} blocks the dc content, R_{57} and C_{28} together integrate the ac content to near sinewave. The resistor R_{59} and R_{60} are to provide forward bias and R_{61} determines collector current. The output develops across a reasonant circuit formed by L_4-C_{31} which is also tuned to 7.8 KHz to obtain high gain at this frequency. The tuned circuit also provides desired selectivity. The waveform developed at the upper end of tuned circuit is used as a burst phase identification reference by the 180° PAL switch. It has a peak-to-peak amplitude of nearly 3.5V and is fed via R_{58} and C_{32} connected in series.

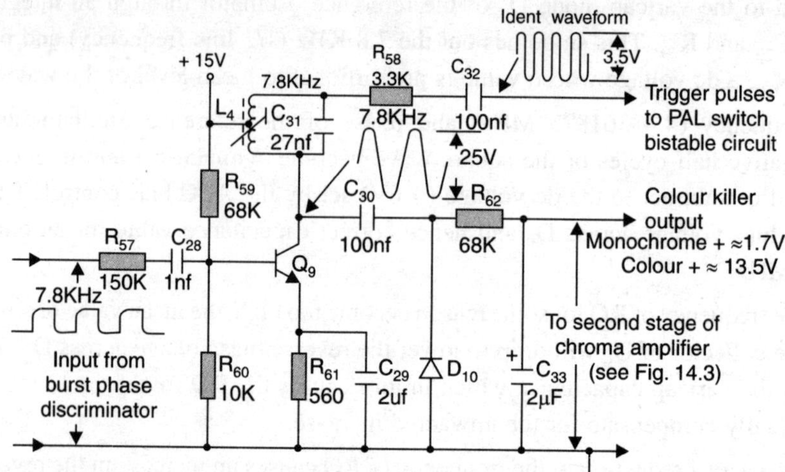

Fig. 14.11 Burst phase IDENT amplifier and colour-killer generation circuit.

Colour Killer

The dc operating voltages to Q_9 is supplied via centre tap on L_4. Such a connection causes L_4 to function as a tuned autotransformer and enables a waveform of about 25V peak-to-peak to be developed at the collector of Q_9. This waveform is fed via C_{30} to diode D_{10} which functions as a half-wave rectifier. The components R_{62} and C_{33} form a low-pass filter which provides a steady dc level of about 13.5V as the ouput. This is the colour killer voltage which is used to control conduction of the second stage of chroma signal amplifier as shown in Fig. 14.3. When a black and white programme is being received, there is no output from the burst phase discriminator and hence, no input to the burst phase ident amplifier. Under this condition the colour killer output falls to less than 2V which is not enough to forward bias Q_2 of the chroma amplifier. Thus the second chroma amplifier stage is inhibited. This prevents application of any signal to the chroma delay line and to the U and V demodulators. Thus any stray colouring signals are prevented from reaching RGB amplifiers and hence, no colour noise appears on the black and white picture during monochrome transmissions.

14.8 THE PAL SWITCH AND 90° PHASE SHIFTING CIRCUITRY

In order to recover U and V signals on demodulation two things must be done so far as the reference oscillator (RO) output is concerned. The oscillator waveforms fed to U and V demodulators must be 90° out of phase with respect to each other. Second, the RO waveform fed to the V demodulator must be phase reversed (switched through 180°) on alternate lines, and then kept in step with similar phase reversals being applied to the V signal at the transmitter. These are the functions of the 180° PAL switch and 90° phase shift shown in the block diagram of Fig. 14.2. The circuitry that is provided in the PAL-D decoder to accomplish these two functions is shown in Fig. 14.12. The transformer T_5 has

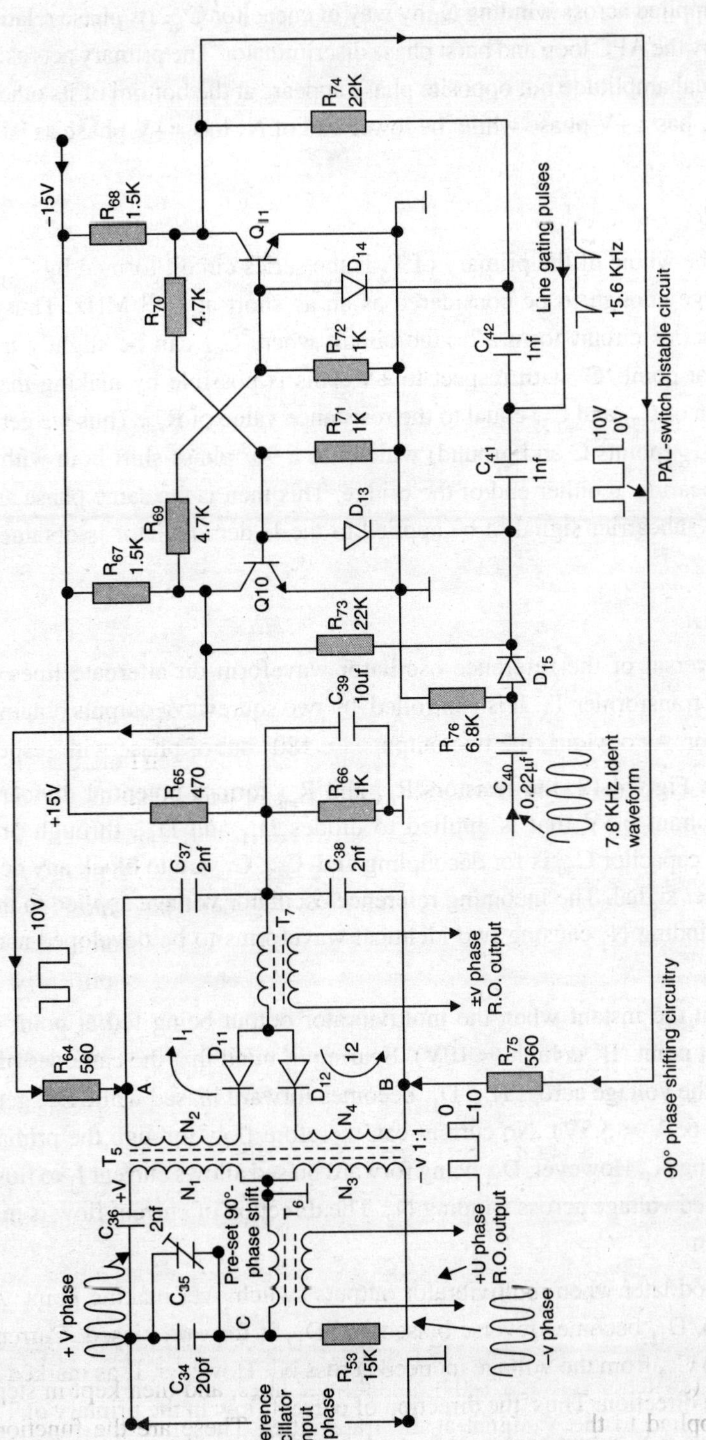

Fig. 14.12 The 180° PAL switch and 90° phase shifting circuits.

CHAPTER 14

identical primary and secondary windings (N_1, N_2, N_3, N_4) with a truns ratio of 1:1. The incoming signal from RO is applied across winding N_1 by way of capacitor C_{36}. Its phase relationship corresponds to +V signal as set by the APC loop and burst phase discriminator. The primary acts as an auto-transformer and a voltage of equal amplitude but opposite phase appears at the bottom of its other winding N_3. Thus the upper end of N_1 has a +V phase while the lower end of N_3 has a –V phase as labelled on the circuit diagram.

The 90° Phase Shift

Connected across the whole of the primary of T_5 is the series circuit formed by $C_{34} \parallel C_{35}$ and R_{63}. The capacitor C_{36} is large enough to be considered as an ac short at 4.43 MHz. Thus the entire primary winding and R-C series circuit form a bridge circuit where C_{35} can be slightly trimmed to obtain a phase shift of 90° at point 'C' with respect to +V. This is possible by making the reactance (Xc) of parallel combination of C_{34} and C_{35} equal to the resistance value of R_{63}. Thus we get a 4.43 MHz signal across the primary T_6 (points C and ground) which has a 90° phase shift both with respect to +V and – V waveforms appearing at either end of the bridge. This then is the same phase as that of +U and as such the 4.43 MHz subcarrier signal to be applied to the U demodulator is obtained at the secondary of T_6.

The 180° PAL Switch

The 180° phase reversal of the reference oscillator waveform on alternate lines takes place in the secondary circuit of transformer T_5. It is controlled by two squrewave outputs obtained from a 7.8 KHz bistable multivibrator. As obvious, the two outputs are 180° out of phase with respect of each other.

As shown in Fig. 14.12, the resistors R_{65} and R_{66} form a potential divider from the 15V dc supply to provide about 6.5V that is applied to diodes D_{11} and D_{12}, through primary winding of transformer T_7. The capacitor C_{39} is for decoupling and C_{37}, C_{38} are to block any dc while providing a short to the subcarrier signal. The incoming reference oscillator voltage applied to the primary of T_5 is developed across winding N_1 causing two identical waveforms to be developed across the secondary windings N_2 and N_4.

Consider first the instant when the multivibrator output being fed at point 'A' is low (≈ 0V) while that feeding at point 'B' is high (≈ 10V). Bearing in mind that the cathodes of both D_{11} and D_{12} are held at 6.5V by the voltage across R_{66}, D_{11} becomes forward biased while D_{12} gets a reverse bias of nearly 3.5V (10V – 6.5V = 3.5V). No current can therefore flow through the primary of T_7 from the voltage across winding N_4. However, D_{11} being forward biased allows current I_1 to flow through primary of T_7 from the induced voltage across winding N_2. The direction of current flow is marked by an arrow on the circuit diagram.

One line period later when multivibrator outputs switch over making point 'A' positive (10 V) and B negative (0V), D_{11} becomes reverse biased and D_{12} is forward biased. Current I_2 now flows in the primary of T_7 via C_{38} from the voltage induced across N_4. However, I_2 as marked by the arrow head flows in the opposite direction. Thus, the direction of current flow in the primary of T_7 reverses on each alternate line enabling ± V output across its secondary terminals. The resistors R_{64} and R_{75} are to limit diode currents to safe values. Thus by operating the mulivibrator at exactly half the line frequency, the phase of reference oscillator output fed to the V demodulator is reversed on alternate lines.

PAL Bistable Switch

The bistable multivibrator built around transistors Q_{10} and Q_{11} is of conventional type and is triggered by line gating pulse to produce two square wave outputs 180° out of phase with respect to each other. Assuming Q_{10} to be 'on' and Q_{11} 'off', the collector of Q_{10} is at near ground potential and that of Q_{11} at about 10V because of drop across R_{68} due to base current flow of Q_{10}. Consider now diode D_{13}. Its cathode is held near ground potential because it is connected via R_{73} to the collector of Q_{10} which is saturated and hence, presents a near short between its collector and emitter. However, D_{13} passes a small current (≈ 20μA) because its anode is at + 0.65V. due to V_{BE} of Q_{10} and is, therefore, slightly forward biased. Diode D_{14}, on the other hand has its cathode connected via R_{74} to the 10V potential appearing at the collector of Q_{11} and its anode to ground via R_{72} and R_{69}. It is, therefore, heavily reverse biased and so at cut-off. D_{15} is zero biased since its anode is connected to ground via R_{76} and its grounded via R_{73} and collector of Q_{10}. The circuit stays in this condition till the next line gating pulse arrives.

The negative going line pulses are applied at the junction of C_{41}, C_{42} and appear at the cathodes of D_{13} and D_{14}, the capacitors being short to signals at line frequency. Since D_{13} is slightly forward biased and D_{14} heavily reverse biased, the line pulse is 'steered through D_{13} to the base Q_{10} which is conducting and kept away by D_{14} from Q_{11} which is at cut-off. The pulse being negative going on reaching at the base of Q_{10}, it causes a small reduction in its base current and therefore in the collector current. The consequent rise in potential at the collector of Q_{10} gets transmitted via R_{69} to the base of Q_{11} to cause a small base current. The usual regeneration action follows to cause almost instantaneous change over of states. Thus, the collector of Q_{10} attains a potential equal to 10V and that of Q_{11} returns to near zero. The next line pulse will get steered to the base of Q_{11} via D_{14} and once again the states will change. The sequence continues if not disturbed by the Ident waveform fed via C_{40} and D_{15}.

The two outputs, as explained earlier, feed at points A and B of the phase shifting circuit to enable ± V output at the secondary of T_7. But unfortunately there is no way of guaranteeing that the swicthing of RO that takes place on application of square wave outputs from the mulitvibrator will always be in synchronism with that occuring at the transmitting end. However, there is a 50:50 chance that the right line will be switched at the decoder. The application of 7.8 KHz Ident voltage along with the line gating pulses makes sure that the steering sequence is always correct. The phase of Ident waveform is the same as that of + V signal in the studio so that all its positive half-cycles reach their peak values at exactly the same time as do the positive half of +V signal.

For correct phasing the negative half cycle of Ident wave should coincide with the state of bistable in which steering diode D_{13} is forward biased with Q_{10} saturated and D_{15} zero biased. Under this condition the Ident signal is of no consequence because a large negative voltage at the anode of D_{15} due to negative half of ident signal keeps it reverse biased during the period of gating pulse and it is steered by D_{13} without hindrance to the base of Q_{10}, where it causes the bistable to be switched.

Should the bistable be operating in the wrong phase, as may happen when the receiver is first switched on, the saturated condition of Q_{10} will coincide with positive half cycles of the Ident waveform and thus with D_{15} conducting, the Ident signal will reach the cathode of D_{13}, where it will oppose the

CHAPTER 14

negative gating pulse and so prevent it from switching to change states. With Q_{10} still in the saturated state, the negative half-cycle of the Ident waveform arrives. This is its wanted phase condition and the next gating pulse will switch the bistable to the normal way. Such a controlling action is illustrated in Fig. 14.13.

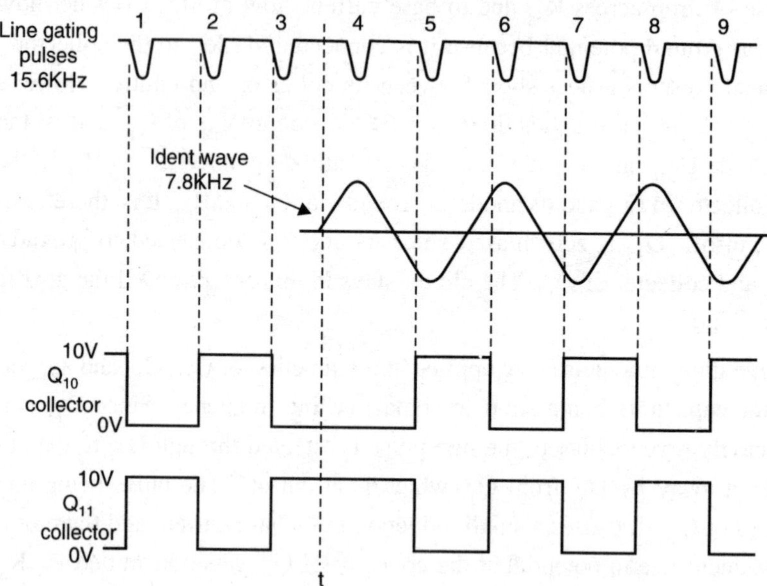

Fig. 14.13 The PAL-Switch bistable operating waveforms.

The waveforms show that the bistable is initially operating in the wrong phase up to time 't' at which moment the positive half-cycle of the Ident is seen to coincide with the 'on' condition of Q_{10}. The ident pulses are applied at instant 't' and its control action does nothing to gating pulse No. 4 and thus the circuit 'waits' for the next gating pulse (No. 5) to arrive and switch states. This then is the correct sequence and the synchronism once attained is maintained to provide correct phasing of U and V chroma sub-carriers that are applied to corresponding demodulators.

14.9 U AND V SIGNAL DEMODULATORS

After learning how the U and V chroma signals are separated and the manner in which phase locked reference subcarrier outputs are obtained, we now turn to U and V signal demodulators. This topic was fully dealt with in chapter 7 (sections 7.12 through 7.15) and hence a detailed discussion here is not necessary. The two demodulator circuits are identical and a typical configuration is shown in Fig. 14.14. In fact, it is the same as given in Fig. 7.15. The diodes D_{16} to D_{19} of the detector bridge function as electronic switches operating at a frequencies of 4.43 MHz and are controlled by the RO waveforms. Every time on conduction they provide a low resistance path between the chroma signal input and output terminals of U (or V) video signals. The resistors R_{77} and R_{78} serve as current limiters for the conducting diodes.

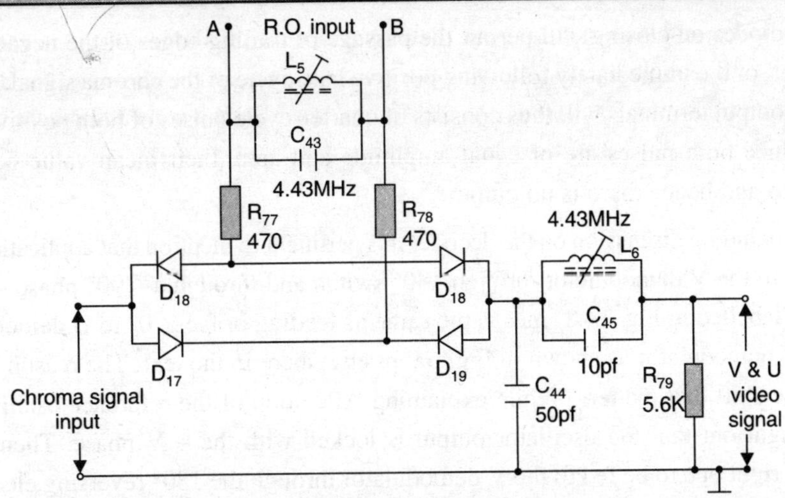

Fig. 14.14 U or V signal demodulator circuit.

The 4.43 MHz parallel tuned circuit formed by L_5-C_{34} and connected across input terminals appears in the V-signal demodulator only. Its functions is to act as a low pass filter to absorb 7.8 KHz fluctuations present in the RO waveform that comes from the 180° PAL switch. Its impedance being very high at 4.43 MHz has no appreciable effect on the subcarrier signal. The capacitor C_{44} and resistance R_{79} have a time-constant equal to the period of 4.43 MHz (subcarrier) but is quite less when compared to the period of highest frequency content (≈1 MHz) of the colour-difference signals. The output is thus able to follow modulating signal variations and reject the subcarrier content. The 4.43 MHz parallel tuned circuit formed by L_6-C_{45} connected between C_{44} and R_{79} acts as a low-pass filter to the colour-difference signals but a rejector circuit to the RO content of the output waveform. The circuit thus functions to reinforce the filtration offered by C_{44} and R_{79}.

Every time the positive half-cycle of RO waveform makes point A positive with respect to point B (see Fig. 14.14) the four diodes become forward biased providing a low resistance path between the chroma signal input and video signal output terminals. During negative half-cylces, all the four diodes get reverse biased and there can be no transmission from input to output terminals. It is usual for the amplitude of RO output to be made considerably larger than that of the chroma signal so that its effect on the switching action of four diodes shall be dominant.

When the RO is locked to the U chroma signal, every half-cycle of the reference oscillator will be in time coincident with similar half-cycles of the sub-carrier content of the appropriate chroma signal. The same is true when RO is locked to the V chroma signal. Thus, either U or V colour-difference signal can be the output depending on the phase relationship of reference oscillator. The output in either case consists of a series of half-cycle pulses which are smoothed by the integrating effects of C_{44}-R_{79} and the 4.43 MHz rejector circuit.

It may be noted that it is essential to keep the phase relationship at the correct value otherwise distortion would occur. For example, when the phase relationship between the reference and chroma signal (U or V) turns out to be 90°, no output would result. This is so because with such as abnormal relationship the diodes on closing will permit the passage of trailing edges of the negative half-cycles and leading edges of the immediately following positive half-cycle of the chroma signal. The waveform appearing at the output terminals will thus consists of quarter-cycle pulses of both positive and negative polarity. Now since both pulses are of equal amplitude and area, their mean value before and after smoothing is zero and hence there is no output.

Before concluding discussion on the decoder it is pertinent to mention that application of reference oscillator output to the V demodulator through 180° switch and through a – 90° phase shifter to the U demodulator (as labelled in Fig. 14.2) means the same as feeding of fsc $\angle 0°$ to U demodulator and fsc $\angle \pm 90°$ to the V demodulator as shown in Fig. 6.9 or elsewhere in the text. The reason being that it is the relative phase shift that matters. While explaining APC loop of the reference oscillator in section 14.6 it was brought out that the oscillator output is locked with the + V phase. Then the oscillator output is just the right one to be fed to the V demodulator through the 180° reversing electronic switch. This signal needs to be delayed by 90° to bring it in phase with the U signal axis and that is what is done. This ensues relative phase relationship which is of utmost importance. However, earlier when the above (locking of subcarrier with +V) was not yet evident the oscillator output was taken arbitrarily to have a phase fsc $\angle 0°$ and then given a 90° phase shift before feeding it to the V demodulator through a 180° phase reversal switch. Thus both the nomenclatures are essentially the same and there should be no confusion about it. Since angles are measured counter clocking from + x axis, the vector +U was designated as $\angle 0°$ and other locations *i.e.*, + (R – Y) and – (R – Y) were referred with respect to this as $\angle +90°$ and $\angle -90°$.

14.10 MATRIXING FOR DRIVE CIRCUITS

On obtaining U and V colour-difference signals as outputs from the decoder and Y signal from the luminance channel (Fig. 14.2) it is now necessary to devise means for driving the picture tube electrodes with appropriate colour voltages. This involves two stage matrixing—first to extract (G – Y) from U and V signals and second to obtain R, G and B signals for feeding corresponding guns in the picture tube. All this is fully explained in chapters 3 and 7 (sections 3.14 and 7.18) and therefore only briefly reviewed here to maintain continuity of the topic.

The two possible methods are:

(*i*) matrixing of U and V to obtain (G – Y) and then use the tube electrodes for getting R, G, B drives.

(*ii*) matrixing to obtain R, G, B directly before feeding the picture tube. The first method is no longer used because it needs high amplitude signal voltages to produce saturated colours. In the R, G, B drive system where matrixing is done outside the picture tube and individual R, G and B signals are applied to the cathode electrodes of their respective guns, the overall sensitivity turns out to be about twice as compared to the colour-difference signal drive. This in turn means easy application of transistors as drive amplifiers and a much lower value of dc power supply.

R, G, B Matrixing

With a view to explaining the technique of obtaining R, G, B signals of sufficient amplitude from the weighted (B – Y) *i.e.*, U and (R – Y) *i.e.*, V signals a simplified circuit configuration is shown in Fig. 14.15. Before proceeding to explain its functioning it is necessary to recollect the following expressions.

(*i*) The weighting factors employed at the transmitting end are: 0.493 for (B – Y) and 0.877 for (R – Y) signals.

(*ii*) Therefore, to obtain (B – Y) the U signal must be multiplied by 2.03 (1/0.493) and V signal by 1.14 (1/0.877).

The transistor Q_2 is an inverting amplifier with 'V' signal input via R_1 and with Y signal input through R_3. Based on the value of associated resistors the 'V' signal is multiplied by the ratio R_7/R_1 = 1.14 and the Y signal by the ratio R_7/R_3 = 1. These two then add to give inverted R signal at the collector *i.e.*,

$$V_R = -1.14\,\frac{(R-Y)}{1.14} + Y = -(R-Y) + Y = -R$$

The transistor Q_3 operates in much the same way to produce:

$$V_B = -2.03\,\frac{(B-Y)}{2.03} + Y = -(B-Y) + Y = -B$$

The transistor Q_5 and Q_6 are connected as emitter followers to avoid any mutual loading effects while combining R and B with Y to obtain G.

As shown in the figure the matrixing of – R, – B and Y signals to obtain G takes plate at the input of inverter amplifier Q_4. A (– R) signal of fixed amplitude is derived from the emitter follower Q_5 and applied to Q_4 via R_{14}. Similarly a (– B) signal of fixed amplitude derived from the emitter follower Q_6 is applied to Q_4 via R_{15}. The Y signal input is derived from Q_1 via resistor R_{13}. With these inputs and appropriate choice of input and feedback resistance (as labelled in the figure) the transistor Q_4 multiples (– R) signal by the ratio R_{18}/R_{14} (16.9K/33K) = 0.51, (– B) signal by the ratio R_{18}/R_{15} (16.9K/90.0K) = 0.186 and Y signal by the ratio R_{18}/R_{13} (16.9K/10K) = 1.69. It then adds the three outputs together to yield inverted G signal at the collector. This operation can be expressed as:

$$-V_G = -(-0.51R - 0.186B + 1.69\,Y) = -G$$

Note that Y = 0.3 + 0.59G + .11 B)

The inverted R, B and G colour signals are fed to the red, blue and green cathode electrodes of the picture tube by way of variable resistors R_{11}, R_{12} and R_{19}. These controls are used to adjust the amplitudes of individual signals to the level which on combining give correct grey scale tracking and therefore correct balance of reproduced colours and black and white shades.

CHAPTER 14

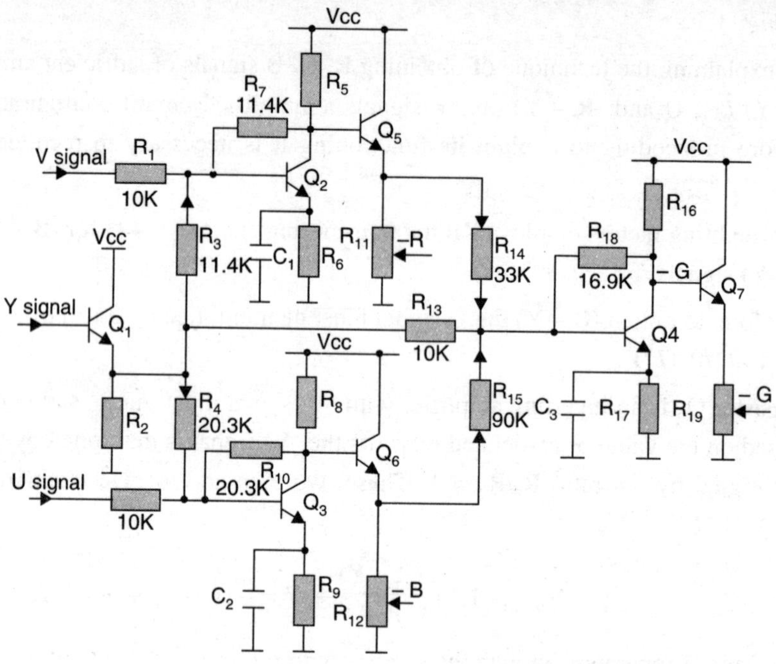

Fig. 14.15 Simplified circuit of RGB matrixing and drive amplifiers.

14.11 PAL COLOUR DECODER IC TDA 3561A

In modern colour receivers nearly all the functions assigned to decoder and luminance channel are performed by a single IC. The TDA 3561A is one such PAL decoder IC that is widely used in various receiver design. It has replaced TDA 3560 which was in use till recently. TDA 3561 is a 28 pin package and its objective block schematic is shown in Fig. 14.16.

The IC performs all the circuit functions shown in Fig. 14.2 plus certain additional features such as external data insertion, output signal clamping and blanking to give added control over the blue, green and red output signals. The purpose of external data insertion facility is to allow the receiver to be used for TV games, to display channels or more importantly to display teletext characters and patterns instead of the normal TV picture. Receivers fitted with this facility use the periods of field blanking pulses to separate and decode the teletext code information transmitted by the TV station. The switching action which decides whether normal or externally-derived signal data are sent to the picture tube is governed by a voltage applied to pin 9, labelled 'video data switch'. In the 'external' mode 'R, G and B matrices are supplied with externally derived RGB colouring signals applied at pins 13, 15 and 17. In the normal operating mode, they receive internally derived colour-difference signals from the R, G and B demodulators. Pulses applied at pin 8 during line and field blanking intervals clamp the externally applied RGB signals to a level set by the 'brightness' control voltage applied at pin 11. In so doing they ensure that both normal and externally derived pictures operate with the same black-and-white reference. Also applied to pin 8 is a two-level sandcastle pulse which is used to gate colour burst pulses and to operate the PAL switch flip-flop (bistable multivibrator).

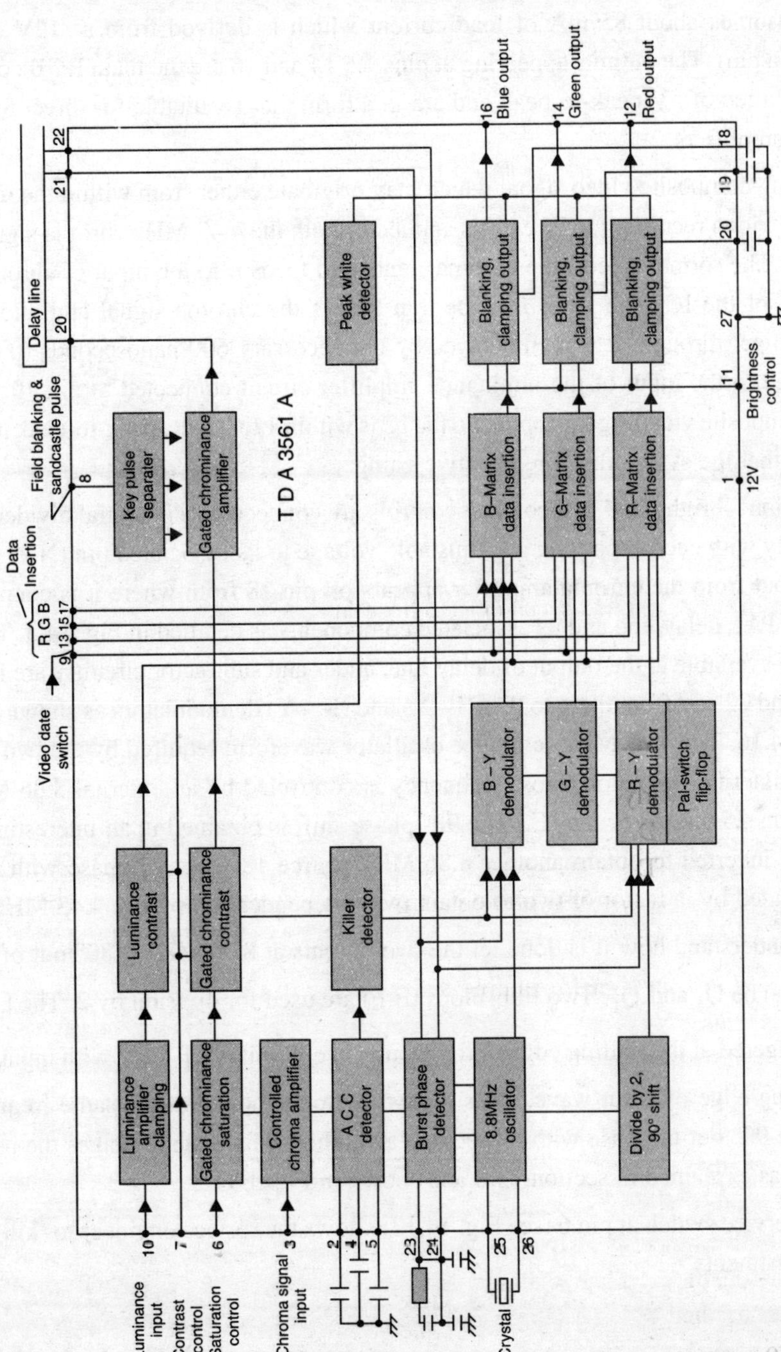

Fig. 14.16 Objective schematic of the PAL colour decoder IC TDA 3561 A.

The much simplified diagram of Fig. 14.17 shows basic details of how the IC decoder is interfaced with the external components needed to complete its decoding functions.

The IC consumes about 85 mA of load current which is derived from a 12V dc operating voltage (applied at a pin). The outputs appearing at pins 12, 14 and 16 are the main RGB colour signals. They have an amplitude of 5V peak-to-peak and are in a form that is suitable for direct feeding to the picture tube drive amplifiers.

The incoming composite video signal which may originate either from within the receiver itself or from an external video recorder (VCR etc.) is applied to both the 4.43 MHz chroma signal filter and the 4.43 MHz trap. The former selects the chroma signal and feeds it to the input of chroma amplifier connected to pin 3 of the IC. The latter *i.e.*, the trap blocks the chroma signal and allows only the luminance signal to get through. It is then delayed by the necessary 600 nanoseconds (0.6 μs) period before being applied to the input of the luminance amplifier circuit connected to pin 10. The overall amplitude of the composite video signal applied to the IC is within 1 to 3 volts peak-to-peak in amplitude. Its polarity is such that the sync pulses are negative going.

The 'saturation' 'brightness' and contrast controls are connected as potential dividers across the common 12V supply with each supplying an adjustable voltage to its associated pin (Nos. 6, 11 and 7) on the IC. The output from the chroma amplifier appears on pin 28 from where it is connected to the externally mounted PAL delay line and its associated components as detailed in Fig. 14.3. The U and V signals that become available at the output of delay line, adder and subtractor circuitry are fedback into the IC via pins 21 and 22 and from there to the (B – Y) and (R – Y) demodulators as shown in the block schematic of Fig. 14.16. The 4.43 MHz reference oscillator waveform required by the two modulators is derived from an internal oscillator whose frequency is controlled by an external 8.86 MHz crystal connected across pins 25 and 26 of the IC. The 90° phase shift is obtained in an interesting way. The 8.86 MHz signal is inverted to obtain another 8.86 MHz source 180° out of phase with it. Both the signals are then divided by a factor of two to obtain two independent outputs at 4.43 MHz.

In order to understand how it is done let the two outputs at 8.86 MHz (180° out of phase with respect to each other) be Q_1 and $\overline{Q}_1$. Two Flip-Flops (F-F) are used for division by 2. The F-F Q_2 with input from Q_1 is triggered at the trailing edge of the input wave. Similarly F-F Q_3 with input from $\overline{Q}_1$ is triggered at the rising edge of input wave. Thus the two outputs though at the same frequency (4.43 MHz) turn out to be 90° out of phase with respect to each other. This method makes the use of a R-C circuit unnecessary as explained in section 14.8 and shown in Fig. 14.12.

The colour service switch at pin 6 (see Fig. 14.17) allows the service engineer to 'kill the colour' for purposes of adjustments.

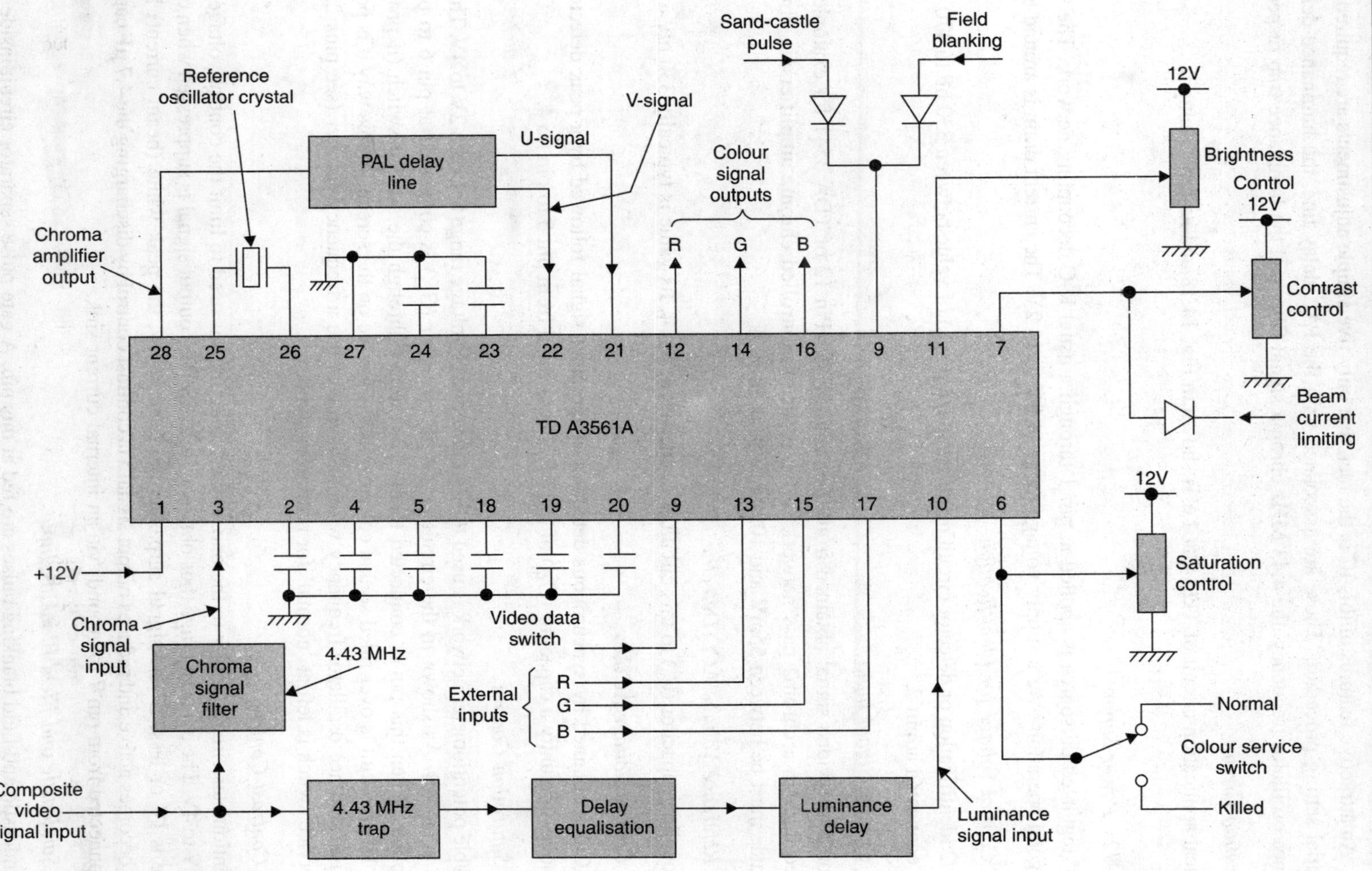

Fig. 14.17 Simplified diagram of IC TDA 3561A showing basic interconnections.

CHAPTER 14

An attractive features of this IC is the fact, that only five simple adjustments are required during the initial setting procedure. These are associated with the PAL delay line, the luminance delay, the reference oscillator frequency, the 4.43 MHz chroma signal filter and the luminance trap respectively.

Application Circuit

A typical application circuit of TDA-3561A is shown in Fig. 14.18 and its functioning (p... ise) is as under:

Pin 1. 12V Power Supply

A 12V regulated dc source is applied at pin 1 through a usual R-C decoupling network. The circuit gives good operation over a supply voltage range of 8V to 13.2V. The current drain is around 85 mA.

Pin 2. Control Voltage for Identification

The H/2 identification or detector circuit needs a capacitor (C_2) of value between 330 nF to 470 nF and this is connected at pin 2.

Pin 3. Chroma Signal Input

The composite video signal obtained from the vision IF IC (Pin 12 of TDA 3541 for example) is ac coupled through a chroma pass network to the input of controlled chroma amplifier via pin 3. Its amplitude must be between 55mV and 1000 mV peak-to-peak.

Pin 4. Reference Voltage ACC Detector

A decoupling capacitor (C_9) to this circuit is connected at pin 4. Its value is typically 330 nF.

Pin 5. Control Voltage ACC

The ACC is obtained by synchronous detection of the burst signal followed by a peak detector. For good noise-immunity a capacitor (C_{10}) of 2.2 μF is connected from pin 5 to pin no 4.

Pin 6. Saturation Control

A suitable potentiometer network is used to limit the control voltage range between 2V to 4V. This is in excess of 56 dB. As shown in the circuit, a 2.2 μF capacitor (C_{14}) is connected at pin 6 to provide decoupling. When this pin is connected to the 12V supply through the service switch (if provided) colour killer circuit is over ruled so that colour signal is visible on the screen. In this way it is possible to adjust reference oscillator frequency without having to use a frequency counter (see pins 25, 26). The service switch is left at 'normal' for routine viewing.

Pin 7. Contrast Control

A potentiometer circuit similar to the one connected at pin 6 is used to limit the control voltage range from 2V to 4V. The contrast range thus obtained is 20 dB. The output signal is suppressed when control voltage is IV or less. IF the signal surpasses the level of 9V, the peak white (beam current) limiter circuit becomes active and reduces output signal via contrast control by discharging the 4.7 μF capacitor C_{15} (connected from pin 7 to ground) via an internal current sink.

Pin 8. Sandcastle and Field Blanking Input

The sandcastle and field blanking pulses are fed at this pin. A gate pulse separator circuit inside the IC separates and routes burst and blanking pulses to corresponding circuits. If the amplitude of sandcastle

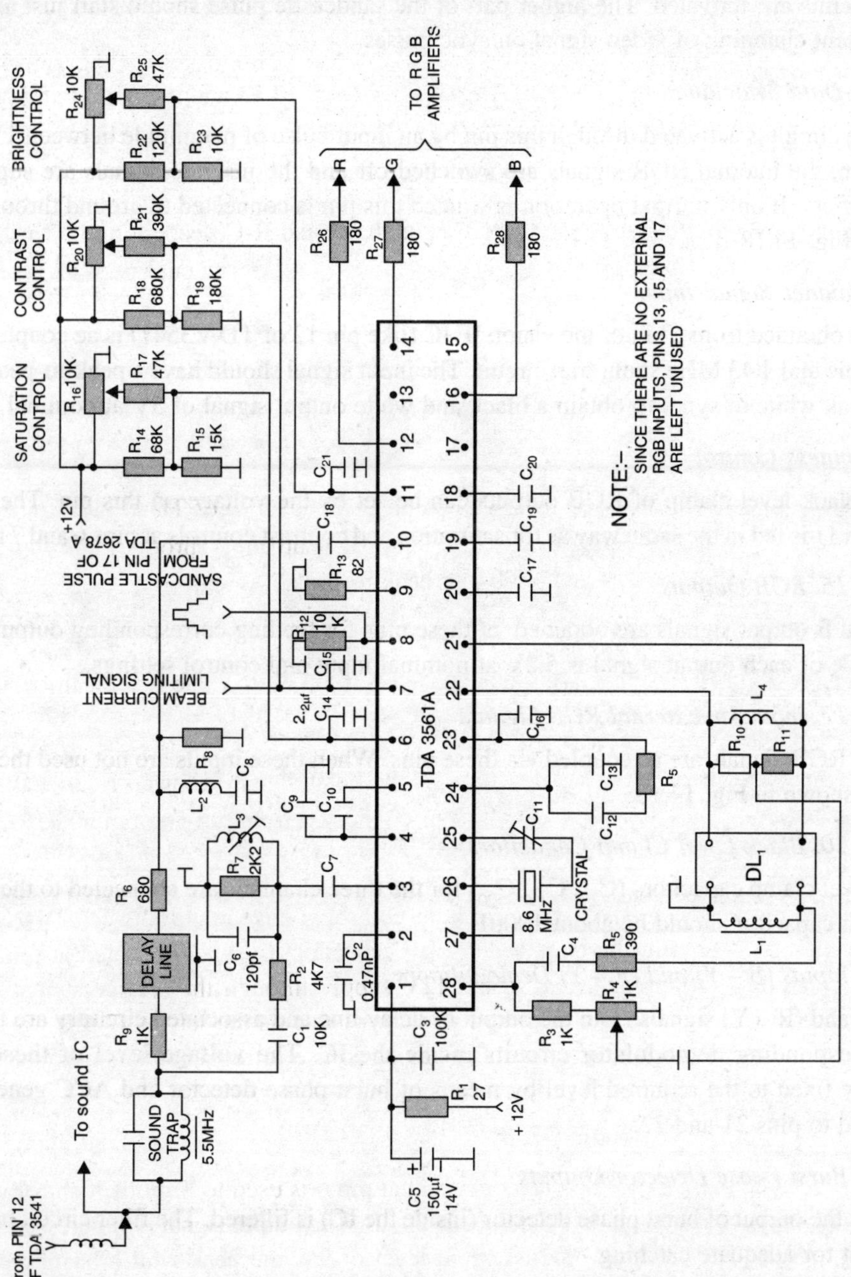

Fig. 14.18 Typical application circuit of PAL Colour Decoder IC TDA3561A.

pulse is between 2V and 6.5V, it indicates that transmission is black and white. Then no RGB signals are available at pins 12, 14 and 16. However, when the pulse amplitude exceeds 7.5V the burst gate and clamping circuits are activated. The higher part of the sandcastle pulse should start just after the sync pulse to prevent clamping of video signal on sync pulses.

Pin 9. Video-Data Switching

The insertion circuit is activated through this pin by an input pulse of magnitude between 1V and 2V. In this condition, the internal RGB signals are switched off and the inserted signals are supplied to the output amplifiers. If only normal operation is wanted this pin is connected to ground through a resistor as shown in Fig. 14.18.

Pin 10. Luminance Signal Input

The Y signal obtained from a pin of the vision IF IC (like pin 12 of TDA 3541) is ac coupled at this pin via a delay line and 4.43 MHz shunt trap circuit. The input signal should have a peak-to-peak amplitude of 0.45V (peak white to sync) to obtain a black and white output signal of 5V at nominal contrast.

Pin 11. Brightness Control

The black level clamp of RGB outputs can be set by the voltage on this pin. The variable dc voltage at pin 11 is fed in the same way as for saturation and contrast controls at pins 6 and 7 respectively.

Pins 12, 14, 16. RGB Outputs

The R, G and B output signals are obtained at these pins for feeding corresponding output amplifiers. The amplitude of each output signal is 5.2V at nominal input and control settings.

Pins 13, 15, 17. Input for External RGB Signals

The external RGB signals are ac coupled via these pins. When these inputs are not used the pins can be left open as shown in Fig. 14.18.

Pins 18, 19, 20. Black Level Clamp Capacitors

The black level clamp capacitors (C_{17}, C_{19}, C_{20}) for the three channels are connected to these pins. The value of each capacitor should be about 100μF.

Pins 21, 22. Inputs (B – Y) and (R – Y) Demodulators

The (B – Y) and (R – Y) signals from the output of delay line and associated circuitry are fed via these pins to corresponding demodulator circuits inside the IC. The voltage level of these signals is automatically fixed to the required level by means of burst phase detector and ACC generator which are connected to pins 21 and 22.

Pins 23, 24. Burst Phase Detector Outputs

At these pins the output of burst phase detector (inside the IC) is filtered. The filter circuit has a suitable time constant for adequate catching.

Pins 25, 26. Reference Oscillator

The frequency of the oscillator is adjusted by the variable capacitor (C_{11}) that is in series with the 8.86 MHz crystal connected between pins 25 and 26 of the oscillator circuit. As explained earlier the oscillator output is inverted to obtain two 8.86 MHz signals that are 180° out of phase. Both are divided by 2 to obtain two 4.43 MHz outputs that are 90° out of phase with respect to each other.

Pin 27. Negative of DC Supply

The negative end of the 12V dc supply is connected here and the pin is grounded to the common negative rail.

Pin 28. Output of chroma Amplifier

Both burst and chroma signals are available at this pin. As shown in Figs. 14.17 and 14.18 it is fed to the PAL-Delay line circuitry to obtain U and V signals.

Pin Voltages-Typical

Dc and ac voltage data at various pins of IC TDA3561A is as under:

Pin No	*Voltage*		*Remarks*
	DC	*AC*	
1	12	—	+ dc supply (12V)
2	0	—	Identification signal
3	2.4	—	Chrominance signal input (0.5V P-P)
4	4.6	0.2	ACC reference voltage (4.6V)
5	4.7	—	Control voltage ACC
6	3	0.3	Saturation control voltage (2V to 4V)
7	2.6	—	Contrast adjustment (2V to 4V)
8	1.6	—	Sandcastle and blanking pulses
9	0	0.1	Video data switching pin
10	1.6	—	Luminance signal input (0.5V P-P)
11	3.0	0.2	Brightness adjustment voltage
12	9.6	1.2	'R' signal out (5V P-P)
13	6.6	0.2	Provision for external 'R' input
14	9.8	1.6	'G' signal out (5V P-P)
15	6.6	0.2	Provision for external 'G' input
16	9.0	—	'B' signal output (5V P-P)
17	6.4	0.2	Provision for external 'B' input
18, 19, 20	11.0	—	Black level clamps
21, 22	3.0	0.1	Signal from delay line to demodulator
23, 24	9.6	—	Burst phase detector output
25	10.0	0.6	Subcarrier reference
26	2.2	0.1	Oscillator (8.86 MHz)
27	0	—	Ground
28	8.6	0.1	Chroma signal to delay line demodulator

14.12 RGB DRIVE AMPLIFIERS

A typical RGB video amplifier circuit together with associated picture tube circuitry is shown in Fig. 14.19. It consists of three identical video amplifiers for driving the three cathodes of picture tube. The inputs to the amplifiers are obtained from the decoded red, green and blue outputs of the chroma IC (pins 12, 14, 16 of IC TDA 3561A). The devices Q_1, Q_2 and Q_3 are high frequency transistors type BF393 or BF869. The three amplifiers are designed to be identical in all respects so that their frequency response is nearly the same.

Since the three amplifiers are exactly identical only one is considered. The transistor Q_1 of the green signal amplifier is connected in common emitter configuration. The dc supply at 150V is filtered by the network L_2 and C_9 before applying it at the collector through resistor R_{24}. The dc bias at the emitter is provided by an emitter follower or obtained through the 12V dc bus in the receiver. The capacitors C_7 and C_8 are bypass to the emitter supply. The resistors R_{15} and R_{12} connected between collector and base provide negative feedback to improve dc stability and make the amplifier performance independent of device parameters. The inductor L_3 in the collector lead is the series peaking coil to extend the bandwidth. The capacitor C_1 at input to the amplifier is to improve step response. The dc collector voltage, which determines the picture tube cut-off voltage is fixed by adjusting resistor R_{17} in the biasing network. This way it enables easy adjustment of the cut-off voltage for the green gun to obtain standard white output at low beam currents. Similarly, R_1 is varied for monochrome reproduction at highlights.

The voltage gain of this amplifier is about 24 dB with rise time better than 0.18 μs. The resistor R_{26} aids in fast breakdown of the spark-gap in the initial stages of flashover, limits current into the transistor and isolates the video stage from short circuit produced by the spark-gap. Horizontal blanking pulses obtained from LOT are applied at grid No.1 (control grid) through R_{28}.

RGB Amplifier Employing IC TDA 8150

Integrated circuits are now available which combine RGB drive amplifiers thus avoiding any discrete circuitry between the chroma IC and picture tube. TDA 8150 is one such IC, the application circuit of which is given in Fig. 14.20. In addition to three independent video amplifiers, the device features an internal generator for the first grid voltage, flashover protection, spot burn elimination and a common cut-off sensing output for use in sequential sampling applications. The TDA 8150 is supplied in a 15 lead Multiwatt plastic power package.

As shown in the figure pins 1, 14 and 15 are the input points for blue, red and green video signals as obtained from the chroma IC. Similarly pins 10, 5 and 7 are the output points for green, red and blue video signals after due amplification. These outputs drive the cathodes of three guns of the picture tube directly. Feedback from output to input is applied from pins 6, 11 and 9 to stabilize amplifier operation. High voltage (V_h) at about 200V necessary for the collectors of output transistors is fed at pin 12 through a current limiting resistor. Similarly 12V low voltage (V_L) dc supply for the small signal circuitry is connected at pin 3. Both the supplies are suitably decoupled to avoid any feedback effects. Pin 8 is the ground pin for both the dc sources. The reference voltage for the three amplifiers is available at pin 2. Its typical value is 3V. The capacitor connected between pin 2 and ground eliminates ac

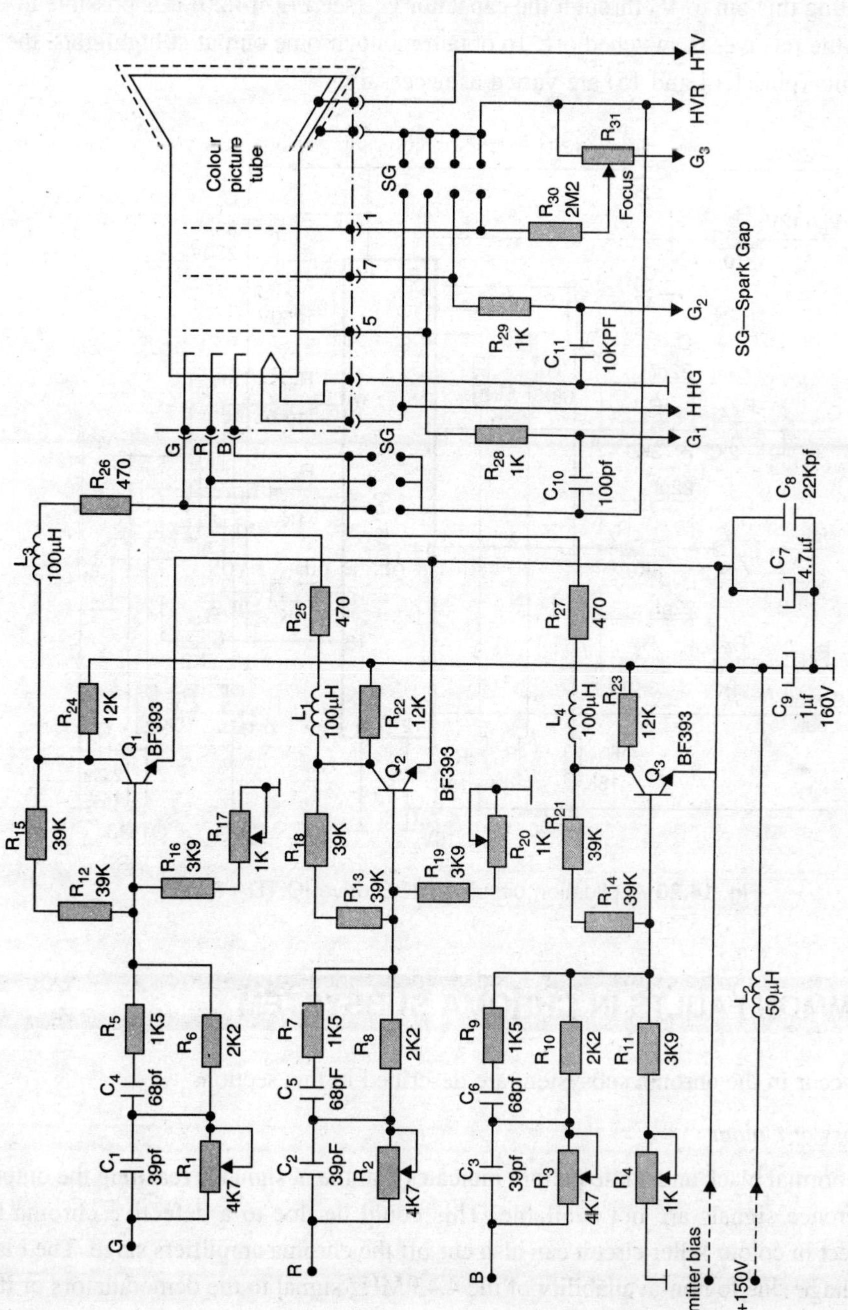

Fig. 14.19 RGB video amplifier circuit and associated picture tube connections.

CHAPTER 14

crosstalk between the amplifiers. Pin 4 provides sum of cathode currents for automatic cut-off adjustments. The dc voltage for the first grid of the picture tube is available at pin 13. It is typically VL + 1V. By connecting this pin to V_h through the capacitor C_8 (see Fig. 14.20) it is possible to eliminate spot burns when the receiver is switched off. To obtain monochrome output at highlights, the resistors in the input circuits (pins 1,14 and 15) are varied as necessary.

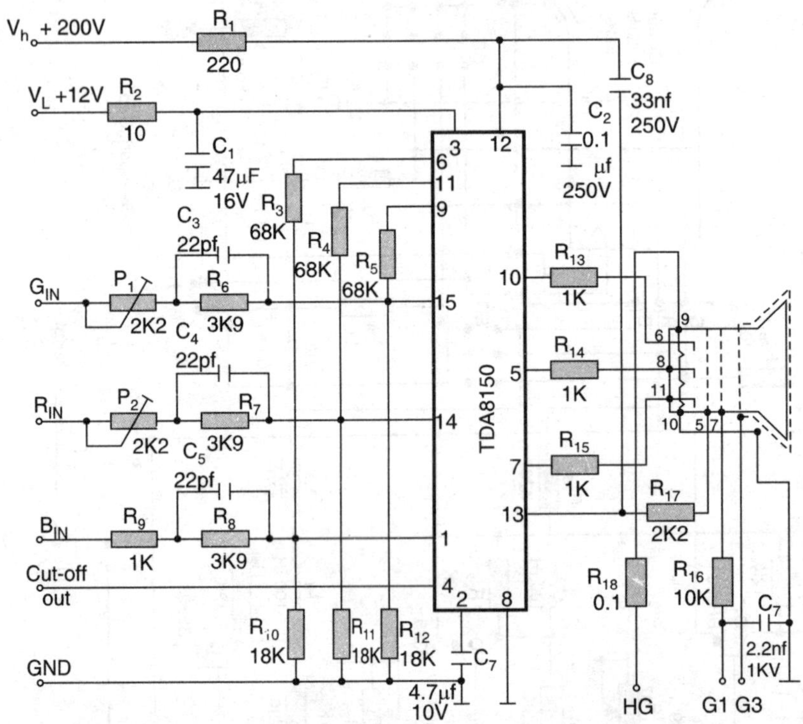

Fig. 14.20 Application circuit of RGB drive IC TDA 8150.

14.13 COMMON FAULTS IN CHROMA SUBSYSTEM

Faults that can occur in the chroma subsystem are described in this section.

(1) Complete Loss of Colour

The presence of normal black and white picture indicates that the Y signal is reaching the output matrix but colour difference signals are not available. This could be due to a defective chroma bandpass amplifier. A defect in colour killer circuit can also cut-off the chroma amplifiers stage. The blanking of colours could also be due to non-availability of the 4.43 MHz signal to the demodulators or it may not be synchronized with the incoming burst pulses.

Before servicing for loss of colour make sure that the channel is properly tuned. Adjust fine tuning control, if necessary. As a double check tune-in another channel to make sure that the composite colour signal is passing through the tuner and hence across the vision IF stage. Also check that the colour (saturation) control is operative and suitably adjusted. If functioning, a dc voltage equal to 3V will be present at pin 6 of TDA3561. Besides make sure that the emitter follower circuit that feeds composite chroma signal from vision IF IC to the chroma IC is functioning and in particular the path that feeds chroma signal at pin 3 of TDA 3561. So far as the IC itself is concerned check if RGB outputs are available at corresponding pins and if not measure signal amplitudes at chroma amplifier input and output points. (pins 3 and 28 of IC TDA 3561A. If proper voltages are not there, check for open or short in the capacitors connected at these pins. For a fault in the subcarrier circuit check APC and ACC circuits and also adjust oscillator trimmer capacitor (C_{11} in TDA 3561A) for minimum beat. In the end check resistors and capacitors at pins associated with de-matrixing circuit blocks. If necessary, replace IC to make sure that the device is in good condition.

(2) Specific Tinted Colour

This can be due to improper amplitude and phase adjustments in the delay circuitry. Readjust associated inductors and capacitors (L_1, L_4, R_4 in TDA 4561A).

(3) Any One Colour Missing

This can happen when one of the demodulators is not functioning or is not receiving the subcarrier signal. However, the most probable cause could be the failure of corresponding drive amplifier. Check RGB outputs at corresponding pins to isolate the fault between the IC and drive amplifiers.

(4) Weak Colours

When weak colours are reproduced even at maximum setting of colour control, the magnitude of U and V signals fed to the R, G, B matrix circuits is quite less. The fault could be a weak chroma bandpass amplifier or defective colour intensity control circuit. Make sure that the dc supply (at pin 1 of TDA 3561A) is normal. Rotate or slide the colour control and see if dc voltage at the corresponding pin varies. If low or not varying investigate the reason. Also check if input and output voltages at the three colour video amplifiers are normal.

(5) Intermittent Colour

The reason for intermittent colour reproduction is non-synchronised colour subcarrier oscillator. Check signals at input and output pins of the chroma amplifier (pins 3 and 28 of TDA 3561A). If normal proceed to check APC and ACC circuits.

(6) Appearance of Venitian Blinds (Hanover Bars) in the Picture

This could also be due to improper amplitude and phase adjustments in the delay line circuitry. Proceed as explained for 'specific tinted colour.'

(7) Darkish Raster

This can occur due to improper voltage at the brightness control pin. Check dc voltage variations at the corresponding pin while rotating (sliding) the brightness control. If necessary check potentiometer connections of this control.

(8) No Picture (Raster and Sound Normal)

This means that video output signals is not available. Check signal amplitudes at the luminance input and RGB output pins of the IC. Also check if sandcastle pulse is available at pin 8 of TDA 3561A.

(9) Erratic Fluctuations in Colours

Such a defect is due to loss of synchronization between the colour burst and subcarrier. The fault is usually in the gated burst amplifier or burst phase detector with the result that demodulator outputs are not phase locked. Sometimes a poor signal from the RF-IF section may also cause loss of colour sync. But in this case, the colours will appear washed out or dim. Check if the sandcastle pulse is present. (pin 8 of TDA 3561). If present check voltage at pin 3 of TDA3561. If low the fault could be in the RF and IF stages *i.e.*, before the chroma IC.

(10) Colour Snow with Black and White Reception

Colour snow appears if the chroma channel is not completely cut-off during black and white reception. Thus, the fault is in the colour killer circuit. Check voltages at the burst phase detector output and reference oscillator pins. Also ensure that the reference oscillator is properly synchronised with the burst phase. Locate corresponding pins on the IC and measure dc voltages to localize the fault.

14.14 COMMON FAULTS IN RGB DRIVE AMPLIFIERS

(1) No Picture but Sound and Raster Normal

The probable causes are (*i*) inoperative video output stages (*ii*) picture tube screen grid (G_2) voltage very low and (*iii*) amplifier transistors not functioning. For these check dc supply voltages at both collector and emitter points of the three transistors. Also check V_{BE} of each transistor and if improper try replacing the defective transistor. Also check dc voltage at the screen grid (G_2) of the picture tube. If less, try varying voltage adjust potentiometer that forms part of the LOT rectifier circuitry.

(2) Colour Tint on Black and White Picture

This shows that white level adjustment is improper. Grey scale tracking should be rechecked for proper adjustments.

(3) Any One Colour Missing

This means that video output is not available at the corresponding cathode of the picture tube. For this check transistor of that video amplifier and also colour video signal at its base input point. In case there is no input, the fault is in the corresponding section of the chroma IC circuit.

REVIEW QUESTIONS

1. Draw circuit diagram of the main luminance amplifier and explain how contrast and brightness controls function to provide luminance output for feeding the matrix circuit. What is the function of line and field blanking pulses in the contrast control circuit and gating pulses in the brightness control section?
2. Draw block schematic of the PAL-D decoder and explain briefly the function of each section. Why is the chroma delay line period chosen to be 63.943 μs?
3. Explain with a suitable circuit diagram how the 1st and 2nd chroma amplifiers function to incorporate colour saturation and colour killer controls. How is the gain of first stage varied to keep the chroma output constant? What will happen if the colour bias is insufficient?

4. Explain with a circuit schematic how the chroma signal is delayed and the two consecutive line signals added and subtracted to obtain U and V signals separately. Illustrate that on all subsequent lines the U signal will continue to be positive but the V signal polarity changes line by line.

5. Describe with a suitable sketch the basic principle of a delay line. Can a SAW filter type configuration be used to introduce necessary delay in the path of chroma signal?

6. Explain with a suitable circuit diagram how the burst pulses are separated from the chroma signal and fed to the burst phase discriminator in push-pull manner. From which point is the burst signal taken for feeding the ACC amplifier?

7. Draw circuit diagram of the ACC amplifier and explain how ACC voltage is developed from the burst pulses.

8. Draw complete circuit diagram of a burst phase discriminator and explain how on comparing the subcarrier waveshape with that of the burst pulses, a control voltage is developed for correcting any frequency or phase deviations in the reference oscillator output. Why such an output has a ripple at 7.8 KHz?

9. Draw a typical circuit of the crystal controlled subcarrier reference oscillator and explain how its output is stabilized by dc voltage (level) variations obtained from the burst phase discriminator output.

10. Explain with a suitable circuit diagram how the 'IDENT' and KILLER' outputs are obtained from the burst phase discriminator output.

11. A PAL 180° switch and a 90° phase shift circuit is employed to obtain two reference oscillator outputs for feeding the U and V demodulators. Explain fully how these are obtained and kept synchronized with corresponding carrier waveshapes at the transmitting end. Draw necessary circuits and associated waveshapes to supplement your answer.

12. Draw basic circuit of a synchronous demodulator and explain how (R–Y) or (B–Y) output can be obtained on demodulation. How is the second signal suppressed?

13. Explain how RGB signals can be obtained on matrixing (R–Y) and (B–Y) colour video signals. Illustrate your answer with necessary equations to prove that the separation holds good for any colour signal. How is deweighting taken care of?

14. A block schematic of the PAL colour decoder IC TDA 3561A is shown in Fig. 14.16. Explain briefly how the luminance and chroma signals are processed to produce RGB outputs.

15. The application circuit of chroma IC TDA 3561A is given in Fig. 14.18. Explain briefly the function of components connected at various pins of the IC.

16. Draw complete of RGB drive amplifiers employing high voltage transistors and explain how their outputs drive the picture tube cathodes. How are grey scale tracking adjustments carried out?

17. The IC version of colour video amplifiers is shown in Fig. 14.20. Explain how the IC TDA8150 functions to produce RGB outputs. What other functions are performed by this IC?

18. Enumerate typical faults that can occur in the chroma subsystem and explain how each fault can be localized and rectified. Take the circuit of TDA3561A as reference to illustrate the trouble-shooting process.

19. Describe how would you proceed to trouble shoot the following faults in the RGB video amplifiers:

   (*i*) Colour tint on black and white picture.

   (*ii*) Green colour missing in the reproduced picture.

   (*iii*) No picture is formed but sound and raster are normal.

20. Explain how sparking is prevented at various electrodes of the picture tube.

15

VERTICAL SCANNING CIRCUITS

INTRODUCTION

Vertical deflection of the electron beam on the face of picture tube is essentially a two step operation. The first stage is a synchronized 50 Hz oscillator or count-down circuit from a reference high frequency source, the output of which is modified with a ramp generator to obtain a sawtooth or trapezoidal waveform. This drives the second stage which is a power amplifier that feeds a linear sawtooth current to the deflection coils for a full raster on the receiver screen.

*The complete scanning circuit is quite complex because of the need to include linearity, height and hold controls; to incorporate circuits for 'S' correction, suppression of undesired oscillations, vertical blanking and voltage boost (pump-up) for a fast retrace of the deflection beam. Initially this meant quite a complicated circuit with several time-consuming adjustments but later, developments in integrated circuit design solved most of the problems. To begin with, ICs became available that contained all the sections from sync separator to vertical output amplifier. Soon after, combination ICs were developed that provided drive signals for both field and line deflection amplifiers. This lead to the development of integrated circuits that contained only the vertical output stage to match different yoke assemblies designed for various sizes of picture tube screens. However, many receiver manufacturers still prefer a discrete vertical output stage for economic reasons with input from a combination IC. Thus, in order to discuss all such variations in the frame deflection system, various representative circuit combinations have been included in the chapter.

15.1 FIELD SCANNING GENERATOR—TDA 2652

This represents a circuit combination where input is vertical sync pulses (derived from the composite colour video signal) and output directly drives vertical deflection coils in the yoke. In particular TDA

---

*Refer Chapter 19 of the book "Monochrome and Colour Television" by R.R. Gulati ; for an in-depth appreciation of the need to provide an elaborate vertical deflection circuitry.

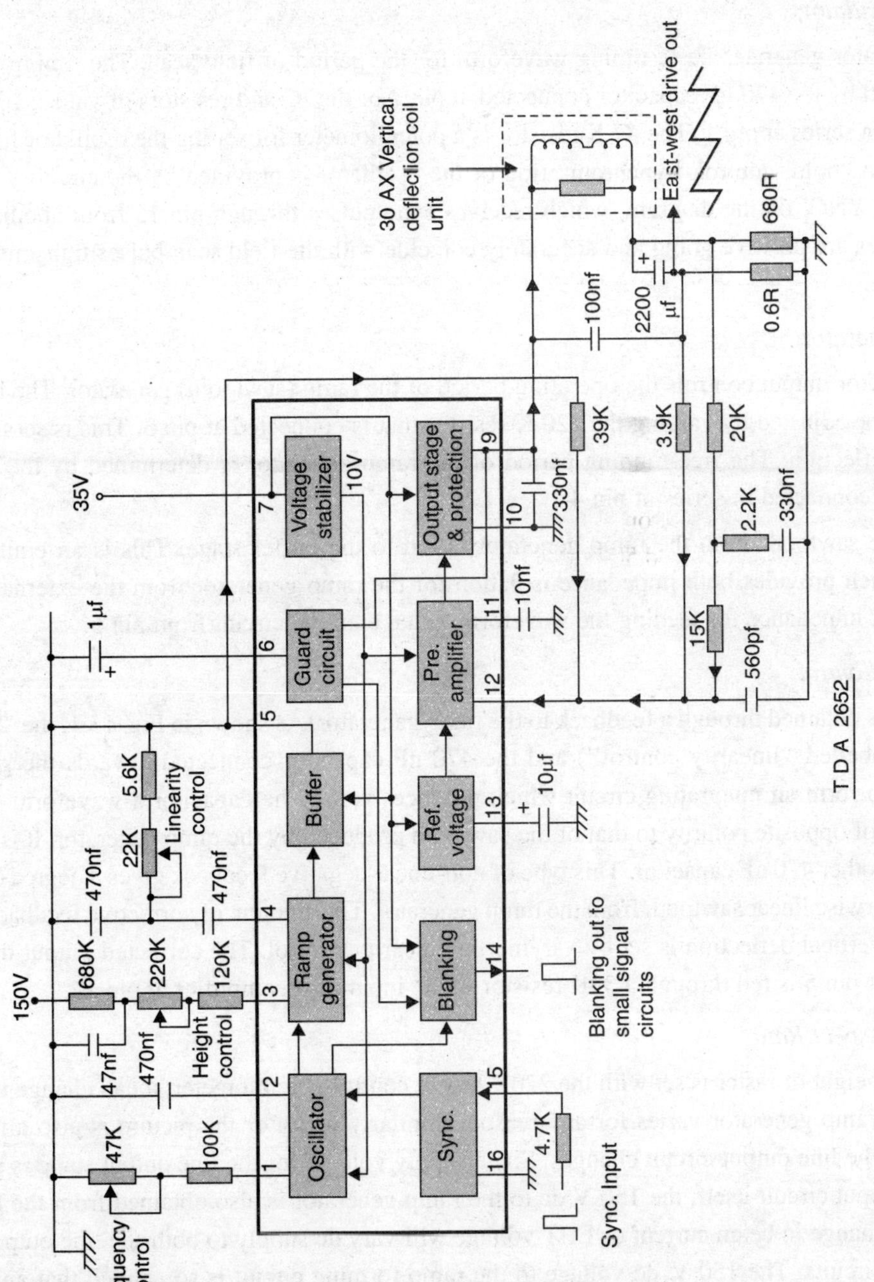

Fig. 15.1 Block diagram and essential exteral circuitry of the feild scanning generator TDA 2652.

CHAPTER 15

2652 is meant for 110° 20AX and 30AX deflection systems. Some external circuit changes are necessary to match each system. The circuit shown in Fig. 15.1 is for the 30 AX system.

50 Hz Oscillator

The oscillator generates basic timing waveform for the period of field scan. The timing waveform is determined by the 470 nF capacitor connected at pin 2 of the IC and resistors of values 100K and 47K provided in series at pin 1. The 47 K resistor is a potentiometer for setting the oscillator frequency and thus acts as 'hold' control. Synchronisation of the oscillator is provided by the circuitry in the block labelled 'SYNC' on the diagram, which receives sync pulses through pin 15 from another IC. These input pulses are positive going and accurately coincide with the field scan pulses transmitted from the TV station.

Ramp Generator

The oscillator output controls the operating period of the ramp (sawtooth) generator. The height of the ramp can be adjusted by varying the 220 K resistor that is connected at pin 3. This is set to obtain full vertical deflection. The free-running period of the ramp generator is determined by the two 470 nF capacitors connected in series at pin 4.

The sawtooth from the ramp generator is fed to the buffer stage. This is an emitter follower circuit which provides both impedance isolation for the ramp generator from the external load and a low output impedance for feeding the waveform to the external circuit from pin 5.

Linearity Control

Linearity is obtained through a feedback to the ramp generator. As shown in Fig. 15.1, the 22K variable resistor (labelled "linearity control") and the 470 nF capacitor connected towards the ground point combine to form an integrating circuit which produces across the capacitor a waveform of parabolic shape and of opposite polarity to that of the sawtooth produced by the ramp generator. It is fed at pin 4 through another 470 nF capacitor. This type of non-linear negative feedback gives a degree of concavity to the otherwise linear sawtooth from the ramp generator. The amount of corrective feedback necessary for linear vertical deflection is set by varying the linearity control. The corrected output that becomes available at pin 5 is fed through a 39K resistor to the input of preamplifier at pin 12.

Constant Aspect Ratio

While the height of raster is set with the 220K height control potentiometer, it can change if dc voltage fed to the ramp generator varies for any reason. Similarly width of the picture is also affected when voltage to the line output circuit changes. Since supply voltage for the line output stage is provided by the line output circuit itself, the 150 V dc to the ramp generator is also obtained from the L.O.T. side. Thus any change in beam current or EHT voltage will vary dc supply to both-the line output and ramp generator circuits. The 150 V. dc voltage to the ramp forming circuit is so applied that any change in width is reflected in a proportionate change in the ramp height. Thus aspect ratio is maintained at its correct value of 4 : 3 despite voltage fluctuations in the line output stage.

Pre-amplifier

The pre-amplifier is a differential configuration with two input terminals. One of these is connected to a steady dc potential derived from the reference. The second input terminal is connected at pin 12 and receives inputs from external sources. The main input, as stated earlier, is from pin 5, the output of buffer amplifier to the sawtooth generator. The second input, also ac, but of much lower amplitude and of opposite polarity to the sawtooth itself, is developed across a 0.6 ohm resistor connected in series with the deflection coils and 2200 μF coupling capacitor. This signal is applied to the preamplifier as negative feedback through a 3.9 K resistor. Its purpose is to assist in maintaining steady gain of the amplifier. The third input is a small dc voltage proportional to variations of voltage across the 2200 μF capacitor. Its purpose is to help compensate for any change which may occur in the dc operating conditions of the pre-amplifier. It is filtered by the low-pass network consisting of 20K, 2.2K and 15K resistors together with 330 nF and 560 PF capacitors. The filter removes any ac content from this essentially dc voltage. The 10 μF capacitor at pins 13 and 11 are decoupling capacitors to corresponding circuit blocks.

It may be noted that a 680 ohm resistor is connected to ground from the upper and (marked +) of the 2200 μF capacitor. The purpose of this resistor is to allow a very small dc current to flow through the deflection coils that is sufficient to give a slight upward movement to the raster which compensates for the downward shift given to it by the exponential shape of the flyback waveform.

Output Amplifier

The preamplifier output is directly fed to the output stage of the field scanning generator. The amplifier provides necessary gain to provide large sawtooth current (typically 2 amp) to the vertical deflection coils. The output that becomes available across pins 9 and 10 of the IC is ac coupled to the coils through the 2200 μF capacitor. It is not possible to use dc coupling because any flow of unipolar current would cause a vertical shift of the picture.

Instability of the output stage at high frequencies is prevented by the 330 nF capacitor connected across the output terminals. Similarly the purpose of 100 nF capacitor across the coils and coupling capacitor (see Fig. 15.1) is to lessen the possibility of line ripple on the raster caused by mutual coupling (cross-talk) between the line and frame scanning fields.

Protection and Guard Circuits

The power transistors in the output stage are protected against damage from overheating by an internal temperature scanning circuit which comes into operation as soon as the substrate temperature exceeds 150°C. The circuit senses base current and collector-to-emitter voltage (V_{CE}) to automatically limit the amount of collector current and hence the power dissipated by transistors. The 1 μF capacitor at pin 6 is a decoupler to the guard circuit.

Blanking pulses are available at pin 14 and are routed to either the video amplifier or on the pin of another IC for generating sandcastle pulses for the chroma processor. The drive for E-W connection circuit is obtained across the 0.6 ohm resistor and fed to the line output stage.

The voltage stabilizer circuit, as in other ICs, regulates the 35V input dc and distribution it to various sections at 10V as shown in the figure.

15.2 VERTICAL OUTPUT IC-TDA3651A

This is a widely used vertical output IC in modern colour receivers in conjunction with the combination IC TDA2578A. Its block diagram is shown in Fig. 15.2 where each section is suitably labelled. The functions performed by various sections are as under:

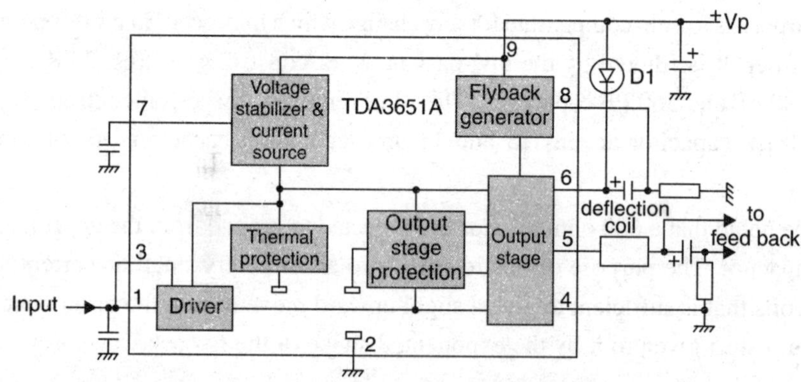

Fig. 15.2 Block schematic of the IC TDA 3651A.

Output Stage & Protection Circuit

Pin 5 is the output pin. The supply for the output stage is fed to pin 6 via diode D_1 and the Output stage ground is connected to pin 4. The output transistors of the class-B output stage can each deliver 1A maximum. The 'upper' power transistor is protected against short circuit currents to ground, whereas during flyback, the 'lower' power transistor is protected against too high voltage which may occur during adjustments. A thermal protection circuit is incorporated to protect the IC against too high dissipation. The circuit is 'active' at 175°C and then reduces deflection current to such a value that the dissipation cannot increase.

Driver and Switching Circuit

Pin 1 is the input for the driver of output stage. The signal at pin 1 is also applied to pin 3 which is the input of a switching circuit. When flyback starts, this switching circuit rapidly turns off the lower output stage and so limits the turn-off dissipation. It also allows a quick start of the flyback generator. Pin 3 is connected externally to pin 1, in order to allow for different applications in which pin 3 is driven separate from pin 1.

Flyback Generator

The capacitor at pin 6 is charged to a maximum voltage, which is equal to the supply voltage V_p (pin 9) during scan.

When flyback starts and voltage at the output pin (pin 5) exceeds the supply voltage (pin 9), the flyback generator is activated. The Vp is connected in series (via pin 8) with voltage across the capacitor.

The voltage at the supply pin (pin 6) of the output stage will then be maximum (twice Vp). Lower voltages can be chosen by changing the value of external resistor at pin 8.

Voltage Stabilizer

The internal voltage stabilizer provides a stabilized supply of 6V for drive of the output stage so the drive current to it is not affected by supply voltage variations. The stabilized voltage is available at pin 7. A decoupling capacitor of 2.2 μF can be connected to this pin.

Application Circuit

A typical application circuit of this nine pin single-in-line IC is shown in Fig. 15.3. The vertical drive signal from pin 1 of the IC TDA-2578A is applied via R_{17}-C_{18} to the driver and thermal protection circuits at pins 1 and 3 respectively. The amplified output is obtained across pins 5 and 4 where the latter is grounded for return path of the sawtooth current flowing in the deflection windings. The output to the coils is ac coupled to avoid any vertical shift of the raster. As in the case of IC TDA2652, both ac and dc feedbacks are used from the output to the preamplifier input at pin 2 of TDA2578A. This ensures linear deflection and stable operation of the output stage. The pre-sets R_{21} and R_{22} are adjusted initially to obtain a distortion free raster.

The dc supply (Vp) to the output stage at pin 6 is through D_3 where C_{20} charges equal to supply potential during the trace period. However, when retrace commences, the self-induced voltage across coil and windings is large enough to reverse bias diode D_3. But, at the same time, flyback generator is activated to connect pins 8 and 9 inside the IC. Thus supply voltage and potential across C_{20} add in series and a voltage equal nearly to 2Vp is applied at pin 6. This is enough to cause a smooth and fast retrace of the beam. The cycle repeats field after field to produce linear deflection of the beam in the vertical direction.

15.3 DISCRETE VERTICAL OUTPUT CIRCUIT

Earlier when integrated circuits for the vertical output stage had not been developed or were quite expensive, many receiver designs employed a discrete vertical deflection amplifier with drive from a combination IC. One such circuit that was widely used in conjunction with combination IC TA7609P is shown in Fig. 15.4. The driver transistor Q_1 receives input from pin 7 of TA 7609P and feeds it to the complementary pair formed by *p-n-p* transistor Q_2 and *n-p-n* transistor Q_3. The 42V dc supply for the amplifier is obtained by rectifying flyback pulses across winding a-b (on L.O.T.) with diode D_1 and filtering with capacitor C_1. The Vcc supply during trace period is fed to Q_3 via D_2 and D_3 and to Q_2 from the charge stored across C_4 during retrace interval. Since Q_3 is off during trace period, C_3 is charged to nearly 42V. The base current for Q_3 is through R_3, R_4 and that for Q_2 from the collector of transistor Q_1 when it conducts. The collector voltage to Q_1 is through R_3, R_4 and diodes D_6, D_7. The diode D_4 and D_5 are to protect base-emitter junctions against shorts when drive pulses are applied to Q_2 and Q_3. When Q_3 is inoperative (during trace period) flyback pulses across winding c-d on the L.O.T. are rectified by D_3 to change C_3 to a higher voltage equal to about 90V. The diode D_2 is then reverse biased. The capacitor C_2 is also charged to potential that exists across it.

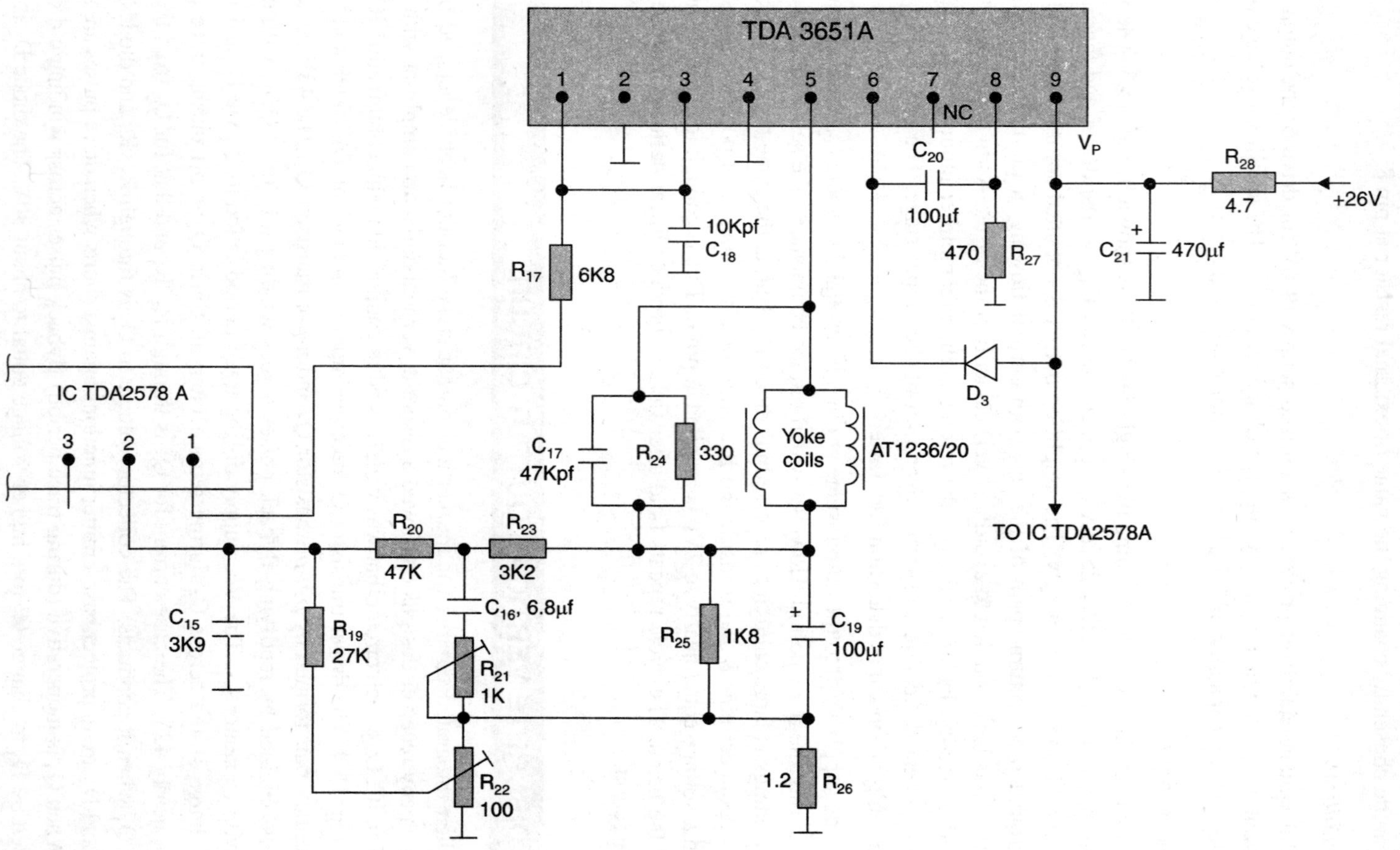

Fig. 15.3 Application circuit of the vertical output integrated circuit TDA 3651A.

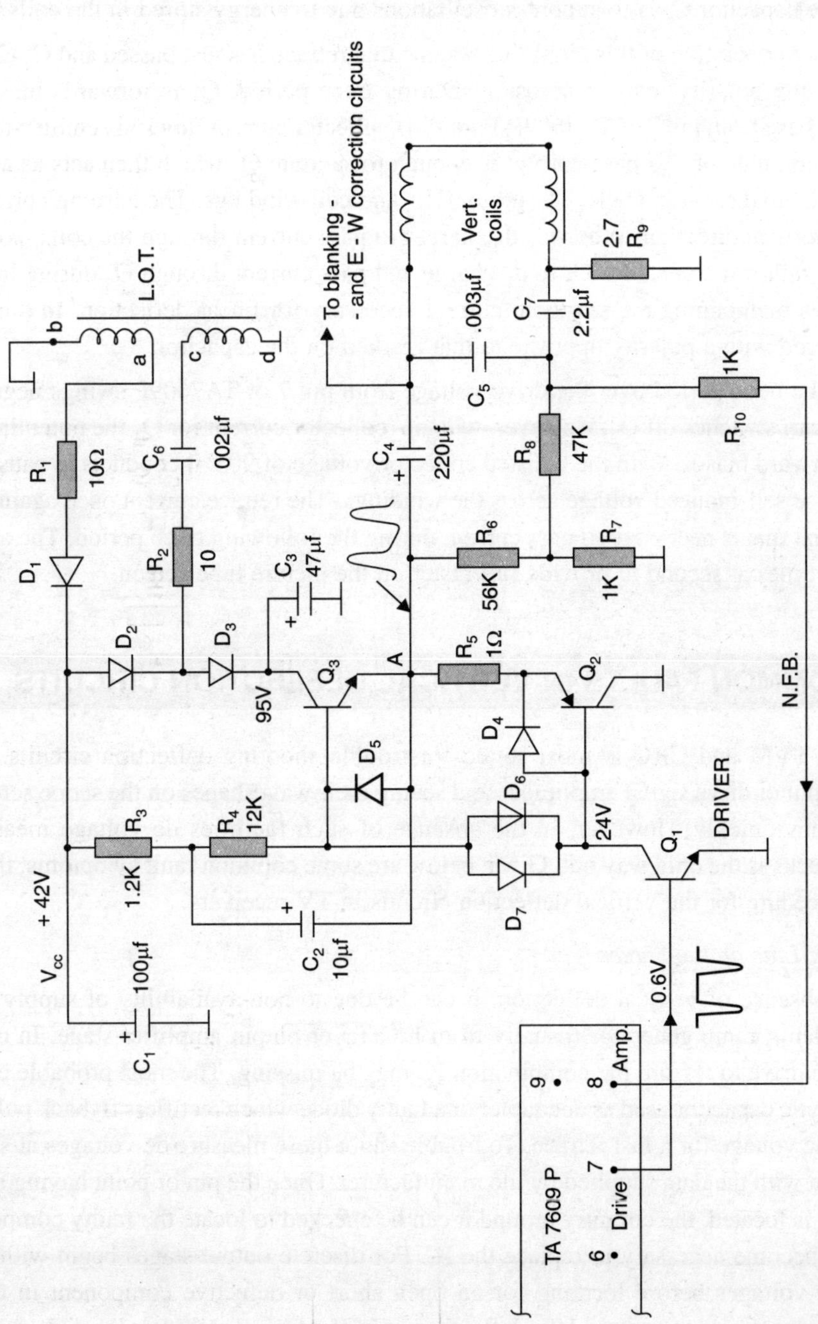

Fig. 15.4 A typical discrete vertical output amplifier.

As in previous circuits, the coils are ac coupled and a feedback voltage proportional to deflection current is developed across the 2.7 ohm resistor R_9. Negative feedback, both dc and ac is applied to the preamplifier in the IC at pin 8. As shown in the figure, blanking pulses are obtained from one end of the windings. The capacitor C_5 is to suppress oscillations due to energy stored in the coils.

To follow operation of this amplifier assume that retrace has just passed and C_4 (220 μF) is fully charged with the polarity marked across it. During trace period. Q_1 is forward biased with 0.6 V applied at the base from pin 7 of TA7609P. Part of its collector current flows via emitter-to-base junction of Q_2. The magnitude of this base current is enough to saturate Q_2 which then acts as a closed switch. This enables C_4 to discharge via R_5, Q_2, ground, R_9 and coil windings. The ensuing current deflects the beam in the vertical direction. When C_4 discharges to pass current through the coils, potential at point 'A' gradually falls which enables C_2 to discharge and pass current through Q_2 during latter half of the trace time thus maintaining the sawtooth current necessary for linear deflection. In this process C_4 is partially charged with a polarity opposite to that marked on the capacitor.

With the trace period over the drive voltage from pin 7 of TA7609P swings negative to cut-off Q_1 which in turn switches off Q_2. However, with no collector current for Q_1 the potential at the base of Q_3 rises to forward bias it. With the boosted collector voltage of 90V it conducts to cause a fast retrace overcoming the self-induced voltage across the windings. The retrace current once again charges C_4 to a high potential that is necessary to pass current during the following trace period. The above sequence continues 50 times a second to provide full raster on the picture tube screen.

15.4 COMMON FAULTS IN VERTICAL DEFLECTION CIRCUITS

The use of VTVM and CRO is most suited for trouble shooting deflection circuits. Measuring of oscillator output or drive signal amplitudes and seeing their waveshapes on the scope screen can lead to faulty locations quickly. However, in the absence of such facilities dc voltage measurements and continuity checks is the only way out. Given below are some common fault symptoms, their causes and servicing procedure for the vertical deflection circuits in TV receivers.

(1) Only Horz Line on the Screen

This means absence of vertical deflection. It can be due to non-availability of supply voltage to the vertical oscillator, ramp generator (usually from L.O.T.) or output amplifier stage. In case of vertical output IC, the drive to it from the combination IC may be missing. The most probable cause could be a bad electrolytic capacitor used as decoupler or a faulty diode which rectifiers flyback pulses or operates to pump-up dc voltage for a fast retrace. To trouble shoot these measure dc voltages at such points and compare them with the data supplied by the manufacturer. Once the pin or point having no or abnormal ac/dc voltage is located, the circuitry around it can be checked to locate the faulty component. In some cases it may become necessary to replace the IC. For discrete output stages begin with checking bias and collector voltages before locating for an open short or defective component in the biasing, dc supply and feedback circuitry around the deflection coils. If no obvious fault is located measure resistance of field coils to ensure that these are not open or there is no partial short in or across them.

(2) Vertical Retrace Visible

This will happen if vertical blanking pulses are not available to cut-off the beam current during retrace intervals. Check for ac voltage on corresponding pin of the IC and if present look for a fault in the circuit that feeds to the video amplifier/R, G, B amplifies or picture tube circuitry. Also check if the sandcastle pulse is formed properly.

(3) Less Height

Assuming there is no partial short in the yoke coils or components in parallel with the windings ; less height could be due to improper adjustment of height control. Very often one/two presets are provided in different locations to obtain optimum height without introducing any non-linearity. Locate these and adjust as necessary.

(4) Non-Linearity

One or two variable resistors are included in the ramp forming and feedback circuits to obtain non-linear deflection. With ageing of components a re-adjustment becomes necessary. Re-adjustment of linearity control pot(s) should restore linear deflection. In case of discrete circuitry defective drive and output transistors can also cause non-linearity. A defective IC can also be the cause for such a fault.

(5) Poor Interlacing or Line Pairing

The most probable cause for it is an open capacitor in the sync pulse integrator circuit. An open decoupling capacitor in the oscillator and ramp generating circuits can also cause line pairing. Locate for the faulty capacitor and try replacing it.

(6) Picture Rolling

This occurs if vertical sync pulses are of low amplitude or not applied to the oscillator circuit. Incorrect frequency of the vertical oscillator can also be the cause of picture rolling. Initially try varying the 'hold' control and later check and adjust oscillator frequency to its correct value of 50 Hz. In case sync pulses are missing, locate their source in the vision IF IC and look for a fault in the coupling circuitry.

REVIEW QUESTIONS

1. Draw block schematic of the IC TDA2652 and briefly explain functions of various sections in the integrated circuit. How are the output transistors protected against overheating.
2. The application circuit of vertical output IC TDA3651A is given in Fig. 15.2. Explain the function of components connected on various pins of this IC. Also trace path of both ac and dc feedbacks to the preamplifier and explain why are these necessary.
3. Draw circuit diagram of a discrete vertical output stage that receives drive from a combination IC like TA7609P and give sequence of operation that lead to produce vertical trace and retrace of the electron beam on the receiver screen.
4. Explain briefly how you will proceed to localize the following faults in the vertical subsystem.
   (*a*) vertical retrace visible
   (*b*) non-linearity
   (*c*) line pairing
   (*d*) picture rolling.

16 LINE SCAN AND HIGH VOLTAGE GENERATION CIRCUITS

INTRODUCTION

The line output stage is designed to perform the following functions:

(1) It generates a stable and repetitive scan current of sufficient magnitude to deflect the beam across the width of picture tube screen.

(2) It provides EHT voltage (typically 25 KV) for the final anode of picture tube.

(3) It generates high voltage (≈ 7 KV) for application to the focussing electrode of picture tube.

(4) It develops a voltage of 550 to 800 V needed for screen grids of electron guns in the colour picture tube.

(5) It feeds low ac voltage to the filaments of three guns in the picture tube.

(6) It produces dc voltage for the vertical scan generator. This indirectly helps maintain aspect ratio at its correct value.

(7) It provides sundry other voltages needed for many other sections of the receiver.

(8) It provides several gating and timing pulses needed in various sections of the receiver. Examples are the colour-burst blanking pulses, black level clamping voltage, bistable PAL-switch gating pulses, line flyback pulses, sync reference pulses and others depending on the design of individual receiver.

The variety and diversity of above noted functions that the line scan generator is called upon to perform accounts for the fact that it is the last remaining section of the modern colour receiver which is not yet fully manufactured in the integrated circuit form.

While the functions of both line and frame scan generators are similar, the design of line output stage is very much different, in fact very complex; on account of high deflection frequency, very short retrace period and need to generate EHT and other dc voltages. It is, therefore, necessary to first grasp the basic technique evolved for providing a linear scan and very sharp retrace by what is known as 'Reaction Scanning'. The efficient method of generating very high voltages is another feat of ingenuity accomplished by receiver designers that needs attention. Therefore, earlier sections of the chapter are devoted to basic techniques of line scan and EHT generation and later sections describe several line deflection circuits found in modern colour receivers.

16.1 ELEMENTS OF LINE SCAN GENERATION

The block diagram of Fig. 16.1 is suitable for an overview of the functioning of various sections of a line scan generator. The starting point is the horizontal (line) oscillator with line sync pulses as its input. It generates a timing waveform, which after further adjustments, determines duration and repetition rate of the scanning current waveform fed to the line scan coils on the deflection yoke. In order to keep the picture tube beam exactly synchronised with the frequency and phase of the beam in the studio camera, one or two comparator circuits are provided. Such circuits form part of a complex IC (combination IC), which as explained in Chapter 13, performs several other functions not directly connected with horizontal deflection.

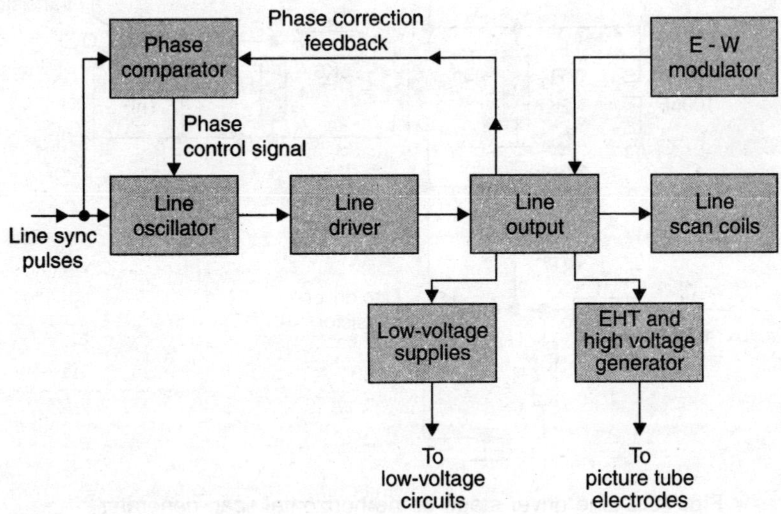

Fig. 16.1 Block diagram of the line scan and high voltage generation circuits.

The essential purpose of line driver is to isolate line oscillator from line output stage and provide necessary power drive to it. The East-West (E-W) modulator carefully distorts the line sawtooth current in such a way so as to compensate for the distortion caused at both ends of every horizontal line due to inherent limitations of the scanning coils, and tube screen not being quite flat.

The EHT and other high and low voltage power supplies are obtained from different sections of the line output transformer (LOT). Other circuits like linearity and 'S' correction (not shown) also form part of the line output stage. As obvious, the line scan generator is then the most important section of the receiver and as such, much effort has gone into its design to keep overall efficiency of the receiver as high as possible.

16.2 LINE DRIVER STAGE

The line driver stage, on receipt of synchronised and phase compared pulses from the line combination IC, generates controlling voltage to be fed to the line output stage. This is shown in Fig. 16.2 where Q_1 and Q_2 are line driver and line output transistors respectively. The line output transistor is operated as an electronic switch, which when forward biased, saturates to close and on reverse biasing cuts-off to cause an open circuit. When in saturated mode it has to deliver large bursts of power to the line deflection coils and also to several power supplies that are derived from it through the line output transformer. Therefore, it is of great importance that the controlling voltage waveform fed to the base of Q_2 shall always be large enough to firmly turn it 'on' yet also sufficiently negative to reverse bias it when required to be cut-off.

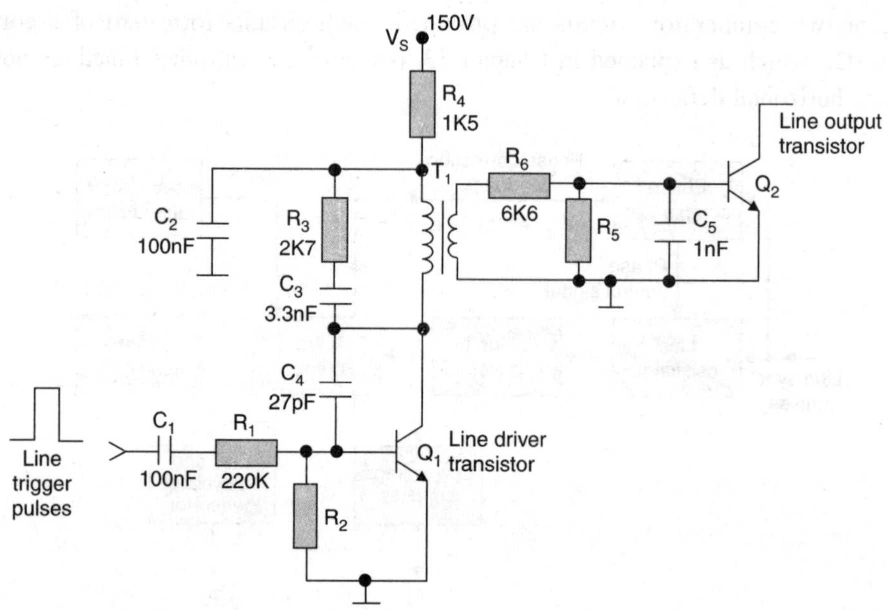

Fig. 16.2 Line driver stage of the horizontal scan generator.

The transistor Q_1 is also operated as an electronic switch and triggered into conduction by positive pulses applied at its base. As stated earlier, these pulses are received from the line trigger pulse output stage of the combination IC (see Fig. 13.3) and repeat at a frequency of 15625 Hz. When Q_1 is saturated its collector is nearly earthed and when turned-off, the collector voltage rises to about the same value as the supply voltage (Vs). Thus the waveform appearing at the collector of Q_1 takes the form of a series of rectangular pulses of amplitude nearly equal to Vs. These are fed at the base of Q_2 through coupling (step-down) transformer T_1. The coupling transformer has a primary-to-secondary turns ratio of about 30:1 designed to match the high output impedance of Q_1 into the low impedance base circuit of Q_2. The high step-down turns ratio also ensures that a negative pulse of about 5V can be applied to the base of Q_2 when it is desired to turn it off. The connections of T_1 are so arranged that the 'turning-on' of Q_1 causes Q_2 to 'turn-off' and vice-versa.

Design Constraints

It is necessary to exercise great care in the design of line driver stage to ensure correct operating conditions for the line output transistor Q_2. The object being to make sure that Q_2 is 'turned on' and 'turned off' at precisely correct moments and at speeds which enable production of good clean waveforms both during scan and flyback periods. Another aim is to ensure that minimum amount of power is dissipated in the transistor itself.

The coupling transformer T_1 that is used to feed switching waveforms to Q_2 is of particular importance. It is designed to have a leakage inductance of some 6μh between the primary and secondary windings. This amounts to as if this inductance is connected in series with the switching waveform fed to the base of Q_2. The effect is to slow down the rate at which base current of Q_2 falls when the edge of negative turn-off pulse appears. A slow turn-off delayed by about 6μs is necessary to allow the excess charge stored in the base of Q_2 to drain off. The excess charge storage in the base of Q_2 occurs because more than enough base current is allowed during turn-on period to make sure that the transistor always saturates when turned 'on' irrespective of any deterioration in its performance caused by ageing or by differences in characteristics between one transformer and another. The excess base current flow result in a surplus of charge carriers (electrons in the NPN transistor) in the base electrode.

While the excess charge storage plays no role in the action of transistor when it is conducting yet it causes delay at the time of switching-off. The excess charge has to be neutralised before the transistor can be turned-off and it takes little time irrespective of the reverse bias applied at the base. The leakage inductance introduced by T_1 causes the arrival of reverse bias waveform fed to Q_2 to be delayed so that adequate time is available for the removal of excess charge. Once this excess charge leaks-off, the transistor can be turned-off cleanly and quickly. In some line driver stage designs a physical inductor is also provided in the base lead for this purpose.

The 2.7K resistor R_3 and the 3.3 nF capacitor C_3 connected in series across the primary winding of T_1 provide just the right amount of damping to prevent ringing (oscillations) and development of excessive voltage at the collector of Q_1. Such voltages could otherwise damage the collector-base junction of the transistor.

The total conduction time of the line driver stage is 22 to 24 μs during which the collector current reaches a peak value of about 150 mA. The conduction period mentioned above is carefully chosen so that forward bias begins to be applied to Q_2 soon after the completion of flyback period and just before the collector current becomes positive.

16.3 LINE OUTPUT STAGE

The elements of line output stage are shown in Fig. 16.3(*a*) where, as explained earlier, rectangular base drive pulses come from Q_1 via transformer T_1 at precisely controlled instants. Such an input pulse waveform is shown in Fig.16.3(*b*). During positive region of the input pulse, Q_2 stays saturated and behaves like a closed switch. The current that flows from Vs to ground through inductive reactance of primary winding of transformer T_2 induces a rectangular back e.m.f. voltage across it. This circulates a sawtooth current (*a-b*) through coupling capacitor Cc in the deflection coils to deflect the beam from centre to right edge of the screen. The diode D_1 stays reverse biased during this interval.

CHAPTER 16

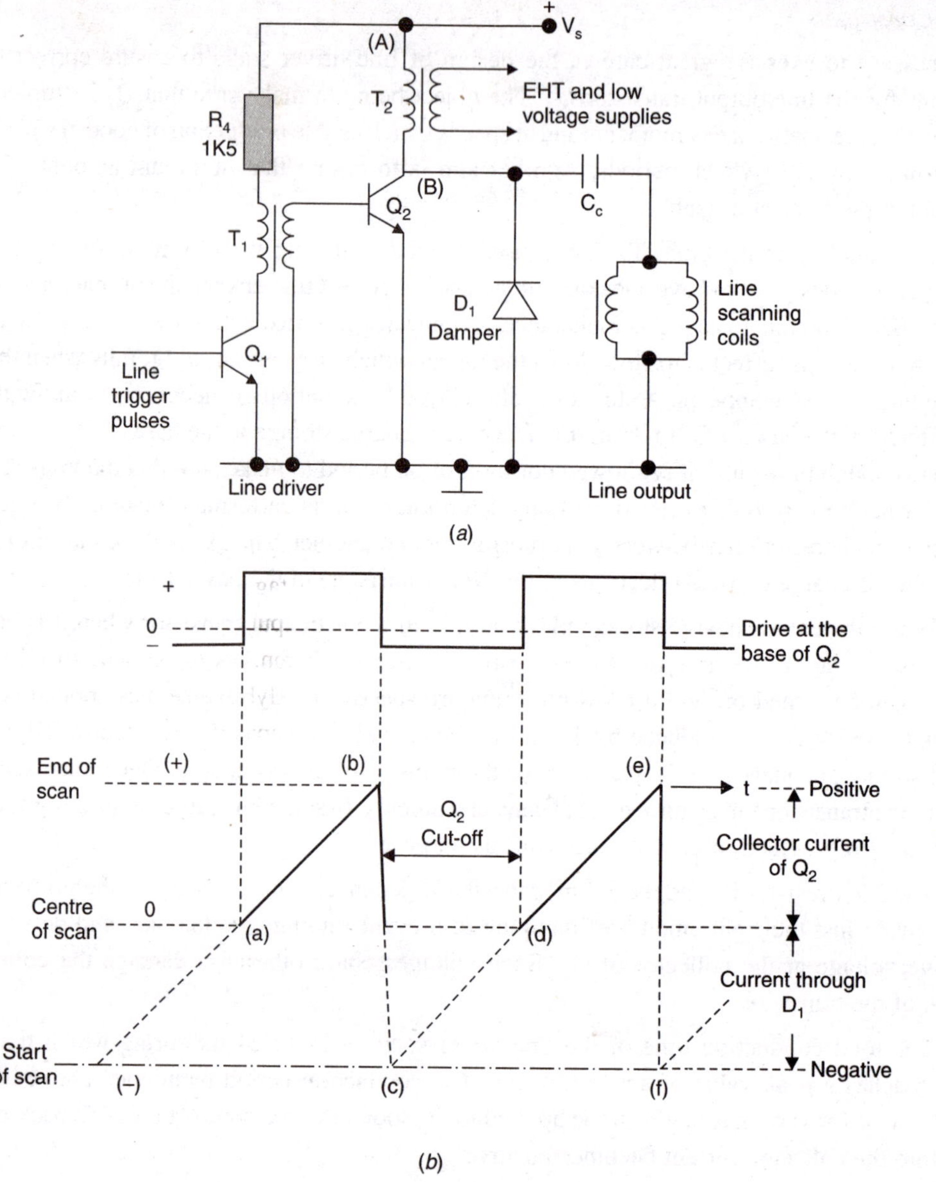

Fig. 16.3 Line output stage (*a*) simplified circuit
(*b*) idealized drive voltage and deflection current waveforms.

Before the collector current of Q_2 reaches its maximum value, it suddenly turns-off because of application of negative region of the drive waveform at its base. The energy stored in the magnetic field around primary of T_2 and deflection coils is thus trapped. This is the beginning of flyback period. The sudden drop in current induces a large voltage pulse which appears across points 'A' and 'B'. The point 'A' is at ground potential from ac point of view because of the filtering and decoupling capacitors connected across Vs. The circuit sets into self oscillations to dissipate the stored energy. The capacitance for the oscillatory circuit is provided by inter-winding and stray capacitances that appear across the inductors. The period of self oscillatory frequency is controlled to be twice that of the allowed retrace period. If necessary, a physical capacitor called the 'flyback capacitor' is connected across the circuit. Self-induced voltage, the magnitude of which is over 1 KV, circulates a lagging current (*b-c*) through the coils. This is of opposite polarity and deflects the beam quickly to extreme left side of the screen. It happens during the first half cycle of oscillations and constitutes the retrace period. As soon as polarity of the induced voltage reverses to become positive, D_1 is forward biased. This enables a flow of sawtooth current (see Fig. 16.3(*b*)) in the deflection coils to deflect the beam back to the centre of screen. The current is labelled as (*c-d*) in the figure. The oscillations are suppressed beyond this moment because the stored energy gets expended in the resistance of its current path. Soon after, Q_2 is again turned 'on' by the positive region of the next drive pulse. The resulting current flow (*d-e*) again deflects the beam to the extreme right. This sequence repeats 15625 times a second to provide a steady raster on the receiver screen.

**Reaction Scanning*

For a better insight into the phenomenon of 'Reaction Scanning' that enables precise control of the retrace interval and prevents any undue oscillations, consider equivalent circuits of the line scan generator shown in Fig. 16.4. This illustrates a step-by-step approach with corresponding voltage and current waveforms. To begin with, an ac equivalent circuit of the line output stage is shown in (*a*) where 'V' is the step input from the driver necessary to turn-'on' and 'off' the output switching transistor Q_2. It is represented by switch S_1 which opens on cut-off and closes on saturation. The inductance 'L' and small series resistance 'r' represent lumped inductance and resistance of the flyback (output) transformer and deflection windings through which the deflection current flows. The capacitor 'C' in parallel with this circuit account for the interwinding, stray and any physical capacitor connected in the deflection circuit. The damper circuit is simulated by switch S_2 and resistance 'R' in series with it. This switch represents damper diode D_1 and thus S_2 closes when the diode is forward biased to allow stored energy to dissipate in resistors R and r when current flows through them.

As already stated, the sequence of operations that cause deflection of beam across the screen is illustrated in Fig. 16.4 with equivalent circuit and waveforms corresponding to each step. The location of beam on the screen is also shown along in each case. The direction of beam deflection has been arbitrarily chosen to be from left to right for the assumed direction of current flow from the voltage source since it can be otherwise for a coil winding that creates opposite magnetic poles.

When switches S_1 and S_2 are open, *i.e.*, when both the amplifier and damper diode are not conducting, no current flows through L and r (see Fig. 16.4(*a*)) and beam is at the centre of the screen. At instant t_0 (Fig. 16.4(b)) when S_1 closes (amplifier is turned on) current i_L flows and rises linearily to deflect the beam towards right side of the raster. The beam reaches at the edge of screen when i_L attains a value equal to I at instant t_1. Since the coil resistance 'r' is very small, the voltage V_L across it = $(i_L r + L\frac{di}{dt})$ is nearly constant during this period. Its polarity is positive and magnitude small because of the relatively slow rate of rise of current. At instant t_1, the sharp negative spike of the drive signal turns-off the amplifier, *i.e.*, switch S_1 opens. With S_1, S_2 both open (Fig. 16.4(*c*)) the C-L-r circuit gets

*Comprehensive details of reaction scanning and basic line deflection circuits are given in chapter 20 of the book 'Monochrome and Colour Television' by R.R. Gulati.

CHAPTER 16

into free oscillations with a period T, very nearly equal to $2\pi\sqrt{LC}$ seconds (since r is small). The initial conditions are such that iL continues to grow further until $V_c = 0$ at $t = t_2$. The entire circuit energy, $1/2L(I_{max})^2$ is now in the inductance. T/4 seconds later at $(t = t_3)$ this energy transfers to C (less a very small loss in r) so that $1/2L(I_{max})^2 = 1/2\ C\ (v_{c\ max}{}^2$ or $V_{c\ max} = I_{max}\sqrt{L/C}$. The capacitance C, being only stray capacitances of the circuit, is very small (a few hundred pico-farads atmost). Therefore $V_{c\ max}$ is usually very high, of the order some thousands of volts.

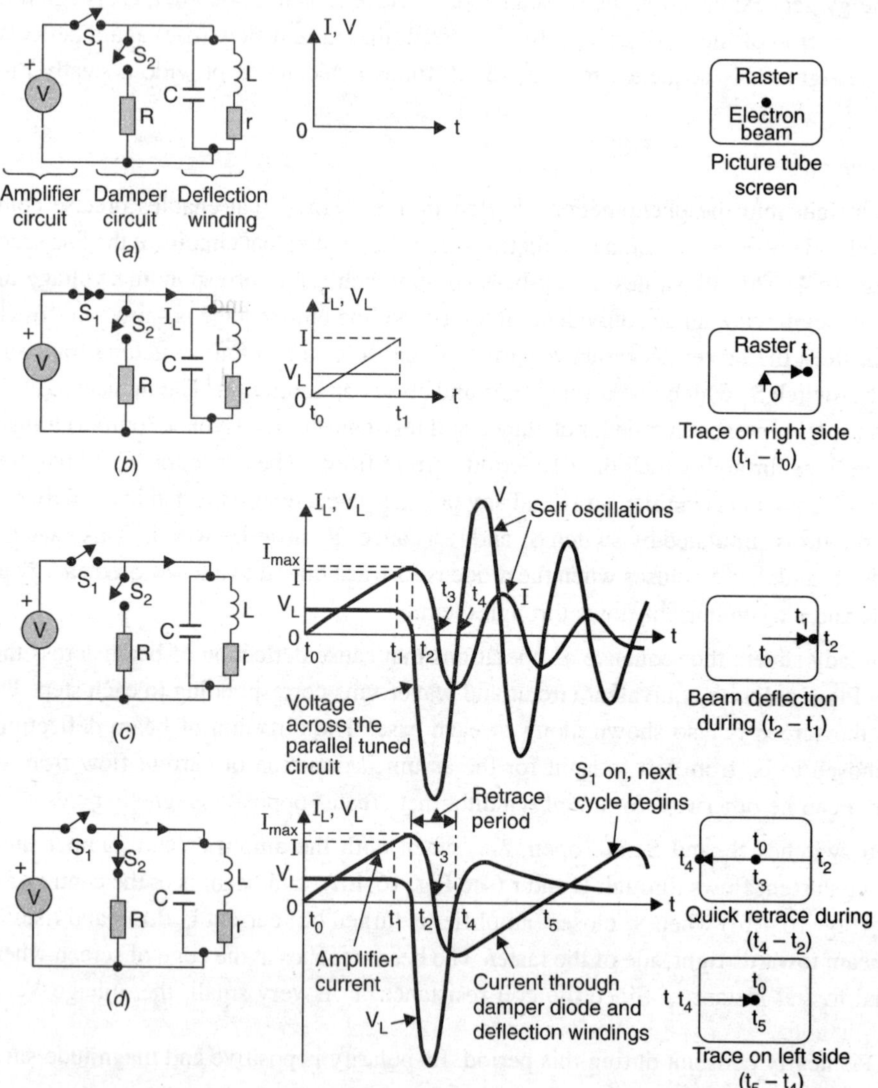

Fig. 16.4 Sequence of operations during scanning of one horizontal line. (*a*) simplified equivalent circuit of the line output stage with no excitation, the electron beam is in the centre of the picture tube screen (*b*) amplifier turns on but damper is open (*c*) amplifier turns on but damper continues to be open (*d*) amplifier remains open but damper diode turns on.

After another T/4 seconds at ($t = t_4$), once more $V_c = 0$ and $i_L = -I_{max}$. In fact, the current varies in a co-sinusoidal manner from its positive peak to its negative peak in a time equal to T/2. If the switches (S_1 S_2) remained open this process of energy exchange would go on for many cycles, the successive values of $V_{c\,max}$ and I_{max} decaying slowly until the whole energy was dissipated in the circuit resistance. The corresponding waveshapes are illustrated in Fig. 16.4(c) with an expanded scale.

However, continued oscillations beyond the first half-cycle are not desirable because the continuing oscillator current in the coil will shift the beam back and forth at the left side of the raster instead of allowing it to trace the next horizontal line at its proper place. Therefore, in order to suppress free oscillations beyond its first half-cycle, switch S_2 closes at instant t_4 to dissipate energy stored in the L, C, r circuit. In fact polarity of the oscillatory voltage changes at t_4 to become positive and thus forward biases the damper diode D_1 (closes switch S_2). With R shunted across L and C, the combination becomes non-oscillatory and current i_L and voltage V_L then decrease to zero with a time constant approximately equal to L/R where R represents resistance of the circuit in series with S_2. The resulting current and voltage waveforms are shown in Fig. 16.4(*d*).

L and C are so chosen that the period of self-oscillations 'T' is twice the horizontal retrace time. Thus, as the coil current decreases from positive maximum to zero and reverses to attain its maximum negative value in time T/2, the beam gets deflected to the left edge of the raster to complete a fast retrace. The resonant circuit is set at about 60 KHz to provide desired flyback time of about 8 μs *i.e.*,

$$\left\{\frac{10^6}{2\times 6\times 10^4}\right\} = 8\mu s$$

Further, the impedance of the circuit is so set, or in the circuit under consideration R is so chosen that current decays to zero in time ($t_5 - t_4$) which is approximately equal to half the trace period. The decaying current from its maximum negative value to zero deflects the beam to the centre of screen without any supply of energy from the voltage source. This amounts to recovery of energy, because if not utilized, this would not only be wasted but also cause undesired oscillations spoiling the raster.

The switch S_1 (amplifier) closes at t_5 to again connect the voltage source to the L-C circuit and just then S_2 (damper) opens to cut-off the damping circuit. The current then smoothly rises to take over deflection of the beam to the right of the raster. This is illustrated in Fig. 16.4(*d*). The cycle described above repeats line after line to trace the entire raster.

CHAPTER 16

16.4 BASIC LINE OUTPUT CIRCUIT

Figure 16.5 is the circuit of a basic line (horizontal) output stage. The corresponding input and output signal waveforms are illustrated in Fig. 16.6. The damper diode D_1 and flyback capacitor C_1 are connected directly across the collector and emitter of the transistor. The value of flyback capacitor C_1 is so chosen that it forms a resonant circuit with the equivalent inductance L of the deflection circuit at a frequency whose period is twice the retrace interval. During the period marked T_1 in Fig. 16.6(*c*), the input signal forward biases the output transistor into full conduction and the resulting current through deflection coils deflects the beam to right of the raster. The damper diode remains reverse biased by V_5 during this interval. At instant t_1 the sharp negative pulse at the base of Q_2 turns it off and its current sharply declines to initiate retrace. The beam returns to left of the raster during first half cycle of self-oscillations. The self-induced voltage across the coil keeps the damper diode back-biased during this (T_2) interval.

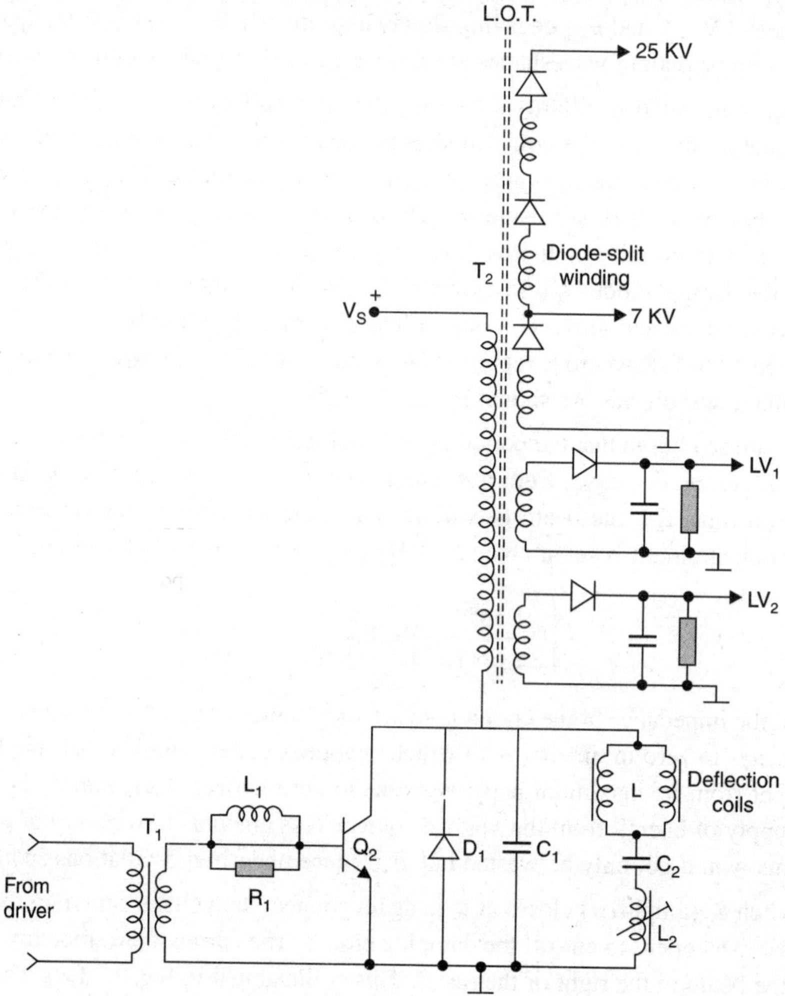

Fig. 16.5 Basic circuit of the line output stage.

The polarity of induced voltage reverses at instant t_3 and forward biases D_1 which then conducts to deflect the beam to centre of the screen. The energy stored in the coil is thus expended and any subsequent oscillations are suppressed. The transistor Q_2 continues to be off for most of the time interval T_3 *i.e.*, (t_4-t_3). A little before the instant t_4 (see Fig. 16.6(*a*)) the sharp rise of base voltage forward biases Q_2 and it conducts heavily to continue deflection of the beam to the right of the raster. Diode D_1 immediately ceases to conduct because of the disappearance of self-induced voltage in the yoke. The Vs supply, however, keeps it off during this interval and the above described cycle repeats to scan the entire raster.

The capacitor C_2 is for 'S' correction. The correction occurs because of the voltage which develops across it when sawtooth current flows through the coils. The capacitor also blocks dc to the yoke coils which would otherwise shift the beam and cause heating of coils due to dc resistance of the windings.

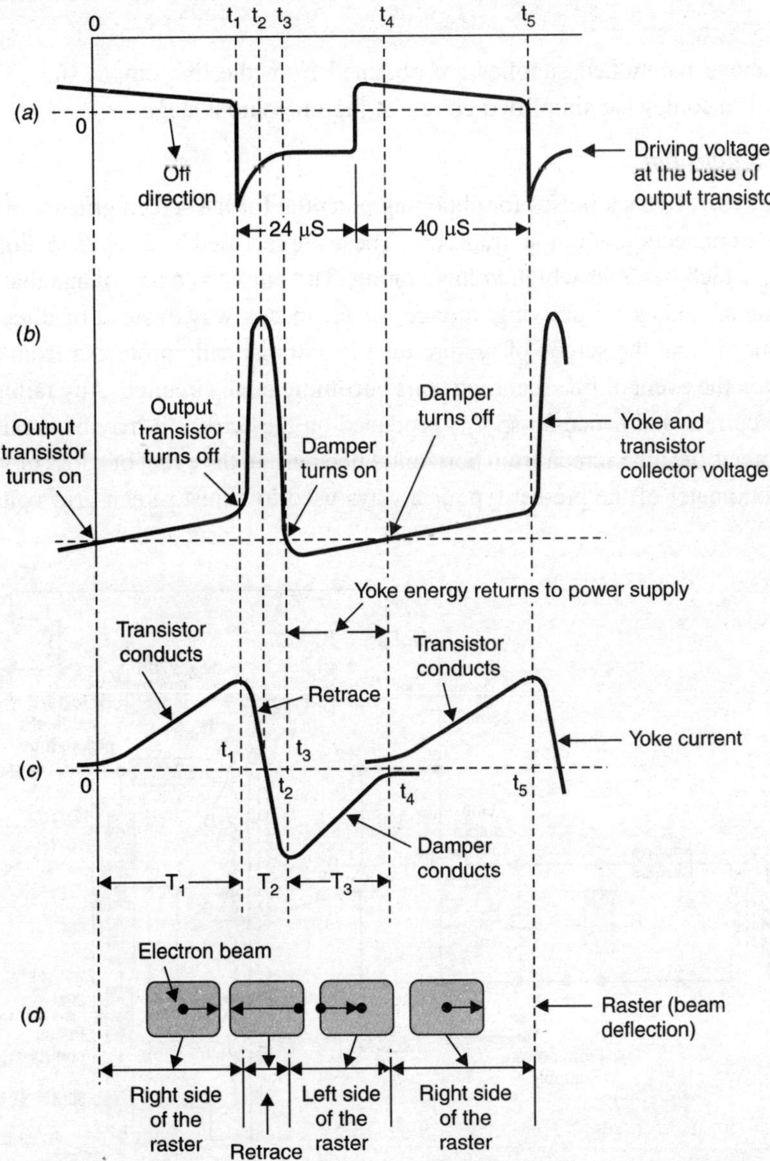

Fig. 16.6 Waveforms corresponding to the line output circuit of Fig. 16.5.

16.5 PICTURE TUBE VOLTAGE SUPPLIES

As explained in chapter 3, in modern in-line colour picture tubes the control (G_1), screen (G_2) and focus (G_3) grids of the three guns are internally connected and only one connection for each is brought out for supplying suitable dc potentials. The final anode or accelerator is the internal conductive coating and EHT is connected to it by way of a metal stud on the outside of picture tube cone. The filament connections of the three heaters are also paralleled and fed externally from a common low voltage ac source.

All the above mentioned supplies are obtained from the line output stage. This is shown in Fig. 16.7 which is a somewhat simplified circuit of the line output stage.

First Anode (G_2) Potential

As shown in Fig. 16.7, flyback pulses for obtaining potential for the screen grid are obtained separately from the collector connection of output transistor. These are rectified by a separate diode D_1 and filtered by capacitor C_{10} which has a very high voltage rating. The output is a dc voltage that ranges from 550 to 800 volts. The advantage of deriving voltage for G_2 in this way, instead of directly from the line output transformer is that the screen of picture tube is automatically protected from having a vertical line burnt into it in the event of line scanning coils becoming open circuited. Any failure of G_2 potential means no beam current and hence no spot is produced on the screen. It may be recalled that a similar guard circuit protects the tube screen from horizontal line burn in the event of a loss of vertical deflection current. A potentiometer of the pre-set type is always used to adjust screen grid voltage for optimum brightness.

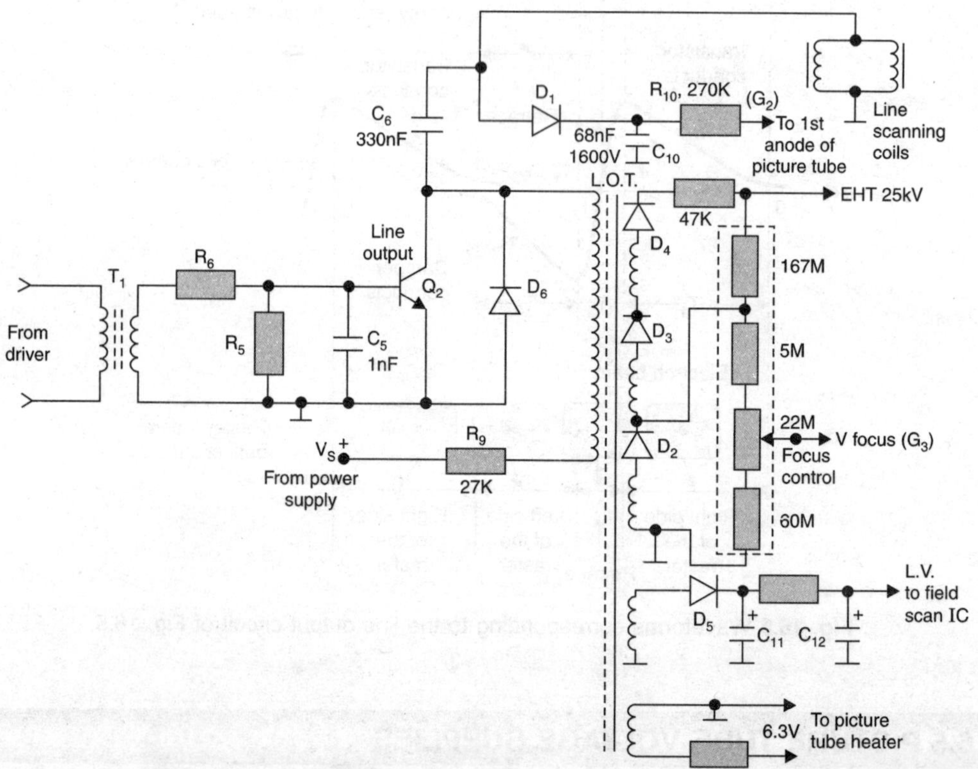

Fig. 16.7 Simplified line output circuit showing generation of supply voltages necessary for a colour picture tube.

EHT Generation

The method of generating EHT potential, as used in monochrome receivers by an overwind on the LOT is not satisfactory beyond 16 KV because of problems like flash-over, high impedance and poor regulation. Till recently a voltage tripler arrangement was used to obtain near 25 KV for the final accelerator. This needs many high voltage capacitors and several diodes. It has now been superseded by what is known as the 'diode split addition' technique. It has many advantages like greater reliability, smaller size and lower cost.

The principle of 'DIODE-SPLIT ADDITION' is illustrated in Fig. 16.8 where three layers of secondary windings are shown wound round the ferroxide core of the L.O.T. While the three sections are shown separately, in actual practice these are wound one above the other and are thus concentric. Each winding is identical to the other and has the same number of turns. The same magnitude of voltage will therefore be induced in each section every time the flyback-derived input pulse gets applied to the primary winding (not shown).

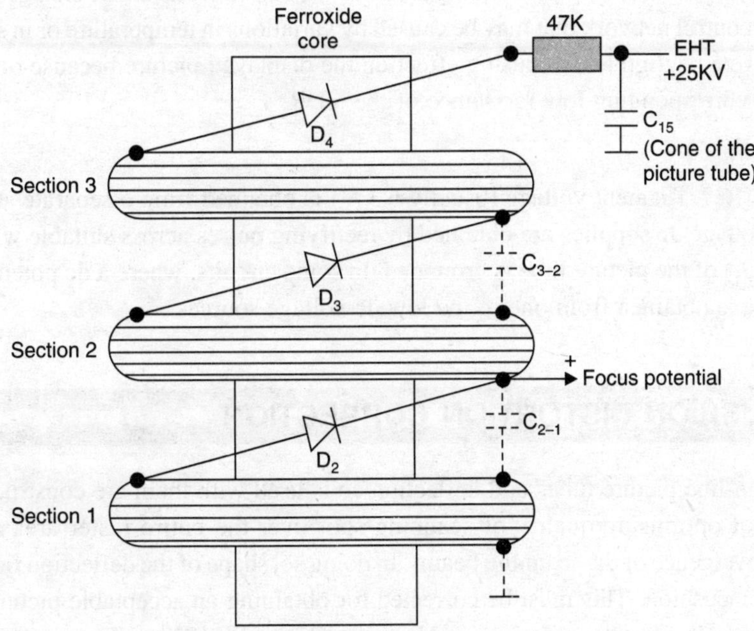

Fig. 16.8 Illustration of the principle of Diode-Split addition to obtain EHT and focus anode potentials.

Because of the close proximity of individual layers, an inter-layer capacitance exists between each of them. It is indicated in the diagram by capacitators in dotted-chain form because these are no physical capacitors. If a diode is connected between the end of the one layer of winding and the start of the next, the ac voltages induced in each layer can be made to charge up all the inter-layer capacitances to the same voltage. Since the capacitances are effectively in series, the total voltage appearing at the output terminal is the sum of all the voltages appearing across all of them. The diodes shown connected in series between the layers are physically embedded in the windings and form an integral part of the transformer.

The three windings are so designed that voltage induced in each layer from the flyback transformer is 8.33 KV. This makes the total potential equal to 25 KV and forms the EHT supply source. In some designs, four layers of windings are used and each layer provides a voltage equal to 6.8 KV.

Focus Anode Potential

The focus potential required for the third (focus) grid (G_3) of a colour picture tube usually lies in the range of 6.5 KV to 7.5 KV. It is obtained from first of the three diodes of the diode-split winding circuit just explained. Since each stage produces potential in excess of 8 KV, it is suitable for feeding a potential divider to obtain necessary dc voltage. The potentiometer as shown in Fig. 16.7 is the focus control. The purpose of using large resistance values in the divider network is to stabilize the magnitude of current flow through the potential divider and thus reduce any voltage changes which might be caused in the EHT and focus potentials by sharp variations in brightness of the scene at the studio. Some TV receivers achieve stability of focus by using a voltage dependent resistance (VDR) instead of a potentiometer for connecting voltage to G_3. As is well known, a VDR offers high resistance to a steady dc current but a low resistance to small changes in such a current. The small changes in current flow through the focus control network that may be caused by variations in temperature or in scene brightness, will therefore, have a negligible defocusing effect on the displayed picture because of the low voltage developed across corresponding low resistance of the VDR.

Low Voltage Supplies

As shown in Fig. 16.7, filament voltage (usually 6.3 V) is obtained from a separate small winding on the L.O.T. Low voltage dc supplies are obtained by rectifying pulses across suitable windings. Usually the control grid (G_1) of the picture tube is grounded. In some circuits, where a dc potential is applied to this grid, the same is obtained from one of the low dc voltage sources.

16.6 PINCUSHION DISTORTION CORRECTION

Electron guns of in-line picture tubes and deflection coils used with them are constructed and aligned with the objects of optimising quality of scanning spot over the entire raster area and of assisting automatic self-convergence of the scanning beams. In doing so, shape of the deflection field gets distorted and looks like a pincushion. This must be corrected for obtaining an acceptable picture on the screen. It involves two separate corrections known as North-South and East-West. The manner in which these are carried out is fully explained in section 3.9 of this text. However, it may be mentioned that because of the in-line structure of modern picture tubes, North-South correction is no longer necessary and thus is not applied. In fact the deflection coil design has been so perfected that in 90° deflection-small screen (51 cm and below) picture tubes, even East-West correction is not necessary and hence not provided for.

East-West Modulation

The manner in which E-W correction is applied in the line output circuits of 110° deflected angle tubes (Philips 30AX) is shown in Fig. 16.9. Out of the many modulation circuits that are possible the one shown is called the diode-modulator. As shown in the circuit, without E-W modulation components, the lower ends of line-scanning coils (L_y) would be connected directly to ground via the dashed-line connection and scanning current would flow through Q_2 and along C_6, L_y and ground. However, with modulation components present, the line coils are connected to ground via L_1-R_7 in parallel and L_2-C_{13} in series, with the whole path shunted by the capacitors C_7 and C_8 connected in series.

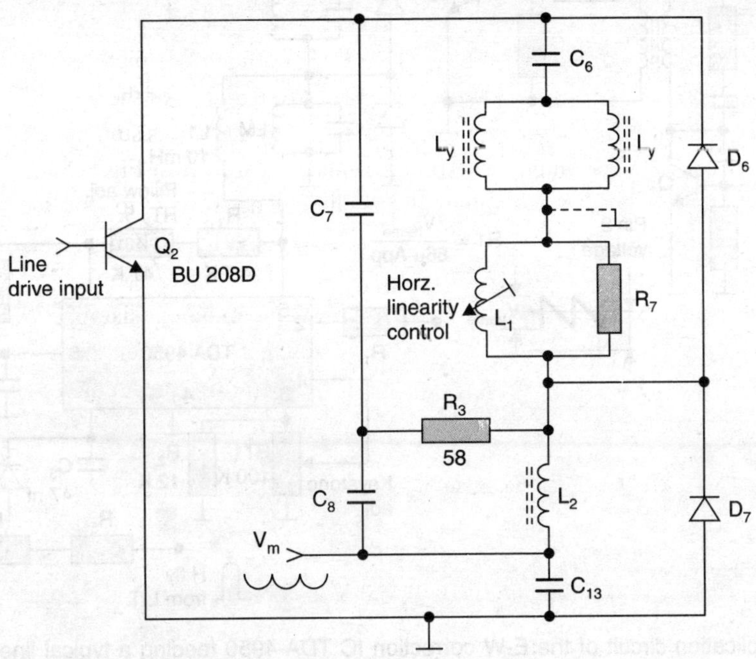

Fig. 16.9 East-West modulation as applied in the line output circuit of 30 AX colour picture tube.

The capacitors C_7 and C_8 form two arms of a balanced bridge, the opposite arms of which are L_2 and the combination L_y and L_1. The diodes D_6 and D_7 function as a bidirectional switch routing the deflection current to different sections of the circuit as the scanning beam moves from the left-half to right-half of the screen. The resistor R_3 reduces magnitude of any ringing currents which may occur due to imbalance of the bridge on account of variations in component tolerances.

The modulating voltage V_m, that is of parabolic shape, is produced by an E-W drive circuit whose input is a linear sawtooth current that repeats at field (50Hz) frequency. The parabolic shape ensures that minimum modulation of the line-scanning current occurs at the top (start) and bottom (end) of each field, with maximum modulation occurring in the vertically central region of the field, just where the horizontal scanning lines need to be most extended.

E-W Pincushion Correction IC TDA 4950

The TDA4950 is a monolithic integrated circuit in an 8 pin minidip plastic package designed for use in the E-W pincushion correction circuit of Fig. 16.9. The modified line output circuit together with connections to TDA 4950 is shown in Fig. 16.10.

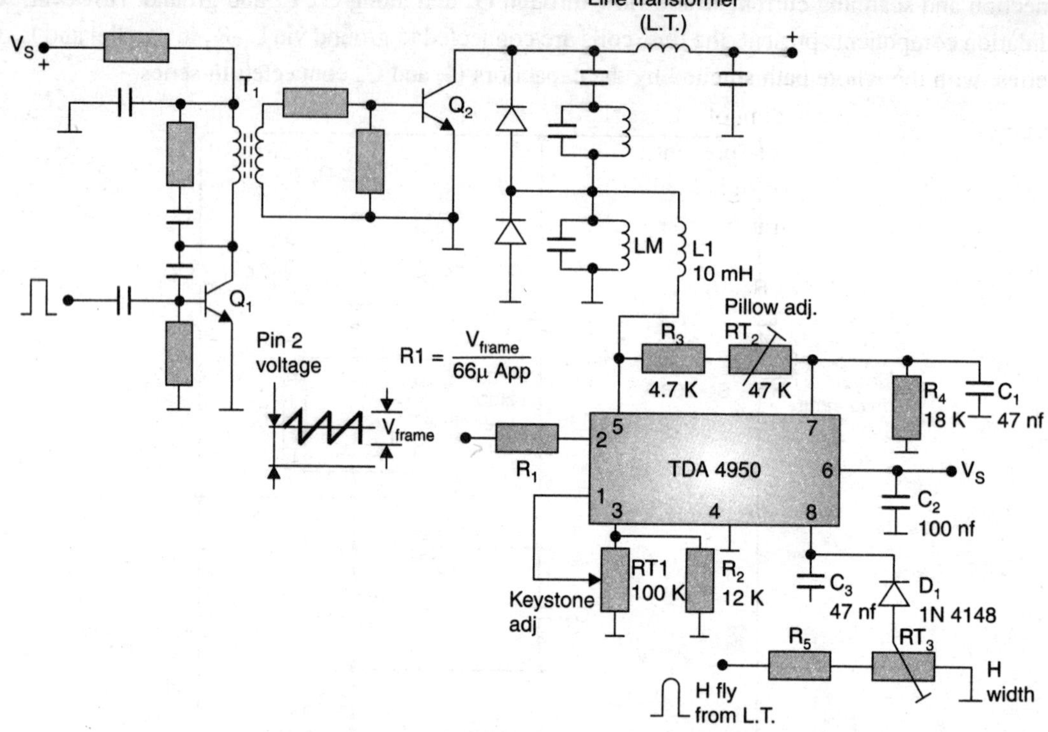

Fig. 16.10 Application circuit of the E-W correction IC TDA-4950 feeding a typical line output stage.

16.7 TYPICAL LINE OUTPUT CIRCUIT

A typical line output circuits is described to gain full familiarity with the operation of a line scan generator. This pertains to 51 cm (90° deflection angle) picture tubes which go into most popular colour receivers. As already explained such tubes do not need any E-W correction and as such corresponding modulation circuit does not form part of the line output stage. Only linearity and 'S' correction components are included. These are predetermined on simulation of the actual output circuit around the deflection yoke and picture tube. Thus only minor adjustments, if any, are necessary in the deflection circuitry.

Horizontal Output Circuit Employing Transistors BF 393 and BU 208D

The circuit of horizontal output stage with BF393 as driver and BU208D as line output transistors is shown in Fig. 16.11. The functions performed by it are the same as enumerated in introduction to this chapter. The input to the scan generator is from pin 16 of the combination IC CA 3223. The driver (Q_1) on receipt of synchronised and phase corrected output of horizontal oscillator in the IC, switches between cut-off and saturation thus developing suitable drive pulses across the secondary of transformer TR_1. The feedback capacitor C_{16} across BF393 (Q_1) is for necessary roll-off (attenuation) at very high frequencies which can cause oscillations in the driver circuit. The step-down transformer TR_1 is for impedance matching between the output of driver and input of transistor BU208D. As explained earlier,

L_1 is for causing slow fall of base current to avoid any excess charge storage effect in the base of Q_2 and to reduce large instantaneous power dissipation at its collector. The collector supply to BF393 through R_3 and primary winding of TR_1 is at 105V from the mains derived dc supply. The capacitor C_2 is for high frequency by-pass to prevent oscillations. The parallel combination of R_2 and C_1 across TR_1 is to waveshape the line drive signal while diode D_1 clips-off any spikes in it which may appear due to inductance of the coupling transformer.

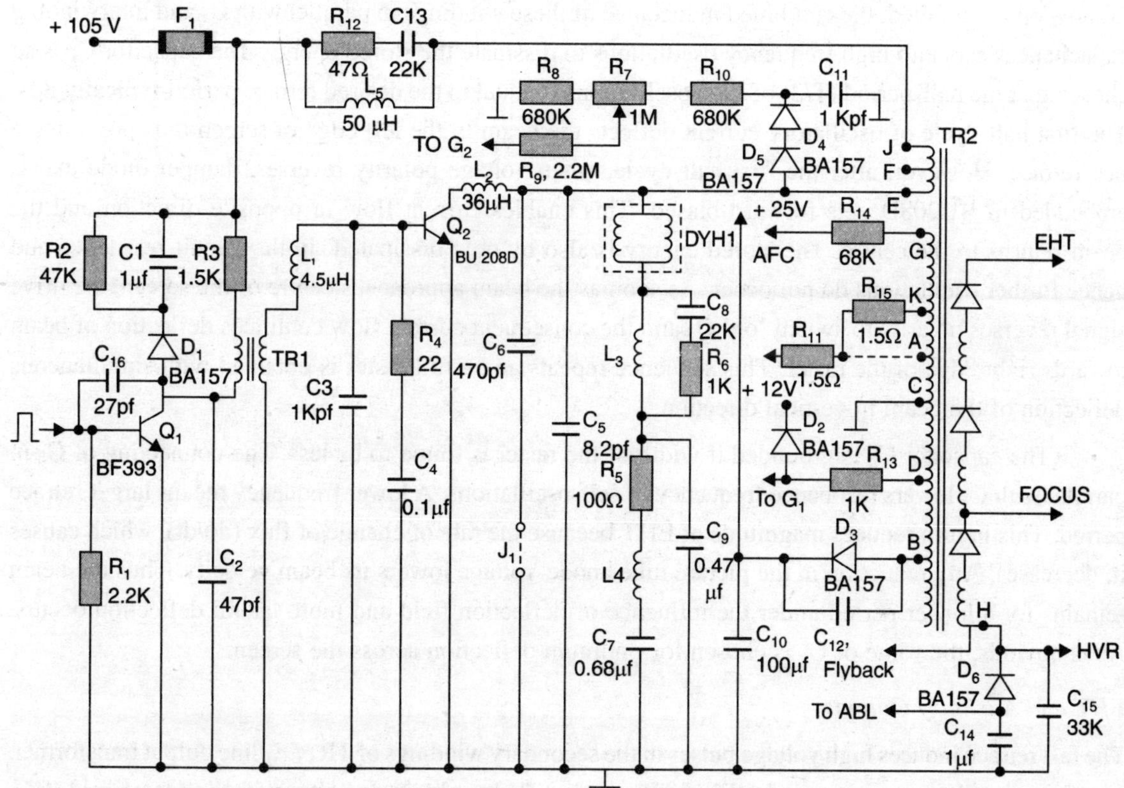

Fig. 16.11 Circuit diagram of the horizontal scan generator employing transistor BF 393 as driver and BU208D in the output stage.

The collector of output transistor BU208D also receives 105 V dc supply from the same source through fusiable resistor F-1, a parallel tuned network (R_{12}, C_{13}, L_5), primary winding J-E of the line output transformer and L_2. The parallel tuned circuit formed by L_5, C_{13} and R_{12} offers high impedance at signal frequencies thus preventing any undesired feedback (coupling) from output to the driver stage. Similarly, inductor L_2 prevents application of any high voltage to the collector of Q_2 from windings on the line output transformer.

The line deflection coils DYH1 are connected directly across the output transistor from point E of primary winding J-E to ground so far as flow of ac current is concerned. The parallel network (L_3, C_8, R_6) and series parallel combination (R_5, L_4, C_7 and C_9) connected between the lower end of DYH1 and ground are for linearity and 'S' corrections respectively. The other windings (secondary) on LOT are for EHT generation, auxiliary supplies and obtaining pulse outputs needed in other sections of the receiver.

The output transistor Q_2 fully saturates when the drive pulse at its base rises to become positive. The rising current in Q_2 develops a near rectangular voltage in the primary winding J-E due to induced back e.m.f. This in turn drives a sawtooth current through the line coils (DYH1) to deflect the beam from centre to right edge of the screen. Soon after, the negative going spike of drive voltage rises to cut-off Q_2. The sudden fall of large current in the primary winding induces high back e.m.f. voltage (typically 1 KV) and lot of energy gets stored in the magnetic field around primary and deflection windings. With Q_2 now open-circuited, the combined inductance of these windings in parallel with C_5 and interwinding capacitances sets into high frequency oscillations to dissipate the stored energy. The capacitor C_5 is so chosen that the half-period (T/2) of such oscillations is equal to the desired retrace period-typically 8μs. The first half cycle of oscillatory current deflects the beam to the left edge of screen thus providing a fast retrace. However, after the first half cycle, when voltage polarity reverses, damper diode that is embedded in BU208D gets forward biased. This enables current flow in opposite direction and the beam returns to the centre. The stored energy is also by now dissipated in the circuit resistance and hence further oscillations do not occur. As soon as the beam approaches centre of the screen, the drive signal reverses polarity to switch 'on' Q_2 and the consequent current flow continues deflection of beam towards right edge of the raster. This sequence repeats and a full raster is obtained with simultaneous deflection of the beam in vertical direction.

The capacitor C_6 is grounded if width of the raster is found to be less. The connecting of C_6 in parallel with C_5 lowers resonance frequency of self-oscillations. A lower frequency means larger retrace period. This in turn reduces magnitude of EHT because the rate of change of flux ($d\phi/dt$), which causes it, decreases. Any reduction in the picture tube anode-voltage lowers its beam velocity. Thus the beam remains for a longer period under the influence of deflection field and more lateral deflection occurs. As is obvious, the value of C_6 is chosen for optimum deflection across the screen.

EHT and Auxiliary Outputs

The fast retrace induces high voltage pulses in the secondary windings of TR_2 *i.e.*, line output transformer. The flyback pulses available at pin 'E' of TR_2 are rectified to obtain dc voltage for the screen grid (G_2). A suitable potential divider is used to adjust it for best results. Similarly, voltage induced across points B and C is rectified to obtain 25V DC for the vertical output stage. Voltage at pin D is applied through R_{13} to G_1 (control grid) of the picture tube. The AC voltage between pins C and K of L.O.T., is the heater voltage for the picture tube. It is fed through current limiting resistors as shown in the figure.

EHT at 25 KV is developed across the diode-split secondary winding and is applied directly to the final anode of C.R.T. High voltage (≈ 7 KV) for the focus grid (G_3) is also derived from one section of the diode-split winding and applied through a potential divider arrangement located at the circuit board of picture tube. The ground end of secondary winding is used to sense average beam current for beam current limiting. This is done by connecting point H of EHT winding to ground through C_{15}. The voltage across this capacitor that varies with beam current is fedback to relevant circuits to obtain average brightness limiting (ABC) and vertical tracking functions.

16.8 COMMON FAULTS IN LINE OUTPUT CIRCUITS

No raster. If the observed fault is 'no raster' the probable causes are:- (*i*) no supply voltage, (*ii*) horizontal driver not working and (*iii*) defective LOT.

For trouble shooting proceed as under:

(*a*) Check supply voltage to the stage and also the fuse in series with it if provided.

(*b*) Measure voltage at the collector of driver transistor and if there is no obvious reason for very high or very low voltage check the transistor.

(*c*) Check drive waveform at the base of output transistor and compare it with the manufacture's data if available. In case drive waveform is unusual or its amplitude is quite less, locate and rectify fault in the coupling transformer and associated components.

(*d*) Check if the flyback or width capacitor at the collector of output transistor is open or short circuited.

Horizontal linearity poor. Assuming circuit functioning is normal, poor linearity (foldover) could be due to disturbed linearity coil. Adjust it to obtain linear deflection.

Less width. It would normally be due to reduced dc supply to the stage. Check supply voltage and if necessary restore it to normal value.

Excessive width. It could be due to excessive dc supply voltage. Check and set it at the specified value.

Excessive blooming. This is caused by poor EHT regulation. Check L.O.T and in particular the diode-split winding and associated circuitry. Excessive blooming will also occur if the picture tube becomes weak. Check picture tube emission by replacing it with a good tube.

CHAPTER 16

REVIEW QUESTIONS

1. Draw block schematic of the line scan generator circuit and label it fully. Also enumerate functions that the line scan stage of a colour TV receiver is expected to perform.
2. Draw basic circuit of the line driver stage and fully explain the need of various components associated with it. Why is it necessary to delay the fall of current in the secondary of coupling transformer?
3. With the help of a simple circuit diagram explain how sharp retrace is obtained by what is known as 'reaction scanning'. Enumerate sequence of operations that go into the scanning of beam across the screen width of a picture tube.
4. A typical line output circuit is shown in Fig. 16.7. Explain how the circuit functions to develop both high and low voltages. In particular explain the diode-split technique of developing EHT. Why are very high value resistors used in the potentiometer employed for feeding the focus grid?
5. Explain with a simple circuit how East-West modulation is affected to correct a particular type of pin-cushion distortion. Why corresponding North-South correction is not necessary in modern P.I.L. type of picture tubes?
6. Dedicated ICs have been developed to produce E-W modulation necessary for correcting pin-cushion distortion. One such IC is TDA 4950. Draw its external circuitry and briefly explain its operation.
7. A complete line scan circuit is shown in Fig. 16.11. Explain briefly how the stage operates to cause horizontal scanning, EHT generation and to provide other auxiliary dc voltages.
8. Enumerate common faults that can occur in the line output stage and describe how these can be traced and rectified.

17

SWITCHED MODE POWER SUPPLY (SMPS)

INTRODUCTION

The dc voltages required for various sections of a TV receiver range in magnitude from above 12V to 160V. Earlier colour receivers employed series regulators to obtain stable dc sources. Such a regulator is a dc to dc converter with provision to keep output voltage constant despite reasonable mains voltage fluctuations and load current variations. Later SCR controlled rectifier and filter circuits became popular. While substantial improvements were made in their design by including such combinations as full wave rectification with transformers designed to ensure proper isolation from mains supply but nevertheless it remained true that the relatively high power dissipation and large physical size of the isolating transformer made it ultimately incompatible with other circuits in the compact modern colour receiver and something better had to be found.

This lead to the development of 'SWITCHED MODE POWER SUPPLY' commonly abbreviated to SMPS which is now used in all modern television receivers. Its main advantages are (*i*) small physical size, (*ii*) greater efficiency because of lesser heat dissipation (*iii*) protection against excessive output voltage by quick-acting guard circuits, (*iv*) isolation from mains supply without the need for a large 50 Hz transformer, (*v*) reduced harmonic feedback into the mains supply and (*vi*) ease of simultaneous generations of both low and intermediate voltage supplies.

The initial sections of this chapter are devoted to basic principle and associated problems of switched mode power supplies. Later sections describe typical SMPS circuits as used in various colour receivers. In the end a typical series regulator is described because a large number of earlier receivers that are still in use employ such a power supply.

17.1 ESSENTIALS OF SMPS

In any SMPS power supply, mains voltage is first rectified and filtered by high voltage rectifiers and capacitors. The unregulated dc voltage thus obtained is chopped at high frequency by switching transistor(s) operated by a control circuit. The chopping frequency is chosen to be 15.625 KHz for supplies to be used in television receivers for reason to be explained later. The chopped dc is applied to the primary of a transformer and voltage induced at its secondary is rectified using fast recovery power diodes and smoothed with a capacitor or inductor-capacitor filter. A fraction of dc output voltage is fed back via a sensing amplifier to the control circuit which adjusts duty cycle of the switching transistor(s) to keep the output voltage constant. The transformer used is very small in size as it works at a very high frequency. Its turns ratio is chosen to obtain desired output voltage. Several windings on the same transformer enable generation of different dc voltage sources simultaneously.

The duty cycle or factor (δ) of the switching device is that fraction of the total period between successive switching operations for which it remains switched 'on'. For example, if the switching rate is 1000 times per second and the device is 'on' for 0.5 ms *i.e.*, half the total period, the duty factor is 0.5 or 50%. Now suppose the 'on' period is reduced to be equal to one quarter of the total period, the duty factor will then be 25% and energy transfer will be reduced to one-half of the previous value. It is thus possible to monitor the duty cycle by what is known as pulse width modulation and vary energy transfer from the unregulated source to keep output voltage constant despite load current changes and input voltage variations.

SMPS Types

Depending on requirements the SMPS can be of the (1) flyback, (2) forward or (3) push-pull type.

Flybacks SMPS

The essential circuit of a flyback type SMPS is shown in Fig.17.1(*a*) to explain how dc output is controlled by means of duty cycle switching. An unstabilized dc input voltage (v_i) is taken from the mains supply through a simple diode rectifier system and connected across the series circuit formed by primary winding (N_P) of transformer T_1 and switching transistor Q_1. The transistor is switched in such a way that when 'on', it is saturated to provide a short and on turning 'off' it becomes an open circuited switch. The circuit is called flyback converter because output power is delivered from it during a period when Q_1 is switched off.

The conduction of Q_1 is controlled by varying width of pulses fed at its base from a pulse width modulator (PWM). This is a voltage-dependent monostable circuit, triggered by pulses supplied to it from the oscillator. The oscillator produces a continuous stream of rectangular pulses of constant amplitude and duration whose frequency is synchronised with that of the horizontal oscillator (15.625 KHz) in the line output circuit. Thus the pulses produced by PWM are also synchronised with the line oscillator. However, the duration of each pulse is determined by the magnitude and direction of control voltage produced by the 'error' amplifier which is also applied to the PWM. As the duration of pulses produced by PWM varies so does the conduction time of Q_1 and hence the power transferred from unregulated dc source changes to be in step with load current demands.

When Q_1 is switched on, the bottom of Np gets connected to ground through the transistor and full input voltage (v_i) causes a linearly rising current in the primary winding of T_1. Because of transformer action a voltage is induced in Ns, the polarity of which is opposite to that of v_i appearing across Np. The diode D_1 is therefore, reverse biased during conduction of Q_1 and no current flows in Ns. Before the rising current in Np reaches its maximum value, Q_1 is cut-off by reversal of base drive from the pulse width modulator. Thus Np is suddenly made open circuit and energy stored in the magnetic field around

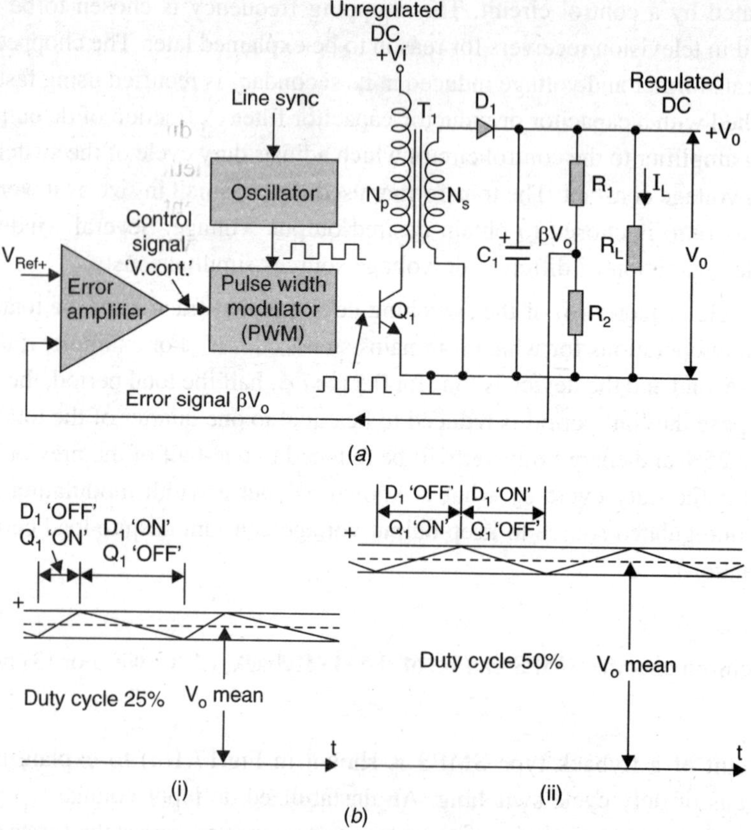

Fig. 17.1 Essential circuit of a flyback switched mode power supply (SMPS) (*a*) circuit (*b*) associated waveforms.

it is trapped. The field at once starts to collapse and its polarity reverses. The collapsing field again induces a voltage in the secondary but now with opposie polarity. This forward biases diode D_1 which conducts to charge capacitor C_1 and also to supply load current (I_L). The value of filter capacitor is so chosen that it can continue feeding current to load even when Q_1 is off and D_1 reverse biased. During intervals when Q_1 is 'cut-off' and D_1 is forward-biased, the charge on C_1 is replenished and output voltage restored to its original value. In this way it is maintained at near constant level and a continous current flows in the external load. The repeated charge and dischrage of C_1 causes a small ripple voltage to be superimposed on the average value of V_o. The larger the value of C_1, smaller is the magnitude of ripple voltage for a given load current. Aslo, for a given value of C_1, the lower the external load current, the smaller is the ripple voltage. The frequency of ripple voltage is same as that of the switching waveform used to control conduction of Q_1.

Variation of V_o by Duty Cycle Switching

The magnitude of output voltage produced by a given input voltage (v_i) is mainly determined by the turns ratio (Np/Ns) of the transformer T_1. With 220V mains supply, the ratio is normally made less than unity for obtaining dc output (v_o) of about 150V. Still lower output voltages can be obtained by putting additional secondary windings on the transformer and providing more rectifier and filter circuits.

Variation of V_o is achieved by changing duty cycle of the fixed frequency switching waveform used to control conduction of Q_1. This is illustrated in Fig. 17.1(*b*) where in (*i*) the output is with a duty cycle of 25% and in (*ii*) it is with duty cycle raised to 50%. As expected, average dc output (V_o) is higher in (*ii*) than in (*i*). This is understandable because at reduced duty cycle, collector current in Q_1 flows for smaller periods and much less energy is stored in the magnetic field around Np and hence less V_o becomes available when D_1 conducts to charge C_1 and pass current in the load circuit. In a typical SMPS the duty cycle ranges from 0.4 (40%) to 0.6 (60%). The output voltage is given by the expression:-

$$V_o = \frac{V_i}{n}\left(\frac{\delta}{1-\delta}\right)$$

where v_i is the dc input voltage, '*n*' the turns ratio of T_1 and δ the duty cycle factor.

Stabilization of V_o

Automatic stabilization of output voltage produced by SMPS is essential because of expected load current variations. For this, V_o is continuously sampled and compared with a stable reference source 'V_{ref}'. As shown in Fig. 17.1(*a*) the resistors R_1 and R_2 are connected to form a potential divider across V_o causing a fraction (βV_o) to be developed across R_2. This fraction, called the error signal, is fed to one input of the error amplifier. This is a differential amplifier where other input is the fixed reference voltage (v_{ref}). The output of this amplifier is a dc voltage, which forms the control signal '$v_{control}$'. The magnitude of this signal is proportional to difference in magnitude of the two voltages present at the input terminals multiplied by the gain of this amplifier. When error signal is larger than V_{ref}, $V_{control}$ moves in a negative direction *i.e.,* becomes less positive. When βV_o is smaller than V_{ref}, the control signal becomes more positive. It is this increase or decrease in the magnitude of $V_{control}$ which controls width of the pulse produced by the pulse width modulator. The PWM in turn, controls duty cycle of Q_1 and therefore, the magnitude of V_o. Thus continuous monitoring of V_o enables to maintain a fixed dc output voltage despite wide variations in the ac mains voltage and dc load current changes.

Forward SMPS. A simplified circuit of a forward converter (SMPS) is shown in Fig. 17.2(*a*). When Q_1 is 'on' V_i is applied to the primary of T_1. The secondary is so wound that when current is rising in Np , the voltage induced in Ns has positive polarity at its upper end as indicated by dots. Thus D_1 conducts during 'on' period of Q_1 to store energy in L_o and C_o and also to feed load current. Diode D_2 stays reverse biased during this interval. As soon as Q_1 turns 'off', the decreasing current in Np induces a voltage across Ns that is of opposite polarity. Diode D_1 is thus reverse biased and no energy is transferred to the load circuit. However, energy stored in L_o and C_o continues to supply I_L which completes its circuit via D_2 which is now forward biased. The pertinent point to note is that in a forward converter, D_1 conducts to store energy when Q_1 is 'on' unlike flyback SMPS where D_1 gets forward biased when Q_1 is turned-off.

Push-pull SMPS

Figure 17.2(*b*) is an oversimplified circuit of a push-pull converter where switches S_1 and S_2 represent two transistors that are necessary for transfer of energy from unregulated source V_i to the load circuit. The two switching transistors (S_1 and S_2) conduct (close) alternately by opposite polarity pulses from the pulse width modulator. Thus D_1 and D_2 conduct alternatey as in a conventional full-wave rectifier circuit. The choke L_o and filter capacitor C_o store energy to maintain steady voltage across the load. The feedback circuit operates in the same manner to actuate the PWM in such a way that conduction periods of switching transistors vary in accordance with variations in load current thus maintaining output voltage constant at V_o.

Relative Merits

Each of the above three methods of dc to dc convertion has a merit that is suitable for a particular type of load. Flyback SMPS is the cheapest but its regulation is poor and output ripple greater. As such, it is not used when the required output power exceeds 100 watts. Butt, where multiple outputs are required with total load power not exceeding 100 watts, this is the obvious choice as it does not require any output choke and this means reduction in cost. Also where relatively higher dc voltages are required, flyback SMPS is more suited because in other types, large inductors would be necessary for high voltage regulation. Thus flyback SMPS is more suited for use in TV receivers where power requirement is less than 100 Watts and more than one dc output voltages are needed to feed different sections of the receiver.

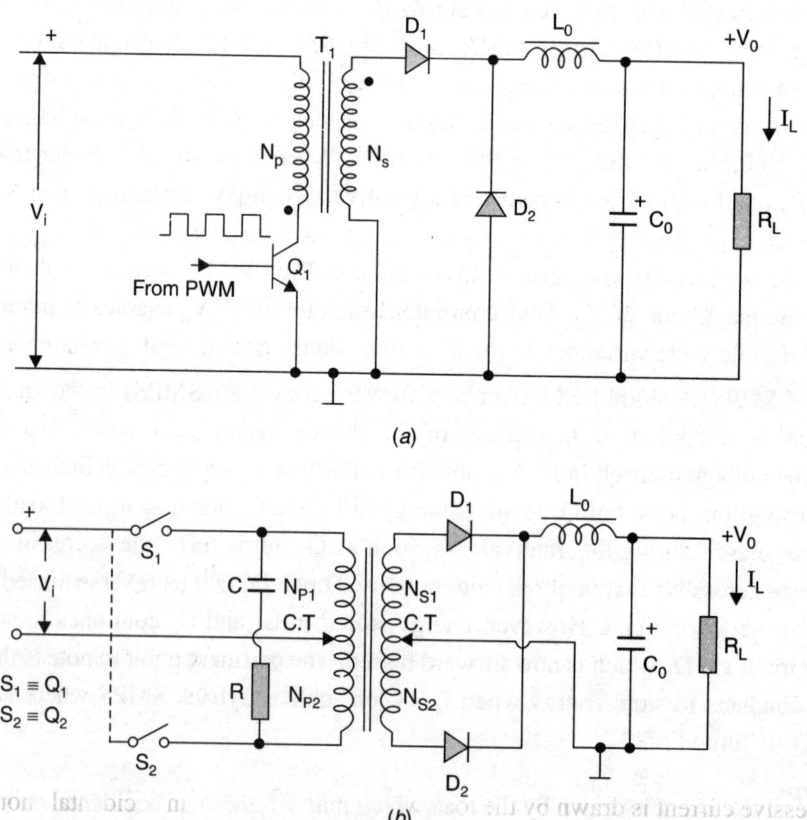

Fig. 17.2 (*a*) Simplified circuit of a "Forward" SMPS
(*b*) Oversimplified circit of a push-pull SMPS.

For higher output power requirements with tight regulation and low ripple content, push-pull SMPS is more suited. It is used when load power exceeds 200 watts. Otherwise, this converter is most complex because with two switching transistors, elaborate drive circuitry becomes necessary.

When power load lies between 100 to 200 watts at low voltage and tight regulation is necessary, forward converter is most suitable. This circuit is not very complex as compared to push-pull type and its regulaton is also good because of choke on the secondary side.

Filter Components

In any filter circuit, the reactance of an inductor is used to store energy and that of the capacitor to act as reservoir of charge and easy bypass to the ripple content. Since $X_L = 2\pi fL$, the higher the frequency the smaller the value of inductor needed in the filter circuit. Similarly since $Xc = ½\pi fC$ higher the frequency smaller is the value of capacitor needed to bypass unwanted harmonic components (ripple) and store energy more effectively. For example, at a frequency of 50 Hz and at a given load, the capacitor needed in a flyback converter is 3000 μF. But when operating frequency changes to say for example 10 KHz, the capacitor value necessary for same regulation comes down to 15 μF. Chopping frequencies used in SMPS supplies of equipments other than TV receivers are generally kept around 20 KHz and hence capacitors with still smaller values serve the same purpose. As stated earlier, flyback SMPS supplies for TV receivers have a chopping frequency of 15.625 KHz and it is synchronised with the line oscillator frequency. The object is to avoid beat frequency interference which could occur if the switching frequency was at some other value. That would need elaborate filtering and close screening of the SMPS transformer and other components.

The very steep edges of switching waveform give rise to high frequency harmonics which, if not adequately screened from the tuner and IF stages, will cause vertical bars and dots to appear on the screen. When the SMPS circuit is synchronised with line flyback, this type of interference appears as a single vertical bar, rather than as multiple bars and dots. It is thus usual to synchronise the SMPS switching frequency with that of the line flyback and to screen the converter circuitry with great care.

17.2 PROTECTION CIRCUITS FOR SMPS

To prevent damage to the switching transister in SMPS circutry, it must be protected from current surges at the instant of switch-on and also against excessive overload currents which may be either intermittent or sustained. Another type of protection is the need to avoid interference to other sections of the receiver because of strong induced voltages when a steep switching current flows in the SMPS transformer.

Automatic overload monitoring. Inrush of current at switch 'on' occurs because the large reservoir capacitor is completely uncharged. It behaves like a short circuit across the output terminals when the initial charging current flows. This current may remain abnormally high for long enough to destroy the switching transistor by feeding it more power than it can handle. Similar damage is likely to occur if a prolonged excessive current is drawn by the load which may be due to an accidental short across output terminals. The damage can also occur if overload transients occur too often.

SMPS protection circits often deal with problems of inrush current by control of the duty cycle. It is arranged that the instant of switch-on, the duty cycle shall be very small. This amounts to feeding bursts of charging current for very small duration to the filter (reservoir) capacitor. As the capaitor acquires more charge, the duty cycle gradually increases until it settles down at its final value.

The existance of either a prolonged short-circuit or of excessive load current is detected by a protection circuit, which reacts by immediately switching-off the SMPS. It does so by a simple method of inhibiting (disabling) the switching transistor. After a short delay, the protecton circuit releases the inhibit and allows the SMPS to operate once more, but with a slowly increasing duty cycle to limit inflow of large current. IF the overload is still present at the output, the SMPS is again immediately turned-off and the slow start-up cycle repeated.

There is another problem that needs attention. Assuming that the protection circuit is designed to operate from a sensing signal derived from the mains high voltage dc circuit (see Fig. 17.1). As such the load currents of other low voltage circuits, if provided, will have to rise to dangerously high levels before protection circuit could be activated to protect them. It is, therefore, usual to include protection fuses in each low-voltage circuit or to arrange for each to be individually monitored by the protection circuit. This is illustrated in Fig. 17.3. The three rectifier circuits numbered 1, 2 and 3 produce 150V, 35V, and 16.5V respectively at their output terminals. Each circuit has a current sensing resistor connected in series with its ground return point. The diodes D_1, D_2, D_3 are for circuit isolation which allow R_1, R_2 and R_3 to be connected at R_4 without mutual interference. The values of R_1, R_2 and R_3 are chosen so that each circuit produces the same voltage drop across R_4 when its load current exceeds some predetermined value. The resistor R_4 is made variable to compensate for small variations in tolerance of resistors and voltage drops across the diodes.

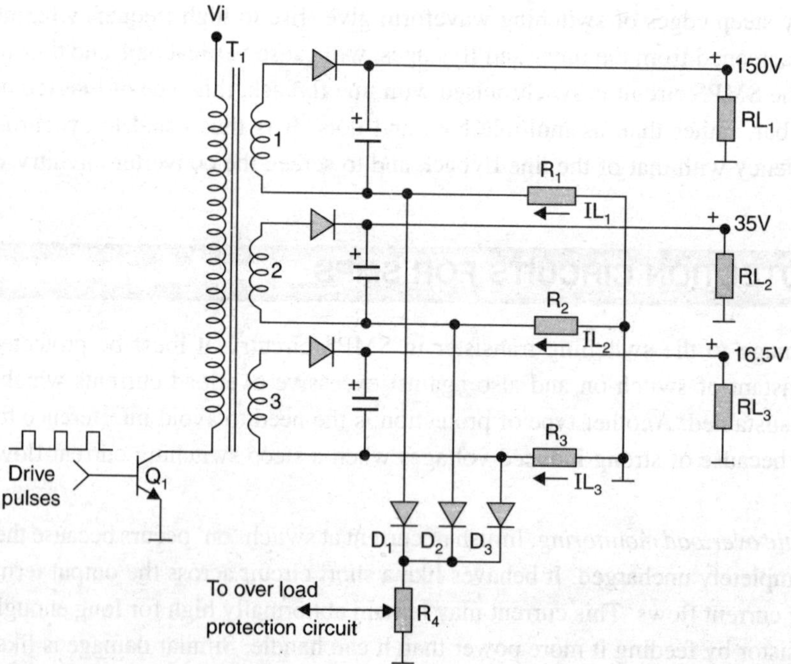

Fig. 17.3 Sensing circuits for automatic overload monitoring in a SMPS circuits.

Suppression of break-down and interfering voltages. The very act of turning 'on' and 'off' causes large current flows with a step-sided switching waveform. These can result in electrical interference and generation of back e.m.fs. large enough to destroy the switching transistor. The interference may also find its way into other parts of the receiver by way of shared earth connections, common supply voltage lines or simply by direct radiation. Careful component layout and earthing, good decoupling and proper screening will usually eliminate such interference but additional components are often necessary to eliminate large back e.m.f surge voltages. The circuit of Fig. 17.4(*a*) shows additional components necessary to suppress breakdown voltages and reducing interference to other sections of the receiver. It may be noted that the switching transistor Q_1 is now located between input voltage V_i and T_1 instead of between T_1 and earth as shown in Fig. 17.1. The advantage is that stray capacitance present in the collector circuit of Q_1 is now maintained at a fixed voltage (V_i) instead of being periodically switched between V_i and ground. This greatly reduces spikes in the current waveform which switching would otherwise produce. The capacitor C_1 connected across the primary of T_1 slows the rate of change of voltage thereby reducing spike in the induced voltage, thus reducing interference from this part of the circuit. The small choke L_1 serves to limit the charging current which flows into C_1 every time Q_1 is switched 'on'.

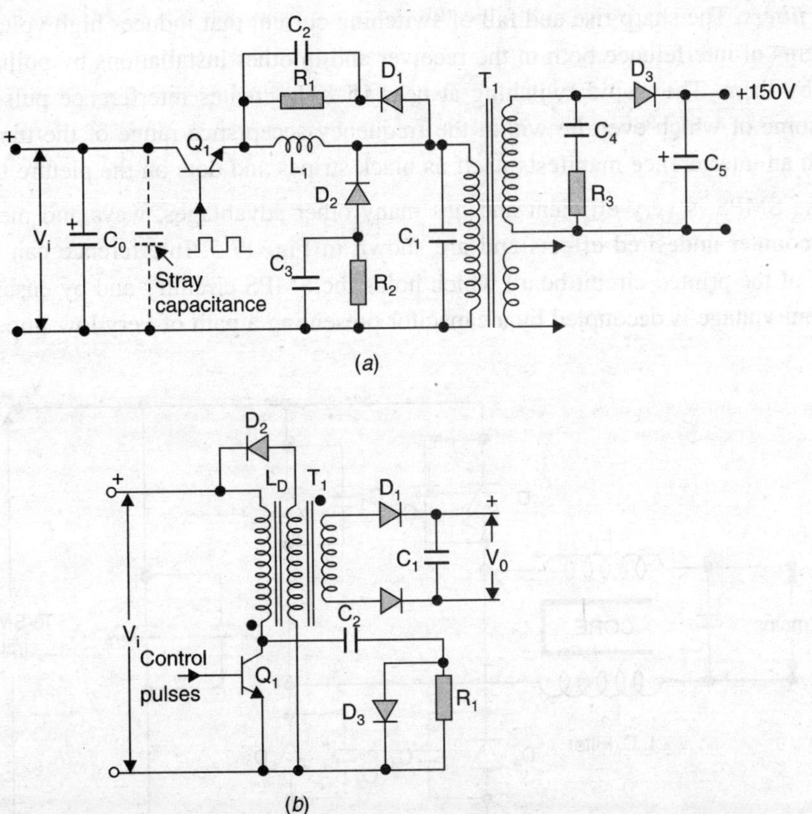

Fig. 17.4 Reduction of electrical interference in SMPS, circuitry, (*a*) with suitable networks (*b*) with demagnetising winding on the transformer core and additional network.

The additional components R_1, C_2 and D_1 connected in parallel with L_1 are to damp any oscillations that would otherwise occur in the series resonant tuned circuit formed by L_1 and C_1. A similar network consisting of R_2, C_3 and D_2 clips any overshoot of voltage caused by leakage inductance of T_1 when Q_1 is switched off. The components C_4 and R_3 connected across the 150V secondary circuit serve to damp any ringing voltages present at this point and so limit harmful reverse voltages affecting D_3.

Another circuit to suppress such undesired effect is shown in Fig. 17.4(*b*). It was used in earlier SMPS designs when transistors capable of switching large currents and having large break-down ratings were not available. To limit maximum voltage across the transistor to twice the value of V_i, a demagnetising winding L_D, having the same number of turns as on the primary is provided. This winding is wound on a bifilar core along with the primary to reduce leakage inductance between the two. In case of lesser load the excess energy which may cause high induced voltage is returned to input dc source when D_2 is forward-biased. Similarly, to keep voltage across and current through the switching transistor to safe values, the network consisting of C_2, R_1 and D_3 is provided. When Q_1 turn-off, the charge stored in the primary winding finds a low leakage path via C_2 and D_3. However, when Q_1 turns on for the next cycle, the capacitor (C_2) discharges through the transistor but the initial rush of current is limited by R_1.

Mains filters. The sharp rise and fall of switching current that induces high voltages can create several problems of interference both in the receiver and in other installations by polluting the mains supply, used by them. The rapid switching at near 15 KHz creates interference pulses having high frequencies, some of which even lie within the frequency-acceptance range of the tuner. The visible effects of such an interference manifests itself as black stripes and dots on the picture tube screen.

Because SMPS is very efficient and has many other advantages, ways and means have been found out to counter undesired effects and are shown in Fig. 17.5. Interference can be avoided by careful layout of the printed-circuit board which holds the SMPS circuitry and by ensuring that every source of output voltage is decoupled by a capacitor presenting a path of very low impedance to the

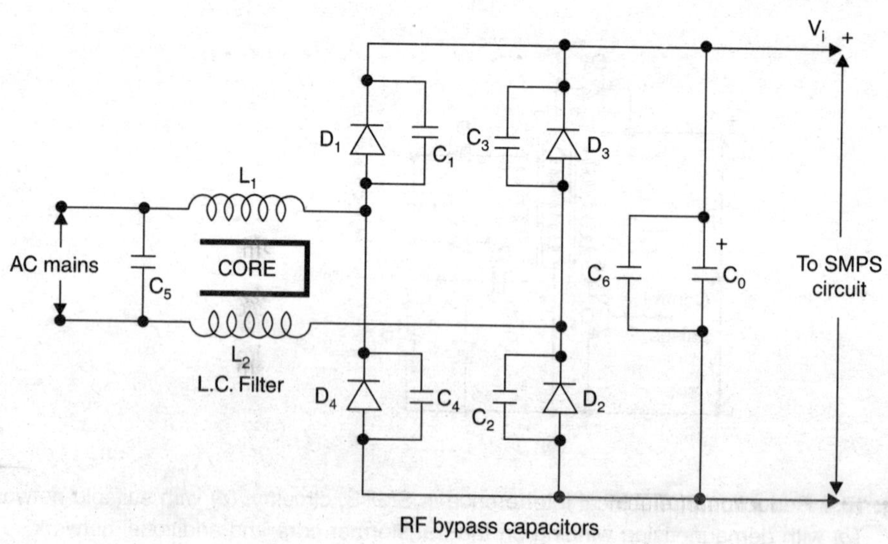

Fig. 17.5 Interference suppression methods at input to SMPS circuits.

passage of radio frequency signals. Electrolytic capacitors of large value perform well at the fundamental frequency of the SMPS switching rate but are ineffective at higher frequencies. The solution adopted is to connect a low value capacitor (C_6) in parallel with the electrolytic type thus offering low reactance, to both low and high frequency interfering components. The type of low-value capacitor used for this purpose is made of polycarbonate and has a capacitance between 0.01μF and 0.02μF. This type of capacitor is also used as an r.f. bypass capacitor in mains rectifier sections of the SMPS, often in conjunction with an L-C filter. Another technique for reducing interference, particularly pollution of the mains supply, consists of an L-C filter connected in series with the mains supply to the SMPS (see Fig. 17.5). The inductive part of the filter consists of two small inductors wound on the same core and mutually coupled in such a direction that magnetic fields produced by each cancel out.

17.3 MERITS AND DEMERITS OF SMPS

After learning the essentials of a switched mode power supply and before proceeding to discuss typical circuits it would be in order to recollect the merits and demerits of SMPS circuits.

Advantages. The main merits of switched mode power supplies are:-

(1) Smaller physical size because chopping rate is kept quite high.

(2) Greater efficiency because the switching transistor is 'on' for periods that are necessary to meet load currents. Thus there is very little heat dissipation in the regulatory transistor.

(3) Isolation from mains supply. This offers advantages like:-

(*a*) No need for a large 50 Hz mains transformer (*b*) antenna can be connected to the tuner directly (*c*) sockets for video, headphones and taperecorder can be incorporated without additional circuitry and (*d*) safety to the user.

(4) Both low and intermediate voltage supplies can be generated simultaneously.

(5) All the outputs are stabilized and protected against open circuit and short circuit conditions at the load terminals.

(6) Interference to other circuits is greatly reduced by synchronising the choping rate with line flyback pulses.

Disadvantages. Despite many advantages the SMPS has some drawbacks. These include:-

(1) The rapid switching of chopping current generates strong interfering signals.

(2) Higher harmonics or oscillatory currents due to trapped energy lie in the radio frequency range and can enter the tuner circuit to cause dot interference on the picture tube screen.

(3) Strong induced voltages get fed back to the mains thus polluting supply to other appliances in use nearby.

As explained in the previous section, the above demerits can be circumvented by elaborate quenching circuits, carefull component layout, mains filter and rigorous shielding.

17.4 SYNCHRONIZED SMPS SUBSYSTEM

The SMPS circuits found in TV receivers employ a relatively simpler method of voltage control. These are either non-synchronous or of line-synchronous type. The functioning of a non-synchronous circuit is first examined with the help of a block diagram (Fig. 17.6).

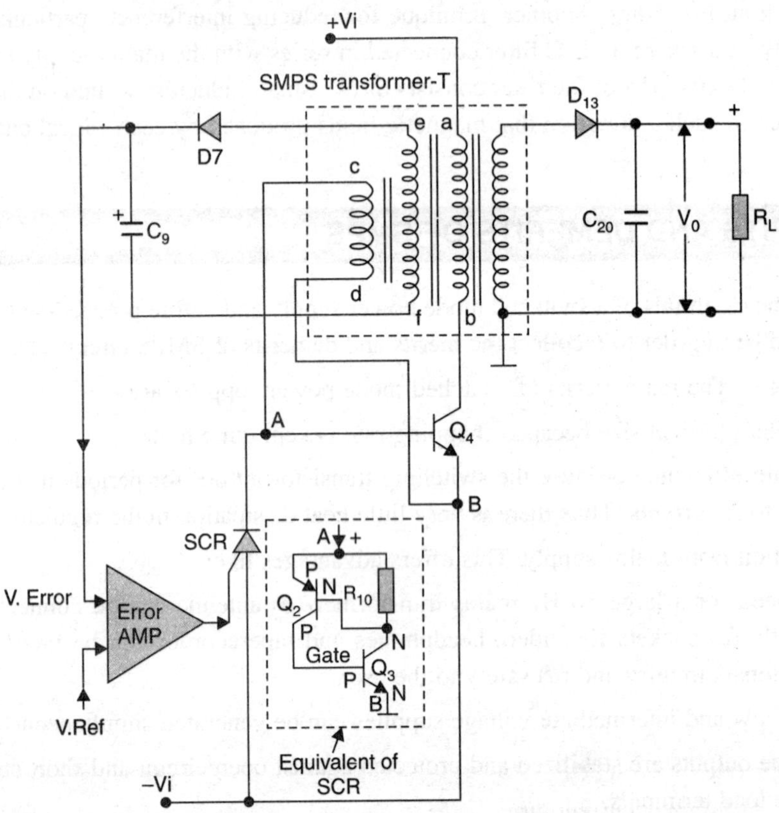

Fig. 17.6 Block diagram of a non-line synchronous SMPS subsystem.

Chopping Mechanism

Any SMPS system is essentially a dc to dc converter where energy transfer is controlled to obtain voltage regulation. The switching transistor conducts to store energy in the primary winding of associated transformer. The amount of stored energy can be expressed as $P_0 = \frac{1}{2I^2 f L}$, where I is the peak current through L the inductance of transformer winding and f the chopping frequency of the switching transistor.

Assuming no losses, $P_0 = \frac{V_0^2}{RL} = \frac{1}{2I^2 f L}$, where V_0 is the dc output voltage and R_L the load resistance.

The above expression can be rearranged to write $V_0 = \frac{1}{I}\sqrt{\frac{R_L}{2fL}}$. It is thus seen that output voltage can be stabilized by varying frequency of an inbuilt oscillator instead of doing so by the expensive method of pulse with modulation.

A blocking oscillator is most suited for this purpose. In the semiconductor version, it may be thought of as a tuned collector configuration designed to produce an extreme case of intermittent oscillations. The feedback winding is polarized to produce such a large amount of feedback. The cumulative action is almost instantaneous and the device switches off immdiately. The reverse bias that develops on account of regeneration is so large that the transistor is driven much beyond cut-off. This prevents continuous oscillations at the rate of natural resonant frequency. The cycle however, repeats when self-bias returns to its conduction region. Thus, the number of times per second the oscillator transistor saturates and then blocks-off is the pulse repetition rate or oscillator frequency. Such an oscillator is represented in the block diagram of Fig. 17.6 by the switching transistor Q_4, which controls current flow from dc source V_i through primary winding (a-b). The winding (c-d) on the SMPS transformer is for driving the blocking oscillator. The winding inductance and associated capacitances (inter-winding and stray) are so chosen that the chopping frequency is close to 15 KHz. It is controlled via another feedback path to regulate transfer of energy on the load side.

Energy transfer control. As seen in the block diagram, there is no pulse width modulator to control duty cycle of the blocking oscillator. The control mechanism is based on varying 'on' time of the oscillator and hence its frequency in a somewhat different way. The control is affected by cutting-off base drive on periodically shorting the feedback voltage. For this, the signal developed across load sensing winding (e-f) is rectified and fed to the base of error amplifier. Here again instead of using an OPAMP, a single transistor is employed as error amplifier. The switching action is obtained by using two transistors connected to simulate the action of a semicondutor controlled rectifier (SCR). The drive winding (c-d) is short-circuited when the transistors are fully turned on, the instant and duration of which is controlled by the error amplifier.

The manner in which SCR (PNPN switch) action is obtained by using two transistors is illustrated by the circuit shown in dotted-chain box of Fig. 17.6. The *p-n-p* transistor Q_2 and *n-p-n* transistor Q_3 are connected back to back to cause positive feedback. So long as gate voltage is zero, both Q_2 and Q_3 stay in cut-off and there is open circuit between points A and B which are the anode and cathode terminals respectively of SCR configuration. When a positive voltage is applied at the gate, Q_3 turns 'on' whose part collector current turns 'on' Q_2. Due to regenerative feedback action, current in the loop goes on rising till the two transistors go into saturation. The resistor R_{10} is to limit anode current to a safe value. In this state, there is a virtual short across the anode and cathode terminals (points A and B) and hence, winding (c-d) is shorted. The gate voltage looses control once the circuit truns on. However, the switch can be restored to its original state by removing or reversing polarity of the anode voltage.

Complete Circuit

With the above introduction, we can now turn to studying a complete circuit of the non-line synchronous SMPS. This is shown in Fig. 17.7. The ac mains voltage is applied through switch S_1 and protection fuse F_1. The capacitors C_2, C_3 and bifilar inductor L_1 form a 'π' filter configuration to prevent mains side interference entering the SMPS circuit and pollution of ac mains from spiked pulses generated in the switching transistor of SMPS. Similarly capacitors C_1 and C_4 are to guard against accidental application of any high voltage to the power supply circuit. The ac mains is also fed to degaussing coils through PTC thermistor (THI) for demagnetising picture tube structure every time the receiver is switched on.

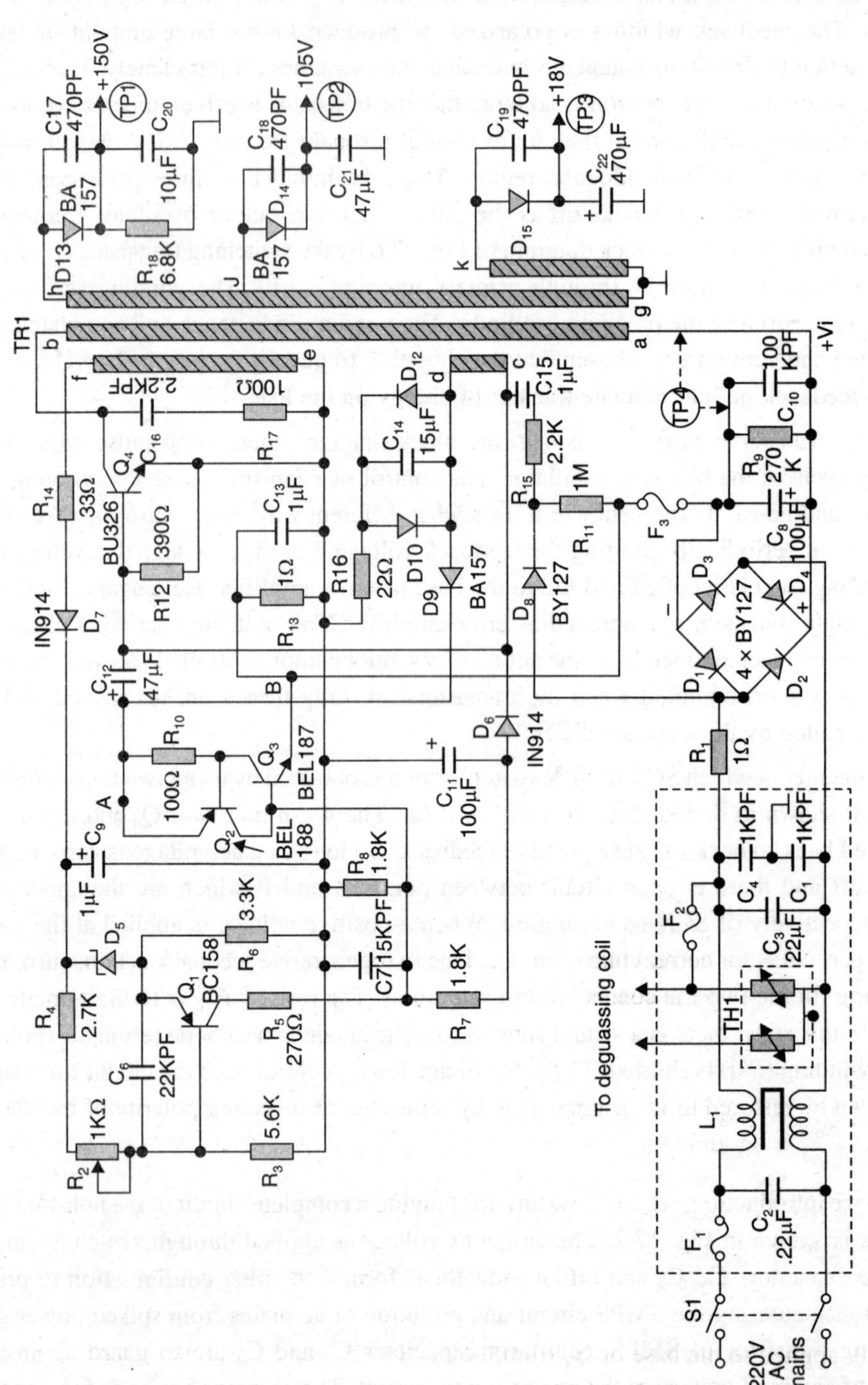

Fig. 17.7 Complete circuit of a non-line synchronous switched mode power supply.

The mains supply is rectified by a bridge circuit formed by diodes D_1 through D_4 and a dc voltage around 330V is developed across the filter capacitor C_8. The resistor R_1 limits inrush current to a safe value at the time of switching on. A discharge path for filter capacitor C_8 is provided by R_9 for safety reasons. Similarly, capacitor C_{10} is for high frequency decoupling.

Switching Action

The switching transistor Q_4 and SMPS transformer TR_1 constitute the blocking oscillator that is set to operate at a frequency close to 15 KHz. Positive feedback for sustained oscillations is obtained on rectifying voltage induced across the winding (c-d). The feedback current completes its path from point 'C' of the winding via base-emitter junction of Q_4, R_{16} and D_{10} to other end (d) of the winding. The resistor R_{12} across the base and emitter is to avoid excessive current through the junction. Similarly, R_{16} is to limit base current to a safe maximum value.

The transistor Q_4 passes a rising current through winding (a-b) from V_i, during its 'on' periods. The current completes its path from V_i positive to its negative end via drive winding (a-b), collector to emitter of Q_4, R_{13} and fuse F_3. The resistor R_{13} produces a small negative feedback to suppress collector current peaks and C_{13} across it is for ac bypass. Similarly, series network (C_{16}, R_{17}) is to quench any high voltage peaks across the switching transistor. The energy stored in the primary winding (a-b), when Q_4 conducts, is transferred to load circuits through secondary windings (g-h) and (g-k). The transfer occurs during 'off' periods of Q_4 when polarity of induced voltages is positive at upper ends of these windings.

Control Action

Regulation is obtained by controlling 'on' period of Q_4 by removing base drive to it for varying intervals. This is done by closing the PNPN switch formed by transistors Q_2 and Q_3 with signals from the load sensing winding (e-f). Normally the gate of this switch (base of Q_3) is held at a negative potential with respect to its cathode (emitter of Q_3) by a dc voltage derived by rectifying and filtering induced voltage across winding (c-d). The rectified current completes its path from upper end 'd' via D_{12}, R_{16}, R_8, R_7, and D_6 to end 'c' of the winding. The capacitor C_7 filters out ac component from voltage drop cross R_8. The dc voltage drop across R_8 applies a steady negative bias to Q_3 since the positive end of this voltage is connected to its emitter via R_{13} and negative directly at the base.

The control (error) signal is obtained from the load sensing winding (e-f) by rectifying it with diode D_7 and filtering it with capacitor C_9. The dc current completes its path via R_{14}, D_7, R_4, R_2 and R_3. The positive voltage developed across R_3 is applied at the base of error amplifier Q_1 (BC158). The zener D_5 passes current through R_6 to develop a fixed voltage across it which charges C_6 to this value. It then serves as reference voltage to the error amplifier. The base to emitter (forward) bias for the p-n-p transistor Q_1 is set by varying R_2 to such a value that it passes only a small collector current under normal output load conditions. The ensuing collector current that completes its path via R_5 and R_8 develops only a small positive voltage across R_8 and net voltage at the base of Q_3 (gate of the SCR switch) continues to be negative. The two transistors, Q_2 and Q_3 remain open circuited and feedback is continuously applied to the base of Q_4 from the drive (feedback winding (c-d)).

During fluctuations in load or line supply voltage, the error voltage changes and this affects base drive to the error amplifier Q_1. With excess energy in the transformer windings, the error signal increases and more voltage is developed across R_6 making the emitter of Q_1 more positive. This forward biases it and increased collector current develops enough voltage drop across R_8 to make the base of Q_3

positive. This triggers the PNPN switch to provide a short across point A and B. The capacitor C_{12} which was earlier charged by diodes D_9 to the peak voltage across winding (c-d) now discharges to pass current through Q_2 and Q_3. The sudden discharges of C_{12} causes two things. The voltage at point 'A' drops to near zero and the PNPN switch is restored to its original state. With near short across C_{12} the feedback winding is almost short-circuited via R_{13}, R_{16} and D_{10} thereby removing base drive to Q_4. This reduces 'on' time of the switching transistor to counteract excess energy stored in the windings. Such a situation can continue for several oscillator cycles to balance energy transfer from primary to secondary. The error voltage continuously monitors output voltage to provide excellent voltage regulation on the load side.

Start of Switching Action

On turning on the receiver, switching action of Q_4 starts with 50 Hz pulses generated by diode D_8 and R-C network R_{15} and C_{15}. As shown in the circuit, these get directly applied at the base of switching transistor. Initially under open circuit and low load conditions the power supply in on intermittent 50 Hz operation. While frequency of switching gradually picks up with feedback from drive winding (c-d) it is interrupted because the 'on' period is reduced with thyristor (SCR) switch remaining closed for longer periods. With normal load high frequency operation takes over and continued monitoring from the error sensing winding (e-f) continues.

Output Voltage

Intermediate dc voltages at 150V and 105V are obtained by rectifying output pulses obtained across secondary winding (g-h) and tapping 'j' on it. Resistor R_{18} is connected across the output at 150V to prevent generation of very high output voltage under open circuit conditions. Similarly low dc voltage at 18V is obtained from winding (g-k). By a proper design of the SMPS transformer and a careful choice of the ratio of R_7 and R_8 that decides initiation of firing of the SCR switch, stored energy in the transformer and hence current from the output diodes can be maintained within permissible limits.

Prevention of Interference

As already explained, input circuit to the ac mains is suitably designed to filter out high harmonic components and prevent passage of interfering pulses either way. To quench high voltage pulses that may damage Q_4 a network consisting of C_{16} and R_{17} is connected between its collector and emitter. This also prevents any high frequency oscillations by dissipating excess energy in R_{17}. Since the SMPS frequency is not synchronised with the line oscillator, careful shielding of power supply components is necessary.

Specifications

The power supply explained above is designed for a 51 cm colour receiver with an eye on cost, efficiency, safety and weight. Its performance characteristics are as under:-

(*i*) Input ac voltage	140 to 250V. AC
(*ii*) Output voltage	150V/40mA, 105V/40mA and 18V/ 800 mA
(*iii*) Line regulation (from 150V-250V AC)	0.3%
(*iv*) Load regulation (25W to 50 W)	2%

Servicing

In the circuit of Fig. 17.7 test point voltages with input equal to 230V ac are as under

TP_1 = 150V dc	TP_3 = 18V dc
TP_2 = 105 V dc	TP_4 = 300 V dc

(1) *For no dc output voltage check:*

(*i*) ac input and line fuse (*ii*) voltage at TP4 (*iii*) Oscillator operation, if not operating check bridge rectifier diodes D_1-D_4, R_{15}, D_8 and collector voltage of Q_4.

Very Low Output

It could be that the SMPS is continuing to operate at 50 Hz as happens at the instant of switching 'on' the receiver. For this condition check (*i*) all outputs for any short, (*ii*) dicde D_{12} (*iii*) resistor R_{13} and capacitor C_{13} (*iv*) transistors Q_2 and Q_3 and finally (*v*) adjust R_2 to get proper voltages.

17.5 LINE SYNCHRONISED SMPS

Figure 17.8 is the circuit of a line synchronised switched mode power supply where Q_{11} is the SM regulator, Q_{13} the driver to output stage and Q_{15} the switching (output) transistor. The transistor Q_{12} is an electronic fuse which operates only during overload conditions. The circuit also incorporates a safety fuse FIA and a safety fuse diode D_{58}. Winding (1-2) is on the LOT and forms part of the feedback circuit to the blocking oscillator. It serves to synchronise the oscillations frequency with that of the line oscillator operating at 15.625 KHz.

As shown in the circuit, unregulated dc (Vi) is obtained in the usual manner by a bridge rectifier circuit. As necessary, a mains input filter is provided to prevent interfering signals passing either way. There is also provision for feeding ac to the degaussing coil. The secondary winding on the SMPS transformer is optimised to serve both as the primary and output winding. As such the negative of V_i supply is kept floating and hence not grounded to sub-chassis.

Blocking Oscillator

The switching transistor Q_{15} receives dc supply from point 'A' of the unregulated dc via FIA, load across 115V regulated dc output, chassis ground 'G' and secondary winding (c-a) on SMPS transformer. The collector current completes its path from emitter to floating negative of V_i (B) through a small 2.2 ohm resistor R_{24}.

Feedback for sustained oscillations is obtained from winding (d-e) and fed at the base of Q_{15} through winding (1-2) on the LOT, R_{22} capacitor C_{14} and resistor R_{23}. The capacitor C_{14} charges and discharges to develop self-bias. The coil L_{11} in the base lead suppresses any sharp peak in the voltage that is fed back. The voltage across winding (1-2) is at the line scan frequency and serves to synchronise the oscillator with the 15.625 KHz frequency of line oscillator. This makes rigorous shielding unnecessary and a compact chassis assembly becomes possible.

On switching 'on' the receiver, starting pulses to the blocking oscillator are fed from 50 Hz ac mains at the base of Q_{15} via C_{53}, R_{54} and R_{23}. As the oscillatory action picks up, blocking rate rises to near 15.625 KHz and is soon synchronised with the line output circuit frequency with feedback from winding (1-2). Once the oscillator attians its normal frequency of 15625 Hz, starting pulses at 50 Hz become ineffective and hence of no consequence.

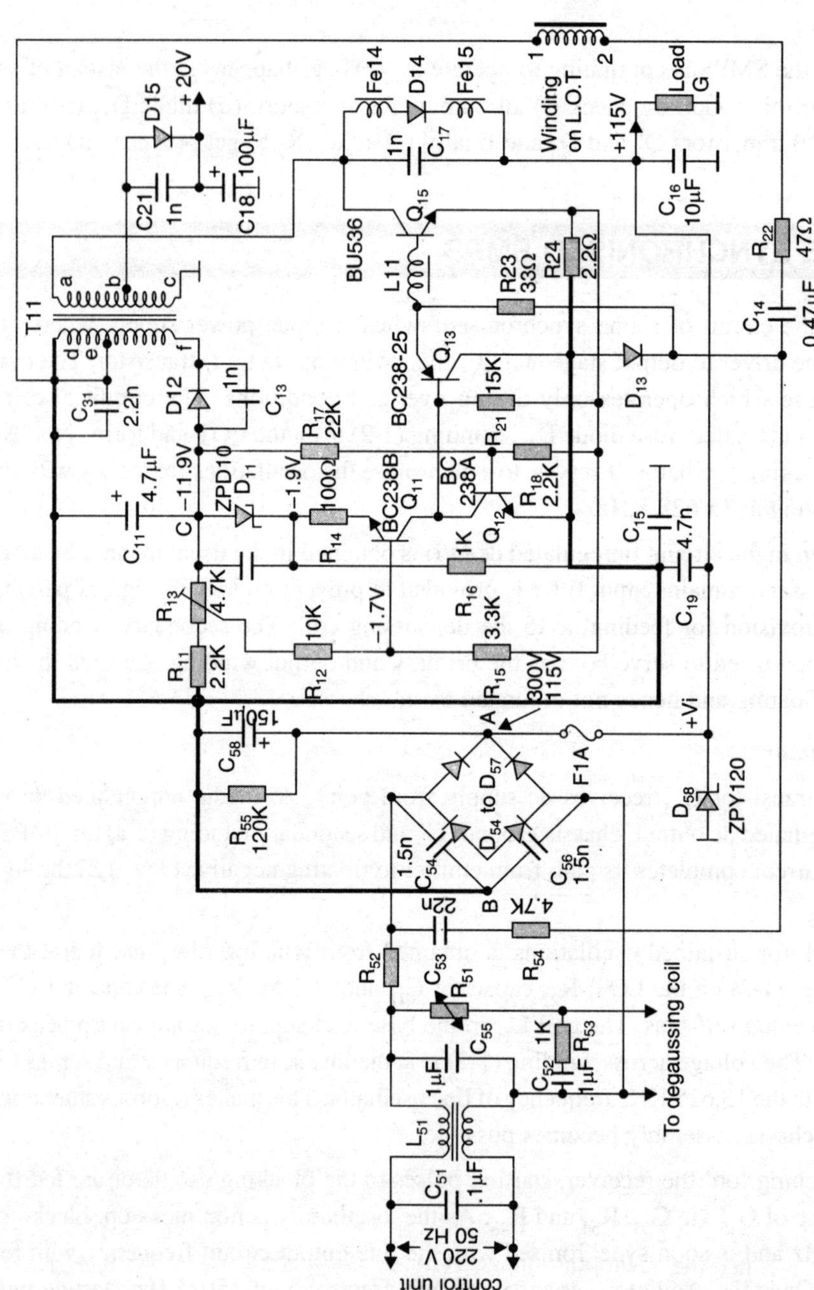

Fig. 17.8 Circuit diagram of the line-synchronised switched mode power supply.

Output DC Sources

As stated earlier the same winding (a-c) carries input and output currents. The diode D_{15} and capacitor C_{18} provide an independent 20V source from tapping 'b' and with respect to ground. Similarly diode D_{14} and capacitor C_{16} generate 115V dc with respect to chassis ground from full winding *i.e.,* (a-c). The capacitor C_{21} and C_{17} are for r.f. bypass and small coils Fe_{14} and Fe_{15} suppress any transients which may otherwise damage diode D_{14}.

As mentioned earlier, positive of supply Vi (point A) and positive of 115V dc are tied together. However, while dc current through load from 115V source completes its path at chassis ground, the switching current from Vi positive flows via load circuit of 115V dc, chassis ground (G), winding (c-a) transistor Q_{15} and resistor R_{24} back to point 'B' on the floating negative of the unregulated dc source. Since the two negative ends are insulated from each other, a common path through the load circuit does not effect the performance of devices fed from 115V dc. The ripple content of V_i that is at a frequency of 15.625 KHz is easily bypassed by associated filter and decoupling capacitors. A dc voltmeter will measure near 300V dc between point A and B while it will measure 115V from point A to chassis ground. In fact it is the common current path through dc load that makes it necessary to keep the negative lead of V_i floating.

Duty Cycle Control

The driver transistor Q_{13} is made to conduct heavily by pulses fed at its base from regulator transistor Q_{11}. Since emitter and collector of Q_{13} are in shunt with base-to-emitter circuit of Q_{15}, each time Q_{13} conducts, the base drive to Q_{15} is bypassed. This happens once during each cycle of oscillation. The duty cycle is controlled by varying time for which Q_{15} is kept turned off. The circuit to do so operates in the following manner.

The voltage induced across the load sensing winding (d-f) of the SMPS transformer is rectified by diode D_{12} to obtain a proportionate dc voltage across filter capacitor C_{11}. The capacitor C_{31} prevents any dc short across winding (d-e). As labelled on the circuit, the dc voltage thus obtained is typically 11.9V and point 'C' can then be said to be at – 11.9V potential in a relative sense. The positive end, as shown, is tied up with the floating negative of V_i. From this low voltage dc source, the emitter of Q_{11} is maintained at a fixed potential of – 1.9V by potential divider formed by the 10V Zener D_{11} and resistor R_{16}. Similarly a potential divider connects – 1.7V at the base of this transistor. Thus Q_{11} attains forward bias of 0.2V. In fact, as explained later R_{13} is varied to obtain correct dc output voltages. It may also be noted that collector of regulator transistor Q_{11} is connected to positive end of 11.9V dc through load resistor R_{21}, a small 2.2 ohm resistor (R_{24}) located in the emitter circuit of Q_{15} and floating negative line of V_i supply. The capacitors C_{15} and C_{19} are for r.f. bypass purposes.

Every cycle when Q_{15} is switched on, the resulting current develops a sawtooth voltage across emitter resistor R_{24}. This voltage is fed to thc base of Q_{11} through R_{15}. Thus net forward bias of this transistor depends on the voltage induced across winding (d-f) and the magnitude of switching current through Q_{15}. Since the sawtooth voltage is positive going, it switches 'on' Q_{11} at a particular instant depending on the magnitudes of two feedback signals. The resulting collector current lowers potential at the base of Q_{13} enabling it to conduct when positive going feedback pulse appears at its emitter from windings (e-d) and (1-2). The shunting effect as explained earlier deprives Q_{15} of its base-drive and it is turned-off. The period for which it tuns-off during each cycle depends on how long Q_{11} is kept in condution. This then amounts to pulse width modulation of the 15.625 KHz waveform that is fed at the base of Q_{15}.

Any increase in current through Q_{15} due to increase in mains voltage and thus in V_i will develop a higher voltage drop across R_{24}. The regulator transistor BC238B will then conduct for a longer period. This will curtail the 'on' period of Q_{15} because Q_{13} will keep its base emitter junction shorted for a longer period. The duty cycle will thus decrease to keep the output voltage stabilized.

With fall in ac mains voltage the switching current will decrease. Analogous arguments will lead to the conclusion that Q_{15} will now conduct for a longer period to maintain supply of energy to the load circuits. Similarly with variations in load current, more or less feedback voltage will be induced in winding (d-f) and dc voltage supply to Q_{11} will vary accordingly. This will also have the same effect of controlling the duty cycle to maintain output voltages constant despite load current variations. The preset R_{13} is initially set to obtain desired output voltages under normal load conditions. Thus it is seen that both mains supply fluctuations and load current variations are monitored continuously to regulate dc output voltages. Since SMPS supplies electric power to nearly all sections of the receiver, a trouble free operation is ensured despite wide variations in the mains supply voltage.

Electronic Fuse

The transistor Q_{12} is so biased by R_{17} and R_{18} from the 11.9 dc source that it stays at cut-off during normal operation of the power supply. However, when an overload occurs, the magnitude of sawtooth voltage across R_{24} increases. This as shown in Fig. 17.8 is also fed at the base of Q_{12}. The net bias is now enough to turn it on. It conducts to lower potential at the base of driver transistor Q_{13} that is enough to keep it 'on' for longer periods during each cycle of switching. The duty cycle is thus lowered to such a value that all operating voltages are decreased. This distrubs receiver operation and it remains inoperative as long as overload condition exists. Once the fault is localized and removed; normal operation is restored with BC238A (Q12) returning back to its cut-off state.

Safety diode. Another protective measure that is provided in this power supply is the safety zener diode D58 (ZPY 120). It is connected between point A and ground and thus has 115V potential across it. Any abnormal increase in this voltage due to faulty regulation will cause more current through D58 and destroy it. This in turn causes more current to flow through fuse FIA which burns to interrupt current flow and thus prevents damage to any receiver circuit.

Trouble shooting. Since negative of the unregulated dc source is kept floating, it should never be allowed to come in contact with chassis ground or ac mains leads. It is thus advisible to use a 1:1 turns ratio isolating transformer between ac mains and receiver input while servicing the power supply or for that matter any sections of the receiver. Typical dc voltage values are labelled at various points on the circuit diagram and should be referred to while servicing the power supply.

(1) If there is no picture or sound on switching 'on' the receiver, the fault could be in the power supply. After connecting an isolation transformer, proceed as follows to locate the fault. Measure dc voltage across C_{16}. It should normally be equal to 115V. However, if it turns out to be zero, measure voltage from point 'A' with respect to floating ground. If this is also not present proceed to locate fault in the unregulated section of the power supply by checking bridge diodes for any short across them. By continuity check or otherwise also check all capacitors for short across them. In addition check C_{16} to

ensure that it is not open or shorted. Remove degaussing coil connections and again check voltages. Thus localize the fault and repair it to restore normal operation of the receiver.

(2) If fuse diode D_{58} is found to be destroyed, the replacing of fuse FIA will restore the power supply with faulty zener disconnected. However, it is not desirable to use the receiver without D_{58} because any excessive voltage due to faulty regulation can destroy ICs that are fed from the SMPS.

(3) If voltage regulation is poor, try resetting R_{13}. If trouble persists check all transistors one by one and also apply continuity test to locate any open circuit in the regulator section.

(4) Though not common, the SMPS transformer may develop a partial short or short to ground from any of its windings. Check if dc voltage across C_{11} is present. If not, check associated components and if necessary try replacing the transformer.

In order to provide rigorous control of the duty cycle and incorporate protective features necessary for successful operation of a switched mode power supply, several ICs have been developed. A representative IC is TDA2581.

17.6 SERIES REGULATED POWER SUPPLY

In colour receivers prior to development of SMPS, the use of a series regulated power supply was quite common. A typical circuit that provides 112V regulated dc output is shown in Fig. 17.19. The mains voltage is stepped down to 110V by autotransformer T_2. This voltage is fed directly to the degaussing coil L_1 through a positive thermistor R_{90} which allows current flow for about one second every time the receiver is switched on.

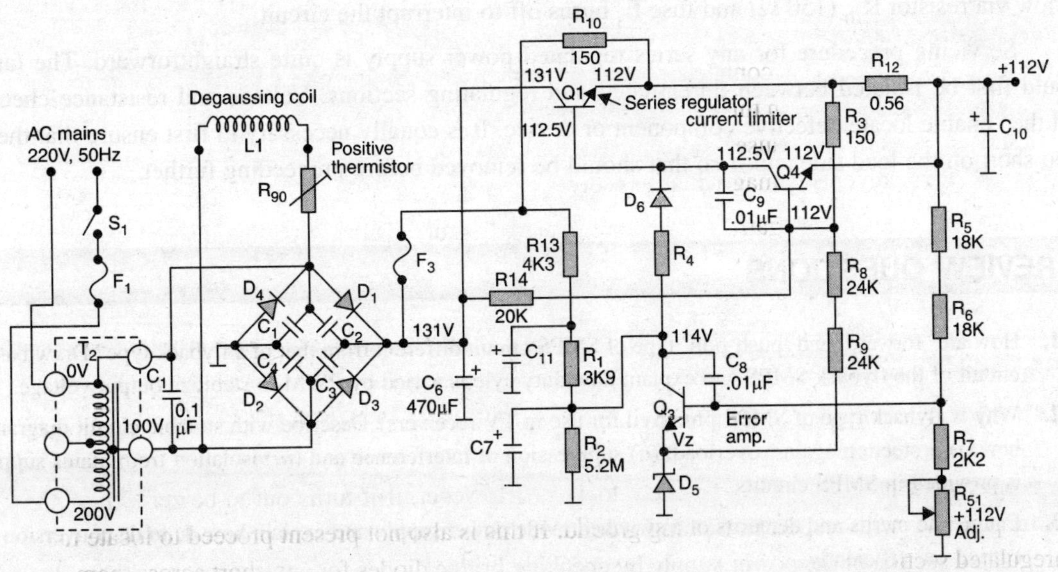

Fig. 17.9 Series regulator power supply circuit with overload protection.

The 100V ac is rectified by a bridge rectifier (D_1 through D_4) and filtered by 470 μF capacitor C_6. A dc voltage of about 130V becomes available for feeding the regulator circuit. The capacitors (C_1 through C_4) connected in shunt with rectifier diodes are for protection from transients for surge voltages.

The regulator employs Q_1 (2SC1829) as series pass transistor, Q_3 (2SC2229) as error amplifier and Q_4 (2SC2120) for current limiting. The error amplifier produces output on comparing a fraction of output voltage V_o with reference voltage Vz obtained across Zener D_5. A potentiometer consisting of R_{13}, R_1 and R_2 feeds decoupled dc at 114V to the collector of Q_3. With any increase in output voltage Vo, current through the error amplifier (Q_3) rises and its collector voltage falls. Since collector of Q_3 is conneceted to the base of Q_1 through resistor R_4 and diode D_6, the drop in voltage gets applied to it to reduce its forward bias. This results in increased voltage drop across Q_1 and output voltage decreases to remain steady at 112V. Any decrease in output voltage will have the opposite feedback action and Vo restored to its normal value.

A short across the load can damage Q_1 because of excessive current flow through it. A current limiting transistor (Q_4) is connected in shunt with Q_1 to prevent such a situation. Under normal load conditions the emitter of Q_4 is at 112V (Vo) and its collector, that is connected to the base of Q_1 is at 112.5V. This is nearly the same potential as at the collector of Q_3 since the two are tied together via R_4 and D_6. The base of Q_4 is also maintained at 112V by potentiometer formed by R_{14}, R_8, R_9 and D_5. Since voltage drop across resistor R_{12} (0.56Ω) is very small, under normal load conditions no additional forward bias gets applied between base and emitter of Q_4 and it stays in cut-off. It is thus of no consequence to the stabilizer circuit. However, when load current increases due to a short in the load circuit or for any other reason, voltage drop across R_{12} becomes sufficient to forward bias Q_4. The resulting collector current that flows through R_{13}, R_1, R_4 and D_6 causes more voltage drop across them and hence base voltage of Q_1 is lowered to the extent that it is cut-off. At this stage load current tends to flow via resistor R_{10} (150 Ω) and fuse F_3 burns off to interrupt the circuit.

Servicing procedure for any series regulated power supply is quite straightforward. The fault should first be isolated between unregulated and regulating sections. Voltage and resistance checks will then enable locate defective component or device. It is equally necessary to first ensure that there is no short on the load side and if so that should be removed before proceeding further.

REVIEW QUESTIONS

1. How are 'forward' and 'push-pull' type of SMPS circuit different from that of a flyback type? Draw basic circuit of the flyback SMPS and explain how duty cyle is varied by PWM to stabilize output voltage.
2. Why is flyback type of SMPS preferred for use in TV receivers? Describe with suitable circuit diagrams how (*i*) protection against overload, (*ii*) suppression of interference and (*iii*) isolation from mains supply is provided in SMPS circuits.
3. Enumerate merits and demerits of a switched mode power supply and explain how dc to dc conversion is carried out efficiently.
4. An economical SMPS circuit is shown in Fig. 17.7. Explain briefly how it operates to regulate various output voltages. How is the blocking oscillator started on switching on the receiver?

5. With reference to the circuit (Fig. 17.8) of a line-synchronized switched mode power supply and in particular explain the following:-
   (*i*) line synchronization of the blocking oscillator
   (*ii*) method of pulse width modulation to vary duty cycle and
   (*iii*) operation of electronic fuse.
   What is function of 'Z' diode D58?
6. Draw circuit diagram of a series regulated power supply and explain how it operates to stabilize dc output voltage. Describe how current overload is prevented.

18

COLOUR TELEVISION RECEIVER'S COMPLETE CIRCUIT, ALIGNMENT AND SERVICING

INTRODUCTION

After studying various sections of a colour receiver, it is instructive to examine a complete circuit to know how various sub-systems combine to give desired performance. In addition, it is essential for a TV engineer or technician to be fully acquainted with the alignment and testing procedure that a receiver undergoes on moving out of the assembly line. Similarly, a service engineer should be well conversant with trouble shooting and servicing of colour receivers. This chapter is written to meet these requirements and for this a popular colour receiver circuit has been chosen. It is of a 51 cm colour receiver based on S.S.P.P. design. A block schematic of receiver is shown in Fig. 18.1 for easy identification of various subsystems in the complete circuit given on the fold out page (Fig. 18.2).

18.1 RECEIVER CIRCUIT DESCRIPTION

The receiver circuit described in this chapter was preselected and its subsystems have already been described in previous chapters. However, the functioning is reviewed here to explain how the RF signal picked up at antenna is processed to recover video and audio outputs.

It may be noted that in the circuit diagram (Fig. 18.2) each component is labelled with a prefix digit to indicate subsystem number. Such a digit was, however, not carried while showing circuits of various subsections in different chapters. As an illustration resistor R302 of Fig. 18.2 is labelled as R_2 in Fig. 14.18 and C418 of the vertical output stage as C_{18} in Fig. 15.4. As such, detailed circuit explanations can be easily referred to and compared with the complete circuit where necessary.

Tuner

The receiver employs a PHILIPS UV 411 electronic tuner. Its internal circuitry is similar to that shown in Fig. 9.4 and described in section 9.3. The program selector is also of the same type as discussed in sections 9.4, 9.5 and shown in Figs. 9.5 and 9.6. The RF signal from the antenna enters at terminal 1 (socket) of the tuner (see Fig. 18.2) and after due processing, IF signals become available at terminal 9 to be fed to the pre-amplifier.

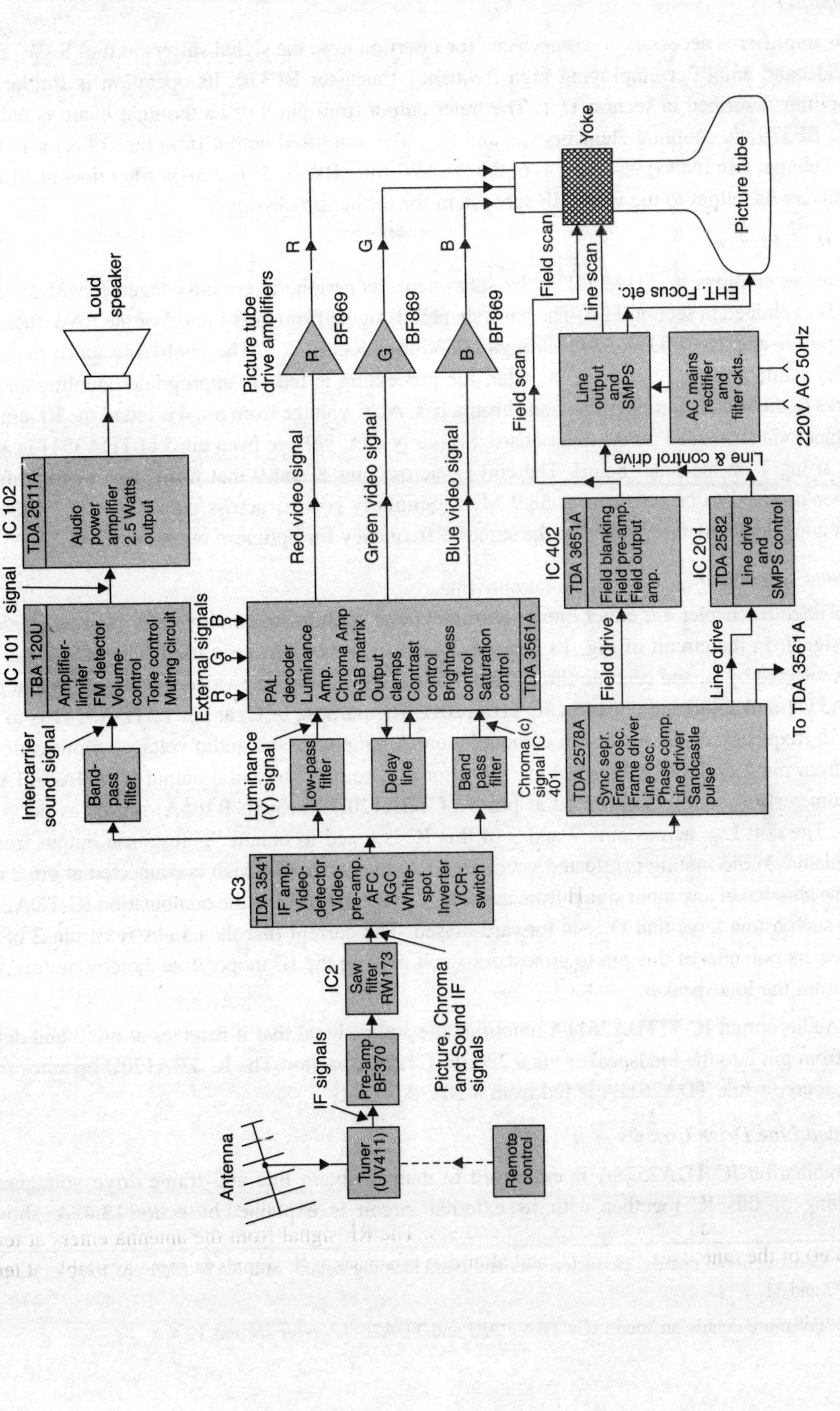

Fig. 18.1 Block schematic of the SSPP colour TV receiver chosen for detailed study.

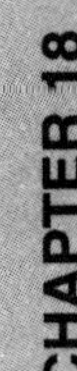

Preamplifier

The preamplifier is necessary to compensate for insertion loss, the signal suffers in the 'SAW' filter. It is a wideband amplifier employing high frequency transistor BF370. Its operation is similar to the preamplifier described in section 11.1. The tuner output from pin 9 of its terminal board is fed to the base of BF370 via coupling elements C_{13} and R_{10}. The amplified output from the collector is fed via C_{17} to the input terminals (pins 2 and 4) of the '*SAW' filter RW-173. The SAW filter does bandshaping and delivers its output to the vision IF subsystem for further processing.

Vision IF Subsystem

The receiver employs IC TDA3541 in the subsystem. Its peripheral circuitry together with functional details is explained in section 11.4. The band shaped IF signal from pins 1 and 5 of the SAW filter is fed across pins 1 and 16 of TDA-3541 through coupling capacitor C_{19}. The composite video output that becomes available at pin 12 of the IC after due processing is fed via appropriate coupling and filter networks to the sound, combination, and chroma ICs. AGC voltage from pin 4 is fed to the RF amplifier in the tuner via terminal 5 on the tuner board. Similarly AFC voltage from pin 5 of TDA3541 is applied via R_7 at lug 7 on the tuner board. The coil L_4 across pins 8 and 9 that forms part of the reference amplifier is tuned to be resonant at 38.9 MHz. Similarly coil L_5 across pins 7 and 10 of the AFC synchronous demodulator is tuned to the same IF frequency for optimum output at pin 5.

FM Sound Subsystem

As explained in chapter 12, either one or two integrated circuits are employed for processing the FM sound signal. In the circuit of Fig. 18.2, a two IC configuration has been used. The SIF signal at 5.5 MHz is selected by L_2 and ceramic filter (CRF-5.5) from the composite video signal available at pin 12 of TDA3541 and applied at pin 14 of IC TBA120U. The purpose of L_2 at pin 12 TDA3541 is to block higher IF frequency components. On demodulation, de-emphasis and audio voltage amplification, the signal from pin 8 of TBA120U is fed via capacitor C_{108} to pin 7 of audio output IC TDA2611A. The potentiometer R_{104} that is connected at pin 5 of TBA120U (through R105A) serves as dc volume control. The coil L_{101} across pins 7 and 9 of this IC is tuned to obtain 'S' response output from the demodulator. Audio muting is affected via diode D_{101}, the anode of which is connected at pin 2 of the IC. In the absence of any input signal from antenna, voltage at pin 13 of the combination IC-TDA2578A falls to such a low level that D_{101} is forward biased. The current that then sinks from pin 2 of TBA 120U lowers potential at this pin to almost zero volt making the IC inoperative thereby muting sound output from the loudspeaker.

Audio output IC **TDA2611A amplifies the audio signal that it receives at pin 7 and delivers output from pin 2 to the loudspeaker via a 220μF (C112) capacitor. The IC TBA120U operates from a 12V dc source while TDA2611A is fed from a 20V dc supply.

Frame and Line Drive Circuits

The combination IC TDA2578A is employed to mainly obtain line and frame drive voltages. The functioning of this IC together with its external circuit is explained in section 13.4. As shown in

---

* Constructional details, characteristics and band shaping details of a SAW filter are fully explained in chapters 7 and 11.

**For more details on sound ICs TBA 120U and TDA2611A refer section 12.4.

Fig. 18.2, composite video signal from vision IF IC (TDA3541) on passing through the 5.5 MHz sound trap formed by L_3 and C_{27} feeds at pin 5 of TDA2578A through the coupling network–R_{411} and C_{409}. After processing, the combination IC delivers the following outputs:-

(*i*) Drive signal to the vertical output IC TDA3651A is fed from pin 1 through resistor R_{417}.

(*ii*) Line drive pulses from pin 11 are delivered to IC TDA2582 via C_{226} and R_{230}.

(*iii*) Sandcastle pulse that becomes available at pin 17 of the IC is fed to the chroma IC TDA3561A via R_{414}.

(*iv*) When the programme selector switch is set for VCR operation it causes pin 18 of TDA2578A to be connected to ground via resistor R_{418}. This resistor in parallel with C_{413} lowers time constant of the internal circuit connected to pin 18 for optimum reception from the VCR.

(*v*) A mute sensing voltage at pin 13 enables blocking of sound IC TBA120U. It is connected via diode D_{101} as explained in the previous subsection.

Flyback pulses from L.O.T. are brought in at pin 12 via C_{236} and R_{400}. *The R-C network connected at pin 15 is for control of line frequency and is set by potentiometer R_{412} to obtain 15625 Hz output. The vertical oscillator in the IC operates in synchronism with frame sync pulses to provide a 50 Hz sawtooth output. The potentiometer R_{416} connected at pin 3 via R_{415} is varied to adjust the frequency at its correct value. The IC operates from a 12 V dc source obtained from the SMPS supply.

Chroma Subsystem

The IC TDA3561A that is employed in the chroma subsystem of the circuit under discussion (Fig. 18.2) combines all functions necessary to produce R, G, B outputs on accepting composite video signal at two separate pins via filter circuits. The luminance signal input at pin 10 of TDA3561A is obtained from pin 12 of TDA3541 via 5.5 MHz intercarrier trap (L_3, C_{27}), isolating resistor R_{300}, delay line DL-470, R_{306}, 4.43 MHz chroma signal trap (L_{302}, C_{308}) and coupling capacitor C_{318}. Similarly chroma signal input at pin 3 is brought from the same source via C_{301}, R_{302}, chroma bandpass filter (L_{303}, C_{306} and R_{307}) and coupling capacitor C_{307}. The sandcastle pulse from pin 17 of the combination IC is fed at pin 8 of the chroma IC via R_{414} and R_{312}.

The PAL delay line necessary for demodulation receives input from pin 28 and feeds outputs at pins 21 and 22 which are input points to U and V demodulators. The reference oscillator crystal is connected between pins 25 and 26 through the trimming capacitor C_{311}. Brightness control R_{324}, contrast control R_{320} and the saturation control potentiometer R_{316} are connected at pins 11, 7 and 6 respectively. The RGB outputs obtained after matrixing and necessary amplification become available at pins 12, 14 and 16 respectively. These are fed to corresponding drive amplifiers via resistors R_{326}, R_{327} and R_{328} as shown in Fig. 18.2. The pins 13, 15 and 17 are for external RGB inputs like those obtained from the teletext decoder. For more details on the functioning of this IC (TDA 3561A) refer section 14.11 and Figs. 14.16, 14.17 and 14.18.

*In Fig. 13.4 (Frame and line drive) R_{403} is labelled as R_{11}, R_{412} as R_{13}, R_{413} as R_{14}, R_{416} as R_{18} and R_{417} as R_{19}.

RGB Amplifiers

The three (R, G and B) drive amplifiers employ transistor BF869 and feed outputs to corresponding cathodes of the colour picture tube via resistors R_{354}, R_{355} and R_{356} respectively. The potentiometers R_{336} and R_{337} are for adjusting input to G and B amplifier to obtain proper colour voltage amplitudes at the cathodes. Similarly potentiometers R_{348}, R_{349} and R_{350} are for adjusting bias voltage (V_{BE}) of the amplifier transistors. The input drive and bias potentiometers are adjusted for 'grey scale tracking' with the aim that no colours get reproduced during reception from monochrome telecasts. The diodes D_{301}, D_{302} and D_{303} between base and emitter of each of the amplifier transistor (BF869) are to ensure that base-to-emitter voltage does not accidently exceed a predetermined value. More details on RGB amplifiers and grey scale tracking are given in chapter 3 and section 14.13 of this text.

Vertical Drive Amplifier

The IC TDA3651A performs as the vertical output amplifier. It receives drive voltage from pin 1 of the IC TDA 2578A and delivers deflection current from pin 5 to the vertical deflecting coils. All necessary detail on how linearity and amplitude controls operate to deflect the beam and fill the raster to its full height are explained in section 15.3.

Line Deflection and SMPS

In the receiver circuit under study (Fig. 18.2) the line deflection and SMPS circuits are combined together in a very interesting way which is both economical and efficient. The two circuits are controlled by a common IC (TDA2582) which feeds timed and synchronised line drive pulses to the line drive amplifier. The feedback loops incorporated in the IC enable over-voltage and over-current protection besides beam current limiting. The IC also includes facilities for automatic re-start after switch-off due to overload etc.

Line deflection circuit. It comprises of the line oscillator, line drive amplifier and line output amplifier. The oscillator circuit is in the IC TDA2582 where external components connected to pins 13, 14 and 15 control its frequency and phase. The frequency of the oscillator is determined by the timing network that comprises of capacitor C_{207} (2k 7PF) connected between pins 13 and 16 (ground) and resistor R_{209} (41k) connected between pin 13 and pin 10 that has an applied reference voltage. The component values are chosen to obtain free running frequency close to 15625 Hz but in no case greater than this. Pin 15 that is one of the inputs to the differential configuration which forms the oscillator circuit receives a fixed reference voltage from pin 10. Line sync pulses obtained from pin 11 of the combination IC are fed at pin 14 of TDA2582 via C_{226} and R_{230} to synchronise the line oscillator with frequency and phase of corresponding oscillator at the transmitting end.

The synchronised line drive pulses that become available at pin 11 of the control IC are applied at the base of line drive amplifier TR_{203} (Q_{203}) via diode D_{210}. A negative feedback voltage obtained from tapping 8 on the secondary of LOT is also fed at the base of TR_{203} via R_{211} for obtaining stable operation of the line output circuit. On amplification, line drive signal is applied to the base of line output transistor TR_{202} (BU508A) via impedance matching and isolation transformer T_{204}. The Vcc supply for BU508A is obtained on rectifying and filtering input mains supply. The transistor's collector current flows via primary winding (12-13) on LOT and through primary of current sensing transformer T_{203}.

The secondary winding (1-6) on the line output transformer drives line deflection coils on the yoke via capacitor C_{231}. The network (C_{234} and R_{238}) across the coils is to suppress any undue oscillations.

SMPS action. Various dc voltages are obtained as in any other SMPS circuit by rectifying ac voltages at 15625 Hz that are obtained from different windings on the LOT. However, the control mechanism is through line drive pulses in such a way that line output transistor BU508A also acts as the SMPS switching device. This explains why such a receiver is called Single-Switch-Power-Pack (SSPP) colour television.

Both over-current and over-voltage protections are obtained via sensing inputs fed to TDA2582 from two different points on the LOT. Pin 5 receives input from tap 17 on the secondary winding that feeds the deflection coils. A voltage proportional to load current from the 12V dc line is sensed via transistor TR_{205} and fed at pin 6 via R_{240}. This serves as over-current protection feedback. Similarly over-voltage protection feedback input is applied at pin 7 of TDA2582. It is derived from two sources. The input via R_{223} is from pin 4 of the 180V dc supply winding on LOT and that via D_{209} is from the secondary of current sensing transformer T_{203}. Thus, any variations in the dc output voltage or excessive current through the line output transistor due to any reason, are fed to the comparator and control circuits in the IC. Whenever voltage at pin 7 exceeds the threshold level, the protection circuit comes into operation and trips it off. The tripping level is about the same as the reference source at pin 10.

A very important factor in stabilizing dc output voltages including the EHT is to maintain a balance between output and input power *i.e.*, volt-amps (V × A). This control as in any SMPS circuit is exercised by varying duty-cycle factor of the switching device. For this, a feedback input is connected at pin 8 of the IC. It is obtained from the ABL transistor TR_{204} and connected (via pin 8) to one input of a differential amplifier functioning as an amplitude comparator, the other end of which is connected to the reference source at pin 10. The differential amplifier output controls timing of line drive pulses that finally get applied to the line output transistor which in effect is also the switching device of the SMPS supply. The dc output voltages can be set as necessary by varying feedback at pin 8 through potentiometer R_{232}.

Since the line output and SMPS switching circuits operate at the same frequency of 15625 Hz and remain synchronised, there is no need to provide elaborate shielding of the SMPS circuit which otherwise is a must.

Starting DC supply. It is necessary to provide dc supply for starting SMPS and allied circuits at the time of switching 'on' the receiver. Once started the SMPS develops various dc sources to feed supply to different sections of the receiver. This, as shown in Fig. 18.2, is provided through a small mains transformer (T_{202}) the voltage on the secondary of which is rectified and filtered to develop a 20V dc supply. This source feeds its output to IC TDA2611A and also to a series voltage regulator (TR_{201}).

The regulated dc output at 12V is fed to pin 9 of TDA2582 (supply control IC) and also to pin 16 of TDA2578A (combination IC). As shown in the circuit diagram, regulated dc is also fed to the line driver amplifier via diode D_{204} and R_{226}. This is enough to start the line deflection circuit and SMPS action. Regenerative action enables these circuits to attain normal operation in a very short time and all dc voltages,including EHT get stabilized at their design values. One of the dc output voltages that now becomes available is at 26V and feeds line driver amplifier via diode D_{216}. This source is also connected to the cathode of diode D_{204} which then gets reverse baised to disconnect 12V dc supply, which, as obvious, is not required any more.

The EHT is obtained as in other circuits by what is called 'diode-split addition' through diodes D_{219}, D_{220} and D_{221}. The supply for the focus grid of picture tube is derived from one of the tapping points of this EHT winding.

Mains input and degaussing circuit. A shown in Fig. 18.2, ac input from mains supply is fed to the bridge rectifier circuit through the usual protection and filter sections. On rectification, it is filtered to provide raw dc to the line output cum SMPS circuit. The line supply is also fed to the degaussing coil via dual PTC (positive temperature coefficient) thermistor R_{202} and R_{203}. After initial inrush of current, the thermistor heats up to develop high resistance to block continued flow of current to the coil. It is kept hot by thermistor R_{202} which is connected across the supply line and placed in close physical contact with R_{203}.

The mains chassis earth is isolated from the ac mains side by magnetic coupling through transformers, T_{202}, T_{203}, T_{204} and the line output transformer. This is indicated in Fig. 18.2 by a dotted line that runs across the chassis area which remains hot. Such a provision keeps rest of the circuit isolated and prevents any shock hazards to the technician both during manufacture and later while servicing the receiver.

Picture Tube Circuit

The picture tube used in the receiver is a A51-590X precision-in-line version and its pin connections are shown (see Fig. 18.2) below the picture tube circuitry. The high voltages to the focus and screen grids are fed via R_{359} and R_{360} on deriving from one section of the EHT winding.

18.2 *ALIGNMENT AND SERVICING EQUIPMENT

Electronic instruments that are commonly used in television industry both during manufacture and servicing include sweep generator with markers, colour TV pattern generator, wide screen high frequency oscilloscope, RF attenuator, multimeters-analog and digital, EHT proble, power supplies, LCR bridge, IC tester, Insulation tester and a Degaussing coil. Additional equipment that is also useful includes Vectorscope, Audio power meter, Cyclic timer, Distortion analyzer, White balance checker, Colour analyzer, Video sweep generator and Microscope of approximately 12 power for phosphor dot observation in the colour picture tube. Many of the above listed instruments are quite expensive and an average manufacturer prefers to buy only those that are essential for routine checks and alignment purposes. These include a VHF and IF Sweep generator with markers (frequency range 1-500 MHz), an oscilloscope designed to handle signals with bandwidth from dc to atleast 20 MHz, and a colour TV pattern generator besides the usual low cost equipment. The sweep generator provides RF output for any channel that can be modulated with composite colour video signal and markers for alignment of tuner and IF amplifiers. The colour TV pattern generator is equipped to generate a wide range of patterns the video signals for

*If necessary refer chapter 28 of the book "Monochrome and Colour Television" by R.R. Gulati for basics of such instruments.

which can be made to modulate IF or any channel in the VHF-I, VHF-III and UHF bands. Sound carrier at 5.5 MHz is separately available with provision for FM modulation with a 1 KHz audio tone. Different makes of Pattern Generators are available. However, for illustrating the use of various patterns, PHILIPS's pattern generator PM 5519 has been chosen.

PATTERN	*FOR CHECKING*
I. CIRCLE (white circle on grey background. It can be combined with all test patterns).	overall picture linearity. When combined with the crosshatch pattern it is useful for checking overall geometry.
II. CHECKER BOARD (of 6 × 8 squares accurately centered)	(*i*) horizontal and vertical synchronization. (*ii*) picture position (deflection yoke) (*iii*) picture height and width (aspect ratio) (*iv*) horizontal and vertical linearity of deflection. (*v*) centering of picture (*vi*) bandwidth and step input response (*vii*) mains hum interference (*viii*) proper suppression of sound intercarrier (*ix*) correct focussing of the picture
III. DOTS (11 horizontal lines of 15 dots)	static convergence and if necessary re-adjusting as per instructions of the manufacturer.
IV. CROSSHATCH (11 horizontal and 15 vertical lines)	dynamic convergence and pincushion correction, E-W, N-S corrections.
V. GREY-SCALE (STAIRCASE) AND DEFINITION LINES (Staircase signal-white with 8 identical steps combined with definition lines 0.8 – 4.8 MHz (step-width 1 MHz)	(*i*) brightness and contrast control circuits. (*ii*) grey scale (on a CRO all the steps can be seen) (*iii*) linearity of video amplifier (*iv*) video bandwidth.
VI. WHITE PATTERN (100% white signal with or without burst)	(*i*) white 'D' (*ii*) brightness control (*iii*) beam current of picture tube (*iv*) luminance writing current
VII. PURITY (red, green or blue signals (rasters) can be individually obtained).	(*i*) brightness and saturation (*ii*) colour purity (individual) (*iii*) interference between sound and chroma carrier. (*iv*) colour AGC.

VIII. DEM PATTERN
Special test pattern of 4 vertical bars–
1st bar:- (G – Y) = 0
2nd bar:- grey (Y)
3rd bar:- (R – Y) NTSC encoded with PAL burst
4th bar:- ± (B – Y) The 2nd bar and lower part of the screen are for reference (see colour plate No. 5)

(*i*) PAL delay line-amplitude and phase error detection.
(*ii*) PAL demodulators-subcarrier frequency (phase) to the two demodulators.
(*iii*) PAL switch (electronic)
(*iv*) matrix

IX. VCR PATTERN

linearity and sensitivity of colour amplifiers.

X. COLOUR BAR WITH WHITE PATTERN
(75% saturation colour bar with white reference field in lower part of screen)

(*i*) overall colour performance
(*ii*) burst keying
(*iii*) PAL identification circuit.
(*iv*) matrix circuit
(*v*) RGB amplifiers.

18.3 COLOUR RECEIVER ALIGNMENT

With the availability of specially designed integrated circuits, the core circuitry of a colour receiver has more or less been standardized and there is no need for rigorous measurements. Routine checks are, however, carried out at each stage as the set moves on the assembly line. The receiver is invariably operational before it is sent for alignment and testing. Therefore, while describing alignment procedure, it will be assumed that there are no constructional faults and all the components of the receiver were pretested and found to be good.

Receiver alignment procedure can vary somewhat depending on its design, availability of equipment and desire to maintain standards. Most medium scale manufacturing units adopt a middle of the line approach and as such the procedure to be described is based on the assumption that atleast a sweep cum marker generator, a colour TV pattern generator and a high frequency large screen oscilloscope are available besides general purpose instruments.

It is often necessary to use an impedance matching transformer between the instrument and receiver to avoid any mismatch and the consequent insertion loss. Similarly, an RF step attenuator is a must for setting signal output levels and to make gain/loss measurements when or where necessary. Almost all signal and pattern generators have an inbuilt calibrated attenuator and a separate unit is seldom necessary. The two commonly used signal-market sweep generators are TELONIC 1011B (U.S.A.) and JSWA3536 (Japan). These have not only in-built attenuators but provide RF output from a 75 ohm impedance source to match the 75-ohm input impedance of the receiver. Similarly, the PHILIPS PM5519 colour TV pattern generator, that is quite popular, has its own attenuator and delivers output from a 75 ohm source. Therefore, while describing alignment procedure, impedance matching transformers and attenuators will not be shown as separate units.

Another point that needs mention is the fact the alignment procedure in general remains the same for all makes of colour receivers and it is only its implementation that varies somewhat as specified by the manufacturer. The procedure described here pertains to the receiver whose circuit is shown in Fig. 18.2. However, it can be easily adopted for other makes of colour receivers which employ equivalent ICs for the otherwise similar circuits or have somewhat different designs.

(1) Channel Bandwidth Alignment at the Tuner Output

The electronic tuner is a purchased item (often imported) and it is reasonable to assume that the unit in conjunction with associated program selector is operational and pretested on all the three bands. It is, however, necessary to tune its IF output circuit because it feeds into a pre-amplifier, the input impedance of which varies from circuit to circuit. A small variable inductor, that forms part of the output stage load circuit is often kept accessible for this purpose. The Philips UV411 electronic tuner used in the receiver of Fig. 18.2 has such a coil mounted on one side of the unit. A test point is provided for IF injection (terminal 4 in UV 411) to align output circuit of the tuner and associated circuits in the IF subsystem. For this, it is necessary to feed 12V dc at pin 4 through an inductor. This is shown in Fig. 18.3 that gives set up for tuner output alignment. After making necessary connections and allowing a warming period of about 15 minutes proceed as under:-

(*i*) Set sweep generator for IF band output with a centre frequency of 36.15 MHz and sweep width of ± 7 MHz.

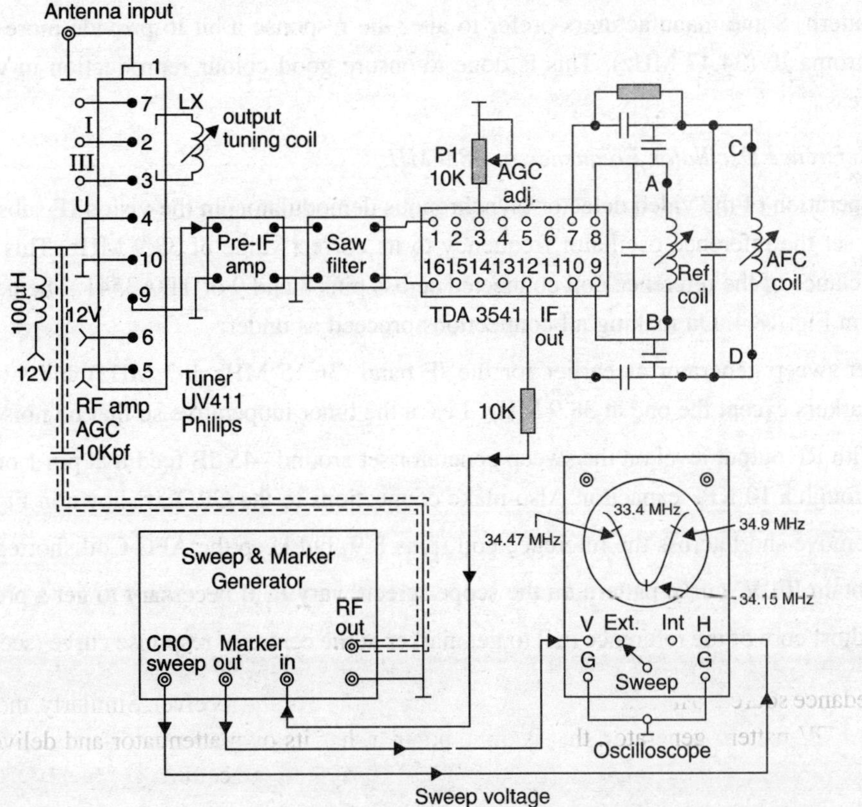

Fig. 18.3 Set-up to align tuner output circuit for the desired bandwidth of 7 MHz.

(*ii*) Set markers for 36.15 MHz, 33.4 MHz (sound IF), 34.47 MHz (chroma IF) and 38.9 MHz (vision IF).

(*iii*) Short both the reference and AFC coils (*i.e.*, A to B and C to D) with short jumper wires to make these circuits inoperative.

(*iv*) To be able to vary IF amplifier AGC, connect a 10 K-ohm potentiometer (P_1) from pin 3 of TDA 3541 to ground.

(*v*) Connect IF output from pin 12 of TDA3541 through a 10 K-ohm isolating resistor to the marker 'IN' terminal on the sweep generator, Also connect marker 'OUT' terminal to the vertical (V) input on the CRO.

(*vi*) Feed 12V dc at pin 4 of the tuner as shown in Fig. 18.3. It is understood that dc supply at pin 6 of the tuner is available.

(*vii*) Set RF output from the sweep generator at about –28dB (reference 500 μV) and feed it through a 10 KPF capacitor to pin 4 of the tuner.

(*viii*) With horz sweep switch on the CRO thrown for 'EXT', vary necessary controls to get IF bandwidth pattern on the screen. If necessary vary P_1 to ensure that the output signal is not distorted due to overloading. Also reduce RF input if necessary.

(*ix*) Adjust core of Lx (see Fig. 18.3) with a small screw driver to obtain markers as shown on the pattern. Some manufacturers prefer to alter the response a bit to provide more gain to the chroma IF (34.47 MHz). This is done to ensure good colour reproduction in weak signal areas.

(2) Setting Reference Oscillator Frequency at 38.9 MHz

For proper operation of the video detector (synchronous demodulator) in the vision IF subsystem, it is necessary to set the reference oscillator frequency to its correct value of 38.9 MHz. This is done by varying inductance of the reference coil connected across pins 8 and 9 of TDA3541. The necessary set up is shown in Fig. 18.4. On making all connections proceed as under:

(*i*) Set sweep generator as earlier for the IF band (36.15 MHz ± 7 MHz) and switch off all markers except the one at 38.9 MHz. Leave the tuner inoperative so that no noise creeps in.

(*ii*) With RF output level on the sweep generator set around –45 dB feed it at pin 1 of TDA3541 through a 10 KPF capacitor. Also make connections to the CRO as shown in Fig. 18.4.

(*iii*) Remove short across the reference coil (pins 8-9) but keep the AFC Coil shorted.

(*iv*) Obtain IF 'V' curve pattern on the scope screen, vary P_1 if necessary to get a proper curve.

(*v*) Adjust core of the reference coil to get marker at the centre of response curve (see Fig. 18.4).

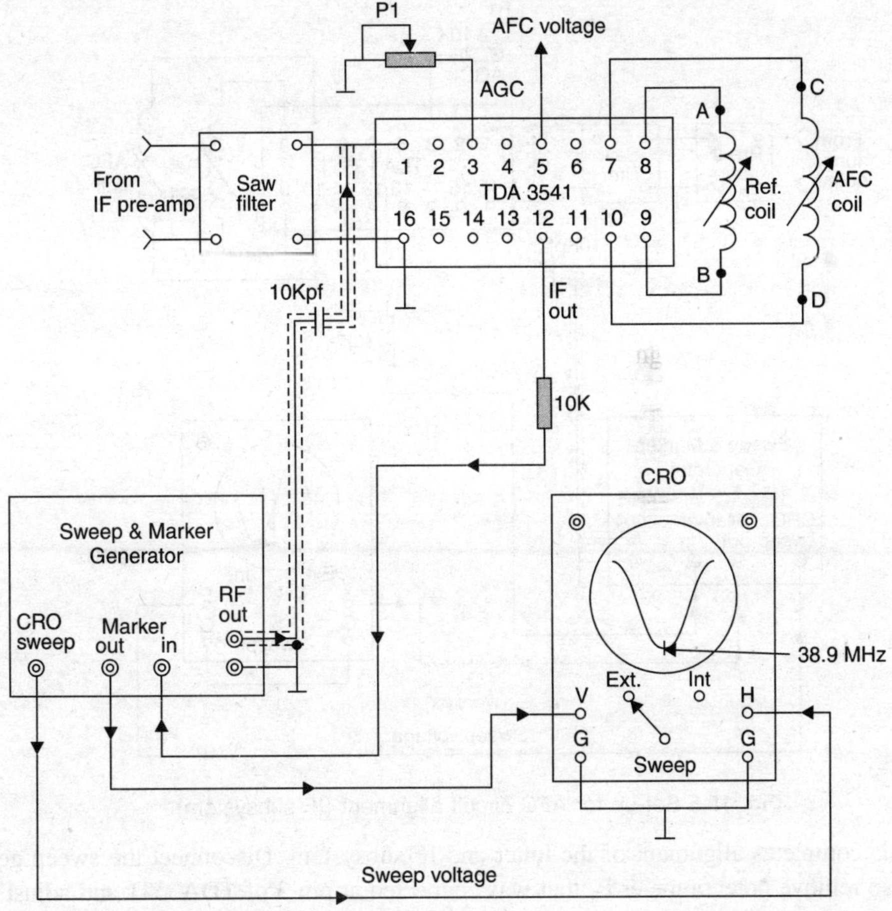

Fig. 18.4 Set-up for tuning reference oscillator frequency to 38.9 MHz.

(3) Setting AFC Circuit in the IF Subsystem

For correct automatic frequency control action, centre frequency of the AFC discriminator must be set equal to picture IF frequency (38.9 MHz). This is obtained by varying core of the AFC coil connected across pins 7 and 10 of TDA3541. The necessary set up is shown in Fig. 18.5 and the procedure is as under:-

(*i*) Remove short across the AFC coil

(*ii*) With sweep RF signal output set for 38.9 ± 7MHz and marker at 38.9 MHz obtain AFC discriminator response curve on the scope screen. Note that output is to be taken from pin 5 of TDA3541.

(*iii*) With AFC switch at OFF, vary core of the AFC discriminator circuit coil to obtain marker as the centre of curve (see Fig. 18.5).

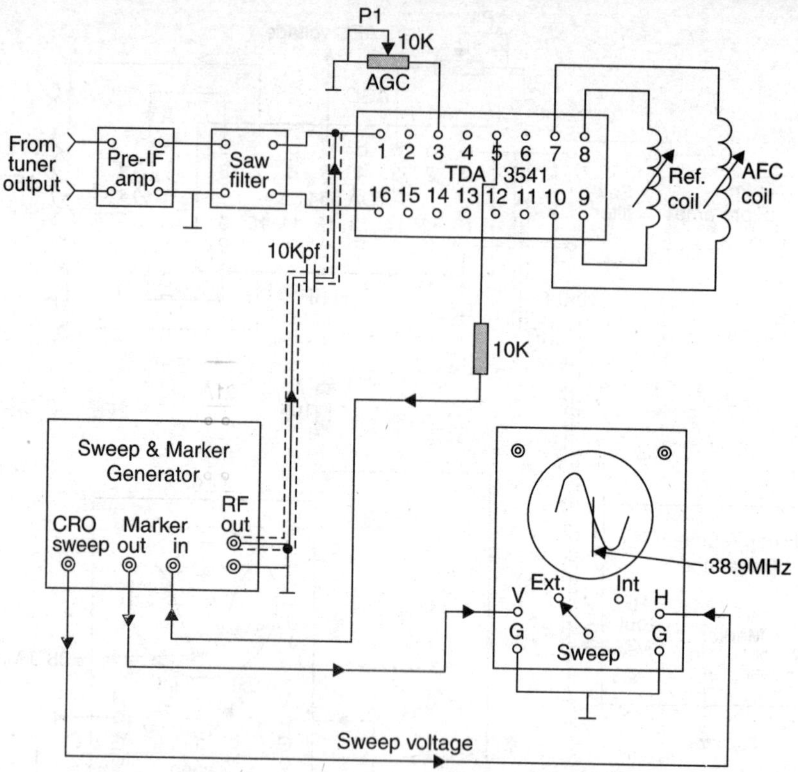

Fig. 18.5 Set-up for AFC circuit alignment (IF subsystem).

This completes alignment of the tuner and IF subsystem. Disconnect the sweep generator. Also remove potentiometer P_1 that was connected at pin 3 of TDA3541 and adjust R_{18} (10 Kohm pot) to obtain 1.2V at this pin. This is the recommended voltage for proper AGC operation of the receiver. Make all connections on the tuner and associated program selector for normal receiver operation. The TV set may now be put in the cabinet if points for remaining alignments remain accessible.

(4) SIF Trap Alignment

The remaining alignments are to be carried out with signal source from the Colour TV pattern generator. The tuner and IF subsystem will duly process it and feed the composite video signal from pin 12 of TDA3541 to other sections of the receiver.

It is necessary to avoid interference from local channels while carrying out these alignments. Therefore, set the tuner to receive an active channel that is not a local one. Set pattern generator switches to supply carrier output for the channel selected on the tuner. Do not disturb these settings till the receiver alignment is over.

The video signal for modulating the chosen carrier on pattern generator can be selected depending on the pattern needed for a particular alignment. Switches are provided for this purpose on the front panel of the pattern generator. It may be noted that the modulating signal also has the sound carrier depending on the selected channel. This appears as 5.5 MHz intercarrier SIF after detection and is available as part of composite video signal at the output of IF subsystem. (pin 12 of TDA3541).

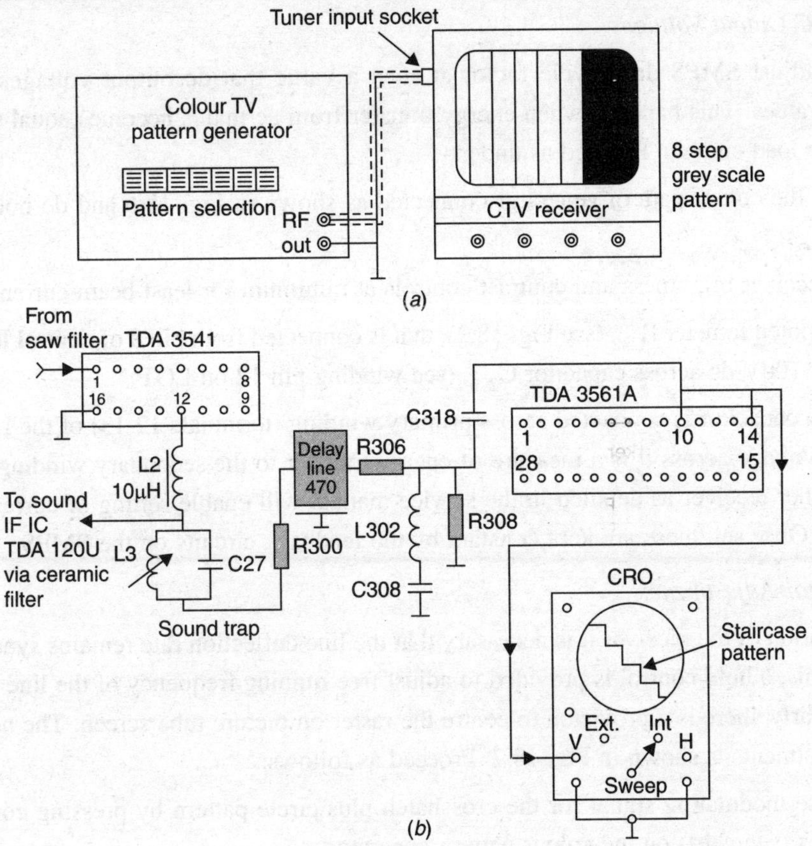

Fig. 18.6 SIF trap alignment (*a*) set-up (*b*) circuit connections.

It is necessary to prevent passage of this intercarrier SIF to circuits other than the sound section. For this, a trap circuit is provided (L_3, C_{27} after pin 12 of TDA3541) that is tuned to block SIF signal to deflection, chroma and luminance circuits of the receiver. The alignment procedure is given below and necessary set-up is shown in Fig. 18.6.

(*i*) Connect colour TV pattern generator RF output to the antenna socket of the receiver.

(*ii*) As explained above select the same idle channel (avoiding local channel) on both the pattern generator and TV receiver.

(*iii*) Choose composite video signal for the staircase pattern to modulate the chosen channel carrier.

(*iv*) With vertical (V) input of the CRO connected at pin 10 (luminance input point) of the chroma IC TDA3561A (see Fig. 18.6), adjust CRO controls (with horz. sweep switch thrown to 'INT') to obtain staircase pattern on the scope screen. For this, set RF output level on the pattern generator as necessary. Any overloading must be avoided.

(*v*) Adjust core of L_3 (see Fig. 18.6(*b*)) for maximum impedance of the tuned circuit at 5.5 MHz. This will occur when no disturbance (hash) is seen along the steps of the staircase pattern.

(5) Setting of DC Output Voltages

The aim is to adjust SMPS duty cycle factor at such a value that dc output voltages attain their predetermined values. This happens when energy transfer from ac mains becomes equal to the power consumed in the load circuits. Proceed as under:-

(*i*) Keep the colour pattern generator connected as shown in Fig. 18.6 and do not disturb any settings.

(*ii*) Set receiver brightness and contrast controls at minimum for least beam current.

(*iii*) Vary potentiometer R_{232} (see Fig. 18.2), that is connected from pin 8 of control IC TDA2582 to get 100V dc across capacitor C_{220}, (see winding pin 12 on LOT).

Since this capacitor is connected across primary winding (terminals 12-13) of the LOT through diode D_{213}, the voltage across it is a measure of energy transfer to the secondary windings. A similar setting in any other receiver as detailed in the service manual will enable setting of corresponding dc output voltages. Once set, these are kept constant by the feedback circuits on the SMPS.

(6) Line Oscillator Adjustments

For proper operation of the receiver it is necessary that the line deflection rate remains synchronised at 15625 Hz. For this, a hold control is provided to adjust free running frequency of the line oscillator at 15625 Hz. Similarly, there is a provision to centre the raster on picture tube screen. The necessary set up for such adjustments is shown in Fig. 18.7. Proceed as follows:

(*i*) Choose modulating signal for the crosshatch plus circle pattern by pressing corresponding buttons (switches) on the colour pattern generator.

(*ii*) Short pin 5 of the combination IC TDA 2578A to ground. The idea is to prevent application of horz sync pulses to the line oscillator.

(*iii*) Vary R_{412} (pin 15 of TDA2578A) to get a stable or slowly floating pattern. It should not have any bending or tearing. Adjust RF output control on the pattern generator, if necessary.

(*iv*) Remove short from pin 5 and check that a stationary pattern is obtained.

(*v*) Next, vary R_{403} (pin 14) on the same IC to exactly centre the pattern. The centering can be checked by counting crosshatch lines or either side of the circle in the horizontal axis. Pin 14 of the IC is input point to the horizontal pulse width modulator. Any change of dc voltage at this pin by varying R_{403} ultimately modifies line drive signal in such a way that a lateral shift of the raster occurs and hence, that of the reproduced picture or pattern.

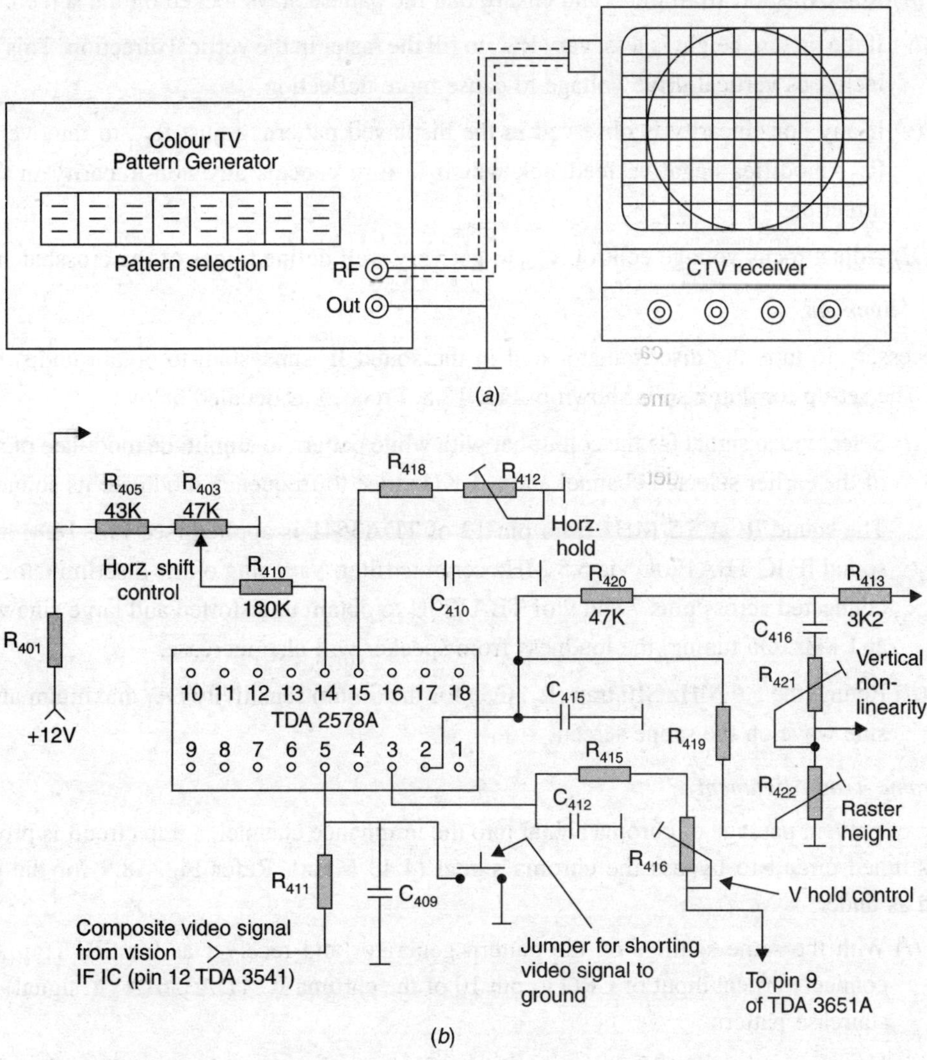

Fig. 18.7 Set-up for adjustments in the line and field deflection circuits.

(7) Vertical Deflection Circuit Adjustments

The three adjustments provided on the vertical drive circuit are (*i*) vertical hold, (*ii*) vertical height and (*iii*) vertical linearity. The set up and initial adjustments remain same as shown in Fig. 18.7 and explained under (6). Proceed as under:

(*i*) Short pin 5 of TDA2578A to ground to prevent application of vertical sync pulses to the field oscillator.

(*ii*) If the pattern rolls vertically, adjust R_{416}, the vertical hold control to stop it. This amounts to adjusting free running frequency of the vertical oscillator at 50 Hz.

(*iii*) Remove short from pin 5 and ensure that the pattern stays locked on the screen.

(*iv*) If the picture height is less, vary R_{422} to fill the raster in the vertical direction. This adjustment increases vertical drive voltage to cause more deflection.

(*v*) If any non-linearity is observed in the displayed pattern, adjust R_{421} to remove it. Varying R_{421} modifies negative feedback which in turn cancels any non-linearity in the vertical direction.

(*vi*) Adjust focus voltage control R_{359} to get very well defined lines of the crosshatch pattern.

(8) SIF Alignment

It is necessary to tune the discriminator coil in the sound IF subsystem to obtain undistorted sound output. The set-up for doing so is shown in Fig. 18.8. Proceed as detailed below:

(*i*) Select video signal for the colour bar with white pattern to amplitude modulate picture carrier of the earlier selected channel and a 1 KHz tone to frequency modulate its sound carrier.

The sound IF at 5.5 MHz from pin 12 of TDA3541 is applied (see Fig. 18.8) at pin 14 of sound IF IC TBA120U via 5.5 MHz ceramic filter. Vary slug of the discriminator coil that is connected across pins 7 and 9 of TBA120U to obtain undistorted and large sinewave output at 1 kHz. On tuning, the loudness from speaker will also increase.

(*ii*) Retune the 5.5 MHz SIF trap (L_3, C_{27}) for maximum sensitivity *i.e.*, maximum amplitude of sine wave on the scope screen.

(9) Chroma Trap Alignment

In order ot prevent passage of chroma signal into the luminance channel, a trap circuit is provided. It is a series tuned circuit to bypass the chroma signal (4.43 MHz). Refer Fig. 18.9 for the set up and proceed as under:-

(*i*) With the same settings on the pattern generator and receiver as for SIF Trap Alignment, connect vertical input of CRO to pin 10 of the chroma IC TDA3561A (Y signal) and obtain staircase pattern.

(*ii*) Vary L_{302} (see Fig. 18.9) to obtain a well defined grey scale pattern on the picture tube screen.

(*iii*) Proper setting of L_{302} will remove any hash like disturbance along the steps of staircase pattern obtained on the CRO. If necessary retouch core of L_{302}.

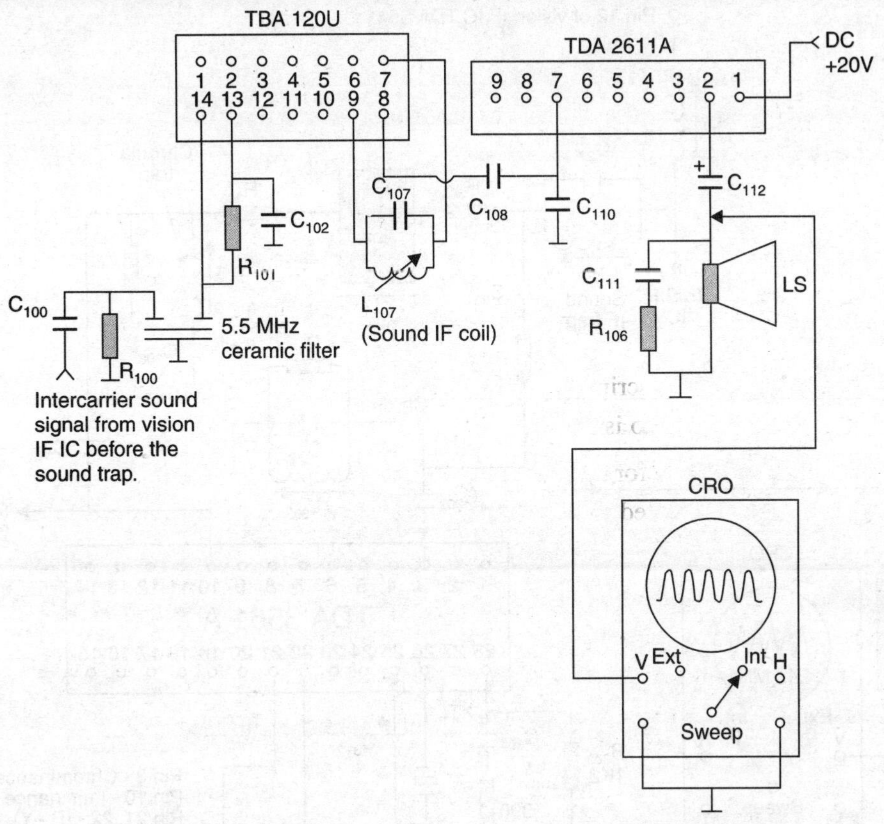

Fig. 18.8 Set-up for aligning discriminator coil in the SIF subsystem.

(10) Chroma Pass Tuned Circuit Alignment

A tuned circuit is provided to select chroma signal from the composite video signal. This circuit is shown in Fig. 18.9 at the chroma signal input pin 3 of the chroma IC TDA 3561A.

(*i*) Without disturbing the setting of chroma trap alignment, shift CRO input to pin 3 through a 1:10 probe to avoid any loading.

(*ii*) Adjust CRO controls to obtain both chroma and burst signals (at least for one line) on the screen.

(*iii*) Vary core of L_{303} to obtain maximum amplitude of the chroma waveform seen on the scope screen. Correct setting will increase saturation of colours in the bar pattern that gets displayed on the receiver screen.

(11) Setting Subcarrier Oscillator Frequency

A crystal controlled oscillator circuit is used for obtaining subcarrier signal voltages for feeding the two demodulators. It must be same as that generated at the transmitting end while encoding colour-difference signals. A small capacitor is provided for setting free running frequency of the oscillator circuit. The crystal and capacitor are connected between pins 25 and 26 of the IC TDA3561A (see Fig. 18.9). Without any change of settings (colour bar with white) at the pattern generator and tuner, proceed as detailed below:

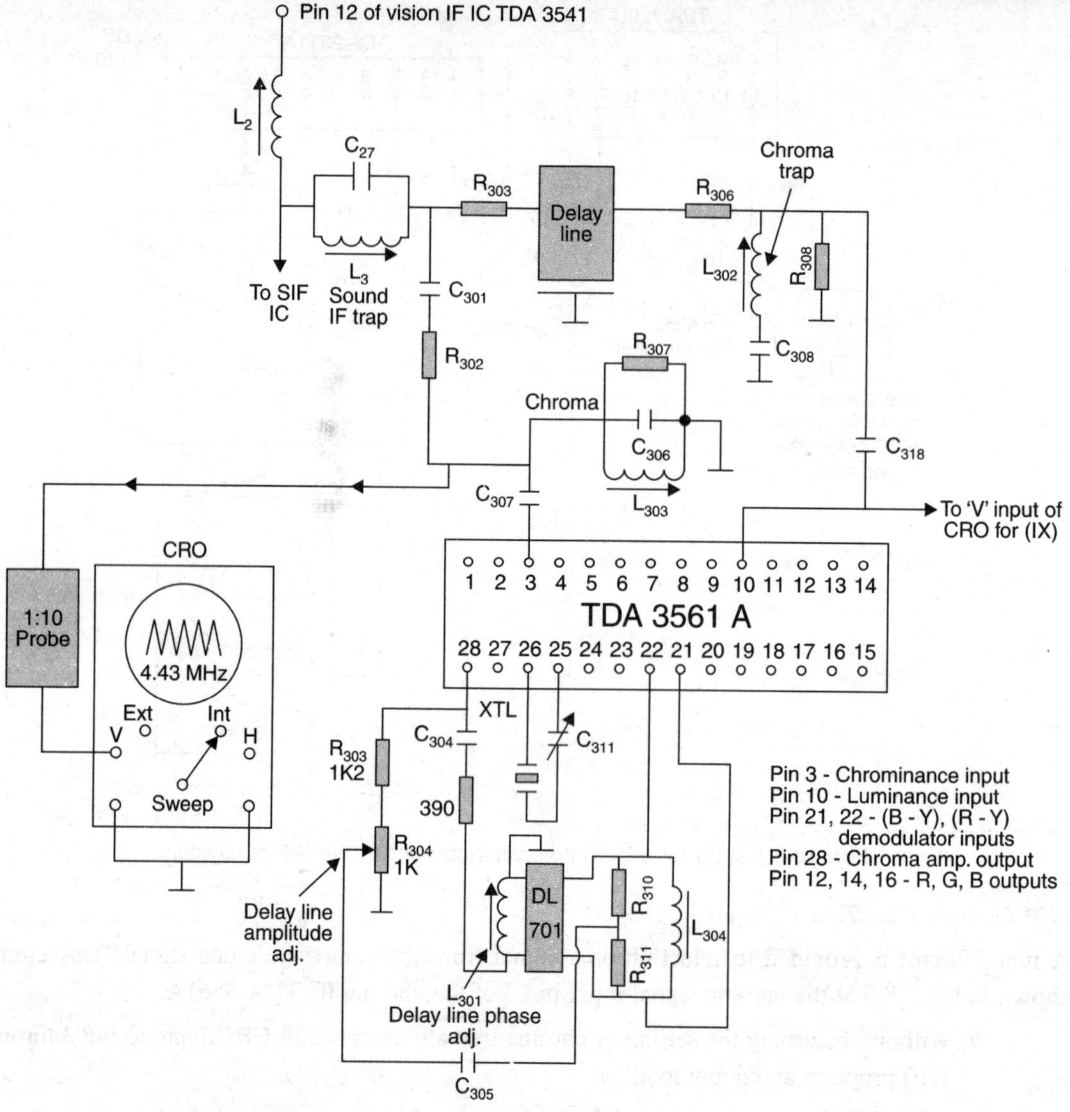

Fig. 18.9 Circuit details around the chroma IC TDA 3561A necessary for the tuning of chroma bypass and chroma pass circuits. Delay line and associated components needed for colour decoder adjustments are also shown.

(*i*) Short pin 1 (dc supply) to pin 6 (saturation control). When this is done, the colour killer circuit is overruled so that colour signal is visible on the screen. In this way it is possible to adjust the oscillator frequency without using a frequency counter.

(*ii*) Interconnect pins 21 and 22. This reduces input signal level at these pins and thus the burst phase detector that receives inputs from these two pins become ineffective thereby enabling the crystal oscillator attain its free running frequency. The colours are still visible because with saturation control connected to full 12V (from pin 1) enough gain becomes available and hence, RGB signals of reasonable amplitude get applied at the picture tube cathodes.

(*iii*) Trim C_{311} to obtain a smooth colour bar pattern on the receiver screen. It should be free of any tearing or streaking.

(*iv*) Disconnect pin 1 from 6 and pins 21 from 22 to restore normal operation. The same smooth colour bar pattern should stay on the screen.

(12) Grey Scale Tracking

The purpose of grey scale tracking is to ensure that R, G, B, signals maintain necessary proportions for correct reproduction of monochrome information both at low and high light levels. The procedure for doing so is as under:-

(*i*) Keep pattern generator connected to the receiver as earlier but remove video modulating signal i.e., feed only RF carrier to the tuner input socket.

(*ii*) Set all the five presets (R_{336}, R_{337}, R_{348}, R_{349} and R_{350}) located around R, G, B amplifiers (see Fig. 18.2) at middle points.

(*iii*) Set receiver brightness and contrast controls at minimum.

(*iv*) Adjust the three cut-off presets (R_{348}, R_{349} and R_{350}) to obtain 140 V dc at the collectors of R, G and B amplifier transistors. Since only carrier is applied, there are no signal voltages at the input of R, G, B amplifiers while voltages are set as explained above.

(*v*) Modulate channel carrier (pattern generator with video signal for white circle and grey background.

(*vi*) Adjust V_{G2} of the picture tube ($_{360}$) to a level so that the circle is just visible.

(*vii*) Change modulating video signal on the pattern generator for a black circle on white background.

(*viii*) Increase brightness control and keep contrast at minimum. Vary cut-off presets (R_{348}, R_{349} and R_{350}) to remove any colour if present on the white background.

(*ix*) Increase contrast control to maximum and adjust drive presets (R_{336}, R_{337}) to remove any colours on the pattern.

(*x*) Repeat the above on maximum and minimum contrast control settings to ensure that no colour is reproduced when there is no colour input.

18.4 TESTING OF CTV RECEIVER

For obtaining ISI (Indian Standards Institute) approval the receiver has to comply to a large number of rigorous tests. ISI has dealt with such a test procedure in their standards books IS-454-1968 and reference may be made to it for details. However, six functional tests are necessary and should be conducted. As taking measurements on all the channels is highly time consuming, these are usually confined to two channels in each band. After alignment and testing is over, the receiver is subjected to a soaking test where it is kept operational for about 24 hours to detect failure, if any, under continued use.

The six functional tests are as under:-

Synchronising Sensitivity

The sync limited sensitivity is defined as the input signal applied to the TV receiver, for which synchronization just loses control completely or partly causing the picture quality to become unacceptable. On tuning the receiver to the desired channel, signal for a standard pattern is fed from the pattern generator. The attenuator is set for a stable picture. The attenuation is then gradually increased and receiver controls reset (except tuning) to stabilize the picture. The level at which picture just looses synchronization is noted for computing the sync sensitivity.

Noise Limited Sensitivity

The input required for a standard video output with a S/N ratio of 30 dB is defined as the noise limited sensitivity. Signal to noise ratio (S/N) is expressed as the ratio of P-P video signal (black to white swing) to rms noise at the picture cathode at 50% modulation level. An RF signal source and a CRO at the output of the receiver are used to measure the two levels in a relative sense *i.e.*, number of divisions on the scope screen mesh.

Colour Sensitivity

Colour sensitivity is the level of input signal applied to a colour receiver at which the colour decoding circuits cease to operate causing the receiver to revert to monochrome operation. Using a colour TV pattern generator, the receiver is set for a standard image. The input signal is then reduced in steps (controls reset for optimum performance) till the picture colour quality becomes unacceptable. The level at which receiver reverts to monochrome operation gives the colour sensitivity.

AGC Range

AGC range is defined as the range of input variation required for a 6 dB variation of the output. Ratio between maximum usable input signal level and input level at which video output with reference to the video level of maximum usable input signal level decreases by 6 dB gives the AGC range of the receiver.

Maximum Usable Audio Output (MUAO)

'MUAO' is defined as the output power available at the loudspeaker for a standard input signal (–50dbm) with 1 KHz modulating signal and ± 15 KHz deviation for the sound carrier. The audio power output is adjusted by the volume control for 10% distortion. The corresponding power output is the maximum usable audio output power.

Synchronising Range

The synchronising range is defined as the range over which sync signals are able to control frequencies of the time-base circuits. With the sync control preset rotated either way, lock-in and hold range points are noted. The corresponding free-running frequencies are measured with an electronic counter. The held and lock-in ranges are then calculated from the above for both vertical and horizontal oscillators.

18.5 RECEIVER INSTALLATION

Colour TV receivers are designed to operate from all channel antennas like the one depicted in Fig. 18.10. It is a 15 element antenna having log periodic arrangement for UHF channels. Good colour reproduction requires an adequate antenna installation. Indoor antennas and some outdoor types that are satisfactory for black and white reception may not have sufficient signal pick up to requisite bandwidth for good colour response. Most antennas have a 300-ohm impedance. An antenna is usually connected to the receiver via a 300-ohm twin lead although sometimes a 75-ohm coaxial cable is used. The antenna often needs to be oriented for optimum colour reception from various stations.

It may be noted that excessive signal strength can be as objectionable as insufficient signal strength. In areas which are very close to powerful transmitting stations, resistive pads are used to attenuate the signal strength. For normal locations that are not too far away from the transmitting tower, receiver manufacturers usually provide a balum (HF impedance matching transformer) to match the 300-ohm antenna down lead to the 75 ohm input impedance of the receiver. It is used at the point where the antenna twinlead enters the receiver.

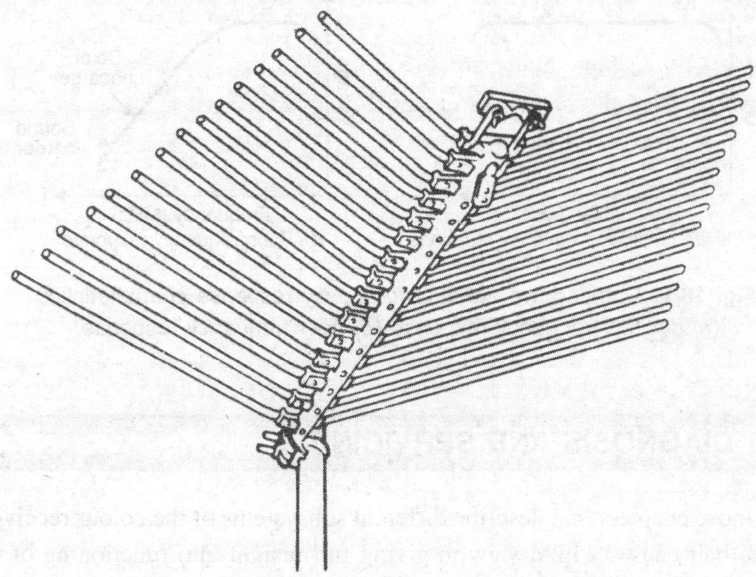

Fig. 18.10 Typical antenna for colour television reception of channels in the VHF and UHF bands.

In weak signal areas, it is often possible to improve reception substantially by installing a preamplifier or booster. It is most effective when it is mounted on top of the antenna mast so that signal-to-noise ratio does not deteriorate due to additional noise pick up by the lead-in between antenna and receiver.

A good colour TV antenna is designed to provide uniform response across each channel as shown in Fig. 18.11(*a*). If the lead-in is properly installed and impedance matching maintained, the same uniform response will be provided at input terminals of the receiver. On the other hand, a poorly designed or a poorly installed lead-in cable with an otherwise good antenna will sometimes have a very non-uniform frequency response as shown in Fig. 18.11(*b*) with the result that, colour and sound signals are considerably attenuated. Reproduction of colour content of the picture will in particular be impaired. It is, therefore, very necessary that a good quality antenna be chosen and installed with a quality twin-lead cable and matching belum. While installing the antenna, it is also necessary that besides proper orientation it is located quite high on the building to avoid any reflections and consequent ghost images. Any compromise on antenna quality and its installation will result in poor colour reception even from an otherwise high quality TV receiver. These days most locations have access to cable TV service and as such no antenna is necessary.

Modern colour receivers do not need any adjustments while installing them. However, the usual precautions of locating the receiver a little away from the wall for proper ventilation (heat dissipation) and from magnetic objects like HI-FI speaker units or other similar objects to prevent wrong beam deflection are necessary. Any magnatization effect due to storage of receiver is taken care of by the degaussing coil when the receiver is switched on for a number of times allowing a reasonable time between successive switchings.

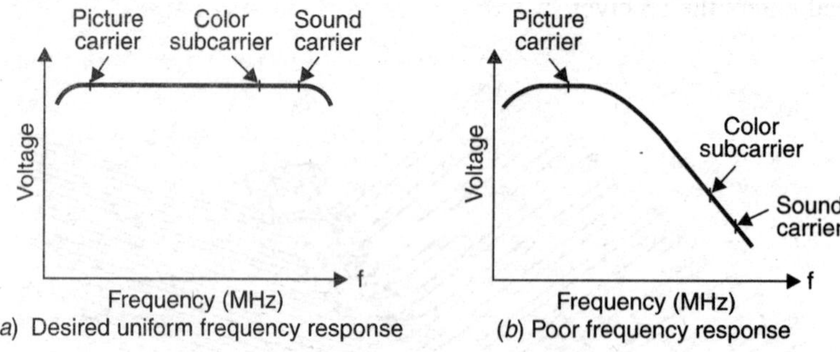

Fig. 18.11 Comparative antenna frequency response characteristics (*a*) desired frequency response (*b*) poor frequency response.

18.6 FAULT DIAGNOSIS AND SERVICING

The last section of those chapters that describe different subsystems of the colour receiver is devoted to probable faults and their causes with a view to giving full insight into functioning of that part of the receiver. It is equally important to gain familiarity with fault diagnosis and to be able to pin-point defective component(s) and/or device(s) from the symptoms observed on switching on a defective receiver. This section is devoted to such an approach where various symptoms together with servicing hints are listed with reference to colour receiver circuit (Fig. 18.2) under study.

Receiver inoperative (dead). As obvious, the fault is in the power supply and associated circuits. Refer circuit diagram (Fig. 18.2) and check ac fuse F_{201}, bridge rectifier diodes D_{205} through D_{208}, low voltage supply rectifier diodes D_{202}, D_{202A}, D_{203A}, low voltage regulator components TR_{201}, D_{201},(12V zener), R_{200}, R_{201} and reservoir filter capacitor C_{213}. Faulty diodes D_{213}, D_{214} and D_{215} across the primary winding of LOT can also lead to failure of power supply and make the receiver inoperative. Check these diodes one by one and make sure that while doing so the receiver is not plugged-in to ac mains.

SMPS not working. Most receivers have a separate SMPS circuit but the one under study (Fig. 18.2) employs TDA2582 which combines SMPS functions with drive for the horizontal output stage. Check IC TDA2582, drive amplifier transistor TR_{203}, associated diode D_{210}, line output transistor TR_{202} (BU508A) and D_{212}–the 6.2V zener at pin 10 of TDA2582.

DC output voltage rises and falls (hunting). Ensure that there is no overloading of 180V, 26V and 12V dc output supplies. Then check feedback pot R_{232}, R_{205} (pin 15 of TDA 2582) and capacitors C_{207}, C_{208}, C_{212}, C_{214}, C_{221}, C_{223} and C_{410} that are in the control circuitry.

EHT not developing. Check driver transistor TR_{203}, line output transistor TR_{202}, flyback transformer T_{206}, associated diodes D_{213}, D_{214}, D_{215} and capacitor C_{215}.

No raster. If EHT and dc supplies are available, no raster means inoperative picture tube or failure of deflection circuits. Check filament supply to CRT, screen grid (VG_2) supply to picture tube, 12V supply to combination IC TDA2578A, vertical output IC TDA3561A, chroma IC TDA3561A, RGB amplifier board besides ground (earth) connections around the picture tube circuitry.

No colours on the picture. Check chroma IC TDA3561A, voltages at pins 3 and 28 of this IC, chroma delay line DL_{701}, XTL (8.866 MHz), sandcastle pulse (pin 8), readjust C_{311} (in series with XTL), C_{300} and C_{308} (pin 1) and retune L_3 (5.5 MHz trap) if necessary. Finally check fine tuning adjustments on the tuner.

Red, green or blue colour is missing. Check colour amplifier transistors TR_{302}, TR_{303} and TR_{304}. Also check voltages at pins 12, 14 and 16 of chroma IC TDA3561A and after isolating resistors R_{326}, R_{327} and R_{328}. In the end, though not common, check CRT by replacing with a known good picture tube.

Dim raster, brightness control not working. Check 12V DC supply to the brightness control pot R_{324} and 8KV supply to the focus control circuit. Also check ICs TDA2578A, TDA3561A, RGB amplifier board and dc supplies to them.

Horizontal line on the screen. Vertical deflection circuit is not functioning. Check 26V dc supply to TDA3651A (vertical output), vertical deflection coils, TDA 2578A (combination IC) and D_{403}.

Vertical rolling (SYNC failure). Check TDA 2578A, readjust frequency, adjust pot R_{416}. Also check R_{411} and C_{412}.

Retrace lines visible. Check ICs TDA2578A, TDA3561A, R_{414}, 180V dc supply line and RGB amplifier board. Adjust R_{360} (VG_2 control) if necessary.

Horz SYNC failure (picture tearing). Check TDA2578A, R_{400}, R_{406}, R_{410}, R_{411}, R_{412}, C_{236} and C_{407}.

Less raster width. Check voltage across capacitor C_{220}, it should be 100V. In the receiver under study this (100V) indicates that all other voltages are correct. For any deviation adjust R_{232} of the SMPS control circuit to obtain 100V across C_{220} (correct dc voltages).

No picture no sound. Check dc supply to tuner and IF subsystem, check TRI (BF370) L_1, L_2 and IC TDA3541.

No picture but sound present. Check L_3, luminance delay line DL_{470}, capacitor C_{318} (pin 10 IC TDA3561A) and ICs TDA2578A, TDA3561A.

CHAPTER 18

No sound but picture normal. Check muting diode D_{101} (pin 2 IC TBA120U,) dc supplies to ICs TDA120U, TDA2611A and loudspeaker circuit. If necessary, check TDA120U and TDA2611A.

Tuning shifts. Check IC, (TAA550-30U). Also check AFT operation for its effectiveness by measuring voltage at pin 5 of TDA3541. It should stay at nearly the same value (= 6V) with or without input signal.

Variations in picture brightness and contrast. Check potentiometers R_{324} (brightness) and R_{320} (contrast) and if continuity is erratic change with good pots.

Negative picture. Check L_1 (collector lead of BF_{370} (TR_1), SAW Filter and TDA3541.

Netting in the picture. Check alignment of L_3 (5.5 MHz trap). Also check antenna lead and tuner operation.

Low DC voltage. Check IC TDA 2582 and readjust R_{232} if necessary.

Resistor in series with AC supply line burns on Replacement (R200A in Fig. 18.2). Check transistor TR_{202} (BU508A) degaussing coil, diodes D_{205}, D_{206}, D_{207}, D_{208}, and capacitors C_{201}, C_{202}, C_{204}, C_{205}, C_{210}, C_{211}, and C_{213}.

DC Voltages high. Check adjustment of R_{232} and R_{235}.

Retrace with picture. Check adjustment of VG_1 (picture tube), preset R_{360} and ICs TDA2578A, TDA3561.

No picture but snow present. Check tuner, channel operating unit, and switching circuits.

18.7 SAFETY PRECAUTIONS IN TELEVISION SERVICING

The following general safety precautions should be observed during operation of test equipment and servicing of television receivers:

(*i*) A contact with ac line can be fatal. Line connected receivers must have isolation so that no chassis point is available to the user. After servicing all insulators, bushes, knobs etc. must be replaced in their original position. The technician must use an isolation transformer whenever a line connected receiver is serviced.

(*ii*) Voltages in the receiver, such as B+ and EHT can also be dangerous. The service man should stand or sit on an insulated surface and use only one hand when probing a receiver. The interlock and back cover of the receiver must always be replaced properly to ensure that receiver's high voltage points are not accessible to the user.

(*iii*) Fire hazard is another major problem and must get the attention it deserves. Technicians should be very careful not to introduce a fire hazard in the process of repairing TV receivers. The parts replaced must have correct or higher power rating to avoid overheating. This is particularly important in high power circuits.

(*iv*) The picture tube is another source of danger on account of possible implosion. If the envelope is damaged, the glass may shatter violently and its pieces fly great distance with force. It can cause serious injury on hitting any part of the body. Though, in modern picture tubes, internal protection is provided but it is necessary to handle it carefully and not to strike it with any hard object.

(*v*) Many service instruments are housed in metal cases. For proper operation, the ground terminal of the instrument is always connected to ground of the receiver being serviced. It should be made certain that both the instrument box and receiver chassis are not connected to the hot side of ac line or any point above ground potential.

(*vi*) All connections with test leads or otherwise to the high-voltage points must be made after disconnecting the receiver from ac mains.

(*vii*) High voltage capacitors may store charge large enough to be hazardous. Such capacitors must be discharged before connecting test leads.

(*viii*) Only shielded wires and probes should be used. Fingers should never be allowed to slip down to the meter probe tip when the probe is in contact with a high voltage circuit.

(*ix*) The receiver should not be connected to a power source which does not have a suitable fuse to interrupt supply in case of a short circuit in the line cord or at any other point in the receiver.

(*x*) Another hazard is that the receiver may produce X-radiation from the picture tube screen and high voltage circuit. It is of utmost importance that voltages in these circuits are maintained at designed values and never exceeded.

REVIEW QUESTIONS

1. With reference to block schematic of the SSPP colour receiver given in Fig. 18.1, explain briefly how the RF signal that enters the tuner input socket is processed by various subsystems to provide a steady colour picture and good quality sound output.
2. Why is the receiver described in this chapter (Fig. 18.2) called a single-switch-power-pack (SSPP) version of a colour receiver? Describe briefly the line output cum SMPS circuit to justify your answer.
3. Colour TV pattern generators provide video signals to produce a variety of patterns for alignment and testing of colour receivers. Enumerate such patterns and briefly describe various alignments for which these can be used.
4. Explain with a suitable set-up and circuit details how would you proceed to align and tune the following.

   (*i*) tuner output circuit

   (*ii*) reference oscillator frequency

   (*iii*) AFC circuit

   (*iv*) SIF trap circuit.
5. Explain with necessary set-up and circuit diagrams how you will proceed to align the following circuits of the colour receiver.

   (*i*) field drive and deflection circuit

   (*ii*) line drive circuit

   (*iii*) SIF discriminator circuit
6. Describe the importance of proper installation of an antenna for a colour receiver. List all precautions which must be kept in mind while installing a colour receiver.

7. **With reference to the given circuit diagram (Fig. 18.2) of a colour receiver, explain how would you proceed to localize the following faulty conditions.**

(*i*) Inoperative (dead receiver)

(*ii*) No raster

(*iii*) EHT rises and falls

(*iv*) EHT not building.

8. **In the colour receiver of Fig. 18.2, localize component(s) and/or device(s) that can cause following faults.**

(*i*) Red colour missing in the reproduced picture

(*ii*) Only a horizontal line appears on the screen.

19 ADVANCES IN COLOUR RECEIVERS AND TELEVISION SYSTEMS

INTRODUCTION

The availability of ICs and microprocessors has brought in significant improvement in colour receivers. Digital technology has also transformed television transmission and reception from analog to digital form of signal processing. Television receivers employing digital circuitry are now available that produce excellent pictures and high quality stereo sound.

Innovative efforts towards achieving further improvements in picture and sound quality have resulted in MAC encoding that eliminates interference between luminance and chrominance components of the video signal. Similarly, High Definition Television (HDTV) that enables almost twice the present horizontal and vertical resolution is gradually replacing the present television systems. Plasma and LCD screen TV receivers are also now available. This chapter is thus written to describe new era colour receivers and introduce advanced television system.

19.1 MODERN COLOUR RECEIVERS

The now available upper range colour receivers are of sophisticated design having features like synthesized electronic channel selection, auto programming, remote control of functions and deliver excellent pictures and sound output. With the application of digital concepts all functions related to video, audio, beam deflection and associated sequences are controlled by VLSI circuits. Specially designed single chip microcomputers control, measure, store and process a large number of functions. All such receivers have provision to store around 100 broadcast station channels in their electronic memory. Also their tuners are of 'Cable Ready Type' meaning have additional provision for receiving all cable TV channels. The receiving system is PAL B/G and AV input for PAL/NTSC 4.43 MHz. Almost all functions are controlled through remote control from the hand-held transmitter. Important functions can also be manually controlled with push-buttons provided on the front panel.

To illustrate receiver function controls, a BPL 51 cm screen colour receiver has been chosen. The remote controlled transmitter unit that is supplied with this is pictured in Fig. 19.1 and various functions that can be controlled with this are briefly explained.

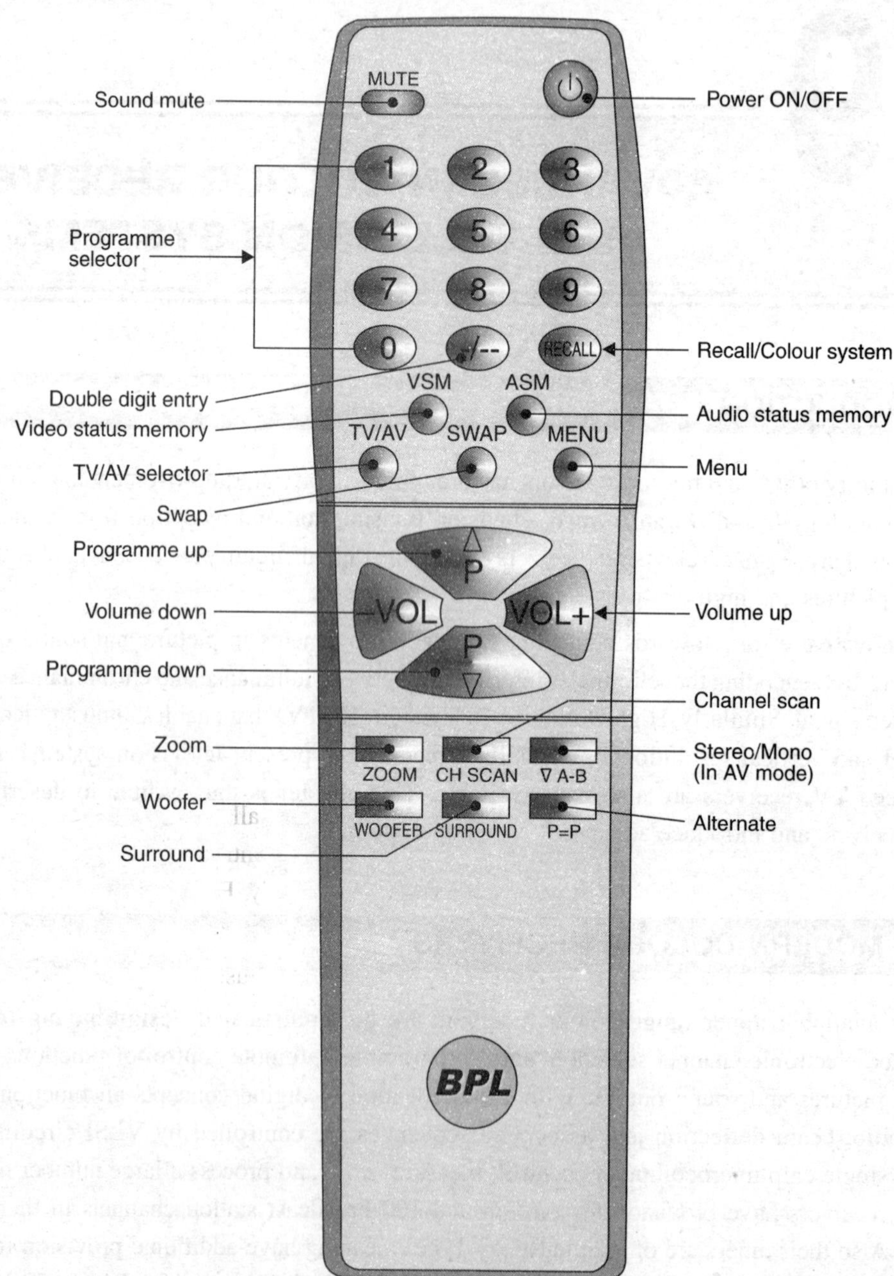

Fig. 19.1 Remote Control transmitter unit of a 51 cm BPL colour receiver (Courtesy BPL, India).

1. **Auto Shut Off Function.** Besides the usual ON/OFF function and the provision to put the receiver on 'Stand-By' mode, the in-built circuitry shuts off the receiver even when left 'ON' after 10 minutes of closing of transmission or cable system failure of the chosen channel.

2. **Channel Scanning.** On pressing the 'channel selection button' on the remote controller, the channel positions are automatically scanned and can be stopped at the desired channel by pressing the same button when that channel appears on the screen.

3. **Sound Mute.** The sound output is muted on pressing the 'sound mute' button and on pressing it again, the sound is restored.

4. **Sound Output Options.** With 'ASM' button on the remote, Personal, Music, Talk and Normal modes can be selected on its successive depression.

5. **Quick Picture Control.** On successive pressing of the 'VSM' button, Personal, Dynamic, Natural, Soft or Game mode can be selected as desired.

6. **Stereo/Mono Sound Selection.** With input from AV terminals, depressing of Stereo/Mono (∇AB) button enables stereo or mono sound output. Similarly on depressing the 'Surround Button' a 3-dimensional sound output effect is created. On further pressing the Woofer button, the Bass effect of sound is enhanced.

7. **Direct Programme Selection.** While channels up to 9 can be selected directly, but for two-digit channels, the double digit button is to be first depressed.

8. **Zoom.** Pressing the 'Zoom' button successively changes the picture display on the screen from 'Normal' *i.e.*, 4:3 aspect ratio, to Zoom (vertically) and 'Wide' in the 16:9 format.

Picture and Sound Adjustments. With receiver 'ON' on depressing the 'Menu' button an Icon showing picture, audio, tune and timer symbols appears on the screen as shown in Fig. 19.2(*a*). For picture adjustments the 'programme' selection button is manipulated to bring the thick arrow head close to the picture symbol as shown in Fig. 19.2(*a*). On pressing the 'menu' button again, picture menu appears on the screen as seen in Fig. 19.2(*b*). Then on repeated pressing of either of the programme buttons, the small arrow (see Fig. 19.2(*b*)) can be brought to point towards the control to be varied out of colour, brightness, contrast and sharpness. In the figure the small arrow is pointing towards colour, then colour intensity can be changed by varying the volume ± control. The strip movement indicates the variation. Similarly other options can be adjusted as necessary. Fine tuning of the chosen channel can also be done in a similar way.

Sound Adjustments. A similar procedure enables audio adjustments by moving the thick arrow towards 'Audio' (see Fig. 19.2(*a*)). On pressing the menu button again an Icon showing Bass, Treble, Balance, Woofer, 3-D surround appears on the screen. On further depressing corresponding buttons any of these choices can be obtained.

Other Adjustments. In a similar way, the built in micro-computer facility in the receiver circuitry enables tuning of incoming channels by either Automatic, Semi-automatic or Manual modes. The additional features that are also available include recall and alternate channels, channel coping, channel swapping, skip channels, programme naming, programme index, volume lock setting, dynamic picture settings, music mode setting, tuning lock setting and storage of often needed telephone numbers.

Colour receivers of other well known brands also provide similar adjustment facilities though the Icons and procedures are somewhat different.

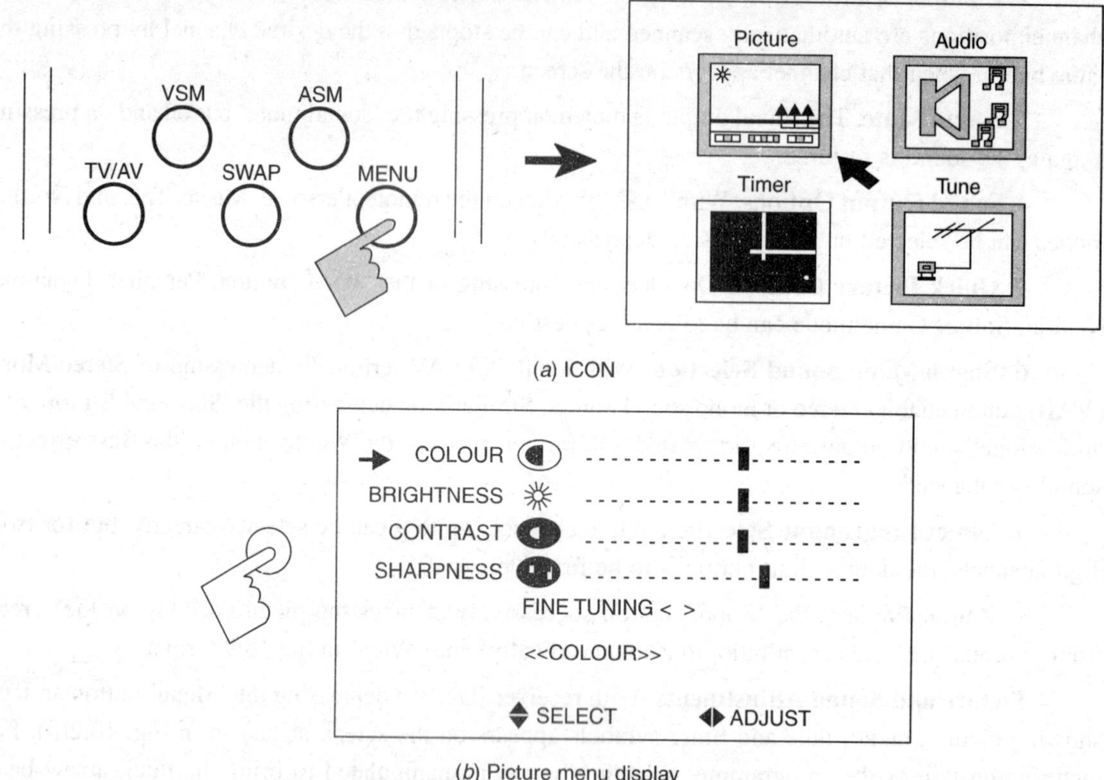

(*a*) ICON

(*b*) Picture menu display

Fig. 19.2 Picture Adjustments (*a*) Icon display (*b*) Picture menu display

19.2 PLASMA AND LCD SCREEN TV RECEIVERS

Over the years demand for bigger screen receivers have been growing to enable better viewing of television programmes and movies. While efforts were made to increase CRT screen size but it has not been possible to go beyond around 100 cm diagonal. However, screen sizes bigger than possible with CRTs are now available. These are based on PLASMA and LCD displays. In plasma screens a solution is coated on the inner side of two glass screen panels. It has millions of phosphor coated miniature glass bubbles containing plasma which is a gas made-up of free flowing ions and electrons. An electric current flow through the screen panels causes certain plasma containing bubbles to emit ultraviolet rays which trigger the phosphor coatings to produce red, green and blue lights. These on scanning and intensity control enable colour pictures on the screen.

In LCD displays the screen consists of a liquid crystal solution inbetween two colour panels. An electric current when applied across very small sections of the panel that are insulated from each other the crystal molecules reorient themselves to allow light to pass through them. With no charge applied across that section, the molecules act as shutters and no light passes through. This forms the basis of creating dark and grey areas on the panel. For obtaining colour display R, G, B colour filters are provided. As for plasma screens, application of controlled charge density and its incidence produces colour pictures on the LCD display panel.

In both plasma and LCD screens, the application of video signal (R, G, B) voltages to the millions of tiny sub-areas called pixels is controlled through a matrix which has insulated conducting lines etched on the panel in the form of columns and rows for reaching each destination.

*The television receivers employing Plasma or LCD screens are otherwise nearly similar. There is no need for EHT and high dc voltages. Similarly no deflection circuitry and yokes are needed. The R, G and B video signals and sync pulses obtained on processing the received RF input, as in conventional TV receivers, are fed into a computer, the control panel of which is programmed to send these signals to the pixels on the screen for obtaining pictures as obtained with CRT screen receivers. Since there is no need for heavy transformers and deflection coils such receivers are very light in weight and occupy less space (≈ 400 mm) as seen in Fig. 19.3. Both LCD and plasma receivers enable good colour reproduction and high contrast.

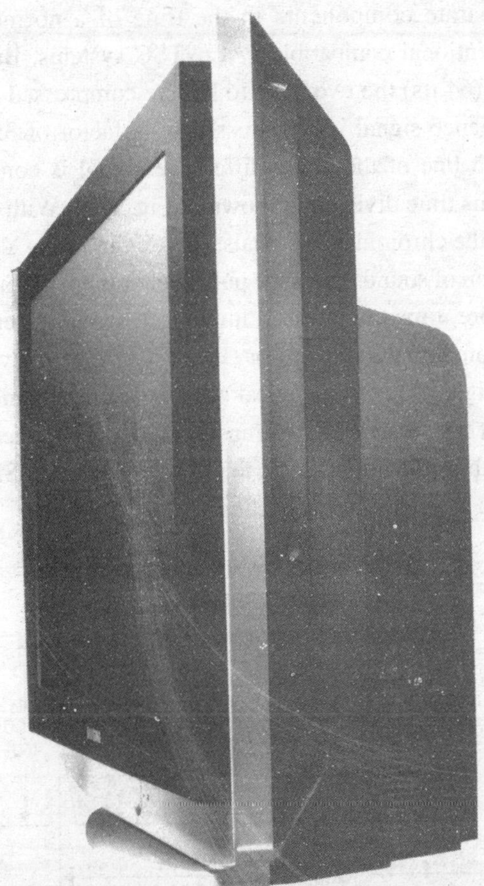

Fig. 19.3 A plasma screen colour receiver (Courtesy SAMSUNG).

*For detailed constructional and action details of plasma and LCD screens refer this author's book "MODERN TELEVISION PRACTICE" 3rd edition published by New Age International Publishers.

19.3 EXTENDED DEFINITION TELEVISION (EDTV)

This term refers to improved performance of TV systems that require different transmission standards but retain the present line number and field rates. EDTV is mainly intended for satellite broadcasting. Normal terrestrial TV broadcasting is by vestigal sideband amplitude modulation. However, satellites cannot use AM-VSB because of non-linearity of satellite amplifiers. Hence, frequency modulation is used which also makes efficient use of the available channel bandwidth.

In the conventional PAL/NTSC systems the luminance and chrominance information is transmitted simultaneously for each line in interleaved form. For sound information there is an additional carrier and there are two separate carriers in the case of stereo sound. This creates problems of interference beats between different signals and hence affects the picture display.

The interference problem is solved by using a new encoding system called "Multiplexed Analog Components" (MAC). In this luminance and chrominance signals are time division multiplexed *i.e.*, sent in time sequence as separate components in the time of a normal horizontal line and not simultaneously as in the conventional compatible PAL/NTSC systems. But, for transmitting the two signals in the same line period (64 μs) the two have to be line compressed *i.e.*, signals are speeded up. In the MAC system, the luminance signal is compressed by a factor of 3:2 and this reduces its time period to 35 μs. Similarly each line of the colourdifference signal is compressed by a factor of 3:1 reducing its time to 17.5 μs. This time division is shown in Fig. 19.4. With these compression ratios, it is not possible to transmit both the chrominance signals, (R – Y) and (B – Y) in every line as this would leave no time for the transmission of sound, line sync pulses and other auxiliary data vital to the business aspect of DSB system. Therefore, time compressed luminance signal is sent on each line with one of the compressed chrominance components (R – Y) or (B – Y). At the receiving end, the signals are de-compressed to obtain original signals. Later, the usual one-line delay technique is used to obtain U and V outputs. It must be noted that the sequencing of transmission as explained above would need a little increase in the base bandwidth but it is not difficult to allow this in the DSB system where bandwidth is much less restricted unlike in terrestrial broadcasting.

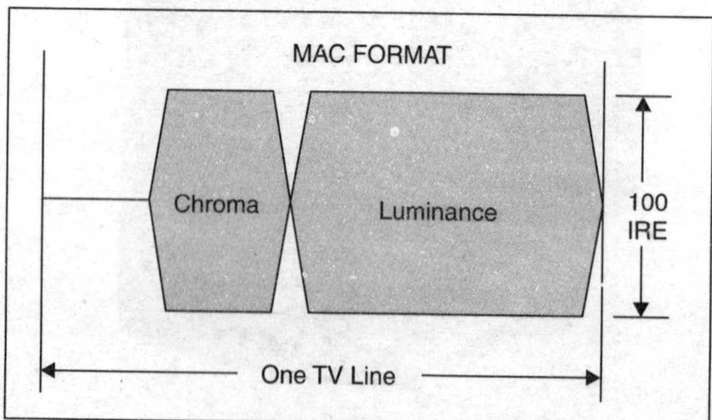

Fig. 19.4 Chrominance and luminance signals are time compressed and transmitted in a sequential format on each television scanning line.

19.4 HIGH DEFINITION TELEVISION (HDTV)

High definition television aims at improving both vertical and horizontal resolution of the reproduced picture by approximately 2:1 over existing standards. Other improvements include improved colour reproduction, higher aspect ratio and stereo sound. These improvements result in a picture quality as clear as obtained from 35 mm movie films and sound as good as from digital audio CDs.

The worldwide pioneers in HDTV were Japan and America. The standards adopted include 1125 scanning lines per frame, 60 fields per second, 2:1 interlace scan and an aspect ratio of 16:9. However, European countries have developed their own HDTV called 'EUREKA' to be compatible with CCIR standards with line numbers equal to 1250/1251, 50 frame rate and an aspect ratio of 16:9. The bandwidth required to transmit one TV channel is between 10 to 12 MHz with prior compression of both video and audio signals.

MUSE Encoding System. The MUSE is 'Multiple Sub-Nyquiest Sampling Encoding' developed by NKH of Japan is used by all the HDTV variants for signal processing to obtain the results stated above. It uses the fundamental concepts of performance exchange in the spatic-termporal *i.e.*, in the transistory transformation domain along with motion compensation to reduce the transmission bandwidth down to about 10 MHz.

Satellite transmission of HDTV signals is carried out in the KU-Band frequency spectrum of 11-13 GHz because more channel bandwidth needed for it can be easily provided in this band. Also C-Band and KU-Band signals do not interfere with each other allowing a single satellite to relay both types of transmissions.

Fig. 19.5 A 42′, 16:9 aspect ratio HDTV receiver with built-in turner (Courtesy SONY).

As of now, these are few satellite HDTV channels because not many countries have yet provided facilities for its reception and onward transmission to subscribers who also are not inclined to buy expensive receivers. However, HDTV transmission is carried out by terrestrial means around big cities where people can afford to buy HDTV ready receivers or converters for reception on existing TV receivers. For best results a 16:9 aspect ratio digital receiver with a built in tuner is the best option though expensive. One such receiver is shown in Fig. 19.5. In India and many other countries HDTV has not yet been introduced.

19.5 ELEMENTS OF DIGITAL COMMUNICATION

Digitization of signal path has gained importance due to many advantages of the digital system. As the digital signal is in 0 or 1 *i.e.*, low and high pulse amplitude form, there is no degradation of the visual and aural components due to noise or distortion while processing it in the studio chain or in equipments at the receiving end. The noise immunity advantage in digital circuits results from the fact that if the noise level is less than the threshold noise margin, the noise signal is sharply attenuated between the input and output, while the desired logic signals get transmitted at full amplitude without error. In addition, the digital bit stream can be stored in logic circuits as long as necessary for signal correction if needed and other related processes. These and other merits of digital processing have resulted in rapid transition from analog to digital form of transmission and reception in almost all electronic communication systems.

Digital Data Processing. Electronic signals may be analog or digital. An analog signal, like the output of a microphone can have any amplitude, within a range, for different frequency components. In contrast, in a digital system, information is represented in discrete or digital form rather than continuous as in analog. For digital processing the most common format is the 'Binary System', where for any signal only two discrete states are possible, which are denoted by 0 and 1 *i.e.*, low land high amplitude. The 0 and 1 as they occur in any digital circuit are called 'bits'.

Based upon only two signal levels, digital signal amplitudes are represented in binary numbers, where the base is 2 *i.e.*, powers of 2 are used unlike 10 in the decimal system. For example, decimal number 19 is expressed in binary system as 10011. This representation is correct because $10011 = 1 \times 2^4 + 0 \times 2^3 + 0 \times 2^2 + 1 \times 2^1 + 1 \times 2^0 = 16 + 0 + 0 + 2 + 1 = 19$. Similarly other numbers can be expressed in binary form as 25 is equal to 11001.

A/D and D/A Converters. As explained, it is advantageous to do signal processing in digital form though the real world is analog. Therefore, analog signals must first be converted to digital form before processing, and digital results converted back to analog form for human consumption.

Digital to Analog (D/A) conversion involves translating digital information to equivalent analog form by converting 'n' digital voltage levels (0 or 1) of the input signal into one analog voltage. Assume that at any instant, the digital input is a 4 bit binary equal to 0011. Since its decimal form is equal to 3 *i.e.*, $0 \times 2^3 + 0 \times 2^2 + 1 \times 2^1 + 1 \times 2^0$, the D/A converter must develop an output voltage equal to 3 amplitude levels, usually mV, in the analog form. Similarly an Analog or Digital (A/D) converter should develop a binary code which represents the input analog voltage. For example an input equal to 11 mV will develop a binary code as 1011.

The converted outputs, either way, vary rapidly depending on the speed *i.e.*, frequency of input signal. Since communication signals are of very high frequency, special high speed converters based on IC technology have been developed and are an essential component in the chain of any digital processing system.

Signal Encoding. Digital data is obtained on quantization and sampling the analog signal at a very fast rate. The input signal is seen divided into a large number of amplitude levels and this is called quantizing of the input signal. This is illustrated in Fig. 19.6 where (*a*) is the signal to be transmitted and (*b*) its quantized form. Only these levels and not the entire analog signal is sampled at a very fast rate to develop corresponding output as a digital data stream. In practice, the sampling is done at a much higher rate than the frequency of the input analog signal, so that the quantized levels fall very close to each other enabling a digital output which in turn will produce at the receiving end an analog output that is nearly the same as the input signal. Note that the quantized levels shown in Fig. 19.6(*b*) have been deliberately exaggerated for clarity of the formation of various levels.

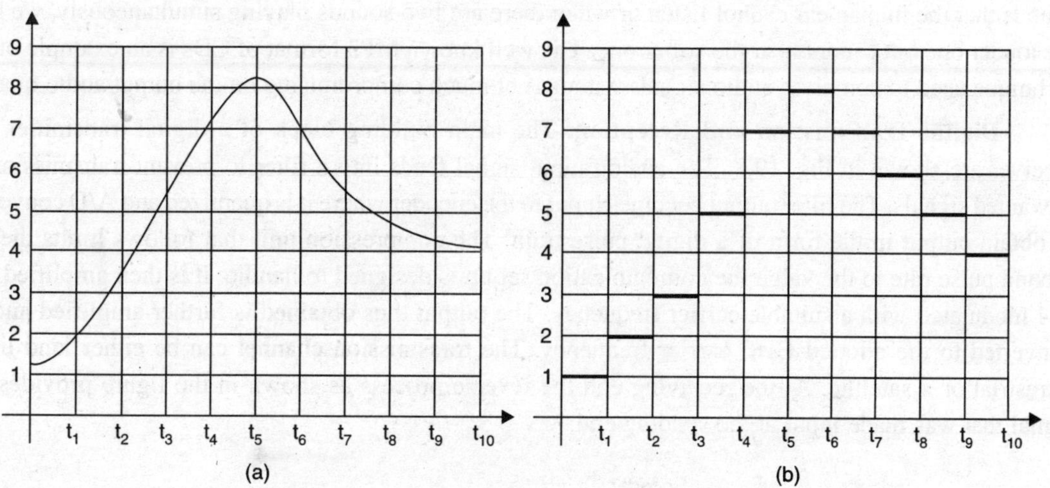

Fig. 19.6 Sampling of analog signals (*a*) original input signal (*b*) same signal quantized into sampling levels.

Encoder. Encoding means development of pulse train in low and high voltage amplitude levels from outputs as obtained on sampling. One type of encoder is shown in Fig. 19.7. As seen the digital clock feeds pulses equal to the sampling rate to both the quantiser and A/D converters.

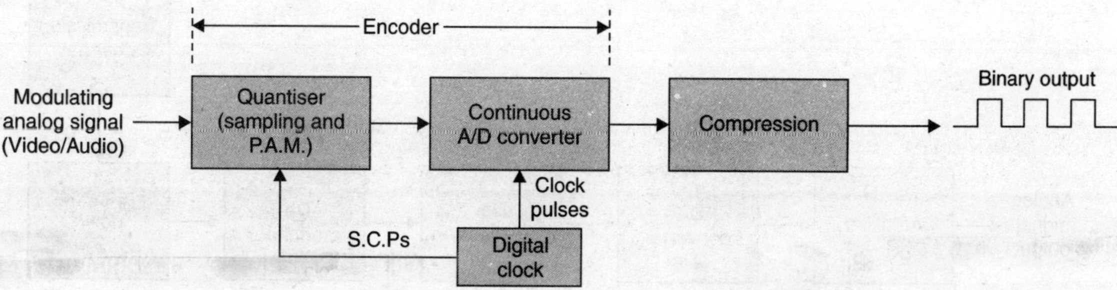

Fig. 19.7 An often used encoder for converting video and audio analog signal to their corresponding binary pulse train.

Data Compression. The binary signal is square wave in nature and to retain its shape during transmission it is necessary that the sampling rate is at least 2 to 3 times the input analog signal frequency. The ensuing digital bit rate will be quite large needing very wide channel bandwidth for transmission. It could be as large as 100 MHz for television video signals. Since the presently allotted channel width per channel is relatively small, it becomes necessary to reduce the per second bit rate. In television where the pulse train rate can be as high as 140 Mega bits/second, a compression of nearly 50:1 becomes necessary. Such a compression rate is possible because in the TV video signal there is lot of redundancy from one frame of video to many that follow, meaning motion and background stays the same for many frames at a time and its repetitive transmission can be avoided thus reducing the data rate. For example, such a situation prevails when news are read by the news reader or for often noticed repetitive scenes in movies. Of course, the data not transmitted has to be added at the receiving end while processing it.

The same applies to audio signals where the sampling rate is reduced to the limit when its effect is noticed in the quality of sound output. Advantage is also taken of the fact that these are certain sounds that the human ear cannot listen or when there are two sounds playing simultaneously, we hear the louder one but cannot hear the softer one. The well known MP3 format of CDs is an example of the technique used to compress audio signals without not much compromising on the output audio quality.

Digital Transmission and Reception. The main building block of a digital transmitter and receiver are shown in Fig. 19.8. The analog input signal feeds into a filter to prevent transmission of unwanted signals. The filter output becomes input to the encoder where it is quantized and A/D converted to obtain output in the form of a digital pulse train. The compression unit that follows limits the per second pulse rate to the value the communication set up is designed to handle. It is then amplified and FM modulated with a suitable carrier frequency. The output thus obtained is further amplified and up converted to the allotted UHF carrier frequency. The transmission channel can be either land lines, terrestrial or a satellite. At the receiving end the reverse process as shown in the figure provides the signal that was made input at the sending end.

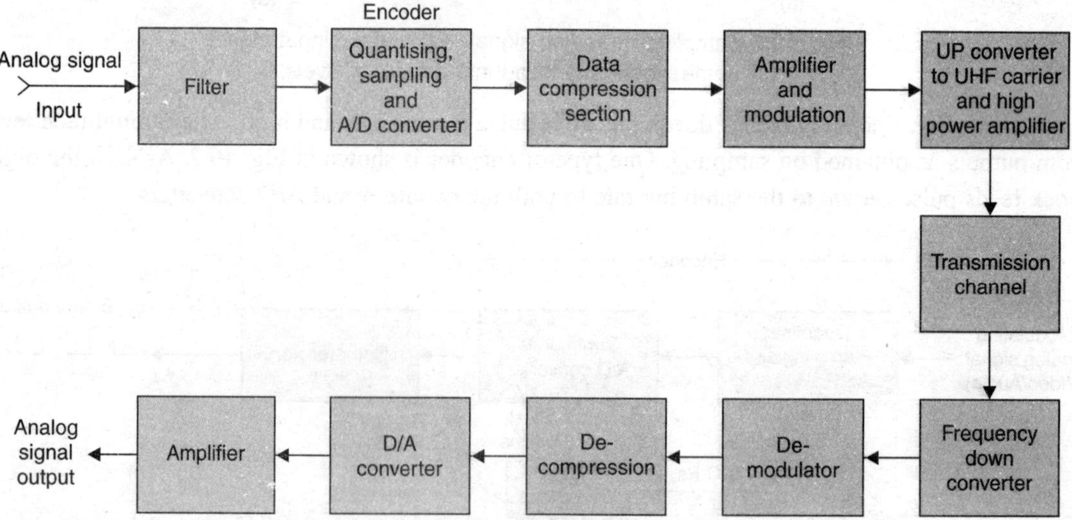

Fig. 19.8 Basic building blocks of a digital communication system.

REVIEW QUESTIONS

1. Explain briefly how in a modern colour receiver its main functions are operated with the provided Remote Control unit.
2. Describe the basic techniques that are used in both Plasma and LCD displays for use in a TV receiver screen.
3. Explain how in a Plasma TV receiver the control computer proceeds to send video signals to the pixels in the display screen.
4. In the EDTV system explain how the MAC encoding system causes separate transmission of luminance and chrominance signals to prevent any mutual interference.
5. Enumerate the main parameters and new features of the HDTV system and explain how it enables excellent reproduction of pictures and high quality stereo sound ?
6. What are the main merits of digital signal processing as compared to the earlier analog system. What is encoding and why is it necessary to use A/D and D/A converters both at the sending and receiving ends of the communication system:
7. What is data compression and why it becomes necessary to do so before transmission? How does compression affect the quality of transmitted audio signals?
8. Draw block diagram showing basic building blocks of a digital communication system and explain the function of each section.

20 DIGITAL SATELLITE TELEVISION

INTRODUCTION

The main drawback of terrestrial broadcasting is in its limiting range because of the earth's curative which eventually breaks the signal path thus preventing reception over long distances. This problem is solved with geo-stationery satellites that orbit the globe at the same speed as the rotation of earth. The curvature of earth thus no longer presents any problem and communication over long distance is carried out through satellites.

At the receiving end, the signals received are very weak and special dish antennas and equipments are needed to process them. The distribution of received signals to the subscribers is either through cables or retransmission over the existing terrestrial network. Direct-To-Home reception is also possible for which each installation needs its own receiving dish antenna and a decoder-receiver. All aspects of satellite transmission, reception and signal distribution are briefly described in this chapter.

20.1 SATELLITE COMMUNICATION SYSTEM

As stated above, the problem of limiting range of broadcast television is solved by communication satellites which move around the earth at the same speed at which the earth revolves around its axis. For this, satellites are launched into space with enough force to enable them to reach 35887 kms (usually referred to be 36000 kms) above the earth. At this altitude the satellite travels around the earth every 24 hours which is the same time it takes the earth to make one full revolution thus keeping pace with it. This way the receiving earth station antenna need to be directed only once towards the wanted satellite and it will continue to receive signals without any further adjustment. Since the down-link signals from far away located satellites are extremely weak special antennas called DISH antennas are used to pick up these signals.

Satellite Signal Path. The basic technique of communication by satellites is shown Fig. 20.1. The earth transmitting station beams signals from a dish antenna as RF waves to the satellite, which in turn retransmits it to the receiving ground station. The frequencies used for this purpose are in the microwave band of 3 to 30 GHz, because at these ultra-high frequencies, the atmosphere no longer acts as barrier and signals are able to travel out into space and back without any obstruction, absorption or deflection. The assigned microwave carrier frequency for sending signals to the satellite is called UP-LINK frequency. As shown in the figure, the satellite has its own dish antenna which receives the up-link signals. It feeds these to a receiver-mixer located in the satellite.

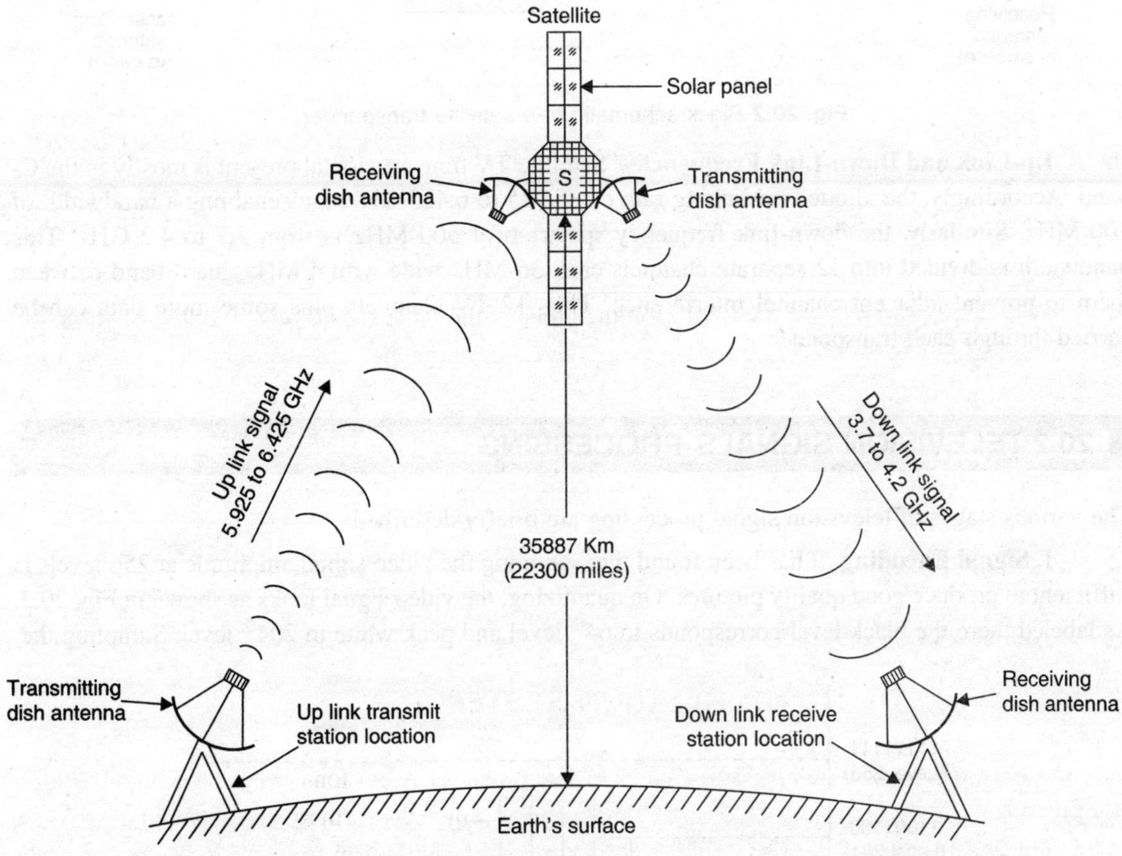

Fig. 20.1 Signal path in satellite communication.

The purpose of this is to amplify the weak signals and convert them to another base frequency called DOWN-LINK frequency. The conversion to another frequency is done primarily to prevent interference between up-link and down-link frequencies. The converted and amplified signal is retransmitted to earth by another dish antenna provided in the satellite. This signal processing unit in the satellite is called TRANSPONDER. Its back schematic is shown in Fig. 20.2.

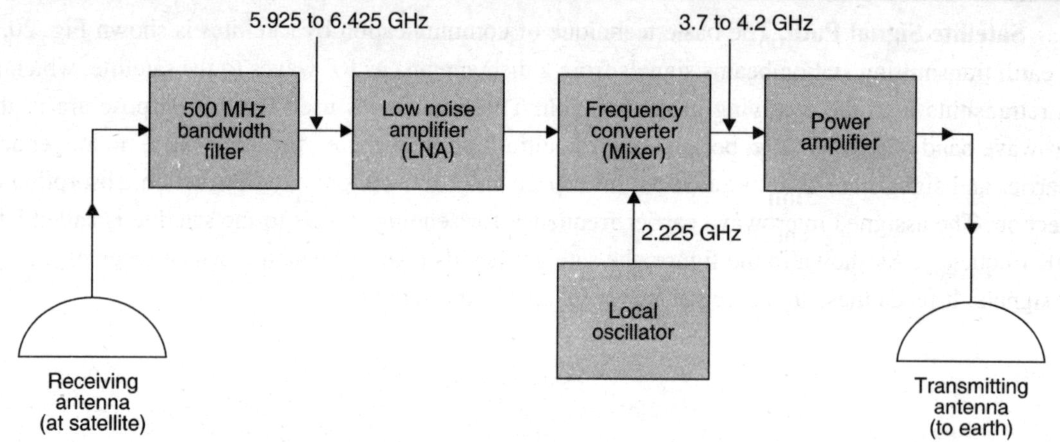

Fig. 20.2 Block schematic of a satellite transponder.

Up-Link and Down-Link Frequencies. Satellite TV transmission at present is mostly in the C-band. Accordingly, the allotted up-linking range is 5.925 to 6.425 GHz, thus enabling a bandwidth of 500 MHz. Similarly, the down-link frequency spectrum of 500 MHz is from 3.7 to 4.2 GHz. This bandwidth is divided into 12 separate channels each 36 MHz wide with 4 MHz guard band between them to prevent adjacent channel interference. Thus 12 TV channels plus some more data can be carried through each transponder.

20.2 TELEVISION SIGNALS PROCESSING

The various stages of television signal processing are briefly described:

1. Signal Encoding. It has been found that sampling the video signal amplitude at 256 levels is sufficient to produce good quality pictures. On quantizing, the video signal looks as shown in Fig. 20.3. As labeled there the black level corresponds to 64th level and peak white to 204th level. Sampling the

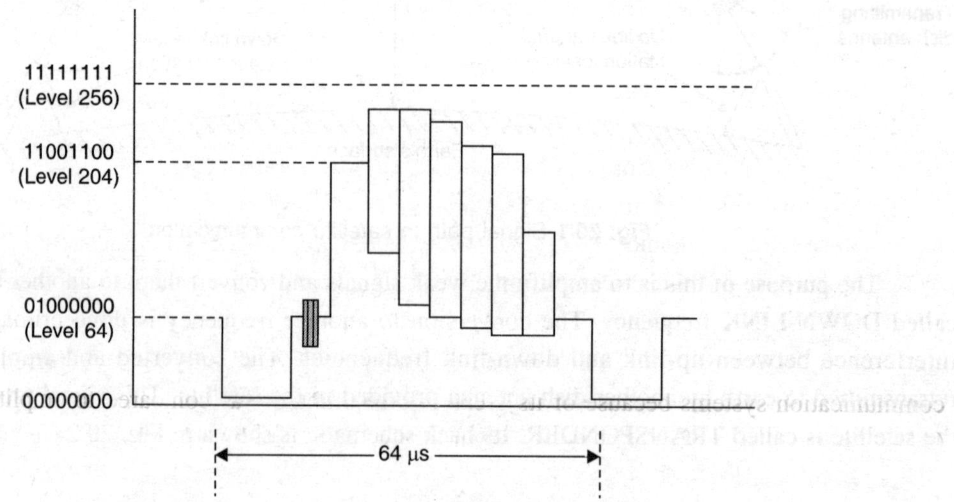

Fig. 20.3 Video signal quantization into 256 levels.

video signal at 256 levels would need an 8 bit A/D converter in the Encoder since $2^8 = 256$. The codes thus obtained for some main levels are noted in the figure. The encoding process is explained in Section 19.5 with Fig. 19.7. For audio signal encoding, a sampling level between 64 to 128 is chosen depending on the required audio quality at the receiver.

2. Data Compression. As explained in the previous chapter, data compression of both video and audio signals becomes necessary because of TV channel width limitations. After great development efforts compression formats have been standardized where the plan MPEG-2 is used for video signal compression and MP-2 for audio signal compression. These are used for all digital TV satellite transmissions.

3. Data Encryption. The TV channels may be divided into Free-to-Air and Pay channels. While there is no special charge for receiving Free-to-Air channels but additional payment is necessary for viewing 'Pay' channels. As such, pay channels are encrypted *i.e.*, scrambled by disturbing video pulse train to the extent that the received pictures are unintelligible. This is done before transmission to restrict access of pay channels to those subscribers only who opt and pay for their viewing. As necessary, de-encryption keys (data) are also transmitted with the channel data to enable authorized customers to descramble *i.e.*, restore correct sequence of video pulse train for normal viewing. This is called de-encryption.

20.3 DATA PACKETISING, MULTIPLEXING AND TRANSMISSION

The UP-LINK setup for digital transmission of one TV channel is shown in Fig. 20.4. The entire data is sent in the form of packets, separately for video and audio. Each packet also carries specific data to register its identity for easy separation at the receiving end. Most programme providers send additional data as labeled in the figure. For each of these separate packets are provided.

Clock System. As shown in the figure, all the data processing blocks are fed with clock pulses for initiation, time duration and sequencing control. For this, timed clock pulses are accurately generated by an electronic digital clock which is driven by a crystal controller oscillator.

Multiplexing. The data stored in all the packets is in parallel form and needs to be serialized in a signal stream before transmission. For this, the packets are time multiplexed, meaning data of each packet is transferred to a single pulse train in a sequential manner. The sequencing mechanism is controlled by the multiplexing control system as seen in the figure. This process though shown by arrows and dots, is electronic in its functioning where logic gates open and close sequentially as timed by the control system. Multiplexing takes place at a very fast rate to ensure steady stream of the contents of each packet at the receiving end.

Modulation and Transmission. Frequency modulation (FM) is preferred in all satellite communication systems because of its greater noise immunity as compared to amplitude modulation (AM). The sequence of operations is shown in Fig. 20.5. The base band video, audio and associated signals are frequency modulated around a centre carrier frequency of the 70 MHz and the resulting output is converted to the 6 GHz range before high power amplification and transmission.

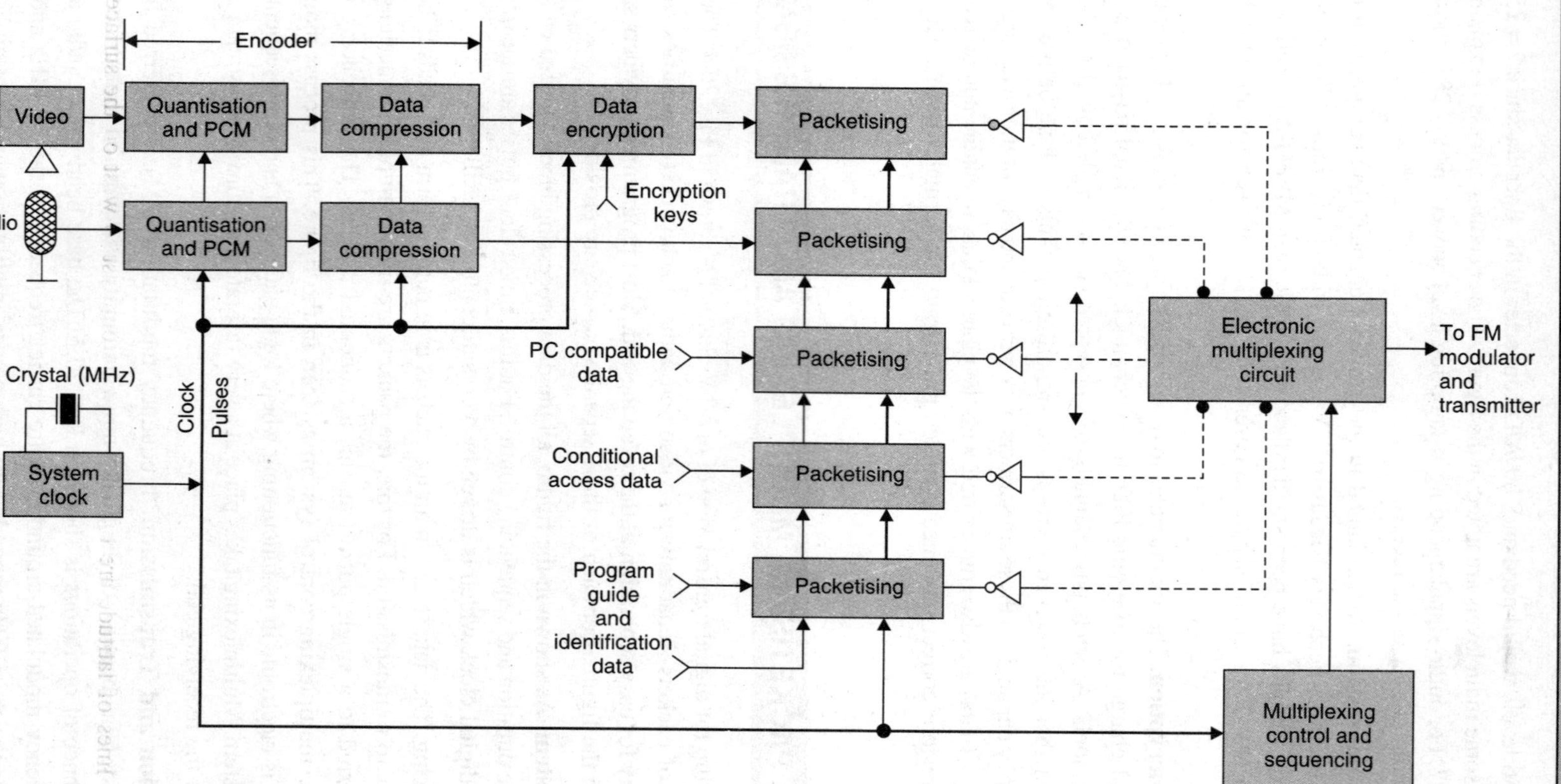

Fig. 20.4 UP-LINK set up for digital processing, packetising and multiplexing of one television channel.

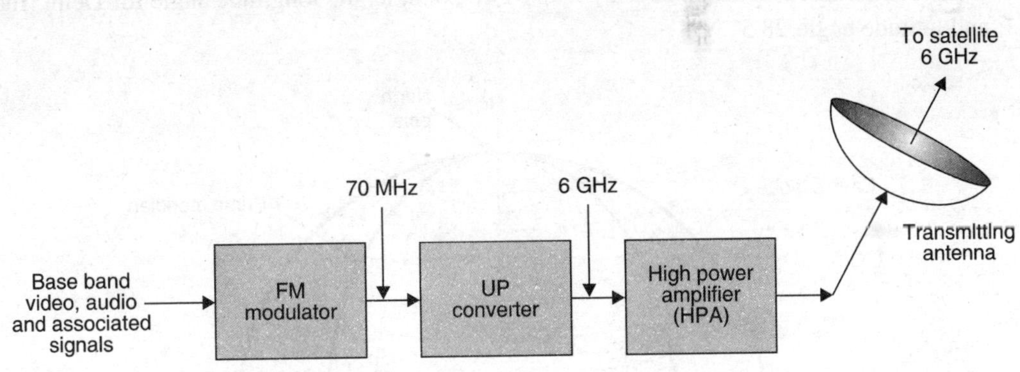

Fig. 20.5 Simplified block diagram of a transmitting earth station.

Transmitting Dish Antennas. Attenuation of microwave signals, both during up-linking and down-linking, is very large. Therefore, special parabolic reflector antennas called dish antennas are used to obtain high gain and directivity. This is achieved by using a horn in conjunction with a parabolic reflector which is a large dish shaped structure made of either screen mesh or metal. The energy radiated by the horn is pointed at the reflector which focuses it into a narrow beam and directs it towards its destination. The same dish antenna works in the opposite way at the receiving end for collecting down-link signals.

The transmitting dish antennas are very big in size because these have to handle large up-link signal power. They are located close to the transmitting site to keep losses to a minimum in coaxial cables that link them. Such dish antennas are very large in diameter and mounted on strong foundations to prevent any shaking by high speed winds. Also there is provision to move the dish horizontally and tilt it up and down for precisely directing it towards the receiving antenna of the assigned satellite.

20.4 SATELLITE ACQUISITION

The first concern at any ground receiving station is to track the wanted satellite. However, satellite tracking from any location on the surface of earth cannot be done at random in the absence of any data for correct orientation of dish antennas. For this, it is necessary to first fix position co-ordinates for both the receiving station and wanted satellite. Longitude and latitude form a system for locating any given point on the surface of earth. The lines drawn (see Fig. 20.6) from North pole to South pole on the earth's surface are called longitudes or meridians. The reference line for longitude measurement called 'PRIME MERIDIAN' is the one that passes through Greenwich in England. The designation east or west (E or W) is usually added to the longitude angle to indicate whether it is being measured to the east or west of the prime meridian.

Similarly lines of latitude are the ones drawn from east to west on the surface of earth and parallel to the equator as shown in Fig. 20.6. The equator that serves as reference for latitude measurements is the central latitude line that divides the earth into north and south hemispheres. Often an N or S is added to the latitude angle to indicate northern or southern hemisphere. Both longitude and latitude angles are often written on country maps. For example, the longitude angle for Delhi (India) is 77° E and latitude angle 28.5° N.

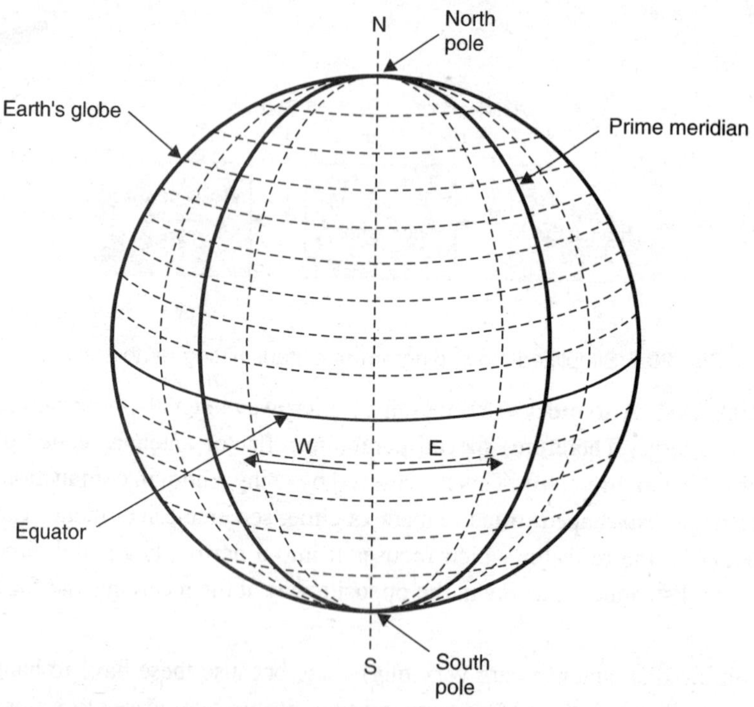

Fig. 20.6 Longitude and Latitude lines for locating any point on the surface of earth.

Location of Satellites. Since geo-synchronous satellites travel at the same speed at which the earth revolves, these are in a relative sense at a fixed point with respect to each other. As such, the location of a Communication Satellite is specified by a point on the surface of earth, directly below it. For geostationary satellites that orbit over the equator, the latitude angle is obviously zero (0̊) and therefore their angle of longitude (E or W) is enough to tell their locations. A large number of satellites are in orbit with different angle of longitude depending on the earth's surface area they are required to illuminate *i.e.*, beam their signals. For example PAS-4 satellite that beams programmes to Middle-East and Asian countries is located at 68.5 degree E. Similarly, India's INSAT-2A that covers nearly the same surface area is parked at 74.0 degree E.

Elevation and Azimuth Angles. For beaming signals from a satellite down to earth, its down linking antenna is inclined at a particular angle which depends on its parking location in the sky and 'Foot Print' (signal receiving) area on the earth. This angle is called 'Look Angle' of the satellite. Therefore, to acquire any satellite, the earth station's dish antenna must point at the 'look angle' of the satellite, meaning orient itself for capturing maximum signal strength. For this, two angles called Elevation and Azimuth are specified. The angle of elevation is the angle which appears between the line from the earth station antenna to the satellite and the line between the antenna and earth's horizon as shown in Fig. 20.7(*a*). The azimuth angle is the angle of direction and is measured clockwise from north pole taken as zero angle point and the plane containing earth station and satellite antennas. It can be better visualized from its schematic shown in Fig. 20.7(*b*), where corresponding representation of the elevation angle is also shown.

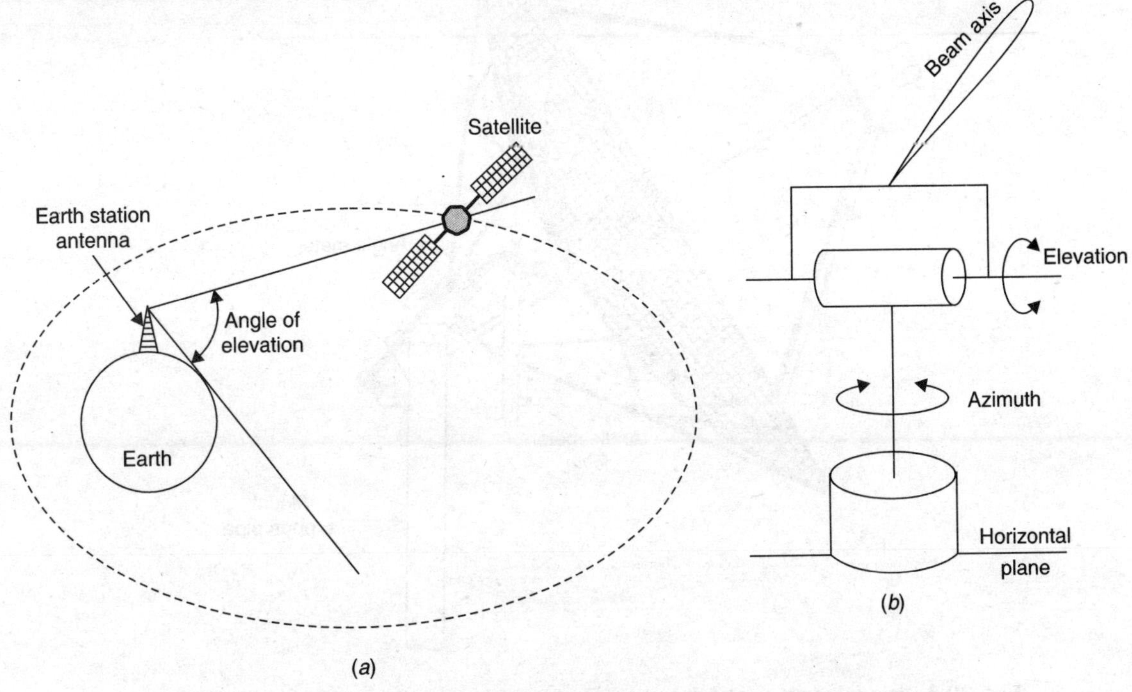

Fig. 20.7 (*a*) Co-ordinates of the angle of elevation (*b*) Schematic illustration of both elevation and azimuth angle.

While the starting point to determine elevation and azimuth angles are the longitudinal and latitudinal locations of both ground station and satellite of interest, their inter-relation is quite complicated because their physical locations are on different axis involving three-dimensional geometry. To make their determination easy computer programmes have been evolved to interpret elevation and azimuth angles for different locations. The results are available in the form of charts which enable immediate determination of the two angles. In India, for receiving SONY's entertainment channel at DELHI from satellite PAS-4, the azimuth angle is 197.76 degree and elevation angle 55.28 degree.

Orientation of Receiving Dish Antenna. For acquiring any satellite from a given location, the dish antenna is first set in the horizontal plane at the corresponding azimuth angle with the help of a common magnetic compass. Since its magnetic needle always points in North-South direction, the angles written on the compass dial serve as a good guide for setting the dish at the desired azimuth angle. For this, the compass is held in hand, brought closer to the dish and rotated to align the needle with the desired angle.

Similarly, for setting the dish at the known elevation angle, an angle meter is often used. As an illustration, in Fig. 20.8 it is shown with its lever set for an EL = 55 degree and then positioned to align *i.e.*, bring it parallel with the mount support pipe thus indicating how much the dish be vertically tilted.

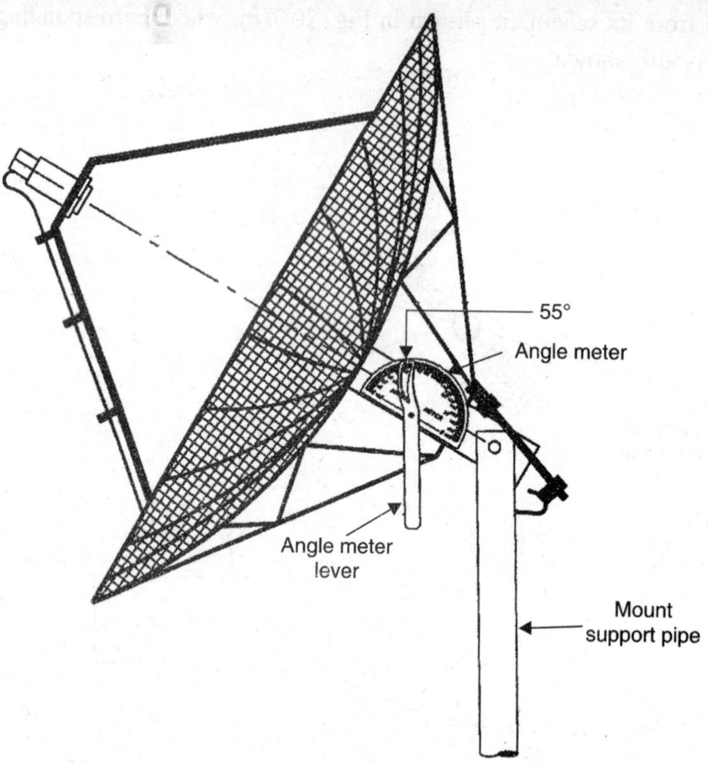

Fig. 20.8 Pictorial view of a dish antenna oriented for an elevation angle (EL) = 55° and azimuth angle (AZ) = 27°.

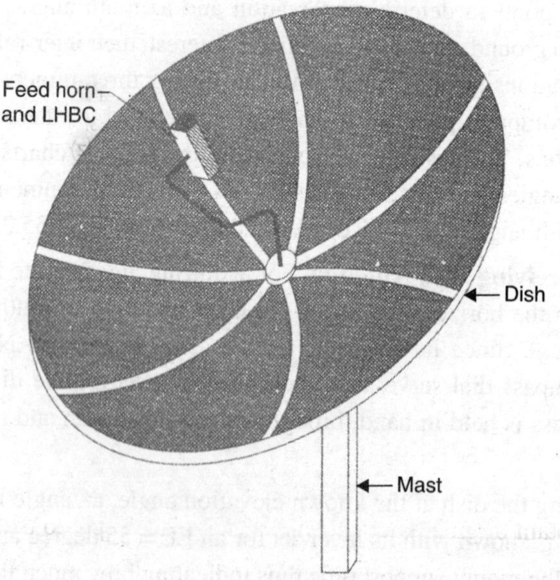

Fig. 20.9 A dish antenna with feed horn and LNBC mounted side-by-side at the focal point

20.5 DOWN-LINK SIGNALS—RECEPTION AND DECODING

The dish antenna when oriented at the look angle of the wanted satellite collects the signal arriving from it and reflects it to a common point called 'Focal Point'. As shown in Fig. 20.9, a feed horn which is actually a small wave guide section is mounted at the focal point to collect the arriving signal. This signal is in the range of 3.7 to 4.2 GHz *i.e.*, the down linking band. It is very weak and will suffer further attenuation if delivered as such for further processing. The collected signal is therefore first fed to a block converter, located next to the feed horn, for amplification and down conversion to a lower frequency range.

Low Noise Block Converter. As shown in Fig. 20.10, the low noise block converter (LNBC) combines a low noise amplifier and frequency converter. Its local oscillator (LO) is set at 5150 MHz and only the difference product is collected. The output will thus be in the range of 950 to 1450 MHz. The bandwidth remains the same *i.e.*, 1450 – 950 = 500 MHz. This output is amplified by a multistage IF amplified and sent through a high grade coaxial cable to the digital receiver-decoder for obtaining video and audio outputs of the tuned-in channel.

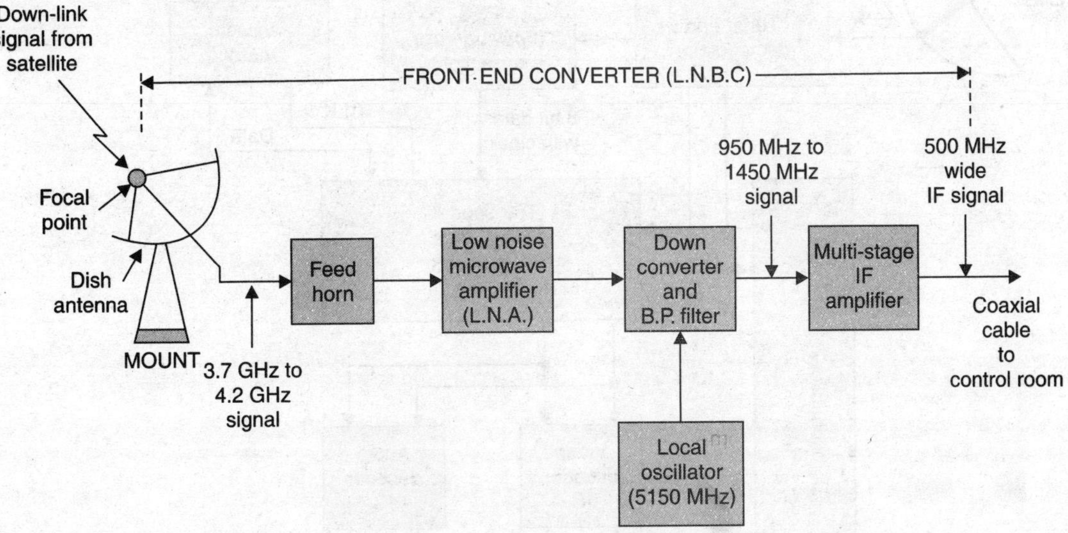

Fig. 20.10 Block diagram of an LNBC.

Digital Broadcast Receiver-Decoder. Since communication by satellites is now all digital, special broadcast receiver decoders are needed to obtain composite video and audio signals. A simplified functional block diagram of the receiver-decoder is shown in Fig. 20.11.

Its various sections are:

(i) **Tuner Demodulator.** The turner is controlled by the microcontroller on receipt of specific code pulses for each channel from the remote control unit. The separated channel signal is heterodyned by the demodulator to obtain original data stream. This is made input to the Forward Error Correction (FEC) section.

(*ii*) **Forward Error Correction.** This section reassembles the received data and corrects it if any error is detected. The error corrected data is forwarded to the Transport IC via an 8 bit parallel interface.

(*iii*) **Transport IC.** It is the heart of the receiver data processing circuitry. It functions under the control of microcontroller unit and isolates video, audio and other data packets. The separated video and audio payloads are sent to respective video and audio decoders.

(*iv*) **Video and Audio Decoders.** The two ICs separately decode the compressed video and audio data and send these to corresponding PAL video and audio encoders.

(*v*) **Video Encoder.** The video encoder converts the digital video information into PAL analog video format.

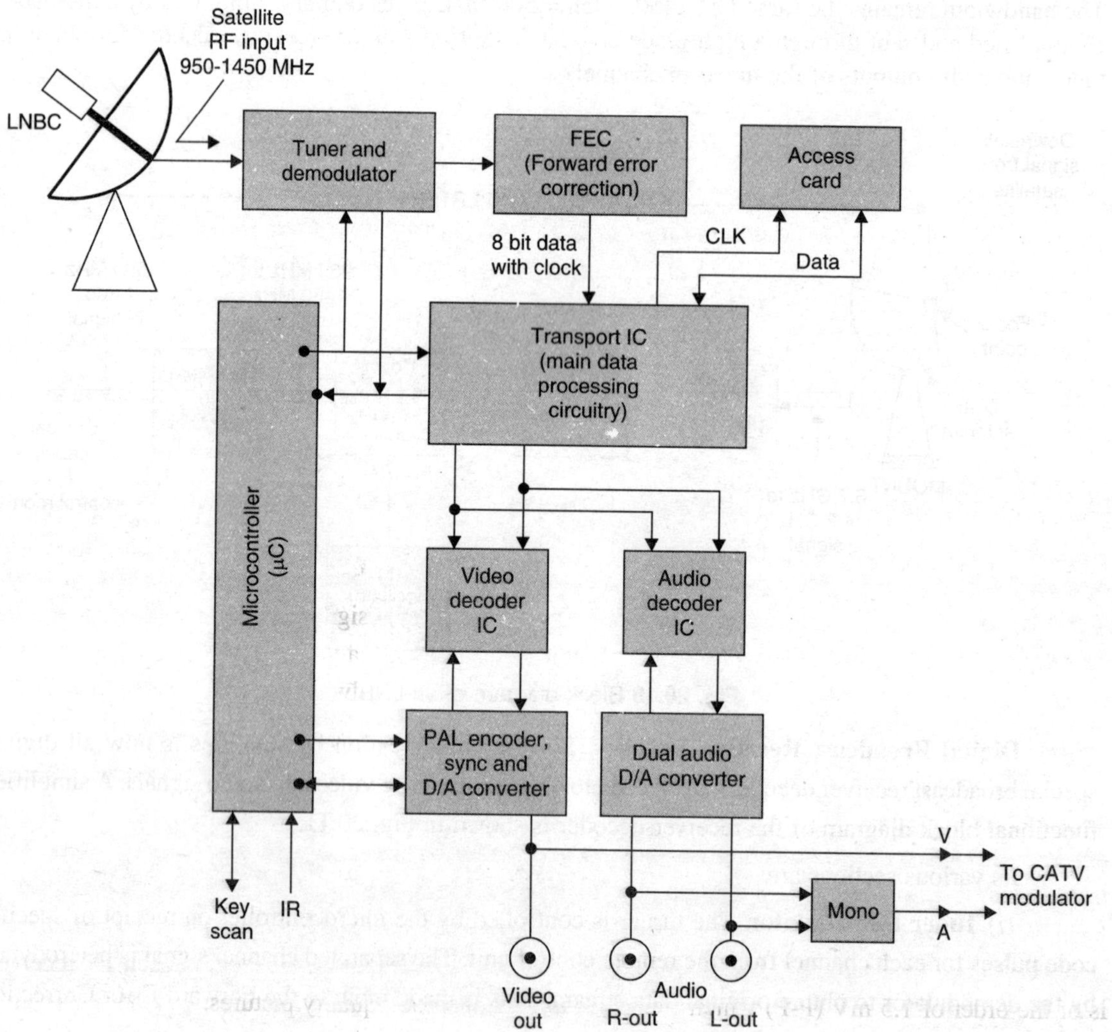

Fig. 20.11 A simplified block schematic of a digital broadcast-receiver decoder.

(*vi*) **Audio Encoder.** This IC is a processor cum D/A converter to obtain audio (mono or stereo) signal in analog form.

The two signals are fed to the input section of equipments used for further processing and distribution of TV channels to subscribers.

20.6 SIGNAL DISTRIBUTION TO SUBSCRIBERS

There are three ways by which television TV channel signals can be made available to subscribers. These are:

*(1) Cable Television (CATV)

(2) Domestic Broadcast System

(3) Direct To Home (DTH) Satellite Television.

All the three modes are separately explained:

1. Cable Television. This is the most common system because it does not have the restriction of channel allocations as is necessary in terrestrial broadcasts. Thus it can offer a variety of programmes on a large number of channels. In big cities, cable TV stations are located in different locations, each independent of each other and deliver programmes on around 60 to 100 channels to their subscribers. For collecting signals from different satellites, cable operators install 20 to 30 dish antennas each with its own LNBC which in turn feed into a large number of tuner-receiver decoders. As explained in the previous section, each decoder gives video and audio signal as its output which become input to the channel modulators.

A CATV channel modulator has all the building blocks of a similar unit at a TV transmitter, though at a very low power level. The audio input from the receiver decoder is amplified, accorded a pre-emphasis of 50 μs and frequency modulated with a carrier of 5.5 MHz. A filter selects the desired output of 5.5 MHz ± 75 KHz and combines it in a combiner with the corresponding composite video signal obtained from the same receiver-decoder. The composite TV signal is amplitude modulated with the assigned cable channel, carrier and the output is passed through a vestigal side band (V.S.B.) filter to bring the modulated output to the standard format of 7 MHz bandwidth as done at PAL system TV transmitters. The output is level controlled to keep it same as for other channel modulators.

The outputs from all the modulators are added in a combining network. The network is a linear mixer where all the signals are simply added together algebraically. The combined output is amplified and sent through a coaxial cable to the Distribution Amplifier for sending it on different routes. The trunk lines have line amplifiers at necessary distances to keep the signal level well above the noise level. From the trunk lines the signal is separated through passive networks called TAP-OFF points and sent to individual subscriber locations. It is ensured that the signal level at the input of each TV receiver is of the order of 1.5 mV (P-P) which is enough to generate good quality pictures.

---

*This author's book "SATELLITE AND CABLE TELEVISION" fully covers all aspects of Cable Television.

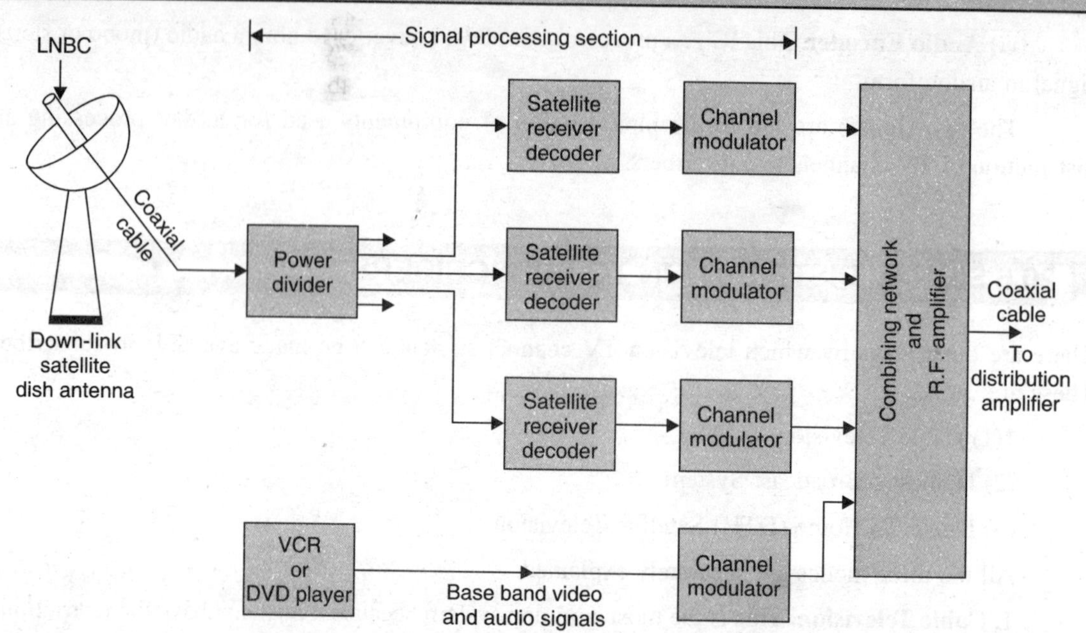

Fig. 20.12 Signal processing block at a cable TV station.

The entire set up at a cable TV station is shown in Fig. 20.12. The LNBC output feeds into a power divider which divides the signal for feeding to many satellite (receiver-decoder) receivers. Also, all cable operators have their own one/two channels for distributing programmes, earlier recorded on VCR tapes and DVD video/audio CDs.

2. Domestic Broadcasting System. Domestic broadcast from geo-stationary satellites is another way that is used for TV signal distribution all over country. In this, signals from the chosen satellite are received by a large number of fairly small and simple earth receiving stations located all over the country. The digital signals thus received are processed to obtain video and audio signals as per PAL TV standards. These are then processed as in conventional analog AM transmitters and rebroadcast terrestrially by low power transmitters (LPT) operating at different channels in the VHF band. In India, there are over 800 such small stations that cover practically the entire populated area of the country.

3. Direct-To-Home (DTH) Satellite Television. The Direct-to-Home (DTH) or Direct Satellite Broadcast (DSB) system enables viewers to access directly many channels of high quality digital video programming over a vast area from one or more high powered KU-Band satellites. There is no need for complex cable networks for signal distribution. KU-Band frequencies are preferred because these are not prone to interference from ground point-to-point communication and also need much smaller dish antennas. These are positioned on top of the building and directed at the Look-Angle of the wanted satellite. Since C-band and KU-band signals do not interfere with each other thus allowing a single satellite to relay both types of transmissions.

The DTH system otherwise functions in the same way for signal encoding, compression, packetising and up linking as explained in earlier sections. The receiving equipment is nearby the same as for digital satellite broadcast reception at cable stations. The video and audio outputs from the receiver decoder directly become input to corresponding jacks provided on present day TV receivers.

However, for billing purposes, the receiver connects through a MODEM to the customer's telephone line for communicating channel viewing details to the service computer. To prevent unauthorized reception a special encryption technique called "Conditional Access System" is used. It has been especially developed for DTH service.

A Smart Card (Access Card) is also provided with the receiver. It receives encrypted keys from the transport IC for decoding a scrambled channel. The keys *i.e.*, data for descrambling is also transmitted with the channel data to enable customers to descramble *i.e.*, restore normal sequence of video data without having to communicate with any external agency. This is achieved by inserting the Smart Card into a slot provided for it on the front panel of the receiver. This operation also activates a coded signal which goes to the billing centre over normal telephone lines via a modem.

When the Smart Card is activated for the first time, the serial number of the receiver is encoded (entered) on it. This prevents the card from activating any other receiver except the one in which it is initially authorized. Also, the receiver does not function with Smart Card removed from it.

In India DOOR-DARSHAN recently launched its DTH service offering many Free-To-Air TV and audio channels. SONY and STAR are also expected to launch their DTH transmission in the near future.

REVIEW QUESTIONS

1. In satellite communication explain why up-link and down-link frequencies are chosen to be in the GHz range and kept far-apart from each other.
2. Describe briefly various stages of digital TV signal processing. Why is it necessary to compress the digital data stream as obtained at the output of Encoder? What is the function of encrypting video signals at the transmitting stage?
3. Draw block schematic of an up-link signal processing and transmission setup and briefly explain the function of each stage.
4. How are dish antennas different from conventional Yagi antennas and why are they preferred for collecting satellite signals?
5. Define angles of longitude and latitude which are used to locate any place or point on the surface of earth. Explain how these angles become reference points for the acquisition of satellites.
6. Explain clearly the significance of Azimuth and Elevation angles which enable correct orientation of dish antenna for acquiring described satellites.
7. Show the location of feed horn together with LNBC on a dish antenna and explain where the horn is located to collect maximum signal from the dish.
8. Draw block diagram of a typical C-band LNBC unit and explain how the RF-IN from the dish antenna is amplified, down converted and further processed to obtain IF output 950 to 1450 MHz.
9. Draw functional block diagram of a Receiver Decoder (digital broadcast receiver) and explain how it decodes the signal fed to it to obtain video and audio outputs.
10. State the three main streams that are used to distribute TV channel signals to subscribers. Describe briefly how it is carried out at cable TV stations.
11. Describe briefly the main functions of the Direct-To-Home satellite television system. Why is this superior to cable television? What is the function of the Smart Card provided with D.T.H. system decoder receivers.